Ultrahigh-Density Magnetic Recording

Ultrahigh-Density Magnetic Recording

Storage Materials and Media Designs

edited by

Gaspare Varvaro

Francesca Casoli

CRC Press
Taylor & Francis Group
Boca Raton London New York

CRC Press is an imprint of the
Taylor & Francis Group, an **informa** business

Ultrahigh-Density Magnetic Recording: Storage Materials and Media Designs

First published in 2016 by Pan Stanford Publishing Pte. Ltd.

Published 2022 by CRC Press
Taylor & Francis Group
6000 Broken Sound Parkway NW, Suite 300
Boca Raton, FL 33487-2742

ISBN-13: 978-981-4669-58-0 (hbk)

British Library Cataloguing-in-Publication Data
A catalogue record for this book is available from the British Library.

Visit the Taylor & Francis Web site at
http://www.taylorandfrancis.com

and the CRC Press Web site at
http://www.crcpress.com

Contents

Preface

A stunning amount of data, of the order of a few zettabites (10^{21} bytes), is presently stored on hard disk drives (HDDs), data storage devices based on magnetic recording technology. HDDs are present in desktop computers, multiple-user computers, digital video recorders, and automotive vehicles. In laptop computers, they are being replaced by solid-state memories, which are lighter, quieter, and faster. However, HDDs remain the desirable option whenever the price per bit, per device recording capacity, and endurance are important requirements, as in massive secondary data storage.

Magnetic recording technology has maintained the lead in data storage since IBM first introduced the RAMAC drive operating at 2 kbit/in^2 in 1956, continuing to the present date, with the latest commercially available modern drives approaching the 1 Tbit/in^2 target. Despite the impressive growth of 500 million times in recording areal density in 60 years, increasing the recording density beyond 1 Tbit/in^2 requires a radical technological improvement, since the current magnetic recording technology is unable to achieve storage densities above this value.

In this book, worldwide experts from university, public research institutions, and industry collaborate to give an exhaustive overview of the technology and deepen its future prospects. A comprehensive approach is followed with the aim of offering a complete guide to a wide readership. Two introductory chapters serve as the basis for a better understanding of the following ones, deepening several aspects related to current magnetic recording technology, to its possible evolutions, and to magnetic memories in general. Four chapters are specifically dedicated to recording media and outline their evolution from the commercialization of the perpendicular technology in 2005 to state-of-the-art drives in 2015 to the most promising alternatives for next-generation recording technologies.

Chapter 1 describes the fundamental aspects of magnetism, which can assist readers new to the topic. The chapter is a complete compendium of magnetism, conceived with an original viewpoint by academy experts on the subject.

The second chapter gives a complete overview of the basic principles and technologies that allowed HDDs to reach today's information densities. This chapter takes advantage of the expertise of industry specialists on the subject in order to highlight recent materials and structure breakthroughs in perpendicular magnetic recording.

The third chapter describes the evolution of conventional media used in perpendicular recording, from the introduction of the technology to the market to date. Written by an industry expert on the topic, the chapter is full of technical details and represents an exhaustive guide on conventional perpendicular media.

Chapter 4 is dedicated to energy-assisted magnetic recording; it includes heat-assisted and microwave-assisted recording, two technologies that modify the media's magnetic properties or switching dynamics to allow higher recording densities compared to conventional perpendicular recording. The chapter is primarily dedicated to heat-assisted magnetic recording (HAMR), which is one of the most promising technologies for the future and will likely appear in HDD products in 2018, allowing an areal density growth up to 4 $Tbit/in^2$.

After the realization that the $L1_0$-FePt alloy is the best candidate for next-generation perpendicular recording media, a big effort has been devoted to the optimization of the morphostructural and magnetic properties of this material for recording applications. The most significant progresses on FePt-based media for HAMR are carefully discussed in Chapter 5.

The following chapter, that is, Chapter 6, focuses on exchange-coupled composite media (ECC media), which exploit the old concept of an exchange-spring magnet to improve the writability of perpendicular recording media. The chapter reviews the results obtained by exploiting the ECC principle on conventional media, FePt-based media, and anisotropy-graded media.

Chapter 7 concentrates on bit-patterned magnetic recording (BPMR), where one single bit of information is stored in one lithographically defined magnetic dot. The chapter deepens the reader's understanding of the fundamental aspects of the magnetization reversal process in BPMR media and considers FePt-based BPMR systems employing the ECC media concept. The extremely challenging requirements for the realization of BPMR media has pushed this technology out on the HDD industry road map to 2020 or beyond.

Chapter 8 makes an exhaustive analysis of the techniques and methods used for magnetic characterization of perpendicular recording granular media, ECC media, bit-patterned media, and energy-assisted recording media. The chapter can serve as a complete and clear guide to the reader approaching the magnetic characterization of magnetic recording media.

Finally, Chapter 9 goes beyond magnetic recording matter, to give a full perspective on magnetic memories for data storage in general—from the different classes of magnetic RAMs, to domain-wall-based memories, and beyond binary static memories, as a final stimulus, to memristors.

Overall, this book presents a thorough treatment of the core principles and future directions of magnetic recording, providing a balanced coverage of theoretical, experimental, and technological aspects. It addresses a broad readership, from final-year undergraduate and postgraduate students to experienced researchers active in the magnetic recording area, both in the academy and industry. Chapters include many illustrations and practical examples, making the book a clear and useful guide for both the experienced reader and the reader approaching the treated topics for the first time.

Finally, we would like to express our sincere gratitude to all the authors: none of this would have been possible without their expertise and commitment. We are grateful to the reviewers, who have rigorously evaluated all the chapters and given helpful feedback to improve the book. Our special thanks also go to the editorial office at Pan Stanford Publishing for their constant assistance in preparing the book and to Nicoletta Marigo and Julie Karel for their help with proofreading. Furthermore, our deepest gratitude goes to the president of the Italian Association of Magnetism, Dino Fiorani, for his continuous support and encouragement.

Chapter 1

Fundamentals of Magnetism

Paolo Allia[a] **and Gabriele Barrera**[b]

[a]*DISAT, Politecnico di Torino, Corso Duca degli Abruzzi 24, Torino, 10129, Italy*
[b]*Chemistry Department, Università di Torino, Via Pietro Giuria 7, Torino, 10125, Italy*
paolo.allia@polito.it, g.barrera@inrim.it

This chapter is intended for readers approaching magnetism and magnetics without a substantial knowledge of the subject matter and/or looking for a summary account of its highlights. A difficulty in mastering the topic stems from the fact that magnetism is a multiscale research area, spanning from basic quantum theory of single atoms (the *Aufbauprinzip* of Bohr and Pauli) to the functional properties of macroscopic magnetic materials. This chapter will touch upon all these aspects by following a slightly unconventional order. The author is aware of the existence of excellent textbooks on magnetism; some are listed as references. However, it is hoped that this effort will be of some interest to some people in some circumstances.

Ultrahigh-Density Magnetic Recording: Storage Materials and Media Designs
Edited by Gaspare Varvaro and Francesca Casoli

ISBN 978-981-4669-58-0 (Hardcover), 978-981-4669-59-7 (eBook)
www.panstanford.com

1.1 Magnetic Fields and Energies: A Macroscopic Approach

1.1.1 *Magnetic Fields*

In spite of the purely quantum nature and origin of ferromagnetism [1–3], an entirely classical framework is often expedient to describe the magnetic properties of magnetized bodies. In a classical approach, use is made of the three magnetic vectors in the continuum approximation. Of course, such an approach loses validity when the size of a magnetized body becomes comparable in at least one dimension to the typical interatomic distance a (or is not exceedingly large with respect to a). One starts with the standard relation among the three vectors **H** (magnetic field), **M** (magnetization), and **B** (magnetic induction):

$$\mathbf{B} = \mu_0\mathbf{H} + \mathbf{M} \tag{1.1}$$

where the International System (SI) of measuring units has been used; μ_0 is the permeability of vacuum. This relation is not unique; for alternate expressions in the SI and in different systems of units see, for example, Ref. [4]. The magnetization vector of a fully magnetized body containing n magnetic dipole moments $\boldsymbol{\mu}$ per unit volume is simply $\mathbf{M} = n\boldsymbol{\mu}$. The difference between the static fields **H** and **B** stems from Maxwell's equations:

$$\begin{array}{ll} \nabla \cdot \mathbf{B} = 0 & \nabla \times \mathbf{B} = \mu_0 \mathbf{j}_\mathrm{T} \\ \nabla \cdot \mathbf{H} = -\frac{1}{\mu_0}\nabla \cdot \mathbf{M} & \nabla \times \mathbf{H} = \mathbf{j}_\mathrm{F} \end{array} \tag{1.2}$$

where it is stated that the field **B** is a solenoidal vector whose source is *all* the stationary currents flowing in the system ($\mathbf{j}_\mathrm{T}$ being the corresponding current density vector, the subscript "T" standing for "total"), whereas **H**, a nonsolenoidal field, admits as sources both the *free* currents flowing in the system ($\mathbf{j}_\mathrm{F}$ being the corresponding current density vector, the subscript "F" standing for "free") and the *magnetic poles* (i.e., sources and termination points of the flux lines of **H**) whose volume density is given by the divergence of the magnetization vector:

$$\begin{aligned} \rho_\mathrm{m}(\mathbf{r}) &= -\nabla \cdot \mathbf{M} \\ \nabla \cdot \mathbf{H} &= \frac{1}{\mu_0}\rho_\mathrm{m} \end{aligned} \tag{1.3}$$

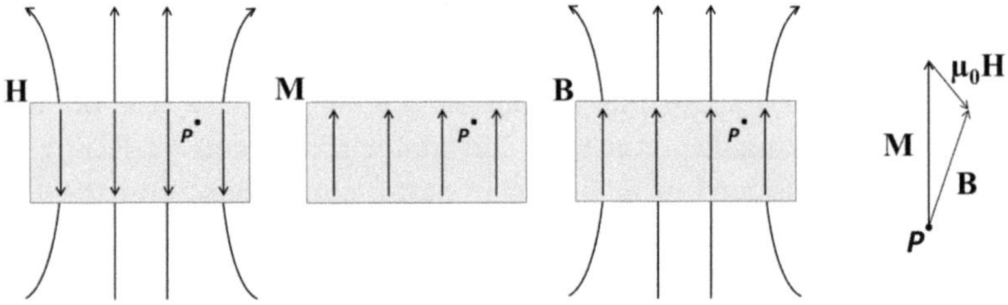

Figure 1.1 Sketch of fields **B**, **M**, and **H** inside and around a permanent magnet. **H** inside the magnet is the demagnetizing field.

Therefore, the field **H** can be generated in two ways: not only by electrical currents under the control of the user—this explains why **H** is more popular than **B** when dealing with electrotechnical applications of magnetic materials—but also by a nonuniform distribution of the magnetization vector, leading to a generally nonzero divergence of **M**. Even if a body is uniformly magnetized (strictly speaking, this occurs in ellipsoidal samples only) the surfaces may act as sources of **H**. The field **H** produced by discontinuities of the normal component of **M** at a surface exists both outside and inside the body. Inside, the predominant direction of the field is opposite to **M**; this is why this quantity is often referred to as the *demagnetizing field*. Owing to the presence of the demagnetizing field, the amplitude of the vector **B** inside an (almost) uniformly magnetized body is generally lower than the one of **M** (Fig. 1.1).

It may be interesting to draw a double analogy among magnetic vectors in magnetic materials and electrical vectors in dielectrics. The electrical counterpart to Eq. 1.1 is of course

$$\mathbf{D} = \varepsilon_0 \mathbf{E} + \mathbf{P} \tag{1.4}$$

D being the electrical displacement, **E** the electrical field, **P** the polarization, and ε_0 the vacuum dielectric constant. The first analogy of course is

$$\begin{aligned} \mathbf{E} &\leftrightarrow \mathbf{H} \\ \mathbf{P} &\leftrightarrow \mathbf{M} \\ \mathbf{D} &\leftrightarrow \mathbf{B} \end{aligned} \tag{1.5}$$

This analogy can be exploited to predict the behavior of the flux lines of the vector **H** generated by a distribution of magnetic poles on a flat surface of generic size and shape (this occurs, for instance, in a uniformly magnetized body of finite size with a surface perpendicular to the magnetization itself). If there are no free currents flowing in or around the material the curl of **H** is zero (exactly as the curl of **E** is zero in static conditions) and the flux lines of **H** must be identical with the flux lines of the electrical field generated by a distribution of point charges on a dielectric surface of exactly the same shape and size.

However, a second, deeper analogy among electric and magnetic vector fields exists:

$$\begin{aligned} \mathbf{D} &\leftrightarrow \mathbf{H} \\ \mathbf{P} &\leftrightarrow -\mathbf{M}/\mu_0 \\ \varepsilon_0\mathbf{E} &\leftrightarrow \mathbf{B}/\mu_0 \end{aligned} \tag{1.6}$$

This analogy arises from considering that the curl of **B** is given by *all* existing currents, exactly as *all* existing charges are the sources of **E**. Therefore, the role of **E** in dielectrics is played by **B** in magnets, the only difference being that **B** is a solenoidal field without poles and **E** (at least in the stationary case) is a conservative field with poles. As a consequence, the expression $\mathbf{H} = (\mathbf{B} - \mathbf{M})/\mu_0$ is—from this viewpoint—only an auxiliary quantity, having a role similar to the one played by the quantity $\mathbf{D} = \varepsilon_0\mathbf{E} + \mathbf{P}$ in dielectrics.

1.1.2 *Magnetostatic Energy*

The magnetostatic energy of a ferromagnetic body derives from the existence of the field **H** produced by the points in a magnetized body where the divergence of **M** is nonzero. The demagnetizing field gives rise to a self-energy, that is, the energy of a magnetized body in its own demagnetizing field. Different expressions can be used to describe such a self-energy. We consider a magnetized body such as a permanent magnet in the absence of applied electrical currents. A *magnetic potential* φ_m can be defined (**H** being here a conservative field). The potential associated to **H** is obtained through the magnetic Poisson equation

$$\Delta\varphi_\mathrm{m} = -\frac{1}{\mu_0}\rho_\mathrm{m} \tag{1.7}$$

where Δ is the Laplacian differential operator, and whose formal solution is

$$\varphi_{\mathrm{m}}(\mathbf{r}) = -\frac{1}{\mu_0}\int \frac{\rho_{\mathrm{m}}(\mathbf{r}')}{|\mathbf{r}-\mathbf{r}'|}dV' \tag{1.8}$$

having set to zero the integration constant; here $|\mathbf{r}-\mathbf{r}'|$ is the distance between the source element of volume dV' in $\mathbf{r}'$ and the observation point in $\mathbf{r}$. The associated self-energy is

$$E_{\mathrm{m}} = -\frac{1}{2}\int \rho_{\mathrm{m}}(\mathbf{r})\varphi_{\mathrm{m}}(\mathbf{r})dV \tag{1.9}$$

where in principle the integral is over all space; however, the density of magnetic poles is nonzero only within the magnetic body, and in many cases it is negligible far from the body surfaces. An alternative expression to Eq. 1.9 is

$$E_{\mathrm{m}} = -\frac{1}{2}\int \mathbf{M}\cdot\mathbf{H}dV \tag{1.10}$$

Here, **H** is the demagnetizing field and the integration is limited to the volume of the magnetized body. In uniformly magnetized ellipsoids a linear relation between **H** and **M** exists:

$$\mathbf{H} = -\frac{1}{\mu_0}\tilde{N}\mathbf{M} \tag{1.11}$$

The quantity $\tilde{N}$ is a demagnetizing tensor that is diagonal when the ellipsoid's principle axes are aligned to the x, y, and z directions of a reference frame; the sum of the three diagonal components N_x, N_y, and N_z is equal to 1. The same relation is usually assumed to hold, although approximately, in bodies with a nonellipsoidal shape.

1.1.3 *Shape Anisotropy of a Magnetized Body*

Magnetostatic energy plays an important role in ferromagnetic materials: As will be discussed in Section 1.4, *magnetic domain patterns* are generated in a magnetized body at equilibrium in order to minimize the magnetostatic energy. Here, we show how E_{m} brings about a particularly significant type of magnetic anisotropy of a ferromagnetic body, that is, *shape anisotropy*. Inserting Eq. 1.11 in Eq. 1.10 one gets

$$E_{\mathrm{m}} = \frac{1}{2\mu_0}\int (N_x M_x^2 + N_y M_y^2 + N_z M_z^2)dV \tag{1.12}$$

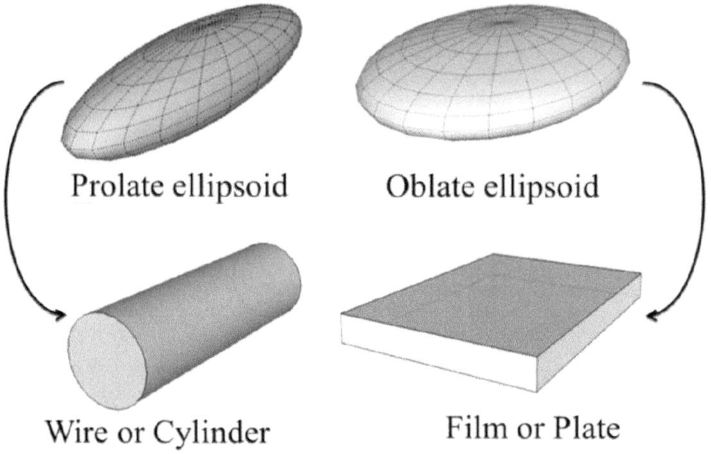

Figure 1.2 Shape anisotropy of magnetized cylinders/wires and plates/films can be approximately evaluated considering these bodies as limiting cases of an ellipsoid.

For an ellipsoid of revolution (Fig. 1.2) with $N_x = N_y = (1 - N_z)/2$ and using $M_z = M\cos\theta$, $(M_x^2 + M_y^2)^{1/2} = M\sin\theta$ the above expression becomes

$$E_{\mathrm{m}} = \frac{1}{2\mu_0}\int M^2\left(N_x\sin^2\theta + N_z\cos^2\theta\right)dV$$
$$= \frac{1}{2\mu_0}\int M^2\left[N_z + (N_x - N_z)\sin^2\theta\right]dV \qquad (1.13)$$

When the magnetization is uniform the energy reduces to

$$E_{\mathrm{m}} = \text{const.} + \frac{1}{2\mu_0}M^2V(N_x - N_z)\sin^2\theta$$
$$= \text{const.} + \frac{1}{4\mu_0}M^2V(1 - 3N_z)\sin^2\theta \qquad (1.14)$$

that is, the energy takes the form of a uniaxial anisotropy energy whose value is dictated by the shape of the sample: For a spherical sample, $N_z = 1/3$ and there is no shape anisotropy energy; for a prolate ellipsoid (elongated along z) $N_z < 1/3$ and the z axis is the direction of minimum energy.

Therefore, in elongated samples the magnetization is typically found aligned with the long axis of the body; instead, for oblate ellipsoids (i.e., squeezed along z) the opposite is true. This situation occurs, for example, in a thin film lying in the x-y plane. In this case $N_z \gg N_x$, N_y so that $N_z \cong 1$, $N_x \cong N_y \cong 0$, and

$$E_{\mathrm{m}} \cong \text{const.} - \frac{1}{2\mu_0} M^2 V \sin^2 \theta \tag{1.15}$$

which is a minimum for $\theta = \pi/2$, that is, when the magnetization vector lies in the film plane. In a circular microwire aligned along z, $N_z \cong 0$, $N_x = N_y \cong 1/2$, and

$$E_{\mathrm{m}} \cong \text{const.} + \frac{1}{4\mu_0} M^2 V \sin^2\theta \tag{1.16}$$

In this case, the magnetization prefers to stay aligned with the wire axis (even if the wire is bent).

Shape anisotropy plays a central role in magnetic nanostructures also. In fact, even in nanosized materials, the classical approach and continuum approximation roughly hold. As a consequence, shape anisotropy is deeply involved in the performance of magnetic dots, antidots, columnar structures, and thin films.

1.2 Basics of Magnetization Dynamics

1.2.1 *Torques*

The dynamic response of a uniformly magnetized body to a constant magnetic field can be treated in a semiclassical way not only in macroscopic bodies but also in magnetic nanostructures where the condition of uniform magnetization is more easily fulfilled. Here the term "semiclassical" is intended to mean that the individual magnetic moments, of quantum-mechanical nature and origin, are effectively aligned and kept parallel to each other by the ferromagnetic exchange interaction (see the next section) in such a measure that one can look for the dynamic response of the classical vector **M** as a whole instead of analyzing the dynamic behavior of each single atomic magnetic moment or spin. As a matter of fact, this approximation generally holds up to microwave frequencies and beyond. However, the treatment cannot be fully classical because the dynamic response of **M** stems from a quantum-mechanical effect:

in the presence of a magnetic field **H** the free magnetization vector coherently precesses around the field axis:

$$\frac{d\mathbf{M}}{dt} = -|\gamma|\tau_{\mathbf{P}} = -|\gamma|\mathbf{M} \times \mathbf{H}$$
$$\gamma = -g\frac{\mu_{\mathrm{B}}}{\hbar} \tag{1.17}$$

where $\tau_{\mathbf{p}}$ is the magnetic torque acting on **M** by effect of **H**, γ is the gyromagnetic ratio, and μ_{B} the Bohr magneton. When **H** is along the z axis, the solution is

$$\begin{cases} M_x = m\cos\omega_0 t \\ M_y = m\sin\omega_0 t \\ M_z = \text{const.} \end{cases} \tag{1.18}$$

$\omega_0 = |\gamma|H$ being the Larmor angular frequency and m the projection of the perpendicular magnetization $M_{\perp} = (M_x^2 + M_y^2)^{1/2}$ on the field axis. Equation 1.17 applies to the lossless case. In a real material with electromagnetic losses one expects that a damping torque appears. A phenomenological expression of the damping torque was given by Landau and Lifshitz [5]:

$$\tau_{\mathbf{D}} = \frac{\alpha}{M}\mathbf{M} \times (\mathbf{M} \times \mathbf{H}) \tag{1.19}$$

α being the dimensionless Landau–Lifshitz damping parameter (typical values of α are in the 0.01–0.1 range). The Landau–Lifshitz equation takes the form

$$\frac{d\mathbf{M}}{dt} = -|\gamma|\mathbf{M} \times \mathbf{H} - \frac{\alpha|\gamma|}{M_{\mathrm{s}}}\mathbf{M} \times (\mathbf{M} \times \mathbf{H}) \tag{1.20}$$

where the first term is responsible for the precession around the z axis and the second term acts to reduce the initial tilt of the magnetization vector with respect to the z axis in such a way that the tip of the vector **M** travels in a spiraling motion around the z direction to finally come to rest when **M** is aligned with **H** (Fig. 1.3).

The dynamic behavior of magnetization is exploited in spintronic devices based on magnetic heterostructures containing two layers of a ferromagnetic metal such as cobalt separated by a diamagnetic metal such as copper. One of the two magnetic layers (named the free layer) hosts an unpinned magnetization **M**, which can be easily rotated by an external or internal torque, whereas in the second

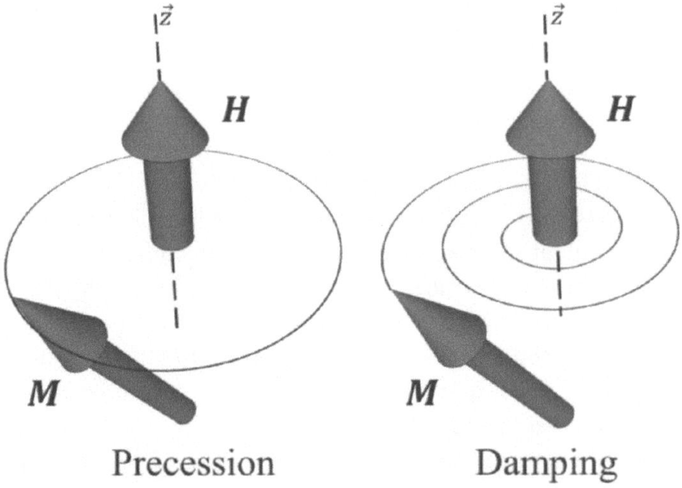

Figure 1.3 Undamped and damped coherent Larmor precession of the magnetization vector (Eq. 1.20).

magnetic layer the local magnetization is kept fixed along one direction (pinned layer of constant magnetization $\mathbf{M_P}$).

If a flow of spin-polarized electrons coming from the pinned layer of magnetization $\mathbf{M_P}$ interacts with the unpinned magnetization **M** traveling across the free layer, a torque effect on **M** is generated: this is the so-called *spin transfer torque* predicted by Slonczewski [6] and expressed as

$$\tau_{\mathrm{ST}} = j\frac{\beta}{M^2}\mathbf{M} \times \left(\mathbf{M} \times \frac{\mathbf{M_P}}{M_{\mathrm{P}}}\right) \tag{1.21}$$

where j is the magnitude of the electrical current density and the parameter β contains the thickness of the free layer and other parameters, including the device efficiency that accounts for all detrimental effects related to the nonideality of the materials. Equation 1.21 means that the free-layer magnetization **M**, initially tilted with respect to the pinned magnetization $\mathbf{M_P}$, exchanges angular momentum with a flow of spin-polarized impinging electrons whose spin is parallel to $\mathbf{M_P}$. In the interaction between electrons and the free layer, the spin of each electron is made to rotate and the electron flow makes **M** turn around.

In spintronic nanostructures of interest for magnetic recording the dynamics of the magnetization vector can be summarized as follows:

- In the absence of any spin transfer torque or damping, the free-layer magnetization **M**, if initially tilted away from the z axis, will precess describing a cone by effect of the torque originated from the external magnetic field.
- If there is damping (but still no applied current) the damping torque will push **M** toward the low-energy direction along the z direction. Consequently, if **M** is perturbed away from the z axis it will precess with gradually decreasing precession angle along a spiral path.
- When a spin-polarized current is applied to the device, the direction of the spin transfer torque can be either parallel or antiparallel to the damping torque, depending on the polarization direction. When the spin transfer torque is opposite to the damping torque and is of greater magnitude, the slightly tilted vector **M** will spiral away from the low-energy direction, increasing the tilt angle. In this case, the current acts to destabilize the orientation of **M** and may excite large-angle dynamics. Depending on the material, **M** may achieve a state of dynamical equilibrium, precessing continuously at some fixed angle in response to the direct current: in this state the energy gained from the spin torque during each cycle of precession is balanced by the energy lost to damping. Or, the precession angle may be excited to ever-increasing values until reaching π, meaning that **M** has been reversed. These effects may be of prominent relevance in the area of magnetic recording, as discussed in further chapters of this book.

1.2.2 *Magnetization Dynamics and Data Writing in Magnetic Recording*

In modern magnetic recording, quick data writing is essential to fulfill the requirements of the computer industry and software developers. Bit magnetization reversal driven by a static magnetic field applied parallel to the bit's easy axis is exploited in current hard disk devices. However, novel faster processes are under investigation in data writing. For example, applying a magnetic field pulse *perpendicular* to the easy axis of a single magnetic bit induces the precessional motion of the magnetization vector and can bring about a notable reduction in the magnetization reversal time; however, the shape and duration of the pulse are both most critical parameters a wrong shape or duration resulting in an unwanted extra precession of the magnetization vector. Current-induced magnetization reversal, as described in the previous section, is also being widely investigated. The magnetization dynamics resulting from injection of a spin-polarized current is a very complex effect, not entirely clarified so far. Finally, more exotic techniques such as nonlinear radio frequency (rf) magnetic resonance methods have been proposed. The reader can find a detailed description of the present state of the art in specialized review papers such as Ref. [7]. The common term in all fast data-writing mechanisms is the widespread use of equations describing the short-time dynamics of **M**, such as the Landau–Lifshitz equation.

1.3 Elements of Microscopic Magnetism

1.3.1 *Stable Magnetic Moments on Free Atoms/Ions*

Stable, intrinsic magnetic moments on free atoms or ions exist as a consequence of quantum-mechanical rules. In many-electron atoms, one starts from the L-S or Russell–Saunders coupling framework [1, 2]. In the presence of a nonnegligible spin–orbit interaction, neither the total orbital angular momentum **L** nor the total spin angular momentum **S** are good quantum numbers, though, instead, the total angular momentum $\mathbf{J} = \mathbf{L} + \mathbf{S}$ is a constant of motion along with $\mathbf{L}^2$, $\mathbf{S}^2$ (Fig. 1.4). When a given atom/ion is in the ground state—magnetism typically involving quantum systems in thermodynamic equilibrium—the existence of an intrinsic magnetic moment μ implies that **J** has a nonzero expectation value. This rules out completely filled electronic shells (whereby Hund's rules [2] $\mathbf{L} = \mathbf{S} = \mathbf{J} = 0$) and therefore all atoms/ions whose electronic structure is

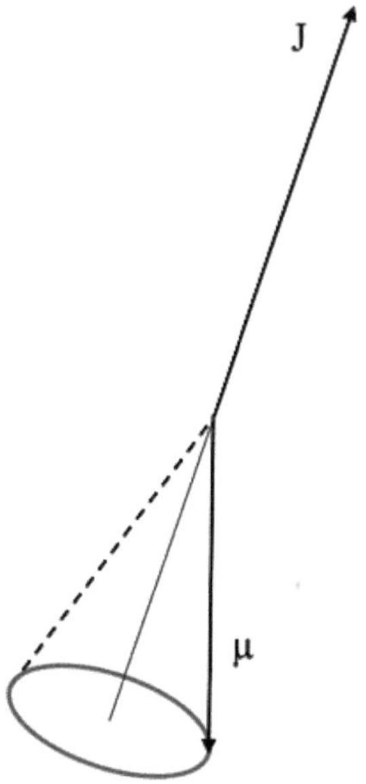

Figure 1.4 In a semiclassical picture, the magnetic moment $\boldsymbol{\mu} = \mathbf{L} + 2\mathbf{S}$ can be viewed as precessing around the $\mathbf{J} = \mathbf{L} + \mathbf{S}$ vector, which is fixed in space. The projection of $\boldsymbol{\mu}$ on $\mathbf{J}$ corresponds to the measured magnetic moment amplitude.

composed of a collection of filled shells. Transition elements of the first family (transition metals proper), second family (rare earths), and third family (actinides) are therefore the species able to bear a stable magnetic moment, which is antiparallel to **J** (because of the negative sign of the electron charge) and has the magnitude

$$\mu = g_J \, \mu_B \sqrt{J\,(J+1)} \tag{1.22}$$

g_J being Landé's factor and μ_B the Bohr magneton ($\cong 1 \times 10^{-23}$ Am^2) [2].

1.3.2 *Van Vleck Paramagnetism: Superparamagnetism*

1.3.2.1 Paramagnetic response of noninteracting magnetic moments

In this framework, the response of a set of independent localized magnetic moments to a magnetic field **H** can be easily derived.

This is Van Vleck's paramagnetism, whose distinctive features can be highlighted as follows. States of given J display a $(2J+1)$-fold degeneracy which is fully removed by a magnetic field that splits the degenerate level by the Zeeman energy:

$$E = -g_J \mu_B m_J H \tag{1.23}$$

where m_J is the quantized projection of J along the **H** axis ($m_J = -J, -J+1, \ldots J-1, J$). The Zeeman energy separation between adjacent levels is $g_J \mu_B H$ and typically is of the order of $k_B T$ at room temperature for experimentally accessible values of **H**. The magnetic response M of n identical magnetic moments (per unit volume) along the direction of **H** is obtained through well-known techniques of statistical quantum mechanics:

$$M = n k_B T \frac{\partial}{\partial H} \ln Z \tag{1.24}$$

Z being the partition function:

$$Z = \sum_{m_J=-J}^{J} e^{\frac{g_J \mu_B H}{k_B T} m_J} = \frac{\sin h\left(\frac{(2J+1)x}{2}\right)}{\sinh\left(\frac{x}{2}\right)}; x = \frac{g_J \mu_B H}{k_B T} \tag{1.25}$$

The result is

$$M(II, T) - n g_J \mu_B J\, B_J(y) \quad ; y = J x \tag{1.26}$$

where $B_J(y)$ is Brillouin function (Fig. 1.5):

$$B_J(y) = \frac{2J+1}{2J} \cot h\left(\frac{2J+1}{2J} y\right) - \frac{1}{2J} \cot h\left(\frac{1}{2J} y\right) \tag{1.27}$$

Explicitly

$$M(H, T) = n g_J \mu_B J \left[\begin{array}{l} \frac{2J+1}{2J} \cot h\left(\frac{2J+1}{2J} \frac{g_J J \mu_B H}{k_B T}\right) - \\ \frac{1}{2J} \cot h\left(\frac{1}{2J} \frac{g_J J \mu_B H}{k_B T}\right) \end{array}\right] \tag{1.28}$$

When either $H \to 0$ or $T >> 0$, $B_J(y)$ reduces to $(J+1)y/3J$ and a linear relation between M and H is found, with a positive magnetic susceptibility χ (quantum Curie's law):

$$M = n g_J \mu_B \frac{J(J+1)}{3} \frac{g_J \mu_B}{k_B T} H$$

$$\chi = \frac{M}{H} = n g_J^2 \mu_B^2 \frac{J(J+1)}{3 k_B T} = \frac{C}{T} \tag{1.29}$$

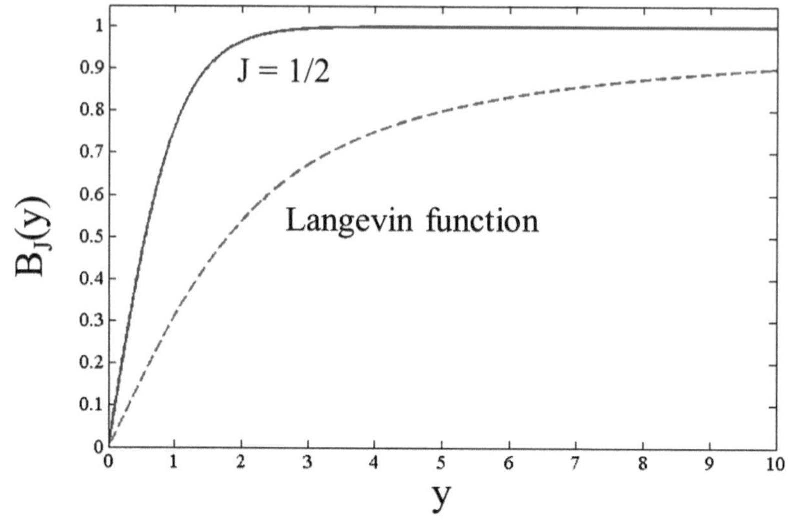

Figure 1.5 Brillouin functions for $J = 1/2$ and for $J \to \infty$ (i.e., the Langevin function). All curves with arbitrary, finite J lie between these extremes.

Paramagnetism can be viewed as the result of a competition between the ordering effect of a magnetic field and the disordering effect of thermal random torques acting on individual magnetic moments. The maximum magnetization of the ensemble of magnetic moments (indicated as the magnetic saturation) corresponds to the limiting case of perfect alignment of all moments along the field direction. If at a given positive temperature T the magnetic field is instantaneously removed, the field-induced orientational order of the magnetic moments is very rapidly lost (within a time of the order of the spin–lattice relaxation time [8]) and the system of magnetic moments spontaneously falls in the state of maximum entropy corresponding to $\mathbf{M} = \Sigma_i \boldsymbol{\mu}_i = 0$ (for $T \cong 0$).

The classical limit of Eqs. 1.27 and 1.28 is obtained by putting $J \to \infty$, $\mu_B \to 0$ in such a way that the product $g_J \mu_B J$ remains a constant ($= \mu$). In such a case

$$M = n\mu L(y); \qquad y = \frac{\mu H}{k_B T} \tag{1.30}$$

where the Langevin's function $L(y)$ is

$$L(y) = \cot h(y) - \frac{1}{y}; \qquad L(y) \cong \frac{y}{3} \quad (y \to 0) \tag{1.31}$$

1.3.2.2 Superparamagnetism

The classical expression, derived from an obsolete view of atomic moments as classical vectors, taking all orientations in space, was definitely abandoned in favor of the quantum-mechanical Brillouin approach; however, the Langevin function has recently enjoyed renewed interest because it is apt to describe the magnetic behavior of an assembly of noninteracting *magnetic nanoparticles*, each carrying a large magnetic moment arising from the sum of individual atomic magnetic moments aligned by ferri- or ferromagnetic interaction. This effect, called superparamagnetism because it involves "supermoments" continuously disordered by thermal effects but able to reorient if submitted to a magnetic field, has many interesting and/or promising applications in various areas such as biomedicine [9] and information and communication technologies (ICTs) [10]. On the other hand, superparamagnetism is a detrimental effect in magnetic recording based upon magnetic media used as memory elements. In fact, a parameter of central importance in a memory element is its stability at or around room temperature. In magnetic recording, this involves the requirement that the magnetization vector of a memory element stay constant in magnitude and direction for a long time (of the order of years). From this viewpoint, a superparamagnetic particle entirely lacks of thermal stability, its magnetic moment continuously changing in direction by effect of random torques of thermal origin. Superparamagnetism is therefore to be avoided in designing and producing media for magnetic recording. These aspects will be considered again in Sections 1.5.2.3 and 1.5.2.5 and in more depth in subsequent chapters of this book.

1.3.2.3 Quenching of the angular momentum

Strictly speaking, the approach to quantum paramagnetism we have just outlined applies to isolated atoms/ions where the most important perturbation term arises from spin–orbit interaction. A very good agreement between theoretical predictions and experiment is often found in rare-earth metals; the agreement is definitely worse for transition metal ions ($3d$) where the very starting point of the model (the L-S coupling scheme) is no longer valid and where the electronic orbitals of $3d$ electrons are affected to a much larger extent by the crystal field generated from the surrounding ions in a lattice [11]. As a consequence, the magnetic moment on a $3d$ metal atom/ion may be different from the prediction (Eq. 1.22). In particular, a crystal field of cubic/octahedral or tetrahedral symmetry has the peculiar effect of totally or partially removing the contribution of the orbital magnetic momentum in such a way that the dominant source of total magnetic moment on a given $3d$ metal atom/ion is the spin angular momentum. This quantum effect called *quenching of the angular momentum* is related to the new form and symmetry assumed by the eigenfunctions of the $3d$ electron orbitals with respect to the free atom case. In transition metal ions, the $3d$ orbitals are affected by the rather strong electric field of cubic/octahedral or tetrahedral symmetry generated by the neighboring atoms. Further details are to be found in textbooks of magnetism [11].

1.3.3 *Ferromagnetic and Antiferromagnetic Interaction*

Paramagnetism implies a uniform, field-sustained ordering of magnetic moments. The ferromagnetic and antiferromagnetic states imply the existence of a *self-sustained*, long-range order of magnetic moments instead; the origin is entirely quantum-mechanical.

The notion of an *exchange integral* was introduced first in the quantum-mechanical description of two-electron systems such as the hydrogen molecule. The electronic states of the system are obtained by means of the canonical first-order perturbation theory; a single and a triplet state, characterized by total spin $S = 0$ or 1, naturally emerge from the standard treatment of the problem (which can be retrieved in elementary texts of quantum mechanics [12]). The important point for our purposes is that the energies of these two orthogonal states differ by an amount proportional to the exchange integral, an energy term of electrostatic origin that stems from the full quantum treatment of indistinguishable electrons. In the H_2 molecule such a quantity is negative and the ground state of the system is the singlet, that is, the two electronic spins always are antiparallel on the ground state ($S = 0$). Calling the one-electron wavefunctions $\psi(\mathbf{r}_{(a,b),(1,2)})$ (subscripts a and b referring to the two nuclei, 1 and 2 to the two electrons), the exchange integral J is written as

$$J_{12} = \int \psi * (\mathbf{r}_{a1})\, \psi * (\mathbf{r}_{b2}) \left(\frac{e^2}{\mathbf{r}_{12}}\right) \psi (\mathbf{r}_{b1})\, \psi (\mathbf{r}_{a2})\, dV_1 dV_2 \quad (1.32)$$

To get $J > 0$ one needs to deal with systems whose electronic wavefunctions are high between adjacent nuclei so that the positive term $e^2/\mathbf{r}_{12}$ in the integral's kernel is large. Such a condition is fulfilled in transition metals and in rare-earth metals for electrons belonging to the incomplete shells ($3d$ and $4f$, respectively). When J is positive, the ground state of a two-electron system is a triplet, that is, the spins of individual electrons are firmly aligned. To excite the system from the ground state (parallel spins) to the higher-energy singlet (antiparallel spins) one should introduce a large energy in the system: the ferromagnetic (parallel) alignment of the spin is robust, particularly at low temperatures, or the ferromagnetic interaction energy is large.

On the contrary, when $J < 0$ the two electron spins are stably antiparallel: this condition may lead to the antiferromagnetic arrangement of magnetic moments in some crystals [2].

It may be interesting to note that the energies of the singlet and triplet states of a two-electron system can be written as

$$\begin{aligned} E_S &= K_{12} + J_{12} \\ E_T &= K_{12} - J_{12} \end{aligned} \quad (1.33)$$

where K_{12} is a constant energy; the values in Eq. 1.33 are coincident with the eigenvalues of the following Hamiltonian for the two-electron system:

$$H_{12} = K_{12} + J_{12} \left[-\frac{1}{2} - 2\mathbf{s}_1 \cdot \mathbf{s}_2 \right] \quad (1.34)$$

where the operator in brackets contains the spin operators $\mathbf{s}_1$, $\mathbf{s}_2$ of the individual electrons and admits the following eigenvalues: -1 when $\mathbf{s}_1$ is antiparallel to $\mathbf{s}_2$ (singlet) and $+1$ when $\mathbf{s}_1$ is parallel

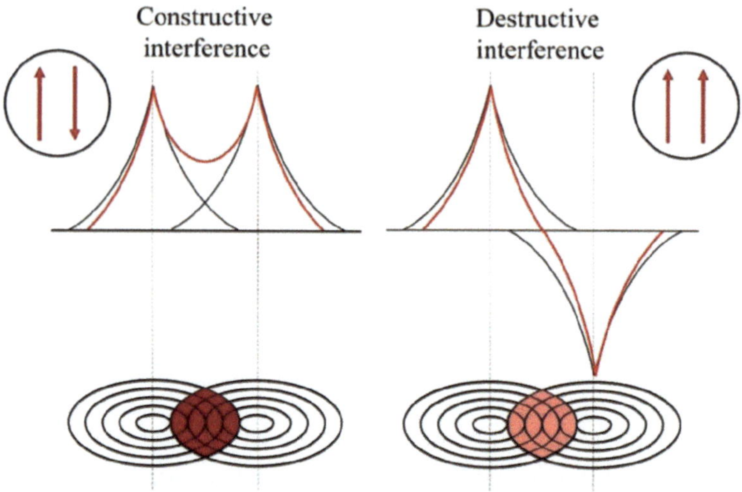

Figure 1.6 The parallel (ferromagnetic) arrangement of the spins of two adjacent transition metal atoms/ions corresponds to *negative* interference of electronic orbitals and hence to a lower electrostatic energy than the antiparallel arrangement.

to $\mathbf{s}_2$ (triplet). In other words, the energy values of the two-electron system can be accounted for by a *spin Hamiltonian*:

$$H_s = \left(K_{12} - \frac{1}{2}J_{12}\right) - 2J_{12}\mathbf{s}_1 \cdot \mathbf{s}_2 \rightarrow -2J_{12}\mathbf{s}_1 \cdot \mathbf{s}_2 \qquad (1.35)$$

Therefore a purely electrostatic, quantum interaction has been reformulated as an interaction between spins. This is not unexpected, because the Pauli principle for fermions implies a strong correlation between the electronic wavefunctions in space and the electronic spin states [12]. We can restate the previous sentence by saying that ferromagnetism/antiferromagnetism can be formally thought of as originating by a strong magnetic interaction (the magnetic moments of the electrons being univocally associated to their spins) albeit its origin is actually electrostatic in nature (Fig. 1.6).

When the two atoms in a bi-atomic molecule have more than one electron each, Eq. 1.35 still holds, with the total spin operators $\mathbf{S}_{1,2}$ substituting for the spins of individual electrons. If an n-atom system is considered, the Hamiltonian is formally written as

$$H_{\text{exch}} = -2\sum_{i<j} J_{ij}\mathbf{S}_i \cdot \mathbf{S}_j \tag{1.36}$$

where the sum is extended to all atom–atom pairs in the system. This is the Heisenberg Hamiltonian, which is a good starting point for more advanced treatments of ferromagnetic compounds (typically insulating materials) and which can be used—although on a much less firm basis—to describe the magnetic properties of $3d$ metals such as Fe, Co, and Ni. In fact, ferromagnetism in these metals arises from itinerant electrons and needs to be described by a different formalism (Stoner/Hubbard models [13]).

When J_{ij} is negative for spins on nearest neighbors and positive for the next-nearest neighbors, a crystal may develop *antiferromagnetism*. An antiferromagnetic crystal can be thought of as the combination of two interconnected sublattices of opposite magnetization. The magnetic moments of each sublattice are all aligned along a common direction (i.e., they are ferromagnetically interacting); however, the magnetization vectors of the two sublattices are identical in magnitude but opposite in sign. As a consequence, an antiferromagnetic material appears from outside as having no net magnetization.

A further important case occurs when the two sublattices in a crystal are magnetically not equivalent (e.g., when different magnetic ions with different moments belong in the two sublattices). In this case the magnetization vectors still have the opposite sign but they are different in magnitude in such a way that a residual magnetization is measured (the effect is called *ferrimagnetism* or imperfect antiferromagnetism). Important types of $3d$ metal compounds such as magnetite (Fe_3O_4), maghemite (γ-Fe_2O_3), cubic ferrites, and magnetic garnets belong to the category of ferrimagnetic materials.

1.3.4 *Mean-Field Approach to Ferromagnetism*

An elementary description of ferromagnetism cannot avoid mentioning the classical derivation of the effect by Weiss [1, 2]. Historically, this model has been the first mean-field description of a physical system characterized by a temperature-dependent order parameter. In fact, Weiss's theory of the molecular field is able to summarily describe not only the behavior of the spontaneous magnetization of a ferromagnetic system as a function of temperature but also the effect of a magnetic field on the magnetization of the system at all temperatures. A great number of mean-field approaches have been subsequently devised in the attempt to describe different physical effects in systems composed of a large number of interacting elements [14]. Generally speaking, the mean-field theory is a rather crude approximation of the actual behavior of any complex physical system (e.g., it completely neglects fluctuations of physical quantities and correlations among the elements of a system); however, it is easy to handle and provides valuable overall information at almost all temperatures. The Weiss approach fails both at very low temperatures because it neglects typical excitations of the ferromagnetic ground state (spin waves) and near the critical temperature for the existence of ferromagnetism (i.e., near the Curie temperature, T_C) [14]. This inadequacy is a common feature of all mean-field approaches.

On the other hand, the mean-field theory is able to predict a most important and general feature of all ferromagnetic systems, that is, the existence of a low-temperature ordered phase of magnetic moments (ferromagnetic phase) followed above a transition temperature by a magnetically disordered phase (paramagnetic phase). In terms of a thermodynamic description, such an order–disorder transition is a phase transition of the second order, characterized by an anomaly of the specific heat without latent heat at the transition temperature (the Curie temperature) [14].

How is this mean field theory linked to Heisenberg's quantum theory? Let us start from Eq. 1.36 modified in the following way:

$$H_{\text{exch}} = -\sum_{i<j} J_{ij} \boldsymbol{\mu}_i \cdot \boldsymbol{\mu}_j \qquad (1.37)$$

where the magnetic moments proportional to the spins have been substituted for the $\mathbf{S}_{i,j}$ operators and the exchange integral has been transformed accordingly (it contains the factor 2 also). Note that in 3*d* metals and their compounds the magnetic moment is mostly given by the spin angular momentum because of the quenching of the orbital angular momentum, as highlighted in the previous section. Formally one can further transform Eq. 1.37 by introducing an effective magnetic field operator for each lattice site i:

$$H_{\text{exch}} = -\sum_{i} \boldsymbol{\mu}_i \cdot \mathbf{H}_i \quad ; \; \mathbf{H}_i = \sum_{j \neq i} J_{ij} \boldsymbol{\mu}_j \tag{1.38}$$

The energy of the system at the temperature T is given by a quantum-statistical average of this Hamiltonian of the type

$$E_T = -\sum_{i} < \boldsymbol{\mu}_i \cdot \mathbf{H}_i >_T \;\; \approx \;\; -\sum_{i} < \boldsymbol{\mu}_i >_T \cdot < \mathbf{H}_i >_T \tag{1.39}$$

The last expression, where the statistical average has been decoupled, corresponds to the mean-field approximation. The average $<\mathbf{H}_i>T = \mathbf{H}_{\text{w}}$ is a quantity identical for all sites (called the *Weiss field*, internal field, or molecular field). According to Weiss, $\mathbf{H}_{\text{W}} = W\mathbf{M}$, where W is a numerical constant much larger than unity. Note that the (electric) Lorentz field existing in a spherical cavity ideally cut within a dielectric material is given by a closely similar expression ($\mathbf{E} = W_{\text{L}}\mathbf{P}$), where $W_{\text{L}} \approx 1$ [15].

In this framework, the temperature dependence of the magnetization $\mathbf{M}$ is obtained by admitting that the total magnetic field acting on a single magnetic moment is not just the applied field but the applied field plus the internal field: $\mathbf{H} = \mathbf{H}_{\text{ext}} + W\mathbf{M}$. Inserting the magnitude of the total field in Brillouin law (Eqs. 1.26 and 1.28), the reduced magnetization is written as

$$m = \frac{M}{M_0} = B_{\text{s}}(x); \qquad x = \frac{M_0 H_{\text{ext}} + W M_0^2 m}{n k_{\text{B}} T} \tag{1.40}$$

where S takes the place of J (this is a reasonable position in 3d metals and compounds), and $M_0 = n g_s \mu_{\text{B}} S$. A solution to Eq. 1.40 is easily obtained by looking at the intersection points between the following two curves:

$$m = B_{\text{s}}(x)$$
$$m = -\frac{H_{\text{ext}}}{W M_0} + \frac{n k_{\text{B}} T}{W M_0^2} x \tag{1.41}$$

For any value of H_{ext}, m is a monotonically decreasing function of T with a maximum at 0 K. The most important result of the model is that the equation admits solutions even in the absence of an external field. When $H_{\text{ext}} = 0$, one speaks of *spontaneous magnetization*; in this case the reduced magnetization m is a quantity ranging between

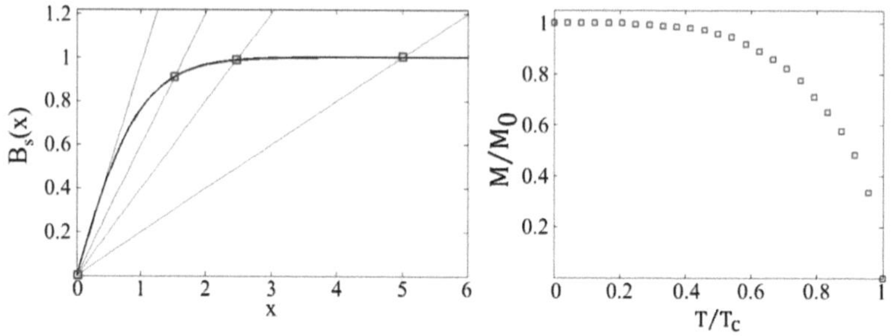

Figure 1.7 (Left) Graphical method to obtain the spontaneous magnetization M as a function of temperature T by exploiting Eq. 1.41 with $H_{ext} = 0$. The straight-line slope increases with increasing T. (Right) Ensuing plot of $M(T)$.

0 and 1 and plays the role of an *order parameter* for ferromagnetism (see Fig. 1.7).

In the high-temperature limit, the magnetic response of the system to an external magnetic field can be obtained by recalling that Brillouin function can be linearized $[B_S(x) \cong (S+1)x/3S]$ so that

$$M = M_0 B_s(x) \cong \frac{S+1}{3S} \frac{M_0^2 H_{ext} + W M_0^2 M}{nk_B T} \tag{1.42}$$

As a consequence, M can be expressed as a linear function of H_{ext}:

$$M = \left(\frac{S+1}{3S} \frac{M_0^2}{nk_B T}\right) \bigg/ \left(1 - \frac{S+1}{3S} \frac{WM_0^2}{nk_B T}\right) H_{ext} \tag{1.43}$$

The magnetic susceptibility is a positive quantity and takes the value

$$\chi = \left(\frac{S+1}{3S} \frac{M_0^2}{nk_B T}\right) \bigg/ \left(1 - \frac{S+1}{3S} \frac{WM_0^2}{nk_B T}\right) = \frac{C}{T - T_C} \tag{1.44}$$

where $C = (S+1)M_0^2/(3Snk_B)$ is the Curie constant (cf. Eq. 1.29) and $T_C = (S+1)M_0^2 W/(3Snk_B)$ is the Curie temperature. Above T_C, the system of spins behaves as a paramagnet and no spontaneous magnetization exists, even in the presence of interactions among magnetic moments. However, the magnetic susceptibility is divergent for $T = T_C$, that is, at the ferromagnetic–paramagnetic phase transition; below T_C a spontaneous magnetization exists (ferromagnetic phase). The spontaneous magnetization can be obtained either numerically or graphically. T_C is a measure of W, which is, in turn, related to the strength of the exchange integral J : the Curie temperature is a figure of the strength of the quantum ferromagnetic interaction in a system of atomic magnetic moments.

1.4 Magnetic Mesostructures in Bulk Materials

1.4.1 *Magnetic Crystalline Anisotropy and Other Anisotropies*

The ferromagnetic state as described either by the Heisenberg Hamiltonian or by Weiss's theory is intrinsically isotropic; in other words, the common direction of all magnetic moments aligned by the exchange interaction can equally point along any direction in space. However, in real materials the experimental evidence always indicates a directional dependence of the large-scale magnetic properties of ferromagnetic materials. One example of magnetic anisotropy, associated with the magnetostatic energy of a ferromagnetic body of anisotropic shape, was previously treated in Section 1.1.3. However, there are a number of other types of magnetic anisotropies, the most important being *magnetic crystalline anisotropy*. In fact, the magnetic response of a ferromagnetic crystal to an applied field typically depends on the direction of the field with respect to the crystallographic axes, which implies that there are preferred directions in space where the spontaneous magnetization of the system (i.e., the common direction of all aligned magnetic moments) is likely to be directed along in equilibrium conditions. These are the so-called *easy directions* for spontaneous magnetization, which are observed at all temperatures below T_C in crystalline ferromagnets. Such an effect is explained in terms of magnetic crystalline anisotropy having the symmetry of the ferromagnetic crystal. For instance, a ferromagnetic crystal with a hexagonal close-packed (hcp) structure, such as cobalt at room temperature, is characterized by a crystalline anisotropy of the type

$$E_{\mathrm{u}} = K_{\mathrm{u}}^{(1)}\sin^2\theta + K_{\mathrm{u}}^{(2)}\sin^4\theta + \ldots ; |K_{\mathrm{u}}^{(2)}| < |K_{\mathrm{u}}^{(1)}| \qquad (1.45)$$

where θ is the angle between the magnetization vector and the hexagonal c axis of the hcp cell. According to Eq. 1.45, the anisotropy energy is represented by a series of even power of $\sin\theta$, with coefficients rapidly decreasing in magnitude; often only the first term is kept. When $K_u^{(1)}>0$ as in cobalt, E_u is a minimum when $\theta = 0$ or π, so that the c axis is the easy axis and E_u has cylindrical symmetry around it. Note that the magnetization at equilibrium can be found to be aligned along the positive or negative easy axis, independent of its sign: this is *uniaxial*, not unidirectional, anisotropy (see Fig. 1.8).

Iron and nickel have cubic crystal structures (body-centered cubic [bcc] for α-Fe; face-centered cubic [fcc] for Ni) and are characterized by a magnetic crystalline anisotropy of the type

$$E_c = K_1\left(\alpha_1^2 + \alpha_2^2 + \alpha_3^2\right) + K_2\left(\alpha_1^2\alpha_2^2\alpha_3^2\right) + \ldots;\ |K_2| < |K_1| \quad (1.46)$$

where the α_i's are the three direction cosines of **M** with respect to the orthogonal axes of the cubic cell. Again, the anisotropy energy is represented by a series whose leading term is the first one. When $K_1 >0$ as in Fe, E_c is a minimum when **M** is aligned along one of the (100) directions of the cubic structure; when $K_1 <0$ as in Ni, E_c is a minimum when **M** is directed along one of the principle diagonals of the cubic cell, that is, the (111) directions (see Fig. 1.8).

The physical origin of the crystalline anisotropy lies in the spin–orbit interaction of magnetic atoms/ions; any rotation in space of the magnetic moment μ corresponds to a slight deformation of the electronic orbitals; the state of minimum anisotropy energy occurs when the direction of **M** (or equivalently the common direction of all atomic magnetic moments) is such that the electrostatic energy arising from the overlap of adjacent electronic clouds is a minimum. Such an effect of quantum nature can be viewed as a first- or second-order perturbation to the Heisenberg Hamiltonian. However, the quantum-mechanical treatment is rather involute, and a phenomenological approach, where the expression of the anisotropy energy is dictated by the crystal structure and the magnitude of the anisotropy coefficients is found by direct measurement, is typically preferred by the experimentalists. Finally, it should be recalled that even antiferromagnetic materials, albeit not displaying a net magnetization, do exhibit magnetic crystalline

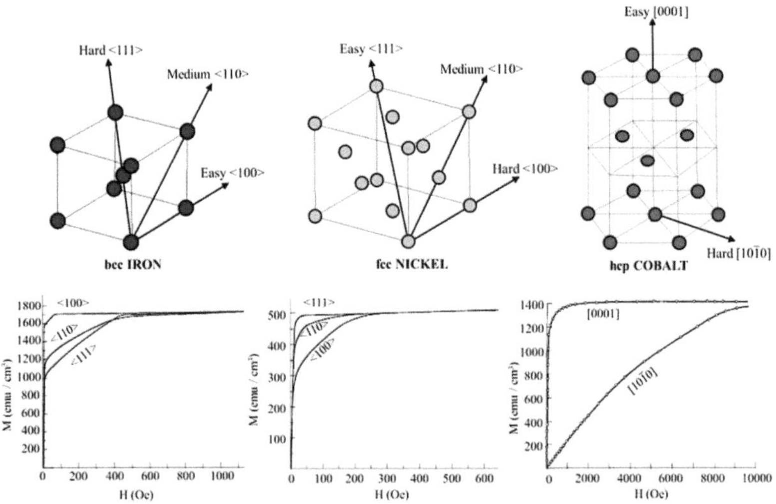

Figure 1.8 (Top) Easy, medium, and hard directions of magnetization in 3*d* transition metals. (Bottom) Corresponding behavior of the measured magnetization curves.

anisotropy. For this reason, antiferromagnetic layers are typically exploited in spintronics as an effective source of magnetocrystalline anisotropy [16]. Antiferromagnetic layers can remarkably modify the response of an adjacent ferromagnetic layer by inducing an *exchange bias field* [17], with an important fallout on magnetic hard disk technology.

In polycrystalline ferromagnetic materials comprised of a number of nonoriented single crystals as well as in amorphous ferromagnetic alloys [18] the crystal anisotropy virtually vanishes and other anisotropies, such as shape anisotropy, play a dominant role.

In fact, there are many other types of magnetic anisotropies in ferromagnetic materials. In bulk materials an important role can be played by the *stress anisotropy* related to magneto-elastic coupling and to magnetostriction [4]. In nanostructured and/or multilayered systems, *surface anisotropy* and *interface anisotropy* play a prominent role. Surface anisotropy primarily stems from the lack of continuity in the crystal symmetry at a surface and typically plays a major role in determining the effective anisotropy of a magnetic nanoparticle or ultrathin film [2]. Interfacial anisotropy typically occurs at the interface between hard-soft ferromagnetic layers or between a ferromagnetic and an antiferromagnetic layer and is basically related to the contact exchange interaction between layers; in FM/AFM systems this anisotropy is typically unidirectional.

1.4.2 *Magnetic Domains*

The magnetostatic energy of a ferromagnetic body (per unit area of the surface) is not negligible, as discussed in Section 1.1.2. To prove this, let us take a thin plate of a ferromagnetic material, of volume V and thickness t, uniformly magnetized *perpendicular to the plate surface*. To achieve this, we admit that the material is a single crystal of cobalt with the c axis directed perpendicular to the plate surface of area A, so the strong uniaxial magnetic crystalline anisotropy overcomes the shape anisotropy (which would cause **M** to lie in the plane of the plate). In this case the demagnetizing field antiparallel to **M** has magnitude

$$\mathbf{H} = -\frac{1}{\mu_0}\mathbf{M} \tag{1.47}$$

the demagnetizing factor being equal to 1 when the length and width of the plate are virtually infinite with respect to the thickness. By Eq. 1.10 the magnetostatic energy of the plate is

$$E_{\mathrm{m}} = -\frac{1}{2}\int \mathbf{M}\cdot\mathbf{H}dV = \frac{1}{2\mu_0}M^2tA \tag{1.48}$$

The energy per unit area of the surface is therefore

$$e_{\mathrm{m}} = \frac{1}{2\mu_0}M^2t \cong 4\times 10^5 M^2t \tag{1.49}$$

which is as large as about 1.4×10^4 J/m^2 for a 1 cm thick plate of Co (taking $M = 1.8T$). Such a strong energy density is reduced if the material's magnetization is no longer uniform. The easy direction of Co being perpendicular to the plane, the magnetization can take either the up or the down direction without changing the anisotropy energy. Let us therefore suppose that the material is divided into a set of *magnetic domains* of uniform width d where the magnetization is pointing up or down in such a way that each

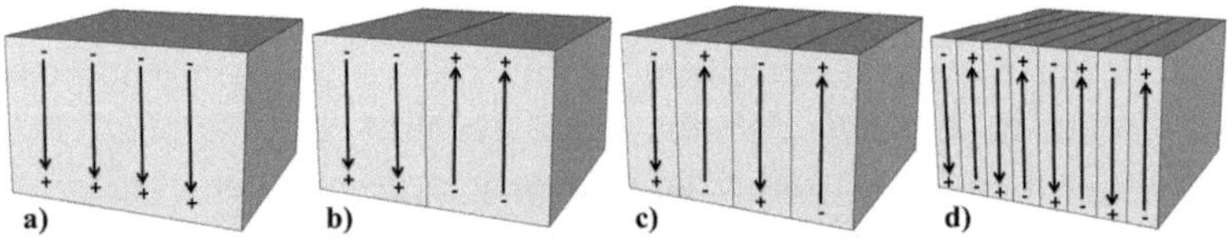

Figure 1.9 Nucleation of magnetic stripe domains effectively reduces the magnetostatic energy of a uniformly magnetized plate (a) by finely distributing magnetic poles of opposite sign on both surfaces (b–d). The magnetostatic energy continuously decreases from case (b) to (d).

domain with up-magnetization is surrounded by two domains of down-magnetization and vice versa. Let us further suppose that all magnetic domains have the same width d and plane boundaries; in this highly idealized case they can be referred to as *magnetic stripe domains* (Fig. 1.9). In this case, a straightforward calculation shows that when $d << t$, the magnetostatic energy per unit area of surface of the Co plate becomes

$$e_m \cong 1.08 \times 10^5 M^2 d \tag{1.50}$$

which is of course much lower than the result of the previous calculation (Eq. 1.49). Magnetic domains of alternating magnetization effectively decreases e_m with respect to the single-domain case, that is, to the uniformly magnetized material. Therefore, in a ferromagnetic material a magnetic domain pattern spontaneously develops at equilibrium in order to lower the magnetostatic energy of the system. In the highly idealized case we have treated, any two adjacent domains have opposite magnetization. Of course, the actual domain patterns of ferromagnetic materials strongly depend on their shape and on all magnetic anisotropies playing a role in the material and can be very complex [19]. In some cases, the magnetization just rotates by 90°, not 180°, from a domain to the adjacent one. However, the origin and purpose of magnetic domains always are the same, that is, they are spontaneously generated in the material in order to decrease the magnetostatic energy.

1.4.3 *The Single-Domain Regime*

Generally speaking, macroscopic magnetic materials spontaneously display magnetic domain patterns at equilibrium (i.e., in the absence of an applied magnetic field). However, granular magnetic media are a notable exception to this rule as they are composed of grains in a *single-domain* state even at zero applied field. Granular media are comprised of a very large number of magnetic nanograins with typical sizes ranging from a few units to some tens of nanometers. In the single-domain state each particle or grain contains an almost uniform, homogeneous magnetization (the only significant nonuniformity of **M** being found at the surface of the particle/grain). An explanation of the onset of the single-domain regime will be given in Section 1.5.2.3, where it will be shown how much dimensions matter in determining the magnetic behavior of a material. Single-domain media have become of notable importance in magnetic recording and will predictably play a central role in forthcoming recording technologies.

1.4.4 *Magnetic Domain Walls*

1.4.4.1 A simple model for 180° domain walls in bulk ferromagnets

In multidomain materials, between any two magnetic domains of opposite magnetization (we neglect here other cases) there is a transition region of finite width. Such a region is called *domain wall,* specifically a 180° domain wall when the magnetization is oppositely directed in adjacent domains A and B (Fig. 1.10).

The domain wall width δ can be estimated by recognizing that the relevant magnetic energies are mainly the exchange energy and the dominant magnetic anisotropy energy, which we suppose to be the magnetic *crystalline* anisotropy in this case. As we shall see, the domain wall thickness is given by a trade-off of antagonist energy contributions. Across a domain wall, **M** rotates by an angle of 180°. Let us suppose that complete rotation is achieved over a distance that is a multiple of the lattice constant of the material, a. Treating the spins as classical vectors of magnitude S, the exchange energy of a pair of magnetic moments at a distance a apart is given by

$$\varepsilon_{ij} = -2J\,S^2\cos\varphi_{ij} = \text{const.} + J\,S^2\varphi_{ij}^2 \tag{1.51}$$

where φ_{ij} is the angle between adjacent spins. The second equality stems from the fact that a complete rotation of **M** is expected to take

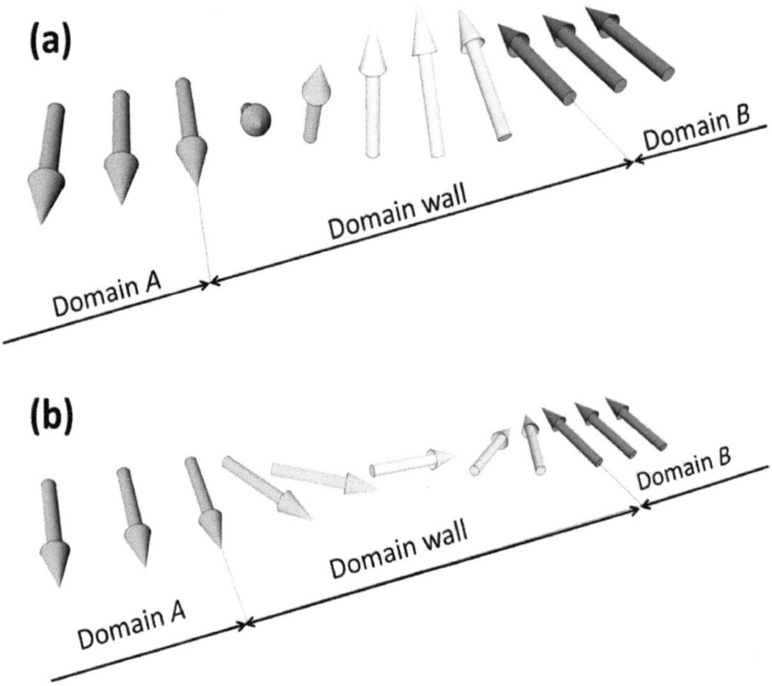

Figure 1.10 Sketch of 180° domain walls in a chain of magnetic atoms between stripe domains A and B of opposite magnetization. (a) The magnetization rotation takes place in a plane orthogonal to the directions of **M** in adjacent domains (Bloch-type wall). (b) The magnetization rotation takes place in a plane containing the directions of **M** in adjacent domain (Néel-type wall).

place over a distance much greater than a, so the angle φ_{ij} is a small quantity. If this thickness is $\delta = Na$, there will be N/a^2 magnetic atoms per unit surface of the wall (this surface corresponds to the material's cross section perpendicular to the axis of rotation of magnetic moments).

As a consequence, exchange energy is stored in the wall where the magnetic energy is higher than within a magnetic domain because of the rotation of **M**, which implies that the condition of parallel magnetic moments, dictated by the Heisenberg interaction, is not respected locally. The extra exchange energy per unit wall surface is

$$\gamma_{\text{exch}} \cong J S^2 \varphi^2 \times \frac{N}{a^2} = \frac{J S^2 \pi^2}{N a^2} = \frac{J S^2 \pi^2}{a \delta} \tag{1.52}$$

where the rotation of **M** has been assumed to be uniform over the domain wall width, so $\varphi_{ij} = \varphi = \pi/N$; γ_{exch} is the energy per unit surface of the wall. We explicitly note that γ_{exch} decreases with increasing δ; the exchange energy stored in the domain wall would become zero for infinite thickness.

On the other hand, a certain amount of magnetic crystalline anisotropy energy is stored in the wall also because there the magnetization vector is generally not aligned along an easy direction, different from what happens inside a magnetic domain. In a first approximation, this energy (per unit wall surface A) is given by

$$\gamma_{\text{an}} \cong |K| \frac{V}{A} = |K|\delta = |K| N a \tag{1.53}$$

V being the volume of the domain wall and A its surface (the domain wall being represented here as a thin foil of surface S and thickness δ). Here, $|K|$ is the magnitude of the dominant anisotropy constant for the considered material. For cobalt, $K \equiv K_u$; for Fe or Ni, $K \equiv |K_1|$ (see Section 1.4.1). The ratio V/S is equal to $\delta = Na$. The anisotropy energy density increases with increasing δ: it would become infinite in the limit of an infinitely thick domain wall. The total energy density for a domain wall in Fe or Ni is therefore

$$\gamma_{\text{tot}} = \gamma_{\text{exch}} + \gamma_{\text{an}} = \frac{J S^2 \pi^2}{a \delta} + |K_1|\delta \tag{1.54}$$

The minimum energy obtained by differentiating Eq. 1.54 with respect to δ can be cast in the form

$$\begin{aligned} \gamma_{\text{tot}}^* &= 2|K_1|\delta^* \\ \delta^* &= \left(J S^2 \pi^2 / |K_1| a\right)^{1/2} \approx \sqrt{J/|K_1|} \end{aligned} \tag{1.55}$$

The domain wall thickness is therefore proportional to the square root of the ratio between the exchange integral and the dominant anisotropy constant. For instance, materials with high Curie temperature (i.e., strong J) and/or low magnetic anisotropy host wider domain walls.

In α-Fe, $\delta^* \cong 4 \times 10^{-8}$ m $= 40$ nm; in Ni, $\delta^* \cong 7 \times 10^{-8}$ m $= 70$ nm. Domain walls are indeed rather thin: in $\alpha-$Fe, the thickness is about 150 times larger than the nearest-neighbor distance between two Fe atoms. In addition, a surface tension (energy per unit wall's surface) is stored in the wall. In iron, for instance, $\gamma^*_{\text{tot}} \cong 4 \times 10^{-3}$ J/m^2. To make a sensible comparison with other surface tensions, let us recall that the surface tension of water is of the order of 7×10^{-2} J/m^2.

1.4.4.2 Thin films: Néel domain walls

The domain walls described in the previous subsection are typical structures of bulk materials; they are usually referred to as *Bloch domain walls.* Across a Bloch domain wall, the magnetization vector rotates in such a way that no magnetic poles are created inside the wall (remember that magnetic poles would generate a magnetic field with an associated magnetic energy). The absence of magnetic poles in the bulk of the material is achieved by the condition that the rotation of the vector **M** takes place in planes parallel to the domain wall surface (in which case there is no component of **M** normal to the domain wall). As a consequence, the divergence of **M** is still zero both within the magnetic domains and in the domain walls—at least inside the material (see Eq. 1.3). However, the domain wall must terminate somewhere at the sample surface. Let us imagine the extremely simple case of a plate of ferromagnetic material with just one domain walls separating two domains where **M** is *parallel* to the plane of the plate (note the difference with respect to the arrangement of Section 1.4.2). The domain wall lies in a vertical plane. Although no magnetic poles are allowed to appear in the plate, they must appear at the plate's upper or lower surfaces (there the divergence of **M** is nonzero) and generate a magnetic stray field.

If the plate is sufficiently thick, the magnetostatic energy associated to these magnetic poles is negligible. In ultrathin magnetic films, however, the energy would be strongly enhanced. To avoid this, the magnetization rotation in thin films will take place *in the film plane,* resulting in a different region of transition between domains of opposite magnetization (named a *Néel domain wall*). In this case, of course, magnetic poles will appear within the film, the magnetization having a significant component in the plane of the domain wall (leading to a nonzero divergence of **M** across the film thickness); under a certain critical film thickness, however, the energy associated to magnetic poles dispersed in the volume of the film is lower than the one associated to magnetic poles appearing at the film surfaces. The critical thickness for the Néel–Bloch domain wall transition is of the order of 30–40 nm in Ni–Fe films at room temperature. Below the critical thickness, Néel walls are more favorable [19].

1.4.5 *Back to Magnetic Domains*

A domain wall stores magnetic energy (per unit wall surface). In a ferromagnetic body containing N ideal magnetic stripe domains, $N - 1$ domain walls exist. Although generating many domains helps reducing the magnetostatic energy (see Eq. 1.50), each added domain wall acts to increase the overall energy. A trade-off is therefore expected. Let us come back to the ferromagnetic plate of Section 1.4.2. Let L, w, and t be its length, width, and thickness, respectively; we suppose that $t \ll L$ and $t \ll w$. Now imagine that the plate is divided into many magnetic domains of width d; the magnetization vector **M** is always directed along t and the domain walls are rectangles of surface area Lt. The total number of domain walls is w/d; the total domain surface area is therefore Ltw/d. The surface area of the plate is Lw. Therefore, the total domain wall surface area per unit surface of the plate is just $(Ltw/d)/Lw = t/d$. As a consequence, the domain wall energy per unit surface of the plate is $e_{DW} = \gamma^*_{tot} t/d$ and the total energy density for a domain wall in Co is

$$e_{tot} = 1.08 \times 10^5 M^2 d + \gamma^*_{tot} \frac{t}{d} \tag{1.56}$$

Therefore, the magnetic domain width will be determined by the balance of two competing energy terms. The total energy can be easily minimized with respect to d; the minimum value e^*_{tot} occurs for

$$d^* \cong 3 \times 10^{-3} \frac{(\gamma^*_{tot} t)^{1/2}}{M} \tag{1.57}$$

and takes the value

$$e^*_{\text{tot}} \cong 6.6 \times 10^2 M \left(\gamma^*_{\text{tot}} t\right)^{1/2} \tag{1.58}$$

For instance, for a large plate of α-Fe with a thickness of 1 cm, the equilibrium domain spacing is $d^* \cong 5.6 \times 10^{-6}$ m and the total energy is $e^*_{\text{tot}} \cong 5.6$ J/m^2, to be compared with a magnetostatic energy as high as 2×10^4 J/m^2 in the absence of magnetic domains. Note finally that in this case the magnetic domain width d^* is orders of magnitude greater than the domain wall width δ.

Although domain walls bring about a moderate increase of surface energy, the saved magnetostatic energy of the sample is so large that a magnetic domain pattern will typically set up in bulk ferromagnetic materials at equilibrium. We shall see how this condition can be violated in fine particles.

1.4.6 *Magnetization Processes*

When a magnetic field is applied to a demagnetized ferromagnetic body, a magnetization along the field's direction appears in the material. The magnetic field applied to the ferromagnetic body generally is the sum of the external plus the demagnetizing field. The term "demagnetized" means that a more or less complex magnetic domain pattern exists at zero field, so the overall *apparent* magnetization of the body is zero. Inside each magnetic domain the magnetization is clearly nonzero; however, the vectorial sum of all magnetization vectors in all domains is zero. In practice, as we will see later, the condition of zero apparent magnetization is not easy to achieve.

When **H** increases, **M** is expected to increase, too, pointing along the same direction of **H**. This occurs through two types of ordering processes for **M**:

- Growth of energetically favorable domains where **M** already is nearly parallel to **H** at the expenses of other (unfavorable) domains; this is typically achieved by *domain wall motion.*
- Coherent rotation of the magnetization vector within each domain.

Generally speaking, both processes coexist during the magnetization process of a ferromagnetic body. For very large values of the applied field, the process will end with the whole body entirely magnetized along the direction of **H** (in which case the magnetization is said to have reached the saturation value). The response of a ferromagnetic material to an applied field, that is, the *technical magnetization* of the material, is expressed in terms of the differential magnetic permeability expressed as $\mu_{\text{diff}} = dB/dH$, B being the magnitude of the magnetic induction vector ($\mathbf{B} = \mu_0\mathbf{H}+\mathbf{M}$).

1.4.6.1 Domain wall motion

Domain wall motion is the main magnetization process at low and intermediate values of the applied field. We have previously seen how a domain wall stores energy. However, in actual materials, the domain wall energy is not uniform in space: in fact, it depends on many local factors, such as internal stresses and/or inhomogeneities of the material, including point defects, line defects, and inclusions of interstitial/substitutional atomic species [4, 18]. To understand the dynamics of a domain wall submitted to a field, it is possible to make use to some extent of concepts borrowed from the dynamics of a particle. At equilibrium, a domain wall will be in a local minimum of the free energy; however, the energy landscape is typically very complicated, containing a very large number of nearly equivalent minima or valleys separated by barriers of higher free energy. The minima are randomly distributed in space.

For simplicity, we consider a one-dimensional case (Fig. 1.11). The specimen is a plate of ferromagnetic material with perpendicular magnetization, containing many planar domain walls, exactly as in Section 1.4.3. A constant magnetic field is applied perpendicular to the plane of the plate. As a consequence, in one half of the magnetic domains the magnetization is already parallel to **H**, whereas in the other half the magnetization is antiparallel to **H**. Let us single out one of the domain walls—it is perpendicular to the plane of the plate, and its initial position is $x = x_0$, which is a minimum of the free energy. For small displacements $s = x - x_0$ (positive or negative) by effect of the applied field, the domain wall energy per unit wall surface changes as $e_W = 1/2\ \alpha s^2$, the

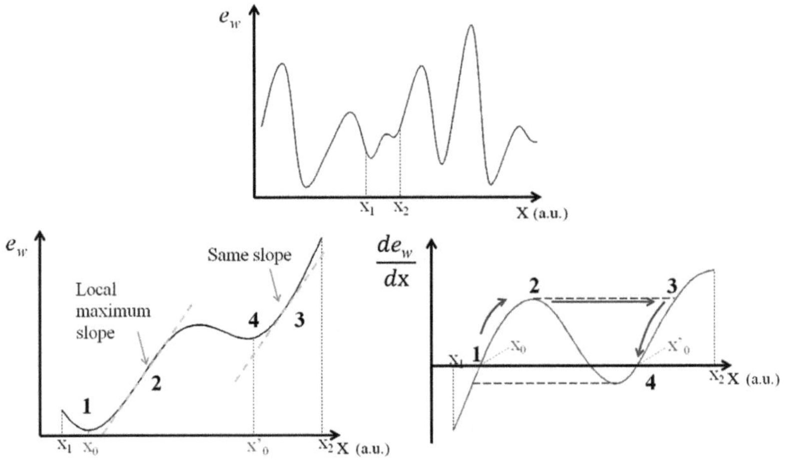

Figure 1.11 (Top) Generic multivalley energy landscape for a moving domain wall; (bottom left) detail of a region with two adjacent local minima; and (bottom right) resulting conditions for domain wall motion; irreversible domain wall jumps are indicated by the two dashed lines.

constant α being a measure of the steepness of the potential well where the domain wall is positioned. On the other hand, the Zeeman energy per unit wall surface varies with s as $e_Z = -2MHs$ because a displacement s of the wall brings about a complete reversal of the magnetization in the region swept by the moving wall. The total energy is therefore

$$e = \frac{1}{2}\alpha s^2 - 2MHs \tag{1.59}$$

By minimizing e with respect to s one gets the equilibrium value of the displacement under a constant field of magnitude H; this is $s* = 2MH/\alpha$, and the ensuing magnetization change is $\Delta M = 2Ms^*/L$, L being the length of the ferromagnetic body. Therefore we have

$$\Delta M = \frac{4M^2}{L\alpha}H \tag{1.60}$$

Assuming S to be the domain wall surface, for a ferromagnetic body of unit volume, $V = SL = 1$ or $S = 1/L$, the reversible magnetic susceptibility takes the value

$$\chi_{\text{rev}} = \frac{\Delta M}{\mu_0 H} = \frac{4M^2}{\alpha\mu_0}S \tag{1.61}$$

The quantity χ_{rev} measures the linear response of the material to a weak magnetic field and is associated to the *reversible* motion of 180° domain walls (if the magnetic field is removed, the domain wall goes back to x_0).

However, domain wall displacements are reversible only when s is very small: it can be proven that it has to be of the order of the domain wall thickness, that is, a fraction of a micrometer [4, 18]. Larger displacements are intrinsically irreversible.

Let us discuss a simple criterion for the irreversibility of the domain wall motion (Fig. 1.11). The restoring force (per unit domain wall surface) acting on a domain wall displaced from a local free-energy minimum in x_0 is $f_R = -de_W/dx$. Typically, the energy e_W is a parabolic function of s for a small s and becomes less steep beyond a certain displacement. This implies that f_R will display a maximum (f^*_R) somewhere (point 2 in Fig. 1.11). On the other hand, the Zeeman pressure (force per unit surface of the domain wall) is $f_Z = 2MH$ and increases with increasing magnitude of H. When $H \geq f^*_R/2M$, the maximum restoring force is no longer strong enough to counterbalance the Zeeman pressure and the domain wall becomes detached from the initial attractor centered in x_0 moving along x until it reaches a new position where the restoring force takes the same value, that is, the $e_W(x)$ curve has the same slope as in point 2 (point 3 in Fig. 1.11). In general, the new position is located around a different attractor x'_0, that is, in proximity of another minimum of the multivalley landscape of the free energy. There, f_R can be locally larger than $f^*_R = f_Z$ (see the region between point 3 and x_2 in the curve in the bottom-right panel of Fig. 1.11) and this stops the domain wall jump. If the magnetic field is removed, the domain wall cannot go back to the initial attractor x_0; instead, it will place itself in the new local free-energy minimum located in the position x'_0. The arrows in Fig. 1.11 (bottom-right panel) indicate the domain wall's movement from x_0 to x'_0 when the field H is first switched on, increased to the value $H = f^*_R/2M$, and finally switched off. Clearly a different position of the domain wall at rest corresponds to a different value of the sample magnetization.

The intrinsic irreversibility of the motion of a domain wall is the ultimate source of the *magnetic hysteresis* of ferromagnetic materials [20]. An example of this is given in Fig. 1.11, bottom-right panel.

Here, the field is first applied in one direction and then changed in sign in order to bring back the domain wall to the initial starting point x_0. Two irreversible domain wall jumps (indicated by the dashed horizontal lines) occur in the process.

By domain wall motion, both reversible and irreversible, favorable magnetic domains can grow at the expenses of the unfavorable ones, in sort of a process of "survival of the fittest." At high fields the ferromagnetic body will eventually contain a single magnetic domain.

Magnetization processes in materials containing domain walls can be divided into two broad classes, (a) processes dominated by domain wall motion and nucleation of new domain walls and (b) processes dominated by the pinning of domain walls [19]. Case (a) corresponds to the main mechanism of magnetization reversal in soft magnetic materials (see Section 1.5.2.1) and in large-grained hard or semihard magnetic materials (see Sections 1.5.2.2. and 1.5.2.4). In this case the material contains many domain walls that can be displaced easily, leading to a large initial permeability. Domain wall nucleation is likely to occur in materials where the domain wall surface energy is comparatively low (see Eqs. 1.54 and 1.55). Case (b) occurs in hard or semihard magnetic materials characterized by a strong interaction between domain walls and pinning centers (precipitations, impurities, defects). These effectively pin the domain walls, leading to a low initial permeability.

1.4.6.2 Magnetization rotation

We consider now a single-domain ferromagnetic crystal with a given magnetic crystalline anisotropy. Imagine that the external magnetic field is not applied along an easy direction of the material; the magnetization will be forced to rotate toward the magnetic field direction and away from an easy direction. Such a rotation may be either reversible or irreversible. In both cases, **M** will take a direction defined by the equilibrium condition between torques related to the magnetic crystalline anisotropy energy and by the Zeeman energy.

Considering a uniaxial magnetic crystalline anisotropy and calling θ the angle between **H** and the easy axis, and φ the angle between **M** and the easy axis, the total energy density is

$$E = K_u \sin^2\varphi - MH\cos(\theta - \varphi) \tag{1.62}$$

A significant rotation of **M** only takes place when $MH >> K_u$, that is, for fields of the order of 10^4–10^5 A/m for most common materials. When the field **H** is removed, **M** may or may not go back to the initial position, depending on the rotation being reversible or irreversible. In any case, at extremely large fields, **M** will eventually be aligned along **H**.

Rotation processes dominate the magnetization reversal of granular media exploited in magnetic recording; deviations from the ideal behavior are related in this case to the interaction existing between or among different magnetic units (i.e., grain–grain contact interactions involving an exchange mechanism, and/or magnetic dipolar interactions).

1.5 Magnetic Hysteresis Loops and Magnetic Losses

1.5.1 *Magnetic Hysteresis*

As anticipated, both irreversible domain wall displacements (or jumps) and irreversible magnetization rotations are the source of a distinctive feature of all bulk ferromagnetic materials, that is, *magnetic hysteresis*. By the very nature of the basically random effects that cause magnetic hysteresis (e.g., the field-induced random displacements of domain walls among quasi-equivalent local attractors) this effect must generally be treated by using rather complex statistical approaches. The interested reader should make reference to the specialized literature [20].

Here, the important parameters of the hysteresis loop are briefly highlighted (see Fig. 1.12). If a previously demagnetized ferromagnetic material is submitted to a very strong magnetic field along one direction, it will eventually reach the positive *magnetic saturation* M_s under sufficiently large fields. This corresponds to having all atomic magnetic moments oriented along the field direction. If the field is removed, the material in general will no longer reach the point (0, 0) in the (H, M) plane. Instead, it will

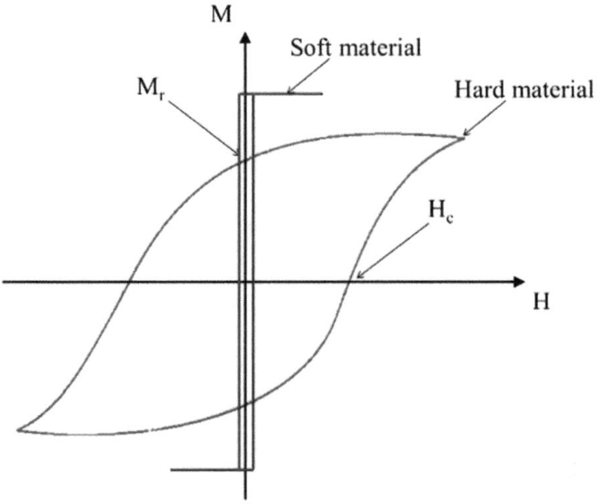

Figure 1.12 Typical magnetic loops of hard/soft magnetic materials. M_r is the magnetic remanence; H_C is the coercive field.

reach a point $(0, M_r)$ on the vertical axis, M_r being the magnetization that remains in the material (positive *remanent magnetization* or *magnetic remanence*). To reach the condition $\mathbf{M} = 0$ a magnetic field must be applied in the opposite (negative) direction. The point $(-H_C, 0)$ on the horizontal axis represents the negative *coercive field* or *coercivity* of the material. If the negative H is further increased in magnitude, the negative magnetic saturation will eventually be reached. Then if H is first brought to zero and then increased again in the positive region, the material will pass through the negative remanence and the positive coercive field, finally reaching the positive saturation state. In a symmetric loop the two values of saturation magnetization, remanence and coercive field, are equal in magnitude and opposite in sign. Under an alternating magnetic field the magnetization will cyclically change describing a closed loop. The demagnetized state (0, 0) is not easily accessed; it can be obtained by submitting the sample to an alternating magnetic field of slowly decreasing amplitude.

Magnetic hysteresis stems from irreversible magnetization processes. Irreversible processes generally imply a loss of energy, so magnetic hysteresis must be related to magnetic losses. Let us briefly discuss this important point. The general expression giving the work (per unit volume of the material) needed to change the magnetization of a magnetized body from a starting value $\mathbf{M}_i$ to a final value $\mathbf{M}_f$ under the effect of an external field $\mathbf{H}$ is

$$W = -\int_{\mathbf{M}_i}^{\mathbf{M}_f} \mathbf{H} \cdot d\mathbf{M} \tag{1.63}$$

During a cyclic magnetization process, the area enclosed by the loop

$$L = \oint \mathbf{H} \cdot d\mathbf{M} \tag{1.64}$$

is the energy lost in the loop (per unit volume). Such an energy appears as heat in the material. In metallic ferromagnets, eddy currents play a dominant role in generating heat. The irreversible domain wall jumps discussed above imply high values of $d\mathbf{M}/dt \approx d\mathbf{B}/dt$; in turn these give rise to strong induced electromotive forces and to high eddy currents flowing around the moving domain wall, leading to a remarkable heating effect. Fast domain wall jumps occur even if the cyclic magnetization of the sample is achieved by applying an almost static magnetic field $\mathbf{H}$. Insulating compounds containing transition metal ions are interesting for high-frequency applications, even if their spontaneous magnetization is lower than in metals, because of their very high resistivity, which helps to effectively reduce the eddy currents.

1.5.2 *Ferromagnetic Materials: An Inventory*

Ferromagnetic materials are customarily divided into at least two broad classes in dependence of their hysteretic properties (see Fig. 1.12): *soft* and *hard* magnetic materials. The most important applications for hard magnetic materials are *magnetic recording* and *permanent magnets.*

1.5.2.1 Soft magnetic materials

Magnetic materials are loosely termed *soft* when they can reverse most of their magnetization under the influence of relatively weak fields, comparable to or lower than the earth's magnetic field. Soft magnetic materials typically have low coercive field, high differential magnetic permeability, and high remanence. Their hysteresis loops are almost rectangular in the (H, M) plane. Soft magnetic materials are looked after for electrical applications as efficient flux amplifiers in motors, generators, transformers, switches, and a variety of light-duty devices.

For decades soft magnetic materials were exploited as sensing elements (read heads) in magnetic recording. They have been definitely discarded in favor of more sensitive nanodevices based on spintronic effects [16]; however, they are still in use as writing elements (write heads). Often soft magnetic materials are exploited in heavy-duty devices such as motors and transformers, which work at a typical magnetizing frequency of 50–60 Hz. The requirement of minimizing the magnetic losses during the cyclic magnetization process implies that the hysteresis loop of a typical soft magnetic material should be as narrow as possible.

1.5.2.2 Hard magnetic materials for permanent magnets

Permanent magnets are passive devices exploited to generate a magnetic field constant in time and uniform without using electrical power and without generating heat. Such a field originates from the magnetic poles of **H** present at the sample's surfaces. The energy needed to keep the magnetic field on has been stored during an initial, single-shot magnetization process and remains frozen in the material for a very long time. In fact, the frozen-in energy—which is mainly the energy associated to the generated field—is not reduced by repeated use of the device, because the magnetic field does not do any work on the surrounding bodies. Materials for permanent magnets are characterized by very high coercivity, very high remanence, and a very high *energy product*, which is a measure of the magnetostatic energy (per unit volume) that the permanent magnet is able to maintain in the air gap of the open magnetic circuit where it is inserted (imagine a horseshoe magnet, for instance). In permanent magnets, the hysteresis loop of the material should be as large as possible. In fact, a typical permanent magnet will *never* be submitted to a complete cyclic magnetization process; it will operate at the magnetic remanence or close to it instead. It should be explicitly noted that the ultrahigh coercivity of a permanent magnet can be achieved by using a material with ultrahigh magnetic crystalline anisotropy. In fact, for a single-domain uniaxial particle, according to Eq. 1.62, the magnetic field applied antiparallel to **M** ($\phi = \pi$) able to flip the magnetization vector from $\theta = 0$ to $\theta = \pi$ is given by $\mathbf{H}_k = 2K_u/M$, where $\mathbf{H}_k$ is the anisotropy field. Different types of materials (either natural or artificial) with ultrahigh magnetic crystalline anisotropy are nowadays commercially available; they include, for example, rare-earth intermetallics such as YCo_5, $SmCo_5$, $Nd_2Fe_{14}B$, $TbFe_2$, and the ordered $L1_0$–FePt alloy, among others [2].

1.5.2.3 Magnetic nanoparticles

Magnetic nanoparticles have typical sizes between 3–4 nm and a few hundred nanometers; below the lower boundary the term "magnetic clusters" is to be preferred; above the higher boundary one speaks of microparticles.

A striking variety of systems containing magnetic nanoparticles can be prepared by many different techniques [21], resulting in a remarkable variety of magnetic properties that depend on nanoparticle size, shape, concentration, and aggregation state. Metal or oxide nanoparticles carrying a permanent magnetic moment can be obtained as dried nanopowders, often surrounded by some passivating shell, or they can be dissolved in fluid hosts or embedded in diamagnetic solids to produce magnetic nanocomposites.

Magnetic nanoparticles below a certain critical size (of the order of tens of nanometers but significantly dependent on the properties of the component material) are in the single-domain state, with the magnetization lying along one of the crystal's easy directions. This can be explained recalling that while magnetostatic energy is a volume energy (see Eqs. 1.9 and 1.10), depending on the particle size L as L^3, domain wall energy is a surface energy (see Eq. 1.58 and the ensuing discussion) with an L^2 dependence. Therefore, for sufficiently small L, the reduction of magnetostatic energy achieved by introducing at least two magnetic domains and one domain wall in a particle would be lower than the increase of the domain wall energy, preventing the nucleation of the domain wall.

If the nanoparticle size is so small that the product KL^3 (where K is the dominant anisotropy energy in the nanoparticle) is less than the thermal energy $k_B T$, the anisotropy energy is no longer able to keep the magnetization aligned along an easy direction indefinitely. As anticipated in Section 1.3.2.2, the vector **M** therefore flips among equivalent easy directions with a temperature-dependent rate that can be high with respect to the typical observation time; in this case, the nanoparticle is said to be in the *superparamagnetic state*, because **M** statistically behaves, in a sense, as a supermagnetic moment of a paramagnetic substance. The magnetic response of a set of independent superparamagnetic nanoparticles can therefore be roughly described by a Langevin function (see Eqs. 1.30 and 1.31).

A common feature of magnetic nanoparticle systems is the existence of low-temperature *blocking* effects. Blocking occurs when $k_B T$ becomes much less than KL^3. In a simple picture an assembly of free, magnetically independent nanoparticles is characterized by a high-temperature superparamagnetic state followed at lower temperatures by a magnetically *blocked* state emerging as a result of the blocking of **M** in each particle by magnetic anisotropy.

Of course magnetic nanoparticles can magnetically interact in a variety of ways, depending on their concentration and state of agglomeration. In systems where contact interactions are minimized, magnetic dipolar interaction plays a central role. At low temperatures, this interaction brings about frozen collective states where the moment directions are dictated by the interplay of energy contributions of random sign [22]. However, significant effects of the dipolar interaction are measured over a wide interval of temperatures [23]. It is worth mentioning that magnetic effects in nanoparticles are basically the same as the ones observed in granular media for magnetic recording. In both cases the size of a magnetic unit (either a nanoparticle or a grain) determines the magnetic response and the thermal stability. In both cases interparticle or intergrain interactions determine the collective behavior of systems comprised of a large number of magnetic units.

1.5.2.4 Materials for magnetic recording

As mentioned, soft magnetic materials are used as write elements in magnetic recording; on the other hand, hard magnetic materials are typically exploited in the recording medium where the information is physically stored. The recording of data is based on the property of ferromagnetic materials to retain a magnetized state after the removal of the external magnetic field. The ability to retain information is therefore measured by the remanence of the material and by its stability in time. Of course the coercive field is an important property of magnetic recording materials, too, because in order to switch a material's remanence from, for example, positive to negative, a field of the order of H_C is to be used. Another parameter characterizing a material for magnetic recording is *loop squareness*. Magnetic memories involve not only hard disks but also magnetic stripes on credit cards, personal documents, and key cards.

1.5.2.5 Meeting the requirements for magnetic recording

The coercive field of a material for magnetic recording must fulfill the requisite of being sufficiently high to prevent undesired loss or change of the stored information (even if H_C must not take extreme values as in permanent magnets, because the recorded information needs to be erased or modified at will by the write head).

Let us briefly recall here a few standard requirements for magnetic recording without entering details, which will be dealt with most authoritatively in subsequent chapters of this book. A recording media must be cheap enough to allow for large-scale production and diffusion; it should have high coercivity in order to keep the stored information for long times; and it should be composed of small particles in order to reduce bit size.

Particulate media comprised of microparticles of low-cost magnetic materials (often Fe oxides such as γ-Fe_2O_3 and Fe_3O_4, or CrO_2) have been used for many years. Because of the low packing density of particulate media, the interparticle interaction could be neglected as a first approximation, and the magnetization reversal was thought of as governed by the intrinsic properties of single particles. However, this technology appears no longer to comply with most of the requirements posed by the modern science of information. Such particulate media are still exploited in a number of low-cost, low-performance applications, for example, in memory cards or flexible magnetic media; they are being increasingly substituted by other low-cost memory elements such as embedded electronic circuits (electronic chips).

Granular thin films with a strong *perpendicular* crystal anisotropy have been the subject of intense research activity and are in use in hard disk memories for more than 10 years [24]. In this case, the perpendicular anisotropy dominates over the natural tendency of the magnetization to lie in the plane of a thin magnetic layer by effect of shape anisotropy; as a result, a higher bit surface density ensues.

As a matter of fact, magnetic recording media able to meet the requirements of the computer industry must have a bit size as small as possible in order to pack more information per unit area (i.e., the *areal* density of bits must be as high as possible). This implies a continuous shrinking of the average in-plane size of the grains, forming a single bit. However, a natural boundary to the continuous decrease of grain (and bit) size is imposed by the constraint of never overcoming the so-called superparamagnetic limit. We recall that a superparamagnetic nanoparticle/nanograin quickly loses any stored information (see Section 1.3.2.2). A countermeasure is to increase the magnetic anisotropy of the nanoparticles/nanograins, while reducing their size: in fact, the parameter determining the onset of superparamagnetism is KL^3 (L being the nanoparticle size; see Section 1.5.2.3).

One disadvantage is given by the fact that stability in data storage implies a difficulty in data recording, because the magnetic field should overcome ultrahigh anisotropy in order to modify the magnetization direction. This problem can be solved, for example, by special writing techniques where local heating of the magnetic support is exploited in data recording (heat-assisted magnetic recording, HAMR, technology) [26, 28].

One further requirement of the magnetic recording industry is the homogeneity of the recording media, which is not automatically ensured by the use of nanogranular materials. A very narrow, almost delta-like bit size distribution would be the optimal solution in order to establish a one-to-one relationship between memory bits and single magnetic units (a wide distribution of particle/grain shapes and sizes implies the need for having more than one magnetic unit in a single bit in order to keep the signal-to-noise ratio sufficiently low [24]; however, this is in obvious contrast with the need of reducing the bit size). A possible solution is to create artificial nanoparticles (e.g., magnetic nanodots) through nanolithography, which is so far a most expensive technique, not yet fully developed at the nanometer scale [29–31]. Regular arrays of magnetic nanoparticles with high perpendicular anisotropy have been actively investigated as possible materials for data storage [25–27]; the array regularity can be ensured by proper self-assembling of nanoparticles, typically mediated by organic shells (self-ordered magnetic arrays, or SOMAs). This technology is an alternative to lithographic bit-patterned recording (see below). Ordered arrays where the single magnetic units are nanoparticles characterized by a ultrahigh perpendicular magnetic anisotropy (such as $L1_0$-FePt nanoparticles) combine the advantages of dealing with a very small bit size with long-term thermal stability at operation temperature.

In any case, even using the best and most suitable magnetic materials and related technologies, superparamagnetism is a limiting factor in the indefinite increase of the areal density of stored information; the upper limit of achievable areal density is estimated to be around 50 Tbits/in^2 and is not very far from the performances of the best products on the market (around 1 Tbit/in^2).

A deeper insight into this subject is beyond the scope of a general introduction. The reader should address subsequent chapters of this book (e.g., Chapter 2) and recent review papers [26, 27, 32–34].

References

1. Nolting W., and Ramakanth A. (2010). *Quantum Theory of Magnetism* (Spinger Verlag, Germany).
2. Coey J. M. D. (2010). *Magnetism and Magnetic Materials* (Cambridge University Press, UK).
3. Skomski R. (2008). *Simple Models of Magnetism* (Oxford University Press, UK).

4. Cullity B. D., and Graham C. D. (2009). *Intorduction to Magnetic Materials,* 2nd Ed. (IEEE Press, Wiley, USA).
5. Landau L., and Lifshits E. (1935). On the theory of the dispersion of magnetic permeability in ferromagnetic bodies, *Physik. Z. Sowjetunion,* **4**, pp. 153–169.
6. Slonczewski J. C. (1996). Current-driven excitation of magnetic multilayers, *J. Magn. Magn. Mater.,* **159**, pp. L1–L7.
7. Piramanayagam S. N., and Srinivasan K. (2009). Recording media research for future hard disk drives, *J. Magn. Magn. Mater.,* **321**(6), pp. 485–494.
8. Slichter C. P. (1990). *Principles of Magnetic Resonance,* 3rd Ed. (Springer Verlag, Germany).
9. Pankhurst Q. A., Thanh N. K. T., Jones S. K., and Dobson J. (2009). Progress in applications of magnetic nanoparticles in biomedicine, *J. Phys. D: Appl. Phys.,* **42**, p. 224001.
10. Sarkar S., Guibal E., Quignard F., and SenGupta A. K. (2012). Polymer-supported metals and metal oxide nanoparticles: synthesis, characterization, and applications, *J. Nanopart. Res.,* **14**, p. 715.
11. Yosida K. (1996). *Theory of Magnetism* (Springer Verlag, Germany).
12. Landau L., and Lifshits E. (1965). *Quantum Mechanics. Non-Relativistic Theory,* 2nd Ed. (Pergamon Press, UK).
13. White R. M. (1970). *Quantum Theory of Magnetism* (McGraw-Hill, USA).
14. Yeomans J. M. (1992). *Statistical Mechanics of Phase Transitions* (Oxford University Press, UK).
15. Kittel C. (2005). *Introduction to Solid State Physics,* 8th ed. (John Wiley and Sons, USA).
16. Hirota E., Sakakima H., and Inomata K. (2002). *Giant Magnetoresistance Devices* (Springer Verlag, Germany).
17. Nogués J., and Schuller I. K. (1999). Exchange bias, *J. Magn. Magn. Mater.,* **199**, pp. 203–232.
18. Chikazumi S. (1997). *Physics of Ferromagnetism,* 2nd Ed. (Oxford University Press, UK).
19. Hubert A., and Schäfer R. (2009). *Magnetic Domains,* 3rd Ed. (Springer Verlag, Germany).
20. Bertotti G. (1998). *Hysteresis in Magnetism* (Academic Press, Elsevier, the Netherlands).
21. Tartaj P., Del Puerto Morales M., Veintemillas-Verdaguer S., Gonzalez-Carreñ T., and Serna C. J. (2003). The preparation of magnetic

nanoparticles for applications in biomedicine, *J. Phys. D: Appl. Phys.*, **36**, pp. R182–R197.

22. Fiorani D., and Peddis D. (2014). Understanding dynamics of interacting magnetic nanoparticles: from the weak interaction regime to the collective superspin glass state, *J. Phys.: Conf. Series*, **521**, p. 012006.
23. Allia P., and Tiberto P. (2011). Dynamic effects of dipolar interactions on the magnetic behavior of magnetite nanoparticles, *J. Nanopart. Res.*, **13**, pp. 7277–7293.
24. Srinavasan K., and Piramanayagam S. N. (2011). Perpendicular recording media, in *Developments in Data Storage: Materials Perspective*, eds. Piramanayagam S. N., and Chong T. C. (John Wiley and Sons, NJ, USA), pp. 52–77.
25. Sun S., Murray C. B., Weller D., Folks L., and Moser A. (2000). Monodisperse FePt nanoparticles and ferromagnetic FePt nanocrystal superlattices, *Science*, **287**, pp. 1989–1992.
26. Weller D., Mosendz O., Parker G., Pisana S., and Santos T. S. (2013). L10 FePtX–Y media for heat-assisted magnetic recording, *Phys. Status Solidi*, **A 210**, pp. 1245–1260.
27. Varvaro G., Laureti S., and Fiorani D. (2014). L10 FePt-based thin films for future perpendicular magnetic recording media, *J. Magn. Magn. Mater.*, **368**, pp. 415–420.
28. Ju G., Challener W., Peng Y., Seigler M., and Gage E. (2011). Heat-assisted magnetic recording, in *Developments in Data Storage: Materials Perspective*, eds. Piramanayagam S. N., and Chong T. C. (John Wiley and Sons, NJ, USA), pp. 193–222.
29. Sellmeyer D. J., and Skomski R. (2006). Introduction, in *Advanced Magnetic Nanostructures*, eds. Sellmeyer D. J., and Skomski R. (Springer Verlag, Germany), pp. 1–12.
30. Weller D., and McDaniel T. (2006). Media for extremely high recording, in *Advanced Magnetic Nanostructures*, eds. Sellmeyer D. J., and Skomski R. (Springer Verlag, Germany), pp. 295–324.
31. Thomson T., and Terris B. D. (2011). Patterned magnetic recording media: progress and prospects, in *Developments in Data Storage: Materials Perspective*, eds. Piramanayagam S. N., and Chong T. C. (John Wiley and Sons, NJ, USA), pp. 256–276.
32. Piramanayagam S. N. (2007). Perpendicular recording media for hard disk drives, *J. Appl. Phys.*, **102**, pp. 011301-1–011301-22.
33. Piramanayagam S. N., and Srinivasan K. (2009). Recording media research for future hard disk drives, *J. Magn. Magn. Mater.*, **321**, pp. 485–494.

34. Weller D., Parker G., Mosendz O., Champion E., Stipe B., Wang X., Klemmer T., Ju G., and Ajan A. (2014). A HAMR media technology roadmap to an areal density of 4 Tb/in^2, *IEEE Trans. Magn.*, **50**, pp. 3100108-1–3100108-8.

Chapter 2

Hard Disk Drives: Fundamentals and Perspectives

Gerardo Bertero, Guoxiao Guo, Shafa Dahandeh, and Anantha Krishnan

Western Digital Corporation, San Jose, CA 94062, USA
gerardo.bertero@wdc.com

This chapter presents a high-level view of the principles and technologies that enable hard disk drives (HDDs) to store massive amounts of digital information, highlighting, at the component level, some of the major innovations or features that made possible the approximately 500 million increase in recording areal density from the introduction of the first IBM RAMAC drive in 1956 operating at 2000 bits/in^2 to the latest commercially available state-of-the-art drives of 2015 approaching the 1 Tb/in^2 mark.

2.1 Magnetic Recording Fundamentals

The last major technological transition in hard disk drives (HDDs) took place in the period spanning 2005 to 2008 when the industry

Ultrahigh-Density Magnetic Recording: Storage Materials and Media Designs
Edited by Gaspare Varvaro and Francesca Casoli

ISBN 978-981-4669-58-0 (Hardcover), 978-981-4669-59-7 (eBook)
www.panstanford.com

shifted from longitudinal magnetic recording (LMR) to perpendicular magnetic recording (PMR). From 2005 to 2015, approximately eight generations of PMR drives have been introduced commercially, delivering a tenfold increase in areal density. The rate of areal density growth has now decreased significantly as PMR becomes a mature technology. While new major technologies such as heat-assisted magnetic recording (HAMR) and bit-patterned recording (BPR) are being developed, the industry is also introducing shingle magnetic recording (SMR) and two-dimensional magnetic recording (TDMR) as ways to extend the range of applicability of current PMR components aiming to deliver higher areal density and storage capacities.

Figure 2.1 shows an exploded view of an HDD, highlighting its major components. The drive may contain one or more media platters. The magnetic heads are mounted at the end of a suspension arm in the head stack assembly (HSA), as shown in Fig. 2.2. The

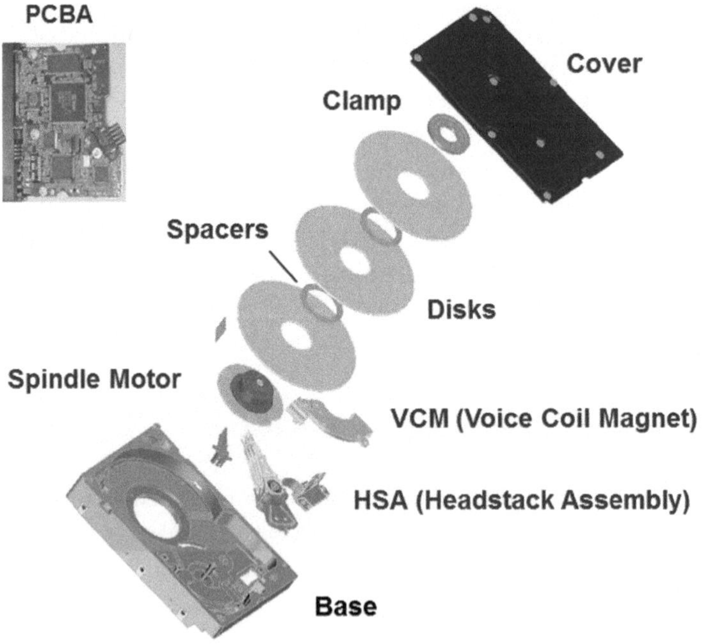

Figure 2.1 Exploded view schematic depicting the main components of a hard disk drive.

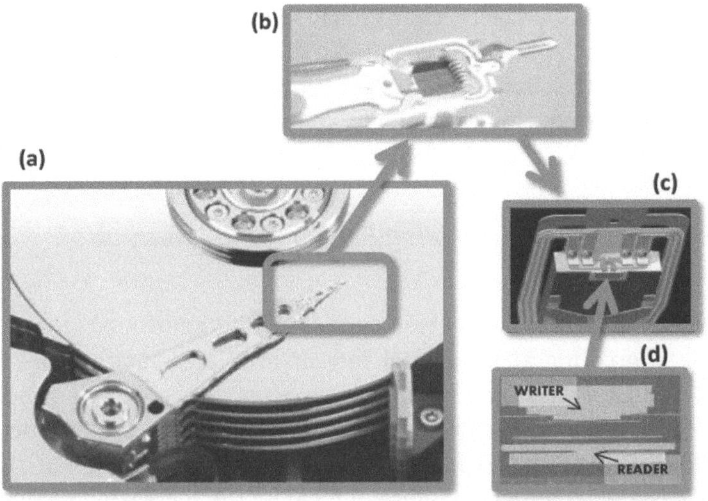

Figure 2.2 Close-up of the inside of a hard disk drive, showing the media platters and the head stack assembly (source: Western Digital corporate website). The sequence of insets (b, c, d) shows in increasing detail the placement of the write and read sensors at the air-bearing surface of the magnetic head slider.

air-bearing surface of the slider is topographically engineered to guarantee stable flying of the slider over the entire surface of the media at an appropriate pitch angle. To maintain stable flying dynamics, even in the presence of vibrations or gentle shock, or changes in humidity, temperature, or altitude, the slider is designed to develop high negative and positive pressure points. Air pressures as high as or higher than 100 atm. are common. Some latest drives are even sealed hermetically in a He or N_2 environment for better control of variables.

In today's state-of-the-art drives, during read and write operations, the thin-film reader and writer clearance to the surface of the disk is adjusted dynamically by means of thermally induced protrusion produced with appropriately placed, built-in solid-state heaters operated to control the flight height at approximately 1.3 nm clearance. Such a small distance from writer and sensor to the media is necessary to achieve high-resolution recording. The magnetic media disk is manufactured with extreme tolerances as well. Typical

R_a roughness characteristic of today's media is ~3.5 Å or less and the entire surface is glide-certified to be practically defect free. This is to ensure that the head does not get damaged by asperities on the media surface during normal operation.

2.1.1 *Magnetic Recording Relevance in Information Storage*

The amount of data created, replicated, and consumed in a single year is growing at an annual rate of over 45%, and by the year 2020 digital data generation is expected to be of the order of 40,000 exabytes. Fueling this phenomenon is the proliferation of mobile devices and social media, resulting in an explosion of digital images and video. This is compounded by the increasing resolution of video and digital cameras and by the advent of new forms of data originating from widespread use of devices such as digital sensors and applications such as surveillance video. In this context, the impact of magnetic storage technology on people's lives and economic activity in general is easily recognized by experts and non-experts alike. While some of this data is redundant and some even transient, most requires long-term storage. The most common method for keeping such volume of information is by magnetic recording stored not only in personal HDDs but increasingly so in large centralized data centers.

Underpinning this impact is the fast and sustained growth in recording areal density delivered by the HDD industry. The material that follows will introduce basic magnetic recording concepts and describe the principles and features that enable HDDs to store information at such incredibly high capacities. Unless otherwise noted, Chapter 2 will focus on conventional perpendicular magnetic recording (CPMR) technology. Occasionally, we will draw comparisons and refer to older technologies when illustrating how innovation was used to overcome hard technical challenges. Since the amount of literature available on HDDs and magnetic recording is already extensive (e.g., Refs. [1–11]), we will give emphasis to more recent materials and structure breakthroughs, which are somewhat less extensively addressed in older textbooks.

2.1.2 *Head Field Basics*

The magnetic recording heads in today's drives are highly specialized devices that operate at the limit of what is technologically possible. The making of a typical recording head takes approximately 5 months and well over 1000 process steps, starting from a blank wafer and ending as a submillimeter-size slider mounted on an electromechanical arm assembly, as shown in Fig. 2.2. A typical dimension, at the moment of writing this book, for the magnetic sensing element of the read head is 4–6 nm in thickness and 20–30 nm in width and height. These sensing elements are part of a complex stack of magnetic and nonmagnetic materials highly engineered to exploit fundamental physical principles that make use of spin polarization, exchange biasing, antiferromagnetic (AFM) coupling, quantum mechanical tunneling, and a myriad more effects to deliver high-quality signals that can be processed practically error free by advanced encoding and decoding channels. The writer element of the magnetic head is also a small wonder in its own. The write structure of the head utilizes the principle of induction to deliver fields on the media capable of writing on high-coercivity materials that can keep the information for at least 10+ years. The most critical materials used in the writer poles have the highest-known saturation magnetization values in nature (~2.4T). A write head structure is designed to deliver not only high-magnitude magnetic fields but also, just as importantly, high magnetic field gradients and high switching rates well into the gigahertz range. For this reason, the main pole in a state-of-the-art PMR writer is properly shielded to maximize field gradients and highly engineered to minimize electrical inductance.

The basic components of a writer are the magnetic core, the electrical coil, and the magnetic gap, g, as shown schematically in Fig. 2.3a. The basic function of the writer is to convert an electrical signal into a magnetization pattern in the media. Simply stated, current through the coil of the inductor head produces a magnetic field in the gap. The fields from a magnetic induction recording head can be estimated with high accuracy by knowing the geometry and surface potential of the magnetized structures [12–16]. Figure 2.3b shows a typical geometry for a longitudinal head having a gap, g, and pole

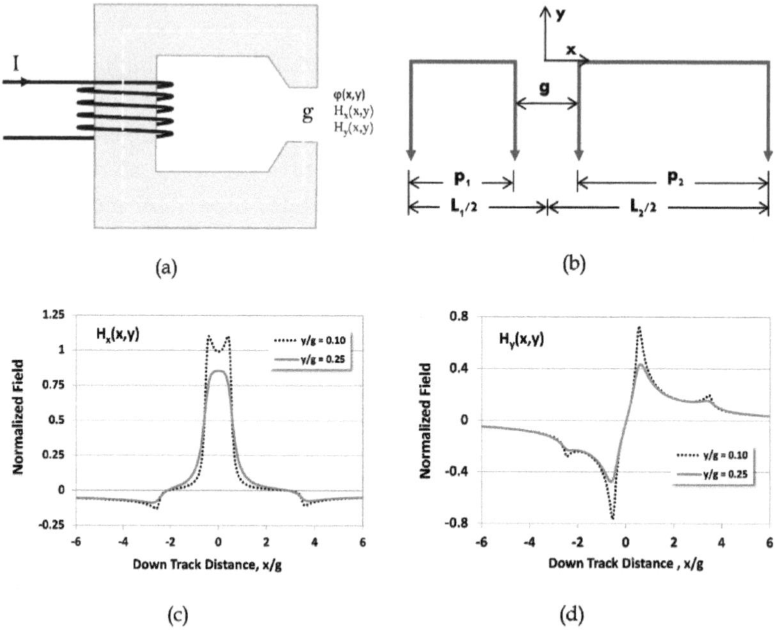

Figure 2.3 Basic components of a magnetic writer (a). Schematic representation of an asymmetric thin-film head (b) and the corresponding longitudinal (c) and vertical (d) components of the fields, where g is the gap length and p_1 and p_2 are the length of the two pole pieces. $L_1 = 2p_1 + g$, $L_2 = 2p_2 + g$. For these field plots, $L_1/g = 5$ and $L_2/g = 7$.

dimensions p_1 and p_2. The corresponding longitudinal and vertical fields calculated for two different height values ($y = 0.1$ g and $y = 0.25$ g) are shown in Fig. 2.3c and Fig. 2.3d. It is readily apparent that both the magnitude and the gradient of the field components deteriorate as one moves away from the head surface [15].

High linear recording density requires a highly spatially varying media magnetization, $M(x)$. In other words, dM/dx must be as high as possible. This spatial gradient of M can be written in terms of two partial derivatives:

$$\frac{dM}{dx} = \frac{\partial M}{\partial H} \times \frac{\partial H}{\partial x} \tag{2.1}$$

It follows from Eq. 2.1 that high field gradients, dH/dx, are necessary to produce high linear densities. Just as importantly, dM/dH also needs to be high. The field magnitude, H, and media properties such

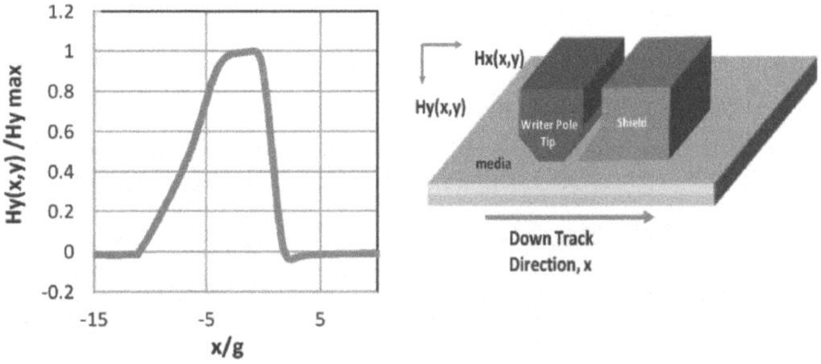

Figure 2.4 Vertical component of the magnetic field of a shielded write pole head, as shown in the schematic diagram on the right.

as magnetic anisotropy field, H_k, grain size, grain size distribution, and anisotropy field distribution, σH_k, amongst others, ultimately determine the slope of M with H. As shown in Fig. 2.2c the air-bearing surface is the part of the slider facing the magnetic media. The inset (Fig. 2.2d) in the picture shows the location of the read element in relation to the write element. Figure 2.4 shows a typical normalized magnetic field profile from a current PMR recording head. Most relevant are the field magnitude, in Oe, and field gradient in Oe/nm. In the figure, the trailing edge of the field profile is shown on the positive x values where the gradient is highest. This is the edge used to form (write) magnetic transitions in the media. As stated above, a sharp field gradient is necessary for sharply written magnetic transitions in the media. In turn, the magnitude of the field is also most relevant since the coercivity of the recording medium needs to be as high as possible to guarantee archivability of the information. To achieve high gradients, magnetic shields are used in close proximity to the writing pole. But these shields do limit the magnitude of the achievable field. Optimization of the field magnitude and gradient as well as other performance metrics such as low inductance for shorter rise time and better frequency response make the design of the writer very complicated.

2.1.3 *Recording Transitions and Signal-to-Noise Ratio*

The recording layer in PMR media is a granular film composed of highly crystallographically oriented magnetic grains separated from each other by a nonmagnetic grain boundary phase typically composed of a combination of nonmagnetic oxide materials (e.g., Refs. [17–20]). The magnetic material in PMR media is based on a disordered hexagonal close-packed (hcp) alloy based on Co, Cr, and Pt. In such a structure, the uniaxial magnetic easy axis is aligned along the crystallographic *c* axis of the hcp cell. In PMR media, magnetic grains are grown with their unique axis oriented in the out-of-plane direction by use of appropriate seed layers normally based on hcp Ru. PMR media also contain a magnetic soft underlayer (SUL) that provides an image to the field of the write head to enhance the writing field. The SUL also provides an image of the recorded transitions in the storage layer boosting the read-back signal, although this effect is mostly observed in long bits (i.e., low frequencies) and is negligible at the highest frequencies.

The function of the write head is to magnetically write or encode areas on the media. The minimum size feature of the written information is called a bit. A digital 1 usually corresponds to the presence of a transition in the direction of magnetization of the media, and a digital 0 corresponds to the absence of a transition at regular timing intervals synchronized to an internal clock. Magnetic bits are typically three to six times wider (cross track) than length wise (down track). The ratio of width over length is called the bit aspect ratio (BAR), and it is a very relevant parameter in magnetic recording. The magnetic grains of (PMR) media are very small in size, typically well under 10 nm in diameter. Because of that, they cannot support magnetization transitions within them. In other words, each grain becomes a uniformly magnetized entity. One recorded bit is composed of a number of uniformly magnetized grains in the media, as illustrated in Fig. 2.5.

Furthermore, when the head is writing on the media, the location where the write head intends to write the transition is stochastic with respect to the loci of the magnetic grain boundaries. For that reason, the actual average location of the written transition as sensed by the reader is usually not exactly coincident with the intended written location. The difference from intended to actual location of a transition, Δs, creates noise in the measurement during read-back. This transition noise is typically characterized by a parameter called transition jitter, representing a statistical

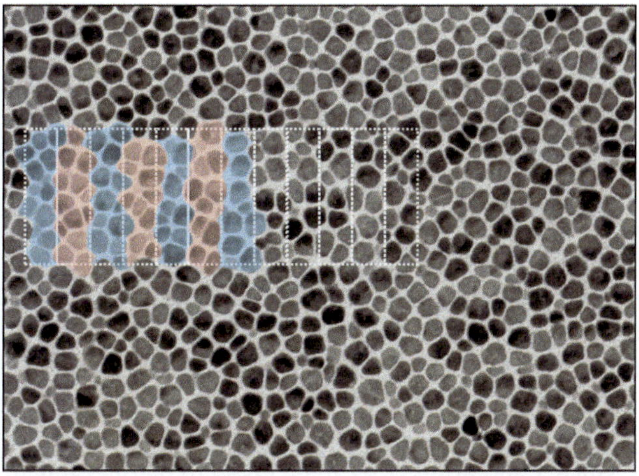

Figure 2.5 Plan view TEM image of state-of-the-art PMR media. The artificially colored areas pictorially represent recorded bits at ~800 Gb/in^2. The faint white traces represent the intended size and location of the bit with a BAR = 4.

standard deviation of all the Δs measured in a large number of transitions. Transition jitter and signal-to-noise ratio (SNR) are inversely related. Because jitter is minimized when the grain size is small, it becomes the major reason for driving media grains to smaller values.

A very useful expression for media SNR is given by [2, 21]

$$\mathrm{SNR} \propto \frac{T_{50} B}{\sigma_{\mathrm{m}}^2} \tag{2.2}$$

where T_{50} is the read-back pulse width, σ_{m} is the write jitter, and B is the bit length. In turn, the write jitter can be expressed as a function of the media magnetic transition parameter, a, the head read width, W, and a media magnetic cross-track correlation length, S, which in well-isolated PMR granular media correlates very well with, and is approximately the same as, the grain size.

$$\sigma_{\mathrm{m}} \approx a\sqrt{\frac{S}{W}} \tag{2.3}$$

Substituting Eq. 2.3 back into Eq. 2.2 and multiplying and dividing by B, the recording SNR can be expressed in terms of well-understood ratios in magnetic recording:

$$\mathrm{SNR} \propto \left(\frac{B}{a}\right)^2 \left(\frac{T_{50}}{B}\right) \frac{W}{S} \tag{2.4}$$

The two terms in parentheses have remained practically constant with drive generation. Scaling requires that the ratio B/a remain $\sim$3 for good resolution of the written transition. Similarly, T_{50}/B is known in the industry as the user bit density (UBD) and generally kept roughly constant at values of approximately 1.1 to 1.4, although in today's most advanced channels, UBDs as high as 1.7 are possible (e.g., Ref. [22]). We conclude that to maintain a high SNR, the ratio of track (read) width to magnetic cross-track correlation length must be kept constant as well from one generation to the next. In other words, since S and grain size can be taken to be the same in today's well-isolated PMR media, the grain size must scale linearly with reductions in read width. A similar analysis [23–25] indicates that the number of grains per bit must remain constant, which also leads to a linear relationship of grain size with track width (as long as the BAR is not changed significantly).

In today's ultrahigh-density recording conditions, the grain size itself becomes a limiting factor to achieving higher linear densities [26–28]. In the grain size limit of magnetic recording, the read width, W, transition jitter, σ, and transition parameter, a, are all interrelated. To illustrate the point, Fig. 2.6 depicts ideally written transitions (limit of infinite write field and gradient) on realistic grain shapes traced from a plan-view transmission electron microscopy (TEM) image of actual media grains. In this exercise a grain was considered switched or not (i.e., marked black or white) according to the location of its geometrical center with respect to the intended transition location. A very narrow read width, W, of 80% of the average grain size was used for digitally acquiring read-back signals from left to right of the image. Multiple read-back traces were acquired at track intervals sweeping the image from top to bottom (microtracks). In the plot of Fig. 2.6b, all the transitions were superimposed. It is clearly seen that each individual transition is very sharp for the narrow read width, W, used here. The resulting transition jitter can be easily estimated from the zero crossing of all the traces. In grain size–limited recording, the grain size ultimately

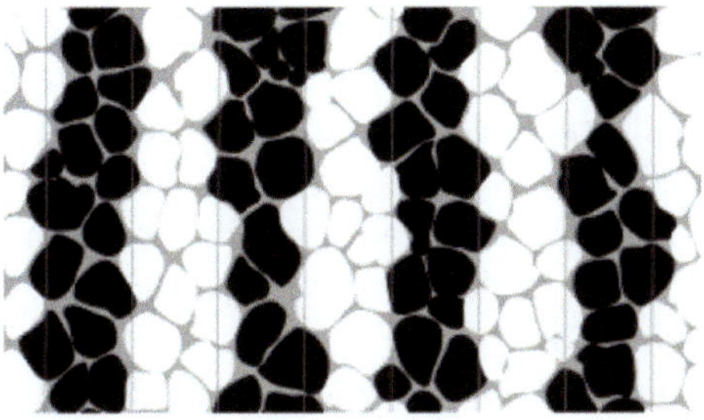

(a)

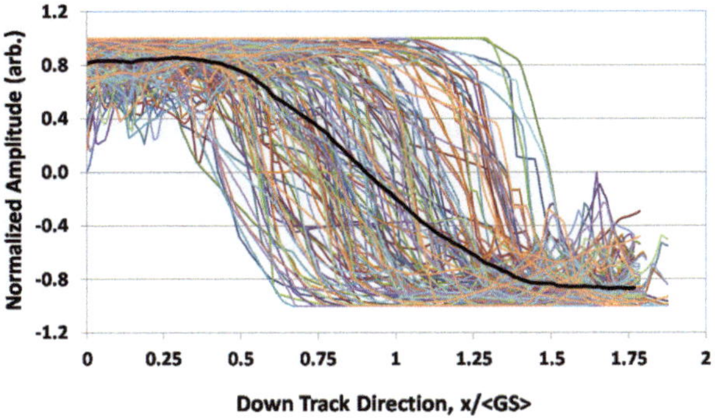

(b)

Figure 2.6 (a) Schematic representation of written transitions on realistic media grains (templated from an actual TEM image). (b) The plot shows the superimposed transition read-back amplitudes. In the image in (a), black corresponds to $M = 1$, white corresponds to $M = -1$, and gray is $M = 0$. For each trace, the read width was 0.8 of the average grain size, <GS>. The darker black line in the plot is the average of all the individual transitions and corresponds to the transition amplitude of a normalized read width, $W/\langle GS \rangle$, of 7.7.

dictates the width of the averaged transition as would be sensed by a read head the size of the height of the entire image (black line in the Fig. 2.6b). It can be readily appreciated from Fig. 2.6b that for the averaged transition, its width from plus to minus saturation is roughly the size of the average media grain size.

In actual recording, that is, not at the limit of grain size recording, the width of the written transition depends on many additional factors but most notably on parameters such as the magnetic spacing defined as the distance between the bottom of the write elements of the head and the top of the recording layer in the media, the writing field gradient, grain-to-grain fluctuations of magnetic properties, media grain size, and size distribution. Because the length of a magnetization transition limits the maximum linear density possible in media, every effort is made to make those transitions as sharp as possible.

Thus, it is desirable to strive for small magnetic spacing, sharp head field gradients, uniform magnetic media properties, and small grain size. As already mentioned, the fundamental limit to transition width is the grain size itself. It is common to approximate the shape of the written transitions based on an arc tangent function or hyperbolic tangent function [2, 29] in which case the transition length can be characterized by a transition parameter, a, where

$$M_x(x) = \frac{2M_r}{\pi}\tan^{-1}\left(\frac{x}{a}\right) \qquad (2.5a)$$

$$M_x(x) = M_r \tan h\left(\frac{2x}{\pi a}\right) \qquad (2.5b)$$

For an infinitely sharp head field gradient, that is, in the grain size limit, the transition parameter, a, cannot be smaller than approximately the average grain size, that is, a media transition parameter of $\sim 1/4$ to $\sim 1/3$ of the grain size, depending on the definition used for a (see Fig. 2.7).

2.1.4 *Thermal Decay and the Recording Trilemma*

The magnetic energy barrier for switching of a sufficiently small particle can be overcome with thermal energy supplied by lattice phonons [30–34]. Thus, to ensure archivability of recorded information, it becomes critical that the energy barriers of magnetic media grains be sufficiently high such that the information will not be thermally erased under normal conditions. The mean energy of phonon modes in a crystal approaches $k_B T$, where k_B is the Boltzmann constant and T is the absolute temperature. This energy must be contrasted with the magnetic energy contained in each magnetic media grain. The magnetic energy of a uniformly magnetized, isolated particle with uniaxial magnetic anisotropy K_u

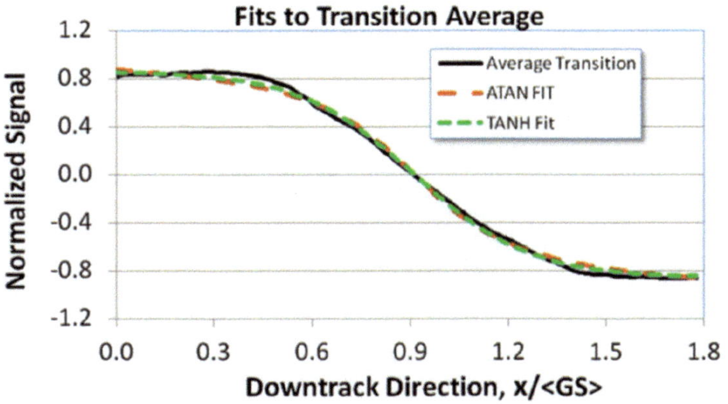

Figure 2.7 Plots of the arc tangent and hyperbolic tangent fits of Eqs. 2.5a and 2.5b against the average transition of Fig. 2.6. In this fit, $a = 0.25$ for the arc tangent transition and $a = 0.35$ for the hyperbolic transition, as defined in Eqs. 2.5a and 2.5b.

and volume V under the influence of an applied field H is given by [35, 36]

$$E = K_u \mathrm{V} \sin(\phi - \theta) - \mu_0 M_S V H \cos\phi \qquad (2.6)$$

In Eq. 2.6 the first term represents the anisotropy energy and the second term is the Zeeman energy, ϕ is the angle between the magnetization vector **M** and the applied field **H**, and θ is the angle between the particle easy-axis direction and the magnetization vector **M**, as shown in Fig. 2.8. A plot of E as a function of angle θ is shown in Fig. 2.9.

Including self-demagnetizing field effects, the energy barrier for switching from one easy direction, I, to the opposite easy direction, II, is given by [32, 34]

$$E_B^{\pm} \cong K_u V \cdot \left[1 \pm \frac{H_d}{H_0(\varphi)}\right]^{m(\varphi)} \qquad (2.7)$$

where H_d is an applied field that includes all demagnetizing fields and H_0 is an easy-axis angle-dependent anisotropy field that for highly oriented media is approximately $2K_u/M_s$ and for randomly oriented media is approximately K_u/M_s. The exponent $m(\phi)$ for highly oriented media is approximately 2. In general, in the absence of external fields, the ratio of $K_u V/k_B T$ must be higher than ~65

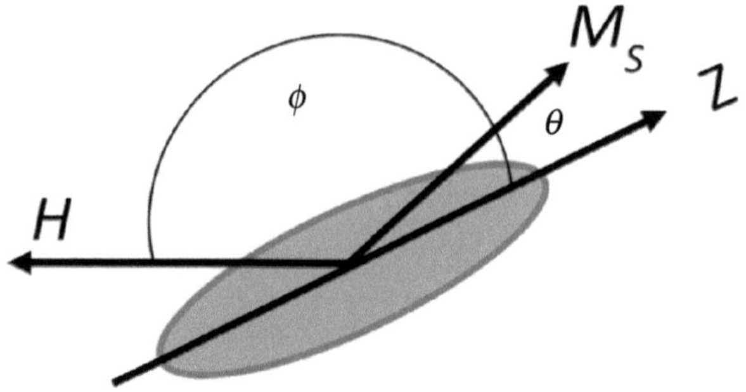

Figure 2.8 Uniformly magnetized particle with uniaxial easy axis along the z direction, magnetization **M**, and applied field **H**.

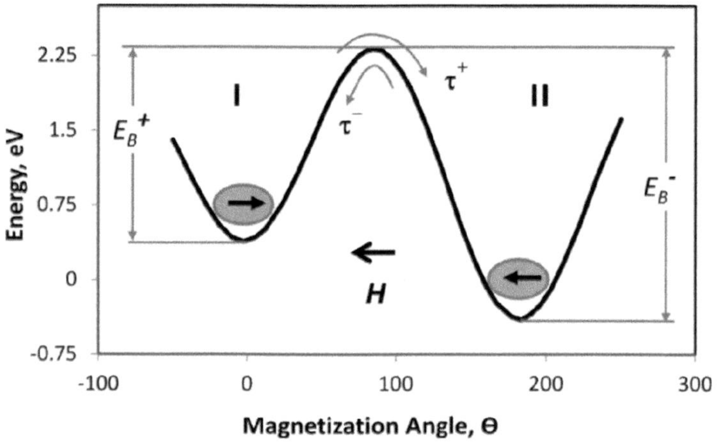

Figure 2.9 Magnetic energy as a function of magnetization angle θ for a uniformly magnetized particle, as shown in Fig. 2.8.

for magnetic media to be thermally stable in commercial drives. At temperature T, the rate of switching from state I to state II and vice versa is given by [30, 34]

$$r_{\pm} = \tau_{\pm}^{-1} = f_0 \exp(-E_{\mathrm{B}}^{\pm}/k_{\mathrm{B}}T) \qquad (2.8)$$

where f_0 is an attempt frequency of typical value 10^9 to 10^{11} (1/s), depending on media properties and structure. Thus the probability of switching after a given time, t, is given by

$$P_{\pm} = \exp(-r_{\pm} \cdot t) \qquad (2.9)$$

Equations 2.8 and 2.9 apply to single particles and show that the probability of switching is a function of both time and temperature. For an ensemble of weakly interacting particles, as present in state-of-the-art media, having an average volume V and magnetic anisotropy energy K_u, Eqs. 2.7 to 2.9 lead to the magnetic coercivity of the ensemble being well represented by Sharrock's equation [33, 36, 37]:

$$H_c(T,\ t) = H_0\left(1 - \left[\frac{k_B T}{K_u V}.\ln(At)\right]^n\right) \qquad (2.10)$$

In Eq. 2.10, the time, t, is the switching time and clearly shows that higher temperatures and longer times cause a decrease in H_c, which may lead to information loss if the media are not designed properly. The need to decrease media particle size to enable higher recording densities conflicts with the need to keep the particle size as high as possible to ensure archivability. In turn, the magnetic anisotropy energy of the media alloy cannot be arbitrarily high since the available field for writing from the write head is limited by the saturation magnetization of the pole material. In current PMR heads the M_S of the CoFe-based write pole has already reached the Slater–Pauling limit of 2.45 T, and while some exotic materials such as FeN have been shown to exhibit higher M_S [36–42], the likelihood of integrating them in real heads is low due to process and synthesis difficulties.

This all leads to the industry recording "trilemma," as shown in Fig. 2.10 [38]. This trilemma of magnetic recording highlights the trade-off between SNR, grain size, and thermal stability and illustrates the fact that areal density progress must take place by clever innovation, making possible a decrease in grain size, without compromising writability of the media or archivability of the written information.

2.1.5 *Sources of Media Noise*

Because transition noise is minimized when the grain size is small compared to the size of the bit, it is always desirable to make the media grain size as small as possible. Figure 2.11 depicts the historical decrease in physical or crystallographic grain size as a

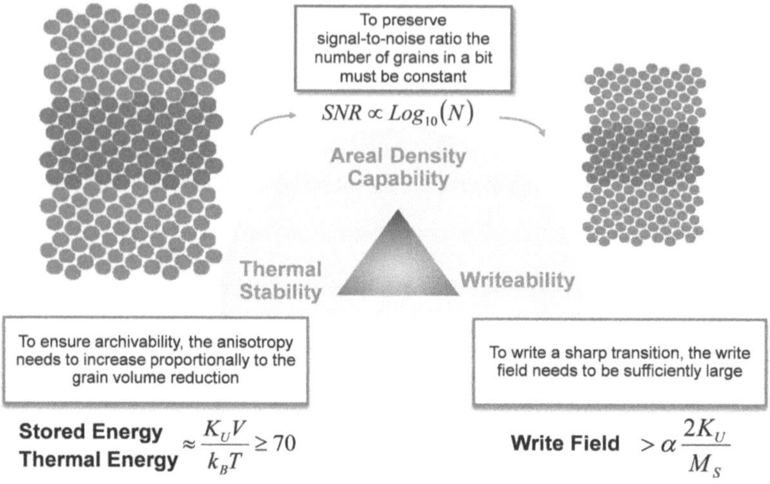

Figure 2.10 The recording trilemma: Areal density increase must take place with no loss of system SNR, writability, or data archivability. All three properties are trade-offs of each other.

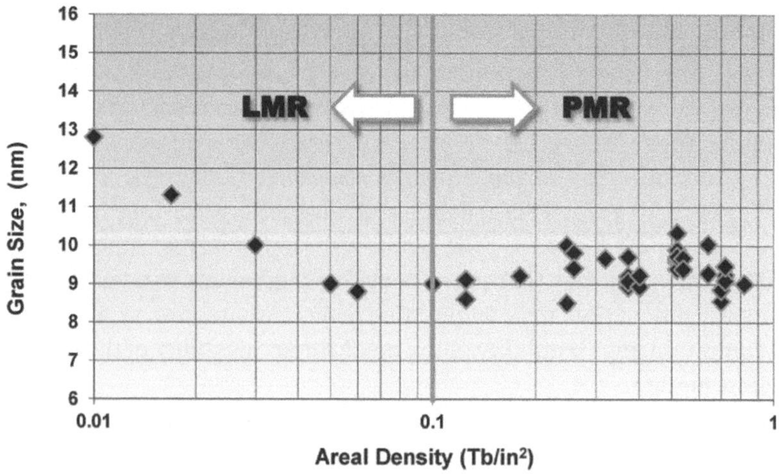

Figure 2.11 Historical grain size trends as a function of recording areal density of HDD products.

function of areal density. It is interesting to see the contrasting situation between the LMR period (1996 to 2005) compared to the PMR period (2005 to present). While grain size experienced a monotonic decrease as areal density increased during LMR recording, the grain size has remained roughly constant during the entire time in CPMR recording. This is so because in PMR media every grain has its magnetic easy axis almost ideally oriented in the perpendicular direction, whereas for LMR media, while the easy axes were in plane, as intended, they were poorly preferentially oriented along the recording down-track direction, as desired. Also, from a demagnetizing energy point of view, decreasing bit length resulted in higher demagnetizing fields for an LMR configuration, leading to the same polarity of magnetic charges at each side of the transition, whereas in PMR recording, increasing linear density results in a lowering of the demagnetizing fields as north and south poles are brought closer together (see Fig. 2.12). Both the much better alignment of easy axes in PMR media and the lowering of the demagnetizing energy as linear density is increased allow for much sharper transitions in PMR recording, resulting in transition parameters of roughly 1/3 of the grain size, as shown in Fig. 2.7. In other words, the linear density in PMR recording has not been to date grain size limited as was the case during the LMR period. In LMR recording the transition parameter, a, was a rather complex function depending mainly on the demagnetizing fields, the physical grain size, the quality of the easy-axis orientation (i.e., the orientation ratio [OR]), the H_c of the media alloy, and the degree of intergranular exchange coupling. The transition parameter played a much bigger role in widening the read-back pulse width and limiting the recording linear density. The presence of intergranular exchange coupling and, to a lesser extent, dipole–dipole interactions in media typically result in magnetization length scales larger than the physical grain size. While the magnetic and physical grain sizes are almost the same in PMR recording, it was not the case in LMR recording, where the magnetic interaction or magnetic cluster size could be a few times larger than the physical grain size.

Another source of noise in magnetic media is that related to the track edges. In magnetic recording, the written track width is basically determined by the width of the write element plus that of

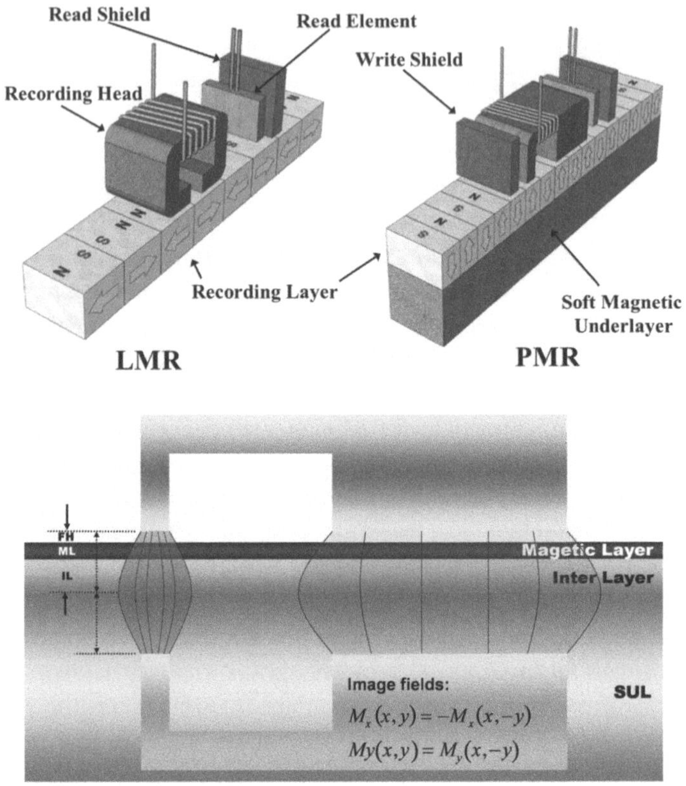

Figure 2.12 Longitudinal (LMR) and perpendicular (PMR) recording configurations. In PMR recording, the presence of a magnetic soft underlayer (SUL) provides for a head field image, boosting the field magnitude and gradient at the recording layer location. Mathematically, the top surface of the SUL layer is the magnetic mirror-image plane. In LMR recording, the demagnetizing fields increased as the linear density increased, thus broadening the recorded transitions and were a major source of thermal decay of information; conversely, in PMR recording, demagnetizing fields are zero at the transition center; thus, closely packed transitions help with thermal stability.

the so-called erase band at the track edge where the quality of the write field magnitude and gradient are very poor. Because of these poorly written track edges and the presence of some curvature in the transitions, the reader element of the head is made to have a smaller dimension than the full track width, for example, 55%–65% of the write element width, in order to read back from the higher-quality portion of the track and avoid the noise contained at the track edges. Because the SNR depends on the width of the read head element (Eq. 2.4) this reduction in read width results in a penalty in both the SNR and consequently the areal density capability.

Aside from transition and track edge noise, direct current (DC) noise, that is, deviation from a track-averaged amplitude (mean amplitude) of a uniformly written track (no transitions) measured as a variance, can also be an important contributor of media noise. DC noise arises because of non-uniformities in the magnetization of the media. Here again, both the granular nature of the media containing magnetized grains and nonmagnetic grain boundaries and intergranular exchange coupling creating magnetic clustering can contribute to most of it. Other contributors to DC noise can be crystallographic defects in the media, dispersion of magnetic easy axes, magnetic voids, and poor, non-saturating writing conditions.

2.2 Magnetic Recording Medium

2.2.1 *Magnetic Recording Media Structure and Evolution*

The structure of magnetic media has undergone both evolutionary and revolutionary changes along the years in response to the industry's need to deliver higher storage capacity. These changes are mostly timed to corresponding changes in related HDD components, mainly those in the write and read elements of the head, and to major technology transitions in the HDD industry, such as the shift from LMR to PMR that took place starting in 2005. Much of these transformations were required in order to deliver higher areal density, while at the same time guaranteeing writability and archivability. The challenging nature of this task is most frequently discussed in reference to the well-known recording density trilemma depicted in Fig. 2.10. In short, areal density and writability scale as inverse functions of the magnetic media grain volume and the media magnetic anisotropy, respectively, while archivability scales as a direct function of the product of media grain volume and its magnetic anisotropy. It would thus seem that one

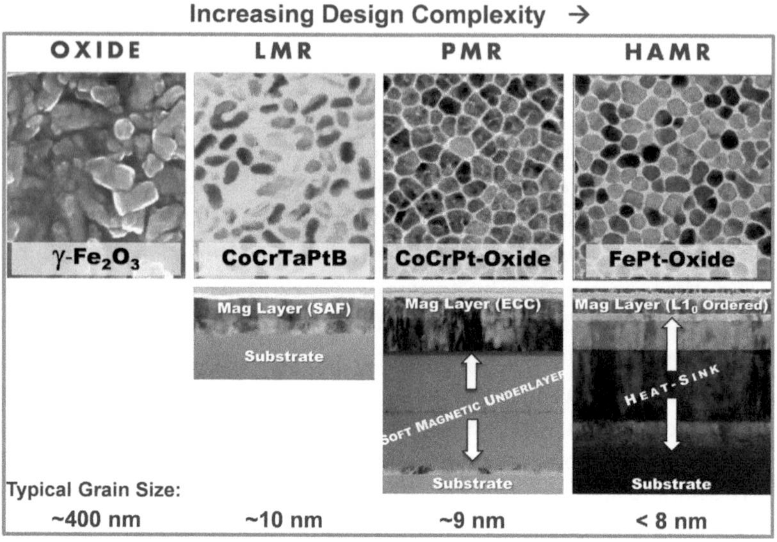

Figure 2.13 Progression of magnetic recording media designs for each of the major technology transitions (HAMR introduction is expected to take place in the 2016–2018 time frame).

cannot optimize one property without trading off for another. Many of the innovations and topics covered in this section find root at the heart of this recording trilemma as ways to deliver higher areal density, while keeping writability and archivability in place.

Figure 2.13 depicts some of the major media structural features from the early days of the technology utilizing oxide magnetic media, through longitudinal recording thin-film media, to today's state-of-the-art PMR and future HAMR media structures under development. The increasing complexity of media can be readily appreciated from these images. In the early days of oxide-coated media, magnetic coercivity, particle size, orientation, and distribution were the main factors determining media performance. With thin-film LMR media, both the process and the structural engineering of the media became significantly more complex. PMR recording demanded magnetic SULs in media for providing an image field to the head and now the interdependence of media and head magnetic designs became ever more critical. With new technologies such as HAMR recording, the media structure must now be designed with magnetic, thermal, and optical functions in mind, becoming even more coupled to the function and design of the recording head.

A good example of process innovation to illustrate microstructural engineering in media is that of LMR recording resulting in preferential orientation of easy axes along the down-track direction of the disk (i.e., the circumferential direction). In LMR, the magnetization of the media is made to lie in the plane of the film and ideally uniaxially oriented along the circumference of the disk. Orienting the *c* axis of the CoPt hcp granular media in the plane of the film was achieved through the use of properly deposited seed layers onto which the magnetic media grains were epitaxially grown. Without any other involvement, the *c* axes would be randomly oriented in the plane of the disk. However, to achieve some degree of preferential alignment of *c* axes along the circumferential direction, the surfaces of the media substrates were topographically grooved, or textured, with very fine circumferential scratches using diamond nanoparticles embedded in advanced tape substrates. When the seed and nucleation layers were vacuum-deposited, the presence of these grooves helped relieve compressive stresses in the cross-track direction (perpendicular to the grooves) but not along the down-track direction (parallel to the grooves). This anisotropic stress state created a distortion in the seed layer crystals, which helped bias the growth of the subsequent CoPt-based magnetic layer such that a higher percentage of *c* axes were oriented along the circumferential direction as opposed to the radial direction.

In other words, the symmetry of an otherwise fourfold symmetric seed layer crystal (bcc grown with [002] fiber texture) was broken by a uniaxial strain distortion created by circumferentially oriented grooves. Since hcp CoPt is a uniaxial crystal there was now a way to bias the CoPt alloy *c* axis preferentially along this direction. Stochastically, a portion of the seed layer grains were oriented with the correct in-plane crystallography to result in a growth-biased CoPt alloy film. This preference for orienting *c* axes along the circumferential direction was characterized by a so-called orientation ratio (OR). To understand how an OR develops, in Fig. 2.14a it can be seen that the spacing between the (00.2) basal planes from the hcp Co alloy crystal is slightly shorter than the spacing between (01.0) planes. A compression of the Cr underlayer grains stochastically oriented

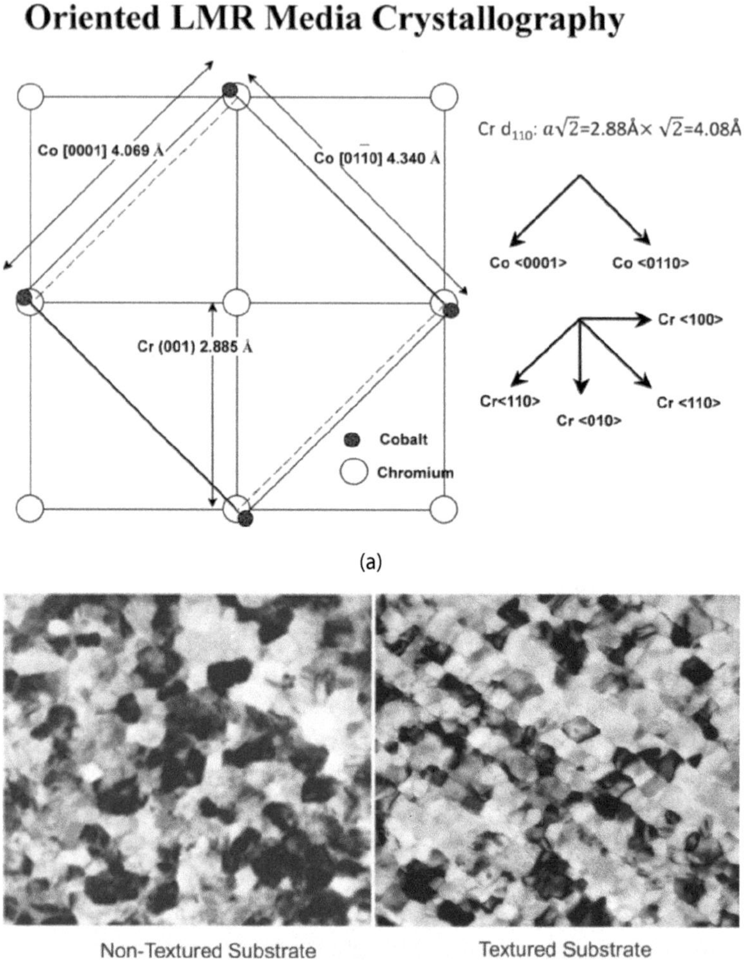

(b)

Figure 2.14 (a) Epitaxial relationship between a (002) oriented crystal and a (11$\bar{2}$0) oriented Co alloy crystal showing a smaller mismatch along the *c* axis direction. (b) Resulting Cr grain morphology in topographically textured vs. non-textured substrates (TEM plan view). The spacing between texture grooves is of the order of the underlayer grain size. Compressive stresses are higher along the grooves where stress relief is less likely. The Co alloy *c* axes prefer to align along the compressed <011> Cr directions, thus developing an orientation ratio (OR) by statistically orienting preferentially along the down-track direction, as desired.

with <110> directions near the circumferential direction also results in a Poisson expansion along the complementary, orthogonal <110> directions. Thus, in those cases a smaller mismatch and corresponding smaller elastic energy expenditure are realized when the hcp Co alloy crystal nucleates and growths with its *c* axis oriented along the compressed Cr underlayer <110> orientations.

Increasing the OR in LMR recording by better substrate texturing and film alloying resulted in a direct reduction of the transition parameter translating into marked SNR improvements. The OR was typically measured as the ratio of circumferential H_c divided by radial H_c. In the most advanced LMR films, an OR of >2.2 was achieved. Figure 2.14b shows two plan-view TEM images of Cr underlayer grains deposited in polished-only and polished-plus topographically textured substrates, respectively. The impact of the grooves on the morphology of the underlayer grains is major. As Pt content was increased on the Co alloy to increase H_c, a corresponding increase in the alloy lattice parameters ensued. It was typical then to add Mo to the Cr underlayer to adjust its lattice constant and maintain the proper amount of mismatch with the CoPt alloy to achieve the largest possible ORs.

In light of the recording trilemma, two great examples of design ingenuity to increase recording areal density come to mind. Synthetic antiferromagnetic (SAF) recording media apply to LMR media structures and exchange-coupled composite (ECC) media as they apply to PMR media. Because of length constrains, we will not dwell here on the SAF innovation. However, the unfamiliar reader is encouraged to learn about this ingenious use of an SAF structure to enable lowering of the demagnetizing fields, increase thermal stability, sharpen transitions, and enable increase of the areal density [20, 38, 43, 44].

Because of its current widespread use, we will shortly describe the ECC media design breakthrough. This innovation arose in response to the realization that if PMR media grains could be grown with the magnetic easy axes at 45° oriented radially, then the switching field for any given thermal stability barrier would be reduced by half [45]. This is because the Stoner–Wohlfarth coercivity of a uniaxial anisotropy particle oriented 45° from the applied filed is half of that when the applied field is co-linear with the easy axis. Orienting the magnetic easy axes at 45° was not possible experimentally, but the concept motivated the realization that the switching field could be also reduced if incoherent switching were to take place in each grain during the recording process. To this end, ECC media and exchange-spring PMR media concepts were almost simultaneously proposed in 2005 and introduced into commercial media in 2007 [46, 47]. The benefit to magnetic recording arises because for the same thermal stability ratio of K_uV/kT, the switching field required is now reduced significantly. This extra margin in design space allows media makers to reduce grain size and raise the anisotropy of the media, for instance, without compromising thermal stability. In practice, this innovation was used to achieve higher write field gradients combined with better overwrite performance. As already discussed earlier, grain size in PMR recording has not limited the areal density; thus, it has not been necessary to decrease grain size in order to achieve higher areal densities.

The principle behind ECC media is to use exchange break layers (EBLs) to diminish the degree of intra-grain exchange coupling, as depicted in Fig. 2.15. The nonmagnetic or weakly magnetic EBL material allows for the spin orientation of each media grain section to rotate semi-coherently with respect to the spin orientation of

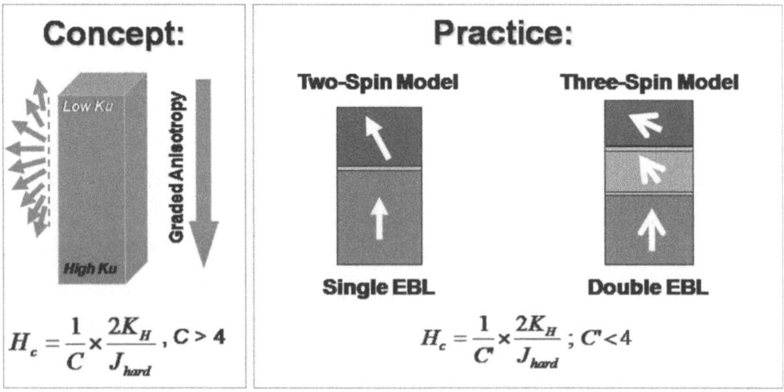

Figure 2.15 Exchange-spring and exchange-coupled composite (ECC) media structures. Incoherent rotation in the column of the grain is enabled by the use of exchange break layers (EBLs).

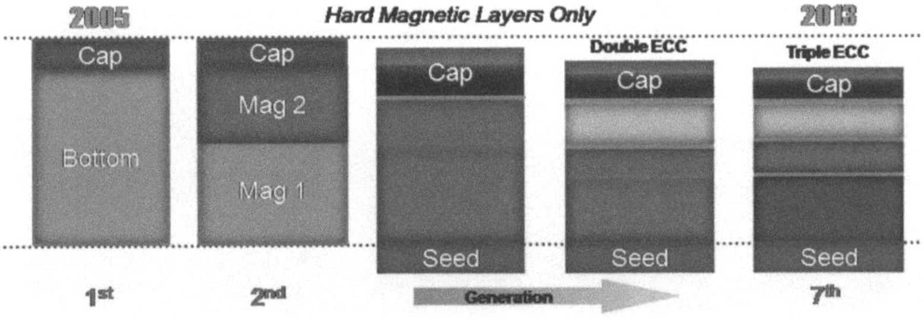

Figure 2.16 Progression of PMR recording layer structure design as a function of drive generation. The ECC media design allows for a reduction in the total media thickness because it enables the use of high anisotropy layers without compromising switching performance.

the other sections. For this scheme to work properly, the layers' thickness, magnetic anisotropies, and degree of exchange coupling across the EBL must all be optimized to provide both high thermal stability and, at the same time, improved switching performance [20, 46–52]. In ECC media, the thermal stability is mostly imparted by the highest anisotropy layer and the switching performance is adjusted by the thickness of both the lower anisotropy layer and the EBL. The first implementation of ECC media utilized one single EBL layer. In today's state-of-the-art designs, two and even three EBL layers through the thickness of the grains are used (Fig. 2.16).

Figure 2.17 shows a schematic and a TEM cross section of a first-generation ECC PMR media structure. In all, approximately 18–20 distinct layers are used to build the media structure, depending on the media maker and application. Starting with an adhesion layer, the next structure up is the magnetic SUL. The main function of the SUL is to boost the magnitude of the write field by providing a low-reluctance flux path to the return pole of the write head (e.g., Refs. [8, 17, 20]). Because soft magnetic materials tend to easily form magnetic domains that are a source of noise upon read-back, the SUL is made as a typically balanced synthetic antiferromagnet (e.g., Refs. [53–55]). In the absence of external fields, the two portions of the structure are magnetized opposite to each other. This not only lowers the number of domains present but also minimizes the domain wall noise upon read-back because there will be opposite magnetic charges present at the wall in each portion

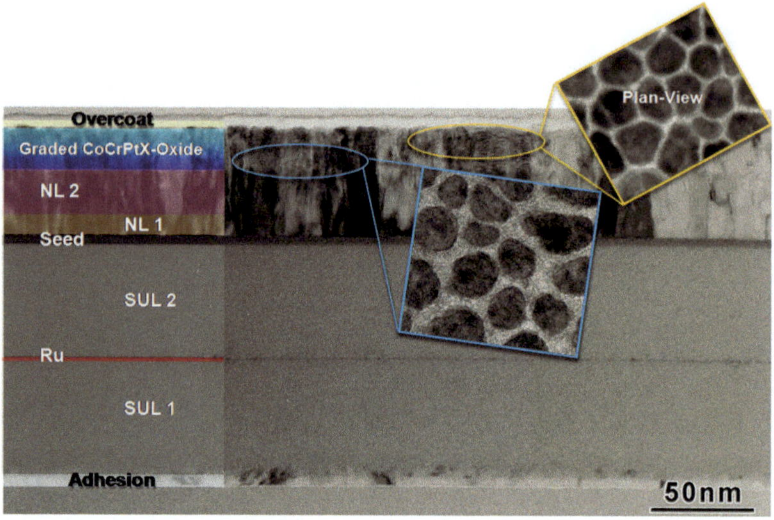

Figure 2.17 Cross section of a single ECC PMR media structure circa 2008. The insets in the figure represent plan-view images at different sections of the hard magnetic layer stack to illustrate the different degrees of oxide grain boundary isolation used from bottom to top of the structure.

of the SAF structure. When a domain wall, mostly Néel type, of a given polarity is present in the top layer of the SUL, a corresponding opposite-polarity wall will be present in the lower portion of the SUL structure, thus minimizing stray fields. The SUL layer is also sputter-deposited with an induced radial magnetic anisotropy such that during the writing process a 90° rotation of magnetization from the easy axis, that is, radial direction, to a hard axis, that is, circumferential direction, takes place rather than switching to and from two easy axes requiring 180° rotation of the magnetization. The SUL film is also amorphous to minimize magnetic domain noise. Next up are the seed and nucleation layers. These layers create the crystallographic and morphological template for the hard magnetic layers to grow with the proper grain size and crystallographic texture.

The seed and nucleation layer structure is typically composed of three to four individual layers, including an Ru layer sputtered at a low Ar pressure to optimize crystallographic texture and a top high-pressure Ru to break down the grain size into smaller sub-grains. Frequently, the seed layer underneath the two Ru layers is made to contain some magnetization to help lower the effective magnetic gap between the recording layer and the SUL. As discussed earlier in this chapter, the magnetic layer structure is a complex ECC stack comprised of high-magnetic-anisotropy layers at the bottom and lower-anisotropy layers toward the top, each spaced with a non- or weakly magnetic exchange spacer layer (EBL) to induce nonhomogeneous rotation of magnetization during the switching process. Finally, a hard coating of a diamond-like carbon, frequently doped with nitrogen, is applied. The nitrogen doping helps provide bonding affinity for the lubricant layer applied separately, which completes the full stack. The thickness of the SAF SUL in today's media is of ~50 nm. The total thickness of the hard magnetic layers is less than 15 nm. The overcoat thickness is less than 2 nm and the seed and nucleation layers add up to approximately 15 nm.

2.3 Magnetic Recording Heads

2.3.1 *Writer Technology and Challenges*

The function of the writer is to produce the highest fields and gradients and be able to operate at high frequencies (e.g., Refs. [1–16, 38, 56]). To deliver high fields, soft magnetic FeCo alloys approaching the Slater–Pauling limit of 2.45 T are used [35]. Gradients are maximized with the use of magnetic keeper layers or shields. In today's most advanced heads, a typical shield gap distance is of 30 nm. The shield enhances the gradient but at the expense of field magnitude; thus, the structure of the writer is extremely complicated with a combination of various M_s materials, layer thickness, and different and carefully designed pole tip angles applied in all three dimensions (see Fig. 2.4). Since the writer works on the principle of magnetic induction, a coil winds the main pole in the yoke region of the writer. A strong magnetic field is generated at the gap between the pole tip and the shield and return pole. When the polarity of the electrical current is reversed, the polarity of the magnetic field in the gap is also reversed. Since the media has a magnetic SUL, flux from the pole tip is returned to the return pole though the SUL during the writing process, as shown in the schematic of Fig. 2.12. Some of the flux is lost to the typically floating shield to enhance the write gradient. When PMR recording was first introduced, the main pole was not shielded. A major breakthrough took place when shielded pole heads similar to those in Fig. 2.4 became available, resulting in much higher linear density performance because of the higher gradients that those heads delivered [9, 56]. The wraparound shield was yet another major development that minimizes interactions with the adjacent tracks, thus allowing for higher track densities (e.g., Refs. [57, 58]) (Fig. 2.18).

The challenge for today's writers is to keep on delivering high saturating fields as the dimensions of the recording bit become smaller, particularly since new materials with higher M_s than those currently in use today are unlikely to become suitable for this application. One positive development from the point of view of the writer is the advent of shingle magnetic recording (SMR), where tracks are written in an overlapped manner. In SMR, the writer width can be larger than the resulting track width, giving more flexibility of design and a welcome respite to writer technology experts, although not all drive applications are accepting of SMR schemes. New recording technologies that greatly impact the writer itself, such as HAMR and microwave-assisted magnetic recording (MAMR), are being developed in response to these limitations and will be discussed in later chapters in this book.

High-frequency performance is also a major writer technology challenge because of the inductance and capacitance present in the head structure and the requirement of high data rates in the drive. The ferromagnetic frequency (FMR) of most magnetic materials is less than 1 GHz; thus the intrinsic damping and H_c of the materials used in the write head are carefully controlled. Associated with this high-frequency performance is the need to utilize materials with minimum saturation magnetostriction. Since the highest FeCo alloys display rather high magnetostrictive responses, their use is minimized to just the portion of the write pole where the highest fields are needed.

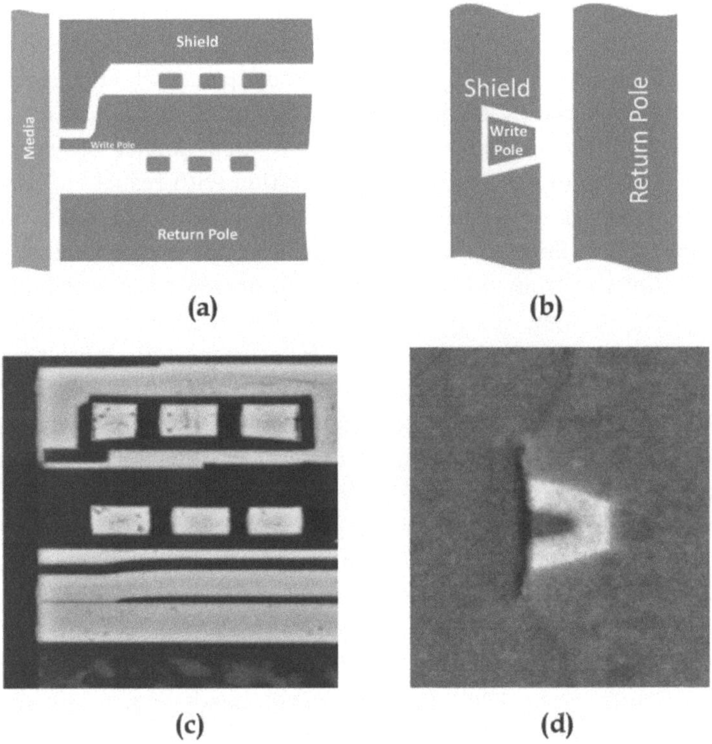

Figure 2.18 (a) Schematic drawing of the cross section of a PMR writer. (b) Air-bearing view of the write pole with a wraparound shield. (c, d) Corresponding SEM images of an actual writer. The write pole trapezoidal shape is to enable the head to operate at large skew angles (e.g., +/−15°).

2.3.2 *Reader Technology and Challenges*

The reader in magnetic recording has been a major enabler and driver of higher recording densities. The first read (and write) heads were bulky, typically FeMn ferrite core devices used up until the late 70s [59–61]. These heads had very limited field performance and thus could not write on high-H_c media needed for higher-linear-density performance. This limitation led to the development of metal-in-gap (MIG) heads with approximately twice the write field capability, and eventually, miniaturization and innovation in lithographic techniques paved the way for the adoption of thin-film heads using mostly Permalloy as the core material but still based on inductance as the main detection principle. A radical paradigm shift indicative of things to come took place when anisotropic magnetoresistive (AMR) heads were introduced in the early nineties. These AMR heads made use of the $\cos^2$ dependence of resistivity as a function of the angle of magnetization with respect to current flow direction. AMR heads worked very well but did not produce large signals. Since the output signal declines proportionally with reductions in track width, these heads presented an extendibility problem in light of the necessary reductions in size required to maintain the aggressive areal density growth over time. A true revolution in magnetic sensing capability took place when giant magnetoresistive (GMR) heads producing significantly higher signals were first implemented in 1995, making use of the spin valve principle, and eventually with the introduction of tunneling magnetoresistive (TuMR) heads in 2005 of widespread use today [59]. Both GMR and TuMR heads are truly quantum mechanical detector marvels pushing the limits of science and technology and widely applied to better the quality of life of people [62–68].

The GMR phenomenon was discovered independently by Fert and Grünberg in 1988 [63, 64]. GMR heads make use of this effect in a spin valve configuration, as shown in the simplified schematic of Fig. 2.19. In a simple spin valve, two ferromagnetic

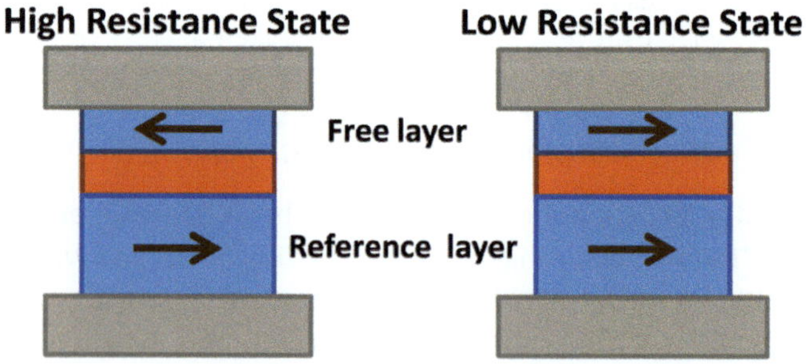

Figure 2.19 Schematic diagram of a simplified spin valve structure. Resistance is lowest when the two ferromagnetic layers (in blue) are parallel and highest when the magnetization of each layer is opposite to the other. The spacer material is a thin nonmagnetic metal (e.g., Cu, Cr) typically 2–4 nm in thickness.

layers are separated by a nonmagnetic, metallic spacer. The resulting electrical resistance in the structure is dependent on the magnetization orientation of each of the two ferromagnetic layers because the probability of electron scattering in magnetic materials is a spin-dependent phenomenon. This effect is observed in both current-in-plane (CIP) and current-perpendicular-to-plane (CPP) configurations [65, 66]. GMR heads resulted in about 1 order of magnitude higher signals than in AMR heads. This allowed for a sharp reduction in the thickness of the LMR media layers, resulting in lower demagnetizing fields and sharper transitions, key to enabling higher linear densities.

The sensing element in today's recording head functions on the principle of TuMR first reported by Jullière in 1975 [68]. TuMR heads resulted in a doubling or tripling of the signal amplitude compared to that of a GMR sensor. The structure of a fully implemented TuMR stack in the head is not simple and we will discuss it in some detail later in this section. However, at the most basic level, a TuMR sensor has two magnetic layers separated by a thin oxide layer. Current flows across the stack in a CPP configuration. Put simply, the electrical resistance is lowest when the densities of states in the two magnetic layers are as similar as possible. This happens when the two magnetic layers are magnetized in the same orientation. In this configuration, the tunneling electrons find the most states available at the other side of the oxide barrier. When the two layers are magnetized in opposite direction, the spin-dependent densities of states of both layers do not match well and the probability of tunneling decreases; in other words, the resistance is highest.

In the TuMR sensor of a recording head one of the layers is made magnetically rigid by exchange coupling it to an antiferromagnetic (AFM) material and referred to as the reference or pinned layer. The other layer across the oxide tunnel barrier, while biased, is allowed to rotate in response to external fields. In current readers, the reference layer is really a stack structure containing a synthetic antiferromagnet, as shown in Fig. 2.20. This is done to minimize magnetic noise arising from magnetic non-uniformities in the AFM layer and to impart the structure robustness against stray magnetic fields.

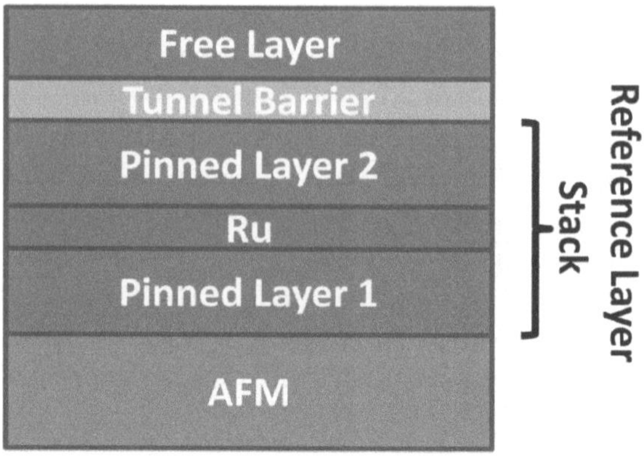

Figure 2.20 The reference layer stack in a TuMR reader is composed of pinned layer 1 exchange-bias-coupled to an antiferromagnetic (AFM) layer. Pinned layer 2 has opposite magnetization to pinned layer 1, imparted by the AFM coupling across the thin Ru layer.

Figure 2.21 shows a schematic and actual low-magnification TEM image of an actual TuMR reader structure. In today's most advanced readers, more than 20 distinct deposition layers are used to create the sensor stack. The free layer itself, a few nanometers in thickness, may be composed of two to four distinct deposition layers in order to optimize its functionality. The dimensions of the reference layer and free layers are typically similar. At 1 Tb/in^2 recording the width of the reader is approximately 25 nm and the thickness of the free layer less than 10 nm. The shield-to-shield spacing is approximately 22 nm and the stripe height (i.e., the height of the reader element) is similar to its width. At any given time, the free layer senses the magnetization of approximately 10 media grains. To achieve sufficient signal and resolution, the magnetic spacing must be of the order of half the shield-to-shield distance and ideally less. This means that the thickness of the head and media overcoats, the media lubricant, and the flight height (read and write clearance) must all add up to less than approximately 10 nm. In fact, in today's state-of-the-art drives, the magnetic spacing is of less than 10 nm and approximately 8 nm at 1 Tb/in^2 recording [69].

The magnetization of the free layer is biased at an angle to that of the reference layer to linearize the read-back response.

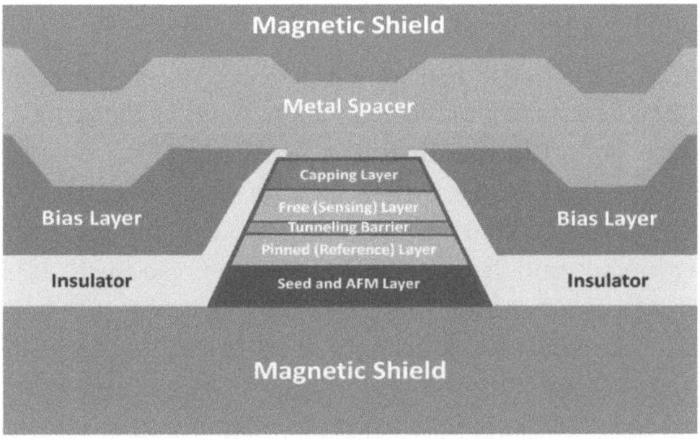

(a)

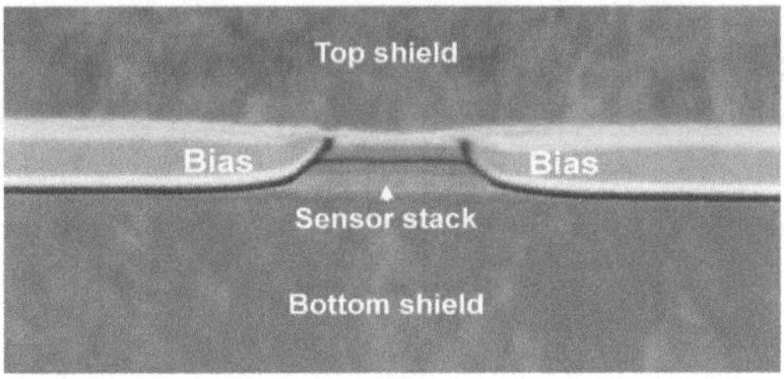

(b)

Figure 2.21 Schematic diagram of a generic TuMR reader stack design (a) and corresponding TEM image of an actual sensor (b). At 1 Tb/in^2 recording the shield-to-shield spacing is of the order of 20 nm or less.

One recent development in current readers is to apply the bias to the free layer by means of soft magnetic materials instead of the more traditional hard magnetic biasing schemes. These soft materials act partially as extra shields providing for improved sensor stability, enhanced screening from adjacent bits, and also extra protection from writer-induced noise the reader. To set the direction of magnetization of the reference layer and the proper bias on the soft bias layers to the correct angles a series of annealing processes under the influence of large, externally applied magnetic fields are used. When the reference and free layer magnetization directions are set appropriately, the sensor responds linearly for the two magnetization states in the media with minimum or negligible asymmetry.

2.3.3 *Sources of Head Noise*

Similar to the situation in media, as we scale dimensions down, reader noise becomes increasingly higher and of changing character. There are many noise categories of relevance in heads, notably Johnson and shot noise, which are mostly resistance driven and of white character; thermal magnetic noise, including spin-torque-transfer noise; and frequency-dependent noise such as $1/f$ and random telegraph noise [62, 70–78]. In general, all components of noise are expected to get worse as recording areal density is increased; however, as shown in Fig. 2.22, thermal magnetic noise has become the dominant noise in readers and is expected to overwhelm other sources of noise past the 1 Tb/in^2 density mark. This noise arises from fluctuations in magnetization as spins interact with lattice phonons. It has a square dependence with electrical current and reader sensitivity ΔR and an inverse relationship to volume and saturation magnetization. As the volume of the free and reference layers are reduced for higher areal density capability, the incidence of all of the main contributors to thermal magnetic noise makes it the dominant noise source [71–74]. Another source of magnetic noise in heads is the spin-torque-induced noise [76, 77]. Here again, as the current density through the device increases, the spin-torque noise will become more relevant. Mitigating these effects through new reader architectures, new materials, and clever design changes is a major focus of attention in the industry.

2.3.4 *Magnetic Head Integration*

Integrating all the components of a head into a working device for use in drives is a very lengthy and complicated process. Starting with AlTiC substrates, the wafers are built layer by layer through multiple process steps. Depending on the device type and application, the wafer process alone ranges from 700 to 1000 steps. Most layers are patterned and defined through lithography. A particular layer can be additive utilizing vacuum processes such as physical vapor deposition (PVD), atomic layer deposition (ALD), chemical vapor deposition (CVD), or plating or subtractive through processes such as reactive-ion etching (RIE) and ion milling. Planarization

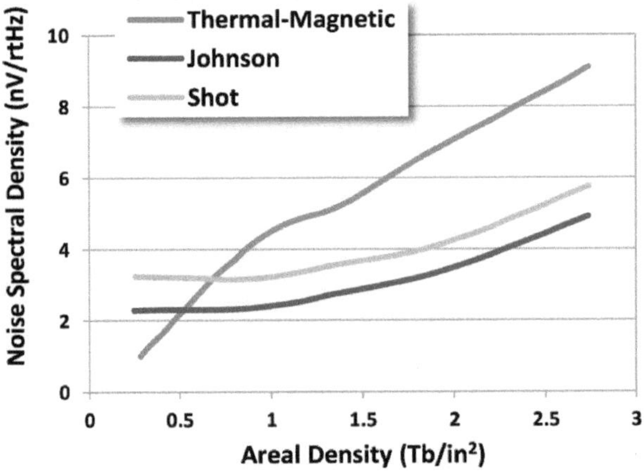

Figure 2.22 Plot of projected reader noise component as a function of areal density. Thermomagnetic noise is projected to dominate the reader noise due to increased sensitivity and fast shrinking volume of the magnetic free and reference layers of the reader.

through chemical-mechanical planarization (CMP) processes is also commonly used throughout the fabrication process. Typically, the wafers are subjected to 50 or so prefabrication steps. The reader process that follows requires 200+ process steps, the heater(s) for flight height control may take 50+ steps, and lastly the writer structure is built requiring 500+ steps. After the wafers are processed the backend steps take place involving slicing of the bars and individual devices, polishing, air-bearing deposition and patterning processes and, mounting on the gimbal and head stack assemblies.

2.4 Mechanics and Servo

In this section, an overview of the HDD servo-mechanical system will be introduced. The servo-mechanical system in an HDD carries the rotating magnetic disk and moves the read/write heads to access the data on the magnetic media. The signal processing system encodes the user data and writes it on the media, reads it back, and reconstructs the original user data.

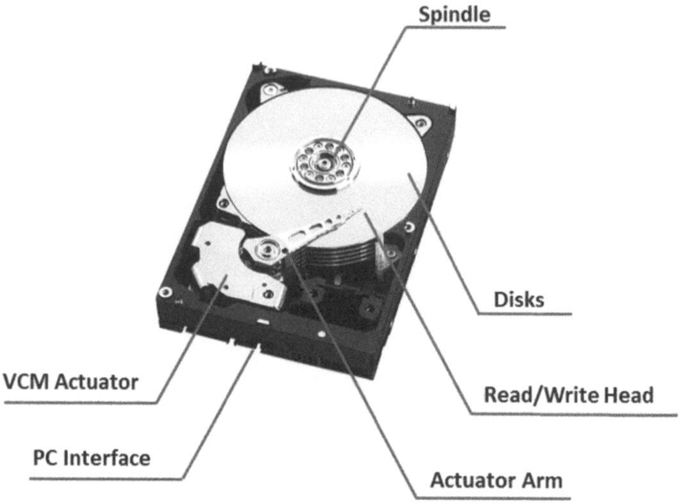

Figure 2.23 Typical hard disk drive.

2.4.1 *HDD Servo Basic Components*

As shown in Fig. 2.23 an HDD contains magnetic disks rotating about a spindle motor at a fixed rotating speed ranging from 4,200 to 15,000 rotations per minute (RPM). The disks are of aluminum or glass substrates coated with magnetic materials where the user data is stored. A voice-coil motor (VCM) moves what is called the HSA, which is composed of the pivot, the arm block (or E block sometimes because of its shape), and the suspensions that carry a slider and head attached at the end of each suspension. The suspension and slider are usually designed to achieve negative pressure required for the head to fly over the disk surface with a desired flying height of about 10 nm over the rotating disk; during a read or write operation, the read/write head will further be lowered to about 1–2 nm from the magnetic disk via thermal fly height control [79]. The servo control system processes the read/write head position information and commands the VCM and milli- or micro-actuators to positon the read/write head to stay near the data track center. The position error, typically referred to as track misregistration (TMR), is measured as 3 standard deviations of the track following position error. A state-of-the-art HDD has about 500,000 tracks per inch

(TPI), with about 300,000 data tracks on a 2.5″ diameter hard drive with 500 GB per platter, or about 500,000 tracks on a 3.5″ disk platter storing over 1 TB user data. TMR is typically less than 10% of the track pitch. On a 500,000 TPI drive, the TMR is close to 2.5 nm or smaller.

2.4.1.1 Disk spindle pack

The brushless DC spindle motor used in HDDs employs a surface-mount permanent magnet and concentric armature windings. The disk rotation allows the user data read from or written to the disk surface in circular tracks at a certain rate. In the meantime, the read/write head follows the radial position of the track center through the servo control system. Move of the read/write head to a target track involves seek/settle delay, whereas reaching the target data on the target track involves rotational delay.

The spindle motor's structure is designed such that the motor has low non-repeatable run-out (NRRO) and low acoustic emissions. This is achieved by employing fluid dynamic bearing in a modern HDD instead of ball bearings, electronic magnetic designs having low cogging torque, low unbalance magnetic force, and sinusoidal driving current to reduce the torque ripple as well as commutation noise [80, 81].

A disk spindle pack is typically balanced to reduce the imbalance-induced vibration [82]. Various disk materials have been studied to achieve lower vibration under windage excitation [83–85]. Similarly, the airflow in disk drive has been extensively studied. Air dams, damping plates, etc., have been reported to reduce the flow-induced vibration as well as acoustic noise [86, 87].

2.4.1.2 HDD mounting and vibration

Regardless of the reduction of a drive's internal vibrations from spindle motor and VCM seeking, the drive will have to be installed in a computer or a storage rack or any other chassis that will introduce environmental vibrations. External vibrations are typically categorized as rotational vibe (RV) and linear vibe (LV).

The HDD mounting design is mainly for shock protection and vibe reduction purposes. Sometimes, a good shock protection mounting condition design could hurt vibe performance and vice versa. Thus trade-offs need to be made to balance shock and vibe performance during the HDD mounting design process. For example, the speaker vibe in a laptop can be reduced by chassis design by moving the vibration generation sources to most insensitive locations and/or reducing the vibration transmission such as using less rigid mounts [86]. Additionally, good shock protection mounting designs need to be matched with the drive internal servo so as to achieve the best performance. Correspondingly, feedback [88–92] and feed-forward controls [93–95] with [96–98] and without additional sensors to reduce the vibrations both internal and external to the HDD will be discussed in the servo control design sections. We only emphasize here that different drive applications have different applicable vibe sources and TMR, as well as performance requirements. For example, in a laptop computer application it is more challenging to achieve the rated full read/write performance when the user plays loud music, which typically is a broadband and time-varying vibration from the speaker. High-level cooling fan vibrations in enterprise/data center chambers is also severe and need to be handled for HDDs to achieve high performance and meet the enterprise-class requirement of a minimal read/write throughput drop.

2.4.1.3 Acoustics

Low acoustic emission is also required in several HDD applications, such as laptops, audio video (AV) boxes, etc. HDD acoustic sources mainly come from spindle motor and VCM motions. Acoustic metrics include both sound power and sound quality. In general, we can sort HDD acoustic requirements into the following categories:

- Idle acoustics: Spindle motion–only sound power
- Seek acoustics: Both spindle and VCM motion sound power
- Loudness: Determined sound quality by the Zwicker method, using human-hearing third-octave bands to mask the tonal component of the noise

- Prominence ratio (PR): High tone-to-noise ratio
- Sharpness: Ratio of high-frequency level to oversound level

HDD acoustics can be improved by reducing the noise source or adding an acoustic cover to prevent noise propagation from the HDD to human ears.

2.4.2 *Servo Fundamentals and Challenges*

The HDD servo system's main function is to process the position error signals (PESs) of the read/write head and to determine the control current of the VCM such that the read/write head reaches the target data sector on a target track just in time, considering the disk rotation latency. This is a constrained time optimal control problem with minimum jerk for acoustic and power considerations. After reaching the target track, the servo steers the head to stay on the target track centerline as close as possible. This can be formulated as a minimum variance control problem [99–103]. Next, we will discuss various aspects of servo control to achieve such objectives.

2.4.2.1 Servo system model and resonance control

Figure 2.24 shows a typical mechanism found inside of today's HDDs [100]. The RRO refers to those repeatable run-out components that are synchronized with disk rotation. All other components that do not repeat with disk revolution are called NRRO. Separating RRO from NRRO is useful in that it allows a specialized control strategy to deal with one of them. NRRO originated from the following sources: air flow due to the rotating disks (often called windage), arm turbulence created by the air flow, the vibration from the spindle motor, disk vibration or flutter, arm sway flutter, analog-to-digital converter/digital-to-analog converter (ADC/DAC) quantization noise and external shock and vibration. These sources are disturbances to the positioning servo loop [104, 105].

One of the major contributors to NRRO components is the one caused by disk vibration modes, that is, the in-plane off-track motion component of the disk vibration in a vertical direction caused by air flow excitation. They are responsible for a number of narrow-band peaks in the PES over the mid-to-high frequency range such as

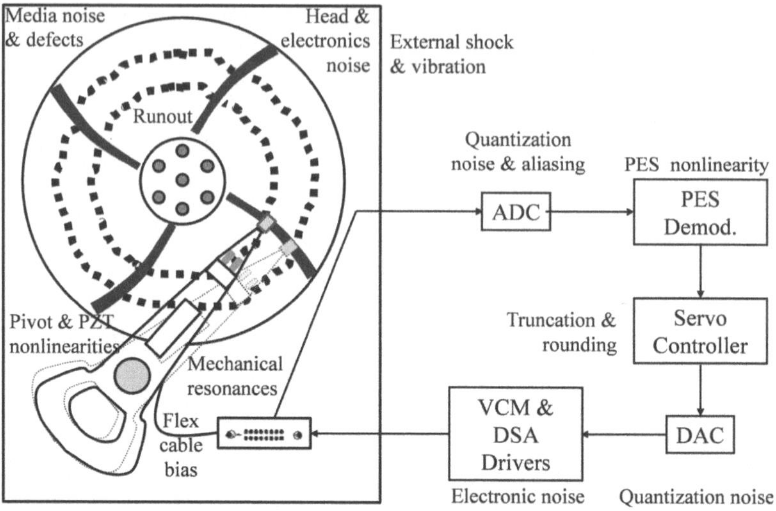

Figure 2.24 An illustration of the servopositioning mechanism in an HDD.

700–3000 Hz [84]. Figure 2.25 shows a typical example of the disk vibration power spectrum.

The main challenge of the track-following servo loop comes from today's ever-increasing track density and the arm actuator's mechanical resonances. Typically, a very-high-bandwidth loop must be designed to reject the effect of disturbances. However, a servo loop cannot be pushed higher than what the actuator resonant mode and parameter uncertainty allow. A sufficient stability margin is needed to account for the resonance modes and uncertainties in their frequency and phase characteristics. Higher-bandwidth actuator designs via optimizing the arm, VCM coil support (bobbin design), arm damper, and suspension damper have been reported [106]. Various filters compensating for the resonances and servo loop-shaping approaches have also been proposed in the literature [107, 108].

2.4.2.2 Servo signal generation and timing control

Typically, the main signal available for feedback in the positioning servo is the PES, which represents the deviation of the head

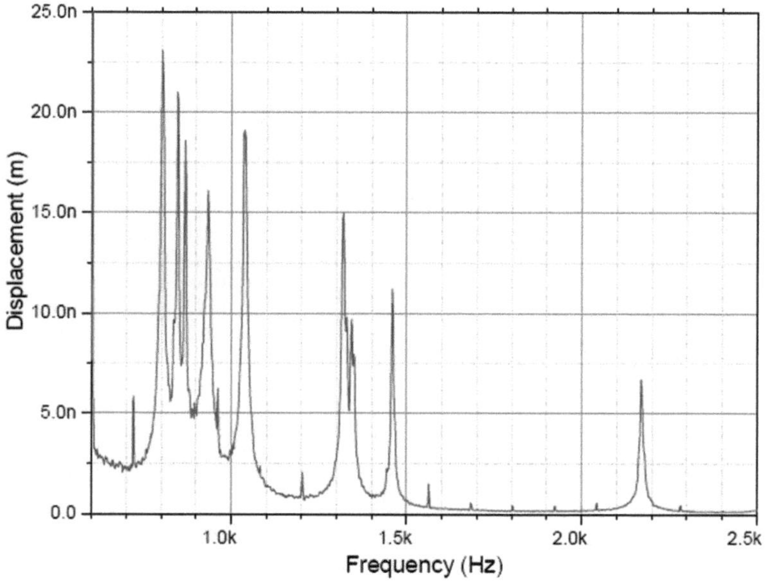

Figure 2.25 A typical example of the disk vibration power spectrum.

position from the center of a target track. The PES is formed via demodulation of the signal read from the servo burst pattern, which is the position information written on the media during the servo track-writing process [109, 110]. The servo pattern is of constant frequency within the same servo zone and hence has a larger size toward the outer diameter direction of the disk and is thus often called a servo wedge. Because the servo pattern is written with a rotatory arm, the servo wedge actually is an arc instead of a straight line on the disk. As shown in Fig. 2.26, each servo wedge consists of a preamble, a sync mark, a wedge identification (ID) and track ID, and bursts. The preamble provides the timing (via its phase) and amplitude reference for the servo channel to decode the current wedge information. The sync marks are special low-error-rate patterns that mark the beginning of the digital fields. The wedge ID identifies the current wedge number within a servo track. A timing loop provides the clock that divides the wedge-to-wedge space to an equal number of clocks, as required by the channel [111].

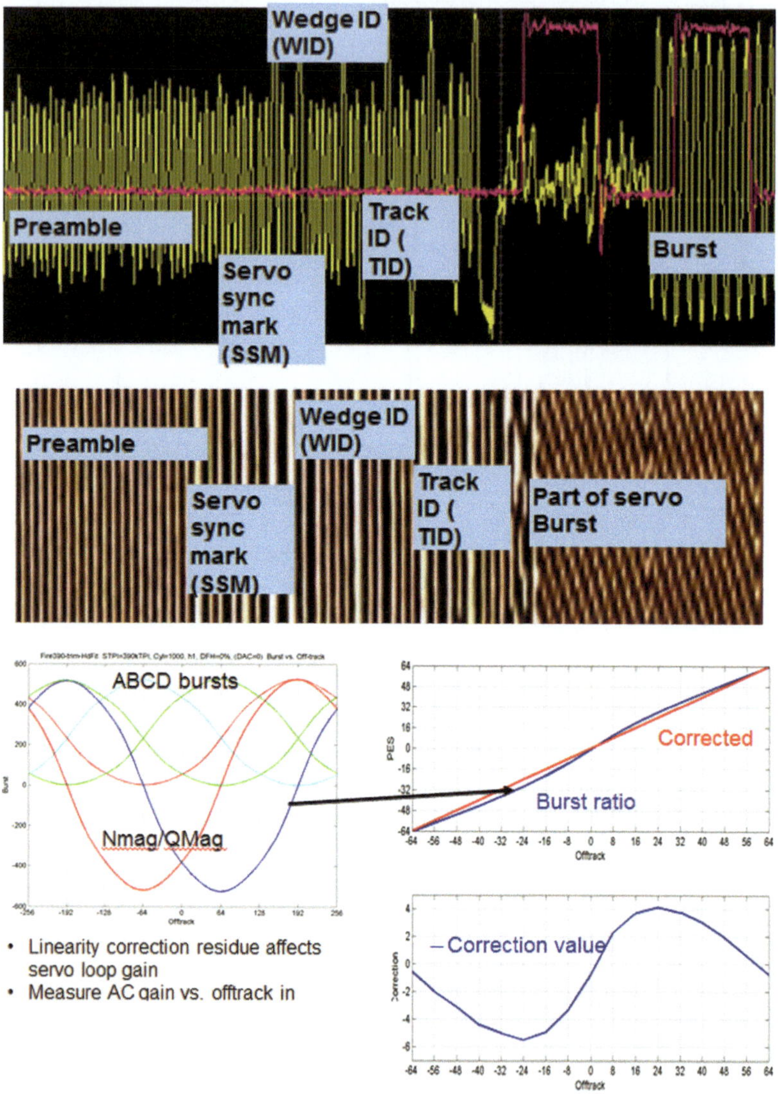

Figure 2.26 PES generation.

The wedge ID and the timing clock provide the down-track position of the read/write head.

In the cross-track direction, the track ID provides the coarse servo track position. The servo bursts provide the fine position information. When the head is moving cross track, the amplitude of the bursts will change, as illustrated by Nmag and Qmag in Fig. 2.26. One can use the Nmag/Qmag or other combinations of the burst ratios to represent the fractional part of PES, while the track ID is the integer part. Because the raw PES is not linear with respect to the true head off-track motion, PES nonlinearity correction has to be done [112]. An appropriate TPI, servo pattern, and servo writing error control scheme has to be chosen such that the PES nonlinearity and various manufacturing considerations such as a low servo writing time and low written-in error can be satisfied [109–111].

The number of wedges times the disk rotating speed gives the PES sampling rate. Noting that the actuators will always have resonances that vary from part to part and these resonances might be excited by mechanical or electrical means, the PES sampling rate needs to be selected such that the high-frequency head motion can be detected through the sampled PES. Correspondingly, there had been reports on how to measure the actuator frequency response beyond the Nyquist frequency (see Ref. [113] and references therein) and control designs considering the above Nyquist frequency behavior [114].

2.4.2.3 Servo control principle: seek/settle control

Seeking is a crucial function in HDD motion control to go from one user data track to another. Faster seeking will result in less track-switching time and higher read/write input/output operations per second (IOPS) performance. In general, HDD seeking algorithms can be divided into two categories, velocity loop seek and position loop seek.

A velocity loop seek is normally applied from a middle-range stroke seek to a full-range stroke seek with bang-bang seek (time optimal seek) type of algorithms. For example, the proximate time optimal servomechanism (PTOS) is one of them [99]. A velocity loop

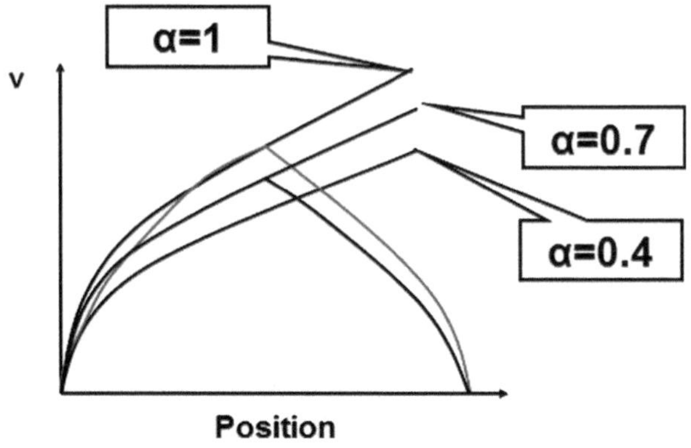

Figure 2.27 Seek profiles.

seek algorithm is independent of the time variable. Only the position error (from current location to target track location) and velocity are used as inputs to compute seeking control current. Assuming maximum constant acceleration and deceleration to achieve time optimal control, we can apply $\mathbf{V}^2 = 2*\alpha*u_{\max}*\mathbf{X}$ to obtain the following velocity profile: $\mathbf{V} = (2*\alpha*u_{\max}*\mathbf{X})^{1/2}$, where $\mathbf{V}$ is the velocity profile, $\mathbf{X}$ is the position error vector, $u_{\max}$ is the maximum available control effort, and α ($<=1$) is the tuning coefficient. As illustrated in Fig. 2.27, the velocity profile becomes less aggressive when α is reduced from 1 to a lower number.

We can apply the following equations to control the VCM to move the read/write head from the current track to the target track:

$$\begin{bmatrix} \dot{y} \\ \dot{v} \end{bmatrix} = \begin{bmatrix} 0 & k_y \\ 0 & 0 \end{bmatrix} \begin{bmatrix} y \\ v \end{bmatrix} + \begin{bmatrix} 0 \\ k_v \end{bmatrix} u$$

where y is the position, v is the linear velocity, and k_y and k_v are the tuning gain,

$$u = u_{\max} sat(k_2(f(y_e) - v)/u_{\max})$$

where $u_{\max}$ is the maximum control limit, sat is the saturation function, f is the velocity profile function, and y_e is the position error,

$$f(y_e) = \begin{cases} \frac{k_1}{k_2} y_e & |y_e| \le y_l \\ \text{sgn}(y_e) \left[\sqrt{\frac{2k_e \alpha v_{\max}}{k_y} |y_e|} - \frac{u_{\max}}{k_1} \right], & |y_e| \le y_l \end{cases}$$

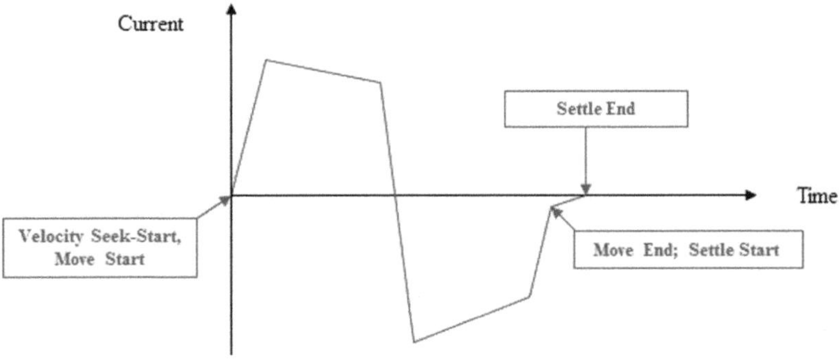

Figure 2.28 Control current profile in a seek operation.

where sgn is a sign function, α is the deceleration profile tuning parameter, and y_l is the distance to target to switch to the linear velocity profile,

$$k_2 = \sqrt{\frac{2k_1 k_y}{\alpha k_v}}, \quad y_1 = \frac{u_{\max}}{k_1}$$

where k_1 is the tuning gain to decide the linear velocity profile slope.

When a velocity seek starts, a move starts and it ends with transiting to the settle phase. The PTOS algorithm can rely on the linear profile to cover the settle phase but some other mode switching control design will switch to a different track-following algorithm to achieve smoother transition and faster settle time, as shown in Fig. 2.28.

The initial value compensation (IVC) algorithm is reported to ensure a smooth change to minimize the position error after switching. The main idea of the IVC is to initialize new controller values with switching time state values with some pre-calculated optimized matrix gain [115]. The algorithm assumes that the full plant states (position, velocity, etc.) are available for feedback. The performance index J (closed-loop state x_c energy) is to be minimized by the selection of the compensator initial state x_D [0]. Q is a nonnegative symmetric weighting matrix.

$$\min_{x_D[0]} J = \sum_{k=0}^{\infty} x_c^T [k] * Q * x_c [k]$$

From linear control theory we have

$$J = \sum_{k=0}^{\infty} x_c^T\,[k] * Q * x_c\,[k] = x_c^T\,[0] * P * x_c\,[0]$$

where P is the solution to the following discrete time Lyapunov equation and A_c is the closed-loop state space A matrix, where $x_c[k + 1] = A_c{}^* x_c[k]$:

$$A_c^T * P * A_c - P = -Q$$

Partition P into 2×2 blocks according to the dimension of $x_G\,[0]$ (plant initial state) and $x_D\,[0]$ to get

$$J = \left[x_D^T\,[0]\; x_G^T\,[0]\right] * \begin{bmatrix} P_{11} & P_{12} \\ P_{21} & P_{22} \end{bmatrix} * \begin{bmatrix} x_D\,[0] \\ x_G\,[0] \end{bmatrix}$$

Take the partial derivative of J with respect to $x_D\,[0]$ and set it to zero. We obtain the optimal $x_D\,[0]$ as

$$x_D\,[0] = -P_{11}^{-1} * P_{12} * x_G\,[0] \cong T * x_G\,[0]$$

where $T = -P_{11}^{-1} * P_{12}$ is the optimal initialization gain matrix. Therefore with the full state feedback, the initialization is simply multiplication by a static matrix gain. When a full state variable is not available, some state estimation scheme is needed.

A position loop seek is also popular in the HDD industry to achieve good seek performance within a short seek distance due to model uncertainty concerns. A position loop seek is usually composed of a position reference and a corresponding feed-forward control profile via the two-degree-of-freedom control concept (Fig. 2.29) [93]. Assume that the VCM real plant model is $G(z)$ with a nominal plant model $G_0(z)$. $D(z)$ is the feedback controller, r is the position reference input, y is the position output, and e is the position error. u_{ffwd} is the feed-forward control, u_{fb} is the feedback control, u_{adapt} is the adaptive control, and A is the adaptation block.

The position error e can be derived as follows:

$$e = \frac{1}{1 + DG}\left(1 - \frac{G}{G_0}\right) r + \frac{G}{1 + DG} u_{\mathrm{adapt}}$$

When $u_{\mathrm{adapt}} = 0$, we can further simplify e, as shown in the following equation, which is composed of feed-forward modelling error part 1

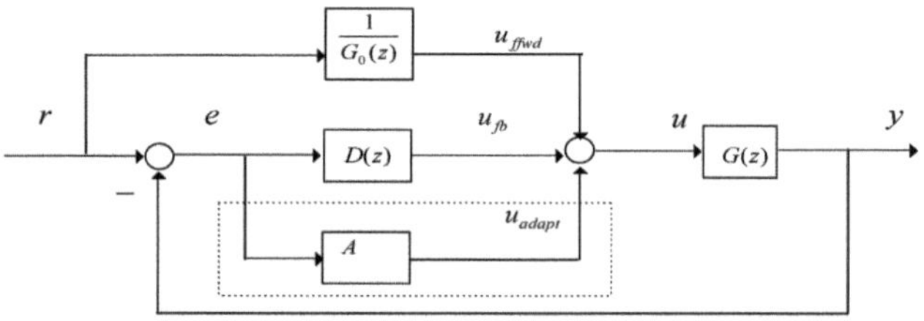

Figure 2.29 Block diagram of two-degree-of-freedom control.

- G/G_0 and feedback error part $r/(1+DG)$:

$$e = \frac{1}{1+DG}\left(1-\frac{G}{G_0}\right)r = \left(1-\frac{G}{G_0}\right)\left(\frac{r}{1+DG}\right)$$

if the VCM model G_0 matches the true VCM plant G exactly and there is no limit to the control signal. The position error is 0 all the time, which is unlikely in real life, so we need the feedback part to cancel the residual error. There are many ways to generate a position reference profile and the corresponding control feed-forward effort using different optimization indexes and constraints. For example, one can constrain the seek control profile to be of certain harmonic frequencies.

2.4.2.4 Servo control principle: track-following control

The primary objective of the track-following control loop is to regulate the position of the read/write head around the center of a target track with as little as possible deviation. Figure 2.24 shows a typical track-following servo mechanism in an HDD. A track-following controller typically includes a lag-lead type baseline compensator to achieve zero steady-state DC position error and desired transient performance, compensator components specialized for dealing with specific types of runout, for example, RRO, and a series of notch filters to precompensate the effect of mechanical resonance modes.

2.4.2.4.1 Laglead control

An HDD actuator response resembles a doubleintegrator model in the lowfrequency region above ~50–100 Hz, with highfrequency resonances. To increase the lowfrequency gain for disturbance rejection, a lag compensation is adopted for the lowfrequency range. To satisfy the 20 dB/dec open-loop Bode cross 0 dB as is required by a typical feedback control system [99, 100, 116], a lead compensator is also necessary. Figure 2.30 shows an example of HDD lag-lead compensator design. As a rule of thumb, a phase margin of ~25–30 degrees is required for an HDD servo

2.4.2.4.2 Advanced loop shaping

Due to everincreasing track density and demanding performance requirements, the track-following servo error rejection transfer function needs to be shaped or optimized in such a way that it will tolerate not only internal disturbances but also the external vibrations and give as small as possible PES variance. Furthermore, this performance must be ensured against variations of electromechanical parameters and resonances characteristics. To achieve this, the following techniques are popularly employed in the industry:

(1) Modern loopshaping and robust design techniques (H_∞, μ-synthesis etc.)
(2) Use of additional serial or parallel connection of peak or notch filters
(3) Disturbance observer (DOB) method with band-pass filter (BPF)-style Q-filters
(4) Numerical search to find controller parameters

In (1), uncertainties in the system are characterized and incorporated into the design; the loop shaping is usually achieved by augmenting weighting functions to the closed-loop system and by synthesizing a robust controller as a solution to an optimization problem. Usually, this approach ends up with a high-order compensator with its order increasing in proportion to the complexity of frequency weighting. The technique in (2) is easy to implement and popularly used when local shaping of the loop is needed. In spite of its simplicity, simply creating a notch in the error rejection transfer function usually amplifies the other area undoing its benefit or even degrading the overall result. Advanced techniques to optimize such a loop [107] will be discussed in a later section.

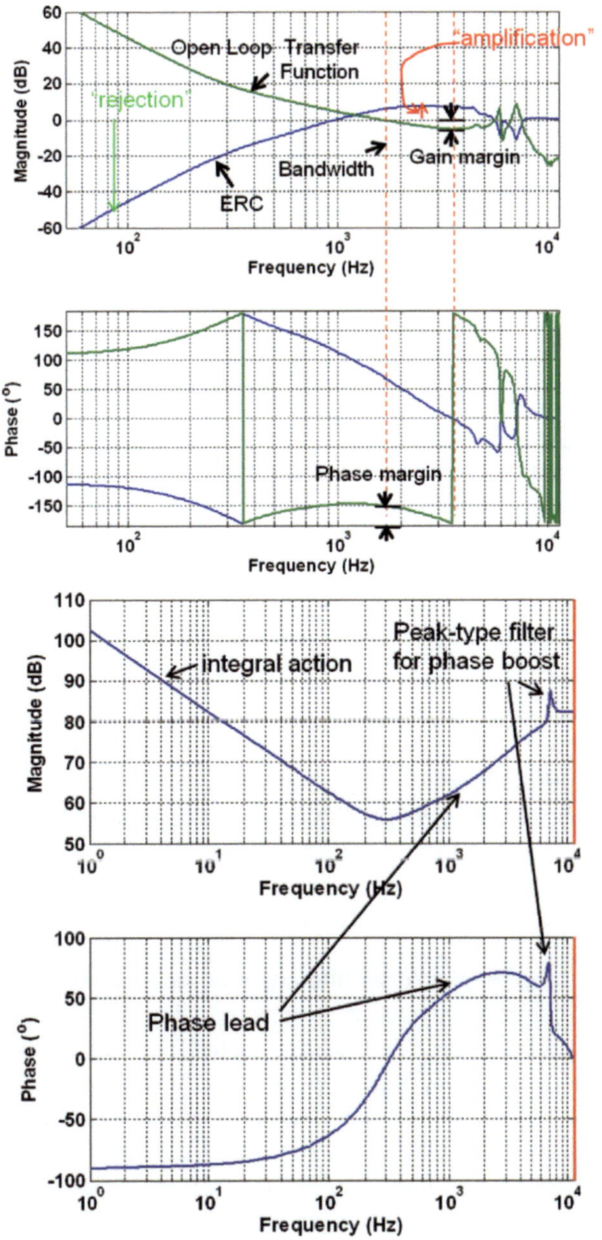

Figure 2.30 Lead-lag compensator design.

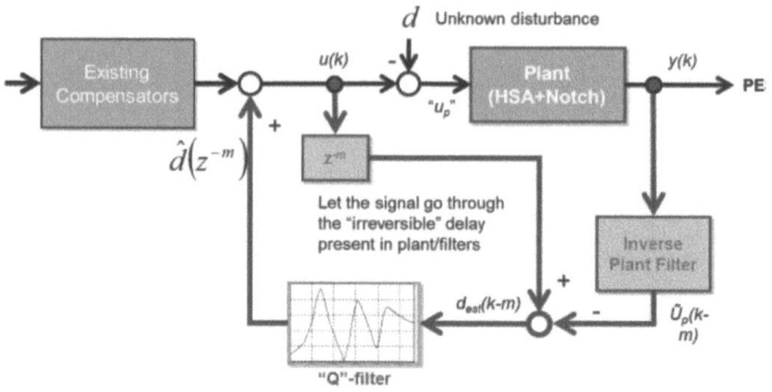

Figure 2.31 A block diagram of a typical disturbance observer with a band-pass-type Q-filter.

The DOB approach in (3) may be adopted as a loopshaping method specialized for vibrations because it can easily provide a localized rejection boost around a given frequency. A typical example of the DOB configuration is shown in Fig. 2.31. The idea behind the DOB is to estimate the disturbance d by subtracting the control effort u from the plant input estimated through an approximate plant inverse along with the measured PES. This raw disturbance estimate is passed through a Q-filter to generate a useable version of disturbance estimate that can be combined into the control effort. However, the presence of the plant delay, which cannot be included in the plant inverse, causes the delay in the estimated disturbance. For this reason, a DOB with a BPFtype Q-filter will create a notch in the error rejection transfer function, whose center frequency is near the BPF's center but does not agree with it exactly.

2.4.2.4.3 Loop shaping/peak filtering: non-repeatable RRO control

A popular approach to deal with narrow-band disturbances is to employ a set of peak filters whose center frequencies are matched to those modes that have the most significant contribution to PES. The advantage of this method is that one can keep the baseline compensator (such as a lead-lag compensator), while adding the desired boost in rejection at selected frequency locations to effectively suppress the narrow-band NRRO. The block diagram of a paralleltype peakfiltering scheme is shown in Fig. 2.32. In practice, multiple peak filters can be added in a similar fashion, where a peak

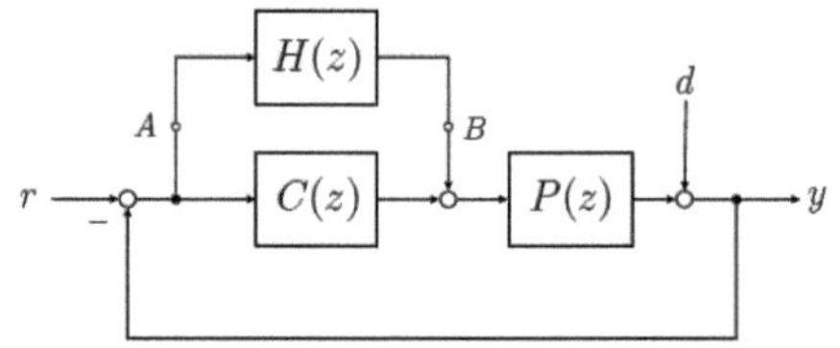

Figure 2.32 A peak filter $H(z)$ connected in parallel with a baseline compensator $C(z)$.

filter of the following form can be chosen:

$$H(z) = \frac{\alpha z + \beta}{z^2 - 2\eta \cos\theta z + \eta^2}$$

Here, the coefficients (α, β) determine the gain and zero location, while η and θ determine the damping and the center frequency of the peak, respectively. In Ref. [107], it is shown that (α, β) can be selected on the basis of the plant characteristic and the filter's damping in such a way that will optimally spread the sideband boosting of the sensitivity function near the peak filter frequency. This result is a nontrivial extension of the optimal zero location method proposed for the harmonic adaptive feed-forward controllers found in Ref. [117].

The efficacy of applying an optimally designed peak filter is shown in Figs. 2.33 and 2.34. Figure 2.33 demonstrates a notch effect created by the application of the peak filter. The PES power spectrum in Fig. 2.34 shows that the filter can effectively suppress the selected narrow-band frequency without significantly boosting the side band.

2.4.2.4.4 Repeatable run-out cancellation

The sources of RROs may come from the disk (eccentricity or clamping deformation), the spindle motor (torque ripples or pulse-widthmodulated driving signals), or any off-track vibration written in during the servo-writing process. There are two types of cancellation schemes for the RRO, depending on the nature of the RRO. One method that has been the most studied in the literature is to control the actuator such that the head "follows" the disturbance [117–119]. The other, a rather unusual approach, is to identify the RRO and modify the position reference such that the controller

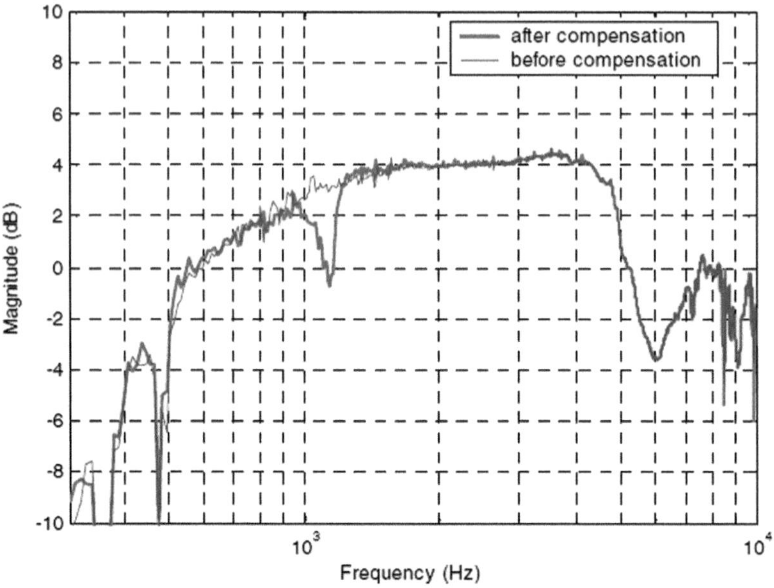

Figure 2.33 Error rejection transfer function before and after the peak filter compensation.

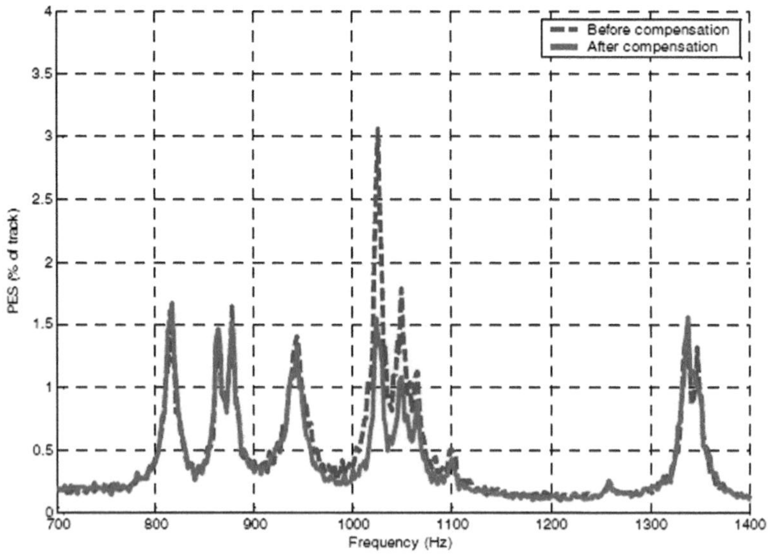

Figure 2.34 PES spectrum before and after the peak filter application.

ignores the RRO [120]. Here, we will briefly introduce the former approach.

Most of the RRO cancellation techniques are essentially based on the well-known internal model principle, where a periodic signal generator or some basis function for periodic functions is incorporated in the design process, to render the closed-loop system track the repetitive disturbance perfectly, while maintaining stability. The harmonic adaptive feed-forward cancellation (AFC) technique in Ref. [117] is a well-known method in conjunction with a simple optimal phase determination method based on its linear-time-invariant (LTI) equivalence. In this technique, the Fourier coefficients (a_i, b_i) corresponding to a selective harmonic frequency ω_i are adaptively estimated on the basis of the PES as follows:

$$a_i(k+1) = a_i(k) + g_i\varepsilon(k)\cos(\omega_i k + \phi_i)$$
$$b_i(k+1) = b_i(k) + g_i\varepsilon(k)\sin(\omega_i k + \phi_i)$$

where $\varepsilon(k)$ denotes the PES at the kth sample and g_i represents the adaptation gain. The cancellation signal to be added to the control effort is generated by

$$u_c(k) = a_i(k)\cos(\omega_i k) + b_i(k)\sin(\omega_i k)$$

It is well-known that the input–output relationship of this AFC is, in fact, equivalent to the following LTI transfer function:

$$H_{AFC}(z) = \frac{\cos(\omega_i - \phi_i)z - \cos(\phi_i)}{z^2 - 2z\cos\omega_i + 1}$$

2.4.2.4.5 Loop-shaping performance limitations

Consider the HDD positioning servo loop shown in Fig. 2.35 with an actuator model $G_p(s)$, a compensator transfer function $G_c(s)$, the input and output disturbances $i_d(s)$ and $o_d(s)$, and the noise $n_p(s)$.

$$\text{PES}_t(s) = -G_p(s)S(s)i_d(s) - S(s)o_d(s) - T(s)n_p(s)$$

where $S(s)$ and $T(s)$ are the output sensitivity function and the complementary sensitivity function, respectively. To meet the HDD recording system TMR requirement, two approaches are usually considered. The first approach is to reduce the disturbance source by improving mechanics such as use of a fluid-bearing spindle motor, disk vibration reduction, air flow optimization, etc. The other approach is to increase the sensitivity transfer function $S(s)$

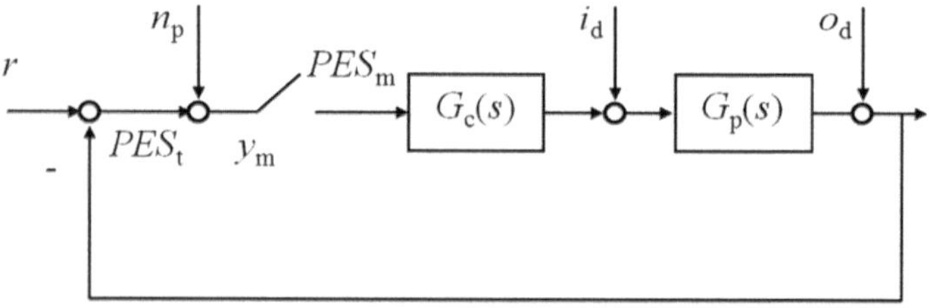

Figure 2.35 Track-following servo loop with input–output disturbances and sensor noise.

attenuation at the disturbance frequencies. The above equation along with the fundamental property

$$S(s) + T(s) = 1$$

tells us that it is impossible to reject both disturbance through reduction of $S(s)$ and measurement noise though reduction of $T(s)$, the complementary sensitivity transfer, at the same time. Reducing the PES measurement SNR is therefore important for servo loop bandwidth boost. Moreover, the well-known Bode integral theorem presents constraints to servo loop design. The theorem suggests that it is not possible to achieve an arbitrary sensitivity reduction [101, 121, 122]. The inherent constraint necessitates design trade-offs that typically increase the attenuation at frequencies having more disturbance and allow some amplifications at frequencies having less disturbances such that the total TMR is reduced.

Another obstacle in achieving higher servo bandwidth is the presence of mechanical resonances and associated phase lag that vary from drive to drive, head to head, and at different temperatures, and even with respect to head fly height changes. A common approach to addressing the effect of resonances is to use notch filtering, which pre-compensates the resonance modes such that the frequency response of the pre-compensated plant resembles that of a rigid actuator. Applying notch filters to suppress the magnitude of resonances is called gain stabilization. Although gain stabilization appears to easily achieve the aforementioned pre-compensation, it inevitably results in the loss of the open-loop phase. This phase loss and the inability to accurately compensate for all the resonances (i.e, plant model uncertainty) limit the achievable servo bandwidth boost.

Finally, the sampling frequency and the achievable servo bandwidth are interrelated. As a rule of thumb, sampling frequency is preferred to be 10 times as high as the bandwidth for effective suppression of vibration. The sampling frequency depends on the number of servo sectors per revolution, which, in turn, is limited by the space on the disk allocated for servo bursts. Increasing the rotating speed of the disk is another way to increase the sampling frequency, but it also increases the level of internal disturbance such as air-flow-induced windage and disk vibration. Besides sampling frequency limitation, the computation delay and transport delay by the ADC introduce a phase lag to the servo loop and thereby reduce servo bandwidth capability [99, 123].

2.4.2.4.6 Dual-stage servo control

Over the past years, continuous efforts have been made to improve the resonances of the VCM actuator. Despites various incremental improvements, the VCM actuator can only offer $\sim$1–2 kHz bandwidth. Another approach adds a secondary actuator placed conveniently between the pivot and the slider in a piggyback fashion [124–128]. Such a configuration, known as dual-stage actuation (DSA), still uses the conventional VCM actuator for long-range seeks, while the secondary stage located closer to the slider can improve tracking performance during track following.

Various methods of designing a controller for the dual-actuated HDD servomechanism have been reported in the literature [129–132]. Commonly used configurations are parallel structure such as the PQ method in Fig. 2.36 and the decoupled master–slave (DMS) structure in Fig. 2.37.

The PQ method [131] is a popular approach for designing a controller in parallel structure. This method is a two-step design approach. In the first step, compensators $C_{\mathrm{v}}(z)$ and $C_{\mathrm{m}}(z)$ are chosen simultaneously to address the issues of stable zeros, relative contribution of the two actuators to the combined output, and the interference between the two contributions. The two parallel branches are then combined to form a single-input single-output model $P_{\mathrm{siso}}(z)$. In the second step, a conventional feedback compensation design scheme for a SISO system can be applied to $P_{\mathrm{siso}}(z)$ to achieve the desired bandwidth and margin.

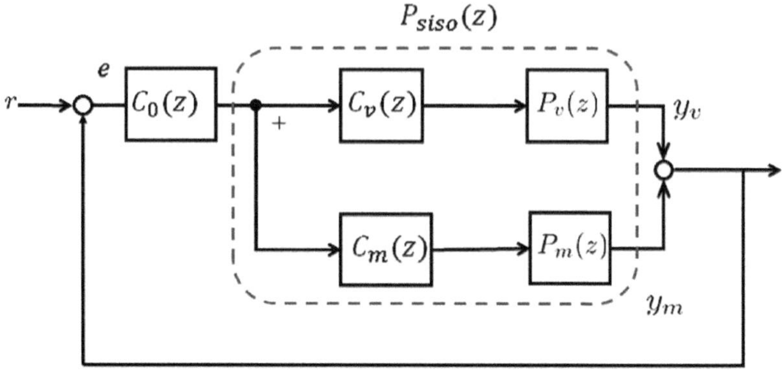

Figure 2.36 Configuration of a parallel-structure DSA controller.

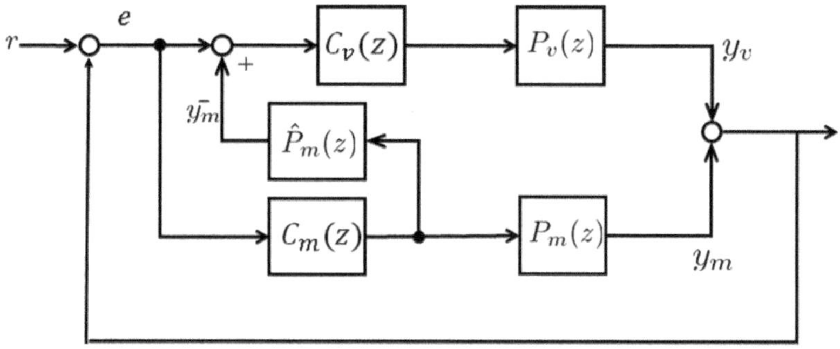

Figure 2.37 Decoupled master-slave-structure DSA controller.

In the DMS structure, the compensators $C_v(z)$ and $C_m(z)$ are designed independently for their individual sensitivity function shaping $S_v(z) = 1/(1 + P_v(z)C_v(z))$ and $S_m(z) = 1/(1 + P_m(z)C_m(z))$. If $\hat{P}_m(z)$ is designed to match the microactuator $P_m(z)$ frequency response, the DSA closed-loop sensitivity function can be decoupled as follows:

$$S_{dms}(z) = S_v(z)\,S_m(z) = \frac{1}{(1 + P_v(z)\,C_v(z))\,(1 + P_m(z)\,C_m(z))}$$

The decoupling property simplifies significantly the design process, which makes this approach attractive to the HDD industry. By defining the VCM compensator as $C_{vp}(z) = \left(1 + C_m(z)\,\hat{P}_m(z)\right)C_v(z)$, the DMS structure is equivalent to a parallel structure. In this sense, the DMS can be viewed as a design procedure for the parallel structure compensator design.

Additional to these basic designs, other enhancements, such as anti-windup control [134], self-sensing and actuation [96], use of one actuator for sensing and one for control [97], the effect of charge versus voltage driver [135], the use of a control algorithm to compensate for hysterysis [98], and tristage control for relaxing the constraint of the secondary stage actuator stroke and bandwidth limitation [133] have all been reported.

2.4.2.5 Current and future servo challenges

Albeit today's HDD is capable of supporting the track density at about half a million TPI, the servomechanical system will need to continue support future recording systems having more disk platters and more complex structures. High-bandwidth actuator systems, low-internal-vibration mechanics, external vibration sources and transmission reduction, and advanced servo designs, including adaptive control for minimum PES variance regardless of actuator uncertainties will continue play a pivotal role in the servomechanical system development, as discussed in previous sections.

In view of the future magnetic recording system moving toward two-dimensional magnetic recording (TDMR), heat-assisted magnetic recording (HAMR), and bit-patterned media (BPM), the following additional aspects will need more research:

(1) Associated with high-bandwidth actuations, low-noise higher-frequency vibration detection and write protection through sampled PES without substantially increasing the number of servo wedges (which will occupy more disk space that is otherwise used for recording user data) are necessary for achieving the best system-level optimization.
(2) For TDMR and BPM designs that need the data being written to the specific down-track magnetic locations, the timing control, that is, the down-track head position relative to the recording bits, becomes critical. In the case of BPM, not only the timing loop but also the cross-track RRO must be followed with a minimum settling time without taking multiple disk revolutions.

In this case, the high-order control system that achieves fast settling is needed in parallel with minimization of media fabrication error.

(3) Position sensing in down-track, cross-track, as well as vertical directions will be vital in providing accurate control of the read/write head position. This includes reduction in the sensing pattern error itself.

2.5 Signal Processing Fundamentals and Challenges

As discussed previously, the first step in retrieving bits recorded onto storage media is to use a read head to read back the bits. The read-back signal obtained thus is noisy due to the presence of non-idealities in the writing and reading process. Sources of noise include process-related issues, for example, electronic noise, inter-symbol interference (ISI), and bit boundary imperfections (also known as jitter), as well as system-related imperfections like TMR, thermal asperity, and media defects. The term "read channel" is used to refer to the collection of signal processing systems responsible for recovery of written bits from the noisy read-back signal. The modern read channel [136] consists of modules that mitigate the effects of these noise sources and non-idealities, for example, equalizer, detector, error-correcting codes and modulation codes, and noise-whitening filters.

The read channel has been a key enabling technology for increasing areal densities, thereby reducing the per capita cost (measured per gigabytes) for HDDs. This is because advances in the field of signal processing have enabled the design of read channels that can tolerate more noise—a direct consequence of increasing areal density—hence paving the way for storage systems with higher areal densities.

The most direct way to assess the performance of a read channel is to measure the maximum operational areal density at which it can support a specified probability of failure to recover bits—also known as sector failure rate (SFR) (typically chosen to be 10^{-14}–10^{-17}, that is, 1 unrecovered block of bits in 10^{14} blocks). It is not usually feasible to measure such small SFRs, so alternative metrics like bit error rates (BERs) prior to application of error correction decoding (also known as raw BERs) are used (refer to Ref. [137] for one such analysis using a raw BER).

There have been numerous improvements in the read channel, leading to better overall system performance. Introduction of partial response maximum likelihood (PRML) detection as an advancement to peak detection and transition from Reed–Solomon (RS) codes [138, 139] to low-density parity check (LDPC) codes [140–142] for error correction remain definitive exemplars of read channel innovations that enabled higher areal density. Interested readers are encouraged to read Chapter 1 of Ref. [136] for a historical perspective on magnetic storage, including the evolution of PRML channels.

2.5.1 *The Read Channel*

Figure 2.38 illustrates the write and read processes for a modern hard drive system. The input data is encoded by the error-correcting code (typically an LDPC code). The error-correcting-code-encoded data is further modulated and parity bits are added for further error protection. The data is then equalized and written to the media.

During the recovery process, the read-back signal is first processed through an analog front end (AFE) that preprocesses the signal. This includes DC correction and analog filtering for noise removal. Subsequently, the signals are discretized and digitized by an ADC sampling them at the channel sample rate. From this point forward in the read channel's path, all operations are done on the digitized and discretized signal. This leads us to term the remainder of the signal processing path the "digital backend" (DBE). Note that in Fig. 2.38, for the sake of clarity, the AFE is not shown explicitly.

Figure 2.38 also shows a timing recovery module. This is an important component of the read channel that works in conjunction with the AFE and the DBE. Although out of the scope of this exposition, it is worth mentioning that the timing recovery module is responsible for controlling the sampling phase of the ADC so as to provide the best BER performance.

The process of data recovery by the DBE of the read channels may be broadly divided into two parts, namely detection and error

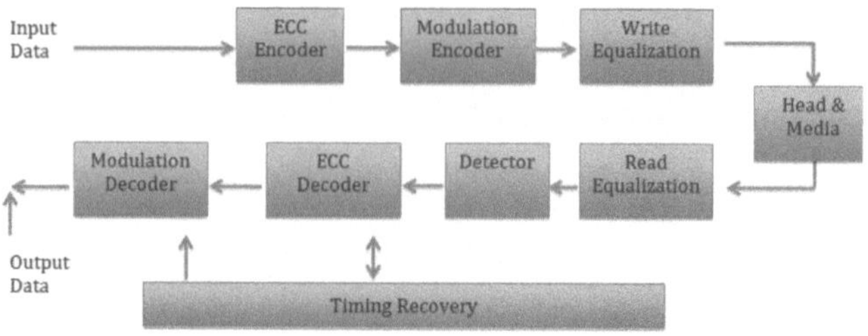

Figure 2.38 Block diagram of the read channel.

correction. Detection may be defined as the process of recovering bits from the ADC samples. The soft-output Viterbi algorithm (SOVA) [143] and peak detection (now obsolete) are examples of detectors. The detected bits are not error free. Indeed, the term "raw BER" refers to the BER of the detected bits.

The next step is error correction decoding that corrects errors made by the detector. Traditionally, RS codes and LDPC codes have been the error-correcting codes of choice.

2.5.2 *Peak Detection*

As the name suggests, the read channel circuit recovers the bits by comparing the amplitude of the read-back signal against a precomputed threshold and then making a binary decision on each bit on the basis fo this comparison. Figure 2.39 [144] describes this process diagrammatically.

It can be seen from the figure that to obtain the best BER performance, it is important that the read-back signal be sampled at its peak. Indeed, it is the case that even small deviations in the sampling instance can have a considerable impact on the raw BER. Control of the sampling phase is achieved by using a timing recovery circuit based on phase lock loops (PLLs).

The main drawback of peak detection is the reduction in tolerance to timing errors at higher areal density. The reduction is primarily due to peak shifts and amplitude loss at these densities. Figure 2.40 explains these phenomena. In Fig. 2.40, the black curve shows the response of the reader to an isolated transition in magnetization—a pulse. The peak detector works best when the

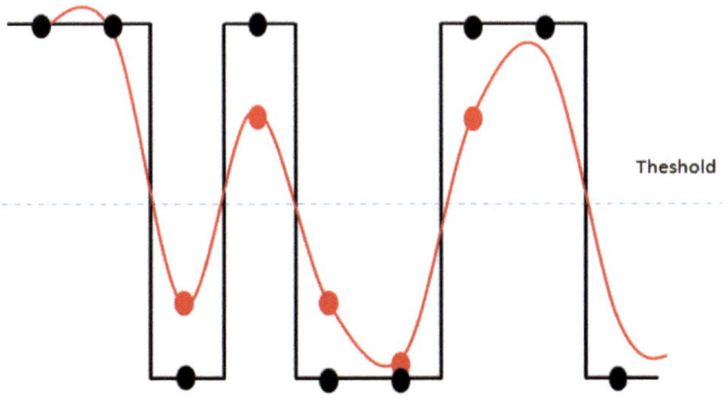

Figure 2.39 Peak detection: The signal marked in black shows the ideal read back and the signal marked in red shows the noisy signal from the reader—the read-back signal is compared against the threshold (indicated by the dashed line) to determine the output bit.

pulses are isolated. However, as bits become smaller (i.e., areal density increases), the pulses are closer to one another (crowding). Pulse crowding leads to consecutive pulses interfering destructively with each other, thereby leading to a loss in amplitude. The blue curve in Fig. 2.40 is the result of two consecutive pulses, namely the pulses marked in red and black, interfering destructively with each

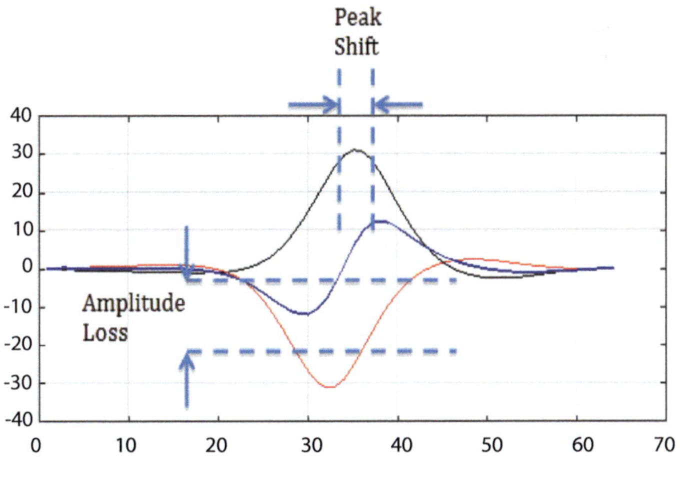

Figure 2.40 Peak shift and amplitude loss during superposition.

other. As a result, the maximum amplitude of the resulting signal is considerably lower. Note also that the peak of the pulses has now shifted from their ideal positions. This leads to a further loss in amplitude. These two phenomena are broadly classified as ISI.

To alleviate the problems of ISI associated with peak detection, run-length-limited (RLL) codes were introduced (see Ref. [145] for a comprehensive review of RLL and other modulation codes). The basic principle of an RLL code was to transform the data written onto the magnetic media so as to ensure that during read-back, the resultant pulses are isolated. A very popular RLL code used with peak detection was the (1,7) code. The (1, 7) RLL code ensured that between any two 1s in the channel bits (resulting in a pulse), there is at least one 0 (no pulse). Further, between two consecutive 1s, there may be no more than 7 zeros. Note that the second condition is for the benefit of the timing recovery module, which relies on the pulses to derive timing error signal for the PLL. The final result was a code of rate 2/3. That is, for every 2 bits of user data, 3 bits of encoded data were stored on the media. This was a significant loss in user data-recording efficiency. It is worth mentioning that other modulation codes, for example, maximum transition run codes based on the (0, 4/4) constraint, have also been devised to improve detection in the PRML read channel.

As the recording density increased, the peak detector became less reliable due to the fact that the ISI became a larger issue impacting the BER and causing data recovery and reliability issues for the system. For these reasons, IBM introduced the PRML detection system in HDDs in the early 1990s.

2.5.3 *Partial Response Maximum Likelihood*

The PRML read channel represented a radical reformulation of the ISI mitigation/bit detection problem. The principle of operation of the PRML read channel may be qualitatively explained as follows: Instead of completely eliminating all ISI, it is controlled. By precisely controlling the amount of ISI and subsequently using the knowledge of this in the determination of each bit, better BER performance can be obtained.

Figure 2.41 Schematic diagram of the PMRL channel. Noise whitening (colored gray) is optional.

The main steps involved in the PRML read channel are schematically shown in Fig. 2.41. As shown in the figure, a PRML channel constitutes (a) equalization and (b) maximum likelihood (ML) detection. Equalization is the process of controlling the ISI, and ML detection is the aforementioned process of using the knowledge of ISI in bit determination.

The equalizer is best understood by viewing the read-back process as an LTI system (see Refs. [146, 147] for mathematical models inspired by this perspective). In particular, the read-back process is visualized as providing an ADC output that is a function of the bits in the large local neighborhood (typically 10–20 bits). The equalizer can then be defined as another LTI system whose output is a function of bits in a smaller neighborhood (typically 2–5 bits).

It may seem at first that making the output a function of only the current bit (instead of a neighborhood of bits), we will be able to detect the bits by simple thresholding. However, it can be demonstrated that such a scheme (often referred to as zero forcing equalization) does not work due to a phenomenon of noise amplification in the signal (the mathematical description of this phenomenon is beyond the scope of this text).

The exact function of the bits that the equalizer output takes (also known as target) depends on the system. In the past, however, some functions have been very successful in magnetic storage. For instance, the PR4 channel characterized by the target function $(1 - D^2)$ (interpreted as the current output is the sum of the current symbol and symbol two time instances before the current one) was used in early systems [148–151]. Another target named EPR4 [152], with a target $= (1 - D)(1 - D^2) = 1 - D - D^2 + D^3$ (interpreted as current symbol is a weighted sum of current symbol and the three symbols, with the weights being +1, –1, –1, and +1, respectively) was also another target function that was very successful in early systems. In general, the construction of target functions is chosen so that its spectral characteristics match the spectral characteristics of the ADC output signal closely. Subsequently, there were efforts to further improve the process of equalization by optimizing the equalizer filter and the targets together. These have led to the socalled generalized partial response targets [153].

The next step in the PRML channel is ML detection. ML detection may be understood as follows: The equalizer output at any point is the weighted sum of symbols in its immediate neighborhood. Since each symbols can be $+/-1$ (for bits 0 and 1, respectively), the total number of possible outputs is finite. For instance, for the PR4 channel, there can only be three distinct possible outputs, namely $+2$, 0, and -2. Moreover, not all combinations of equalizer outputs can occur together. For instance, in the PR4 channel, -2 cannot be followed by a $+2$ (and vice versa). By assuming the equalizer output to be a (noisy) signal that takes on one of a finite number of values, the interdependencies can be used systematically to detect the channel bits. One algorithm that does this is the Viterbi algorithm, which provides the most likely estimates of channel bits.

An enhancement on the Viterbi algorithm is the SOVA [143]. The Viterbi algorithm provides only a string of detected bits without providing a probability metric on each bit (in other words, "how good is the bit estimate"). While this is sufficient input for some error-correcting systems like RS codes, modern coding systems like LDPC codes benefit tremendously from such reliability metrics (also known as soft information). SOVA is an extension of the Viterbi algorithm that provides such a reliability metric.

2.5.3.1 Noise whitening

In the first generations of the PRML read channel, the noise as defined at the output of the equalizer was assumed to be additive white Gaussian. However, this is not true. In particular, noise has nonlinear, colored, and pattern-dependent behavior. Hence, by a more accurate modeling of the noise, better BER performance is possible. Noise whitening refers to a class of techniques that aim to do precisely this. Noise predictive maximum likelihood (NPML) [154] and pattern-dependent noise prediction (PDNP) [155–157] are two algorithms that are examples of this technique.

2.5.4 *Error-Correcting Code*

For HDDs, the error rate of the user data needs to be very low, 10^{-14} to 10^{-17}. It is important that all errors during detection be corrected with a *very* high probability. Decoding of error-correcting codes is the final step in the read channel that is used to enable this.

In HDD systems, data is organized into logical blocks of bits (sectors). Each logical block, containing user data, along with additional bits called parity bits, which are calculated by the error-correcting code encoder, are written together. These parity bits are used to detect and correct errors during decoding.

HDDs historically have used RS codes for error correction. RS codes are widely used for error detection and correction in various computing and communications media such as magnetic storage, optical storage, high-speed modems, and data transmission channels. These codes rose to prominence due to their ability to correct a large number of burst errors typical, in HDDs

In previous generations of HDDs, each sector of data contained 512 bytes of user data and additional parity bits. As HDD areal density continued to increase, the overhead due to error-correcting code parity continued to increase, hence making the error-correcting code less efficient with a 512-byte sector size. This is demonstrated in Fig. 2.42 [158]. The figure shows the net bits per inch (BPI) percentage gain as a function of the number of added parity bits. As can be seen, there is a loss in bit density as error correcting code overhead increases beyond a point.

For this reason, the user data size for HDDs was increased from 512 bytes to 4 KB. This increase enabled improved error-correcting code efficiency and therefore provided an increase in areal density by as much as 6%–10%.

In recent years, many other coding schemes have been proposed to improve the system SFR and reduce the overhead further. LDPC codes are the most widely used codes in magnetic recording systems because of their superior error-correcting capability, especially at lower code rates. Further, decoding of LDPC codes uses a message-passing algorithm, which is highly amenable to practical implementation (e.g., Ref. [159]).

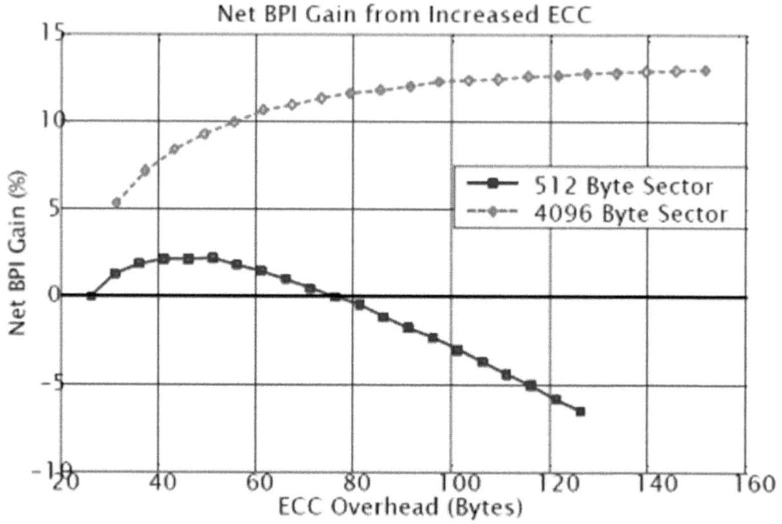

Figure 2.42 Net BPI gain for 512-byte and 4 KB sectors.

LDPC codes are linear block codes that can be represented by parity check matrices with a very small fraction of nonzero entries in them. Gallager introduced them in his PhD thesis in 1960 [140]. However, due to their complexity and difficulties in implementation and the introduction of RS codes, they were mostly ignored until being rediscovered by MacKay and Neal in 1996 [142].

LDPC codes are designed on the basis of a parity check matrix, $H = [h_{i,j}]$. The code is completely specified by H. For binary codes, H is m by n with $h_{i,j} \in \{0,1\}$where m is the number of parity checks and n is the code word length. For example, H defines the parity check equations for a (10,5) code as shown below:

$$H = \begin{bmatrix} 1&1&1&1&0&0&0&0&0&0 \\ 1&0&0&0&1&1&1&0&0&0 \\ 0&1&0&0&1&0&0&1&1&0 \\ 0&0&1&0&0&1&0&1&0&1 \\ 0&0&0&1&0&0&1&0&1&1 \end{bmatrix}$$

LDPC codes are low density because H contains relatively few 1s. The best codes are designed randomly (1s in H are chosen randomly) but typically require a complex encoder. As originally

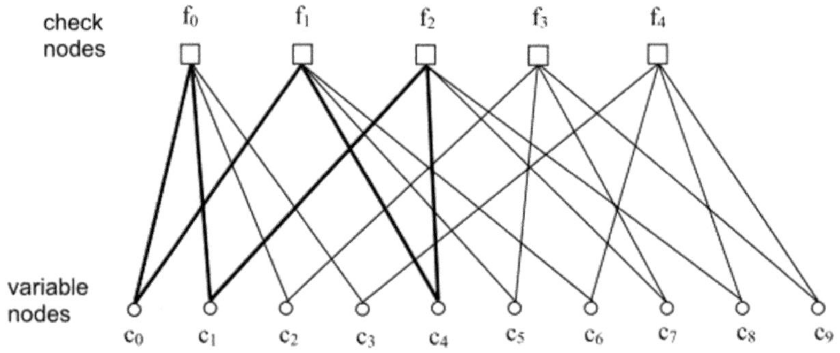

Figure 2.43 Tanner graph for the example H matrix.

described by Gallager, H can be constructed using a set of identity and permutation submatrices.

It is highly instructive to represent LDPC codes using a graphical representation called Tanner graph [141]. In particular, Tanner graphs are a bipartite graphical representation of LDPC codes. There are two sets of nodes, namely variable nodes and check nodes. Variable nodes represent code word bits, and check nodes represent parity check constraints of the codes. Figure 2.43 [160] shows the Tanner graph representation of the H matrix above.

The decoding of LDPC codes is done by using an algorithm called belief propagation. This algorithm is a low-complexity algorithm and is best described as an algorithm that passes messages along the edges of an LDPC code's Tanner graph. In spite of its simplicity, belief propagation is a powerful algorithm and enables LDPC codes to achieve theoretical capacity limits on several channels [161, 162].

2.5.5 *Signal Processing Outlook*

Figure 2.44 [163] represents the signal processing evolution and coding trends. Peak detections were used in HDDs until the early 1990s, when they were replaced by the PRML system, enabling areal density growth to accelerate. Today's the HDD system uses BIC signal processing techniques, such as PRML and LDPC, to further improve the performance and areal density of HDDs.

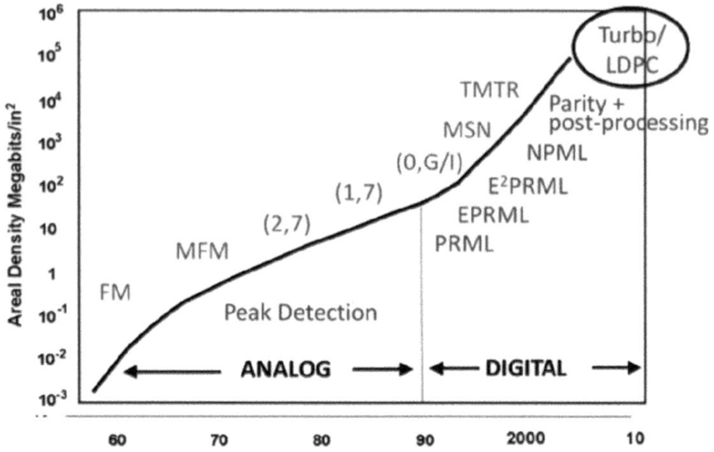

Figure 2.44 Signal processing and coding trends.

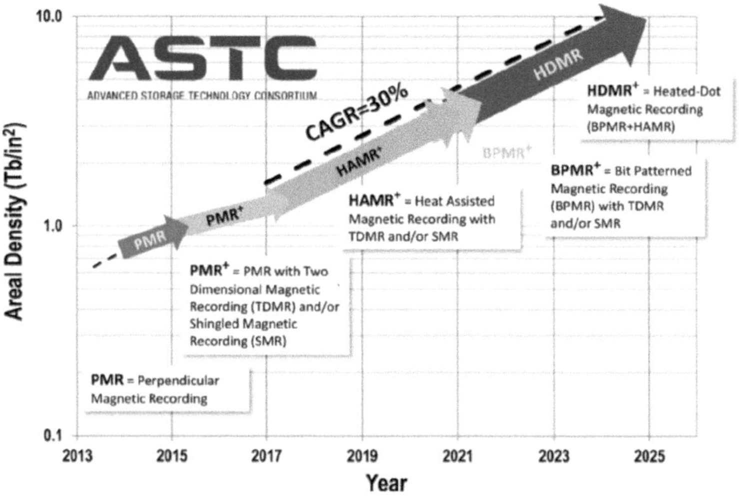

Figure 2.45 ASTC technology roadmap.

The magnetic storage community has been investigating a significant amount of resources to develop other recording technologies, like HAMR [164], TDMR [165], and bit-patterned magnetic recording (BPMR) [166] to further increase areal density. Figure 2.45 shows the Advanced Storage Technology Consortium's (ASTC) roadmap. The storage industry is fully committed to further grow HDD areal density and reduce per capita cost. Further advancements in signal processing and error correction codes are key to enabling these technologies.

2.6 Summary

Magnetic recording technology will continue to be the mass storage medium of choice for many years to come. It has advanced tremendously in the recent years but also faces major challenges going forward. These challenges, however, are not unlike those present in competing technologies. For the most part, they are all rooted in pushing the limits of what is physically known and possible. However, if history is any guide, these challenges will be overcome one by one by continuously perfecting new materials and structures, developing new insights, and making practical use of known and yet to be discovered physical phenomena. The technology roadmap of Fig. 2.45 gives a glimpse of what researchers are working on to deliver higher areal densities. The industry is not only focused on extending current PMR technology but has also been working hard to deliver the next enabler, namely HAMR and, beyond that, heated dot magnetic recording (HDMR). These will be the focus of attention in the following chapters of this book.

Acknowledgments

Guoxiao Guo would like to thank Young Hoon (Jake) Kim and Kuang-Yang (Edward) Tu for their valuable contributions to the Mechanics and Servo section.

References

1. White Robert M. (1985). *Introduction to Magnetic Recording* (IEEE Press, New York, NY, USA).
2. Bertram Neal H. (1994). *Theory of Magnetic Recording* (Cambridge University Press).

3. Wang Shan X., and Taratorin Alex M. (1999). *Magnetic Information Storage Technology* (Academic Press Technology & Engineering).
4. Mee C., and Daniel E. (1996). *Magnetic Recording Technology*, 2nd Ed. (McGraw-Hill).
5. Jorgensen F. (1995). *The Complete Handbook of Magnetic Recording* (TAB Books, an imprint of McGraw-Hill).
6. Daniel E. D., Mee C. D., and Clark M. H., eds. (1998). *Magnetic Recording: The First 100 Years Paperback* (IEEE Press).
7. Mallinson J. C. (2012). *The Foundations of Magnetic Recording* (Academic Press).
8. Khizroev S., and Litvinov D. (2004). *Perpendicular Magnetic Recording* (Springer, the Netherlands).
9. Plumer M. L., van Ek J., and Weller D., eds. (2001). *The Physics of Ultra-High-Density Magnetic Recording* (Springer).
10. Mee C. D. (1987). *The Physics of Magnetic Recording* (North-Holland Personal Library).
11. Hoagland Albert S., and Monson James E. (1998). *Digital Magnetic Recording*, 2nd Ed. (Krieger).
12. Karlqvist O. (1954). Calculation of the magnetic field in the ferromagnetic layer of a magnetic drum, *Trans. R. Inst. Technol.*, Stockholm, Sweden, **86**, pp. 3–27.
13. Ruigrok J. J. M. (1990). *Short-Wavelength Magnetic Recording: New Methods and Analyses* (Elsevier, Oxford).
14. Szczech T. J., and Iverson P. R. (1990). An approach for deriving field equations for magnetic heads of different geometrical configurations, *IEEE Trans. Magn.*, **22**, pp. 355–360.
15. Bertero G. A., Bertram H. N., and Barnett D. M. (1993). Fields and transforms for thin film heads, *IEEE Trans. Magn.*, **29**(1), p. 67.
16. Middleton B. K., Aziz M. M., and Miles J. J. (2000). Accuracy of expressions for the magnetic field of a ring head, *IEEE Trans. Magn.*, **36**(4), p. 2073.
17. Takeo A., Oikawa S., Hikosaka T., and Tanaka Y. (2000). A new design to suppress recording demagnetization for perpendicular recording, *IEEE Trans. Magn.*, **36**(5), p. 2378.
18. Bertero G. A., Wachenschwanz D., Malhotra S., Velu S., Bian B., Stafford D., Wu Y., Yamashita T., and Wang S. X. (2002). Optimization of granular double-layer perpendicular media, *IEEE Trans. Magn.*, **38**(4), Pt. 1, p. 1627.

19. Velu E. M. T., Malhotra S., Bertero G., and Wachenschwanz D. (2003). Low-noise CoCrPtO perpendicular media with improved resolution, *IEEE Trans. Magn.*, **39**(2), p. 668.
20. Piramanayagam S. N. (2007). Perpendicular recording media for hard disk drives, *J. Appl. Phys.*, **102**, p. 011301.
21. Tarnopolsky G. J., and Pitts P. R. (1997). Media noise and signal-to-noise ratio estimates for high areal density recording, *J. Appl. Phys.*, **81**(8), pp. 4837–4839.
22. Chaichanavong P., Bertram H. N., and Siegel P. H. (2006). Design parameter optimization for perpendicular magnetic recording systems, *IEEE Trans. Magn.*, **42**(10), p. 2549.
23. Bertram H. N., and Williams M. (2000). SNR and density limit estimates: a comparison of longitudinal and perpendicular recording, *IEEE Trans. Magn.*, **36**(1), p. 4.
24. Mallinson J. (1974). On extremely high density magnetic recording, *IEEE Trans. Magn.*, **MAG-10**(2), pp. 368–373.
25. Wood R. (2000). The feasibility of magnetic recording at 1 terabit per square inch, *IEEE Trans. Magn.*, **36**(1), p. 36.
26. Hurley G., Amoldussen T., Wood R., Cheng D., and Williams M. (1999). A simulation of magnetic recording on coarsely granular media, *IEEE Trans. Magn.*, **35**(5), p. 2253.
27. Wang X., and Bertram H. N. (2003). Simple transition parameter expression including grain size and intergranular exchange, *J. Appl. Phys.*, **93**(10), Parts 2 & 3, p. 7005.
28. Richter H. J., Supper N., Wilson B., and Terris B. D. (2012). Magnetic recording near the grain size limit, *IEEE Trans. Magn.*, **48**(11), p. 3887.
29. Noyau R. H., Middleton B. K., Miles J., and Williams E. W. (1988). A new theoretical approach to digital magnetic recording, *IEEE Trans. Magn.*, **24**(2), p. 1811.
30. Brown W. F. (1963). Thermal fluctuations of a single-domain particle, *Phys. Rev.*, **130**, p. 1677.
31. Aharoni A. (1969). Effect of a magnetic field on the superparamagnetic relaxation time, *Phys. Rev.*, **177**(2), p. 793.
32. Pfeiffer H. (1990). Determination of anisotropy field distribution in particle assemblies taking into account thermal fluctuations, *Phys. Status Solidi*, **118**, p. 295.
33. Sharrock M. P. (1999). Measurement and interpretation of magnetic time effects in recording media, *IEEE Trans. Magn.*, **35**(6), p. 4414.

34. Weller D. (1999). Thermal effects limits in ultra-high density magnetic recording, *IEEE Trans. Magn.*, **35**, p. 4423.
35. Stoner E. C., and Wohlfarth E. P. (1948). A mechanism of magnetic hysteresis in heterogeneous alloys, *Philos. Trans. R. Soc., A,* **240**, p. 599.
36. Wang X., Bertram H. N., and Safanov V. L. (2002). Thermal-dynamic reversal of fine magnetic grains with arbitrary anisotropy axes orientation, *J. Appl. Phys.* **92**(4), p. 2064.
37. Bertram H. N., Wang N. X., and Safanov V. L. (2001). Dynamic-thermal effects in thin film media, *IEEE Trans. Magn.*, **37**(4), p. 1521.
38. Gao K. Z., Heinonen O., and Chen Y. (2009). Read and write processes, and head technology for perpendicular recording, *J. Magn. Magn. Mater.*, **321**, pp. 495–507.
39. Chen C. W. (1977). *Magnetism and Metallurgy of Soft Magnetic Materials* (North-Holland, Amsterdam).
40. Slater J. C. (1937). Electronic structure of alloys, *J. Appl. Phys.*, **8**, pp. 385–390.
41. Kim T. K., and Takahashi M. (1972). New magnetic material having ultrahigh magnetic moment, *Appl. Phys. Lett.*, **20**, p. 492.
42. Wang J.-P., Ji N., Liu X., Xu Y., Sánchez-Hanke C., Wu Y., de Groot F. M. F., Allard L. F., and Lara-Curzio E. (2012). Fabrication of Fe16N2 films by sputtering process and experimental investigation of origin of giant saturation magnetization in Fe16N2, *IEEE Trans. Magn.*, **48**(5), p. 1710.
43. Abarra N., et al. (2001). Synthetic ferrimagnetic media, *Fujitsu Sci. Tech. J.*, **37**(2), p. 145.
44. Schabes M., et al. (2001). Theory of antiferromagnetically coupled magnetic recording media, *IEEE Trans. Magn.*, **37**(4), p. 1432.
45. Gao K. Z., and Bertram Neal H. (2002). Magnetic recording configuration for densities beyond 1 Tb/in^2 and data rates beyond 1 Gb/s, *IEEE Trans. Magn.*, **38**(6), p. 3675.
46. Victora R. H., and Shen X. (2005). Exchange coupled composite media for perpendicular magnetic recording, *IEEE Trans. Magn.*, **41**(10), p. 537.
47. Suess D., et al. (2005). Optimization of exchange spring perpendicular recording media, *IEEE Trans. Magn.*, **41**(10), p. 3166.
48. Sohn H., and Victora R. (2011). Recording comparison of ECC versus conventional media at equal grain size, *IEEE Trans. Magn.*, **47**(10), p. 4073.
49. Suess D. (2007). Micromagnetics of exchange spring media: optimization and limits, *J. Magn. Magn. Mater.*, **308**, pp. 183–197.

50. Bertero G., Acharya R., Malhortra S., Srinivasan K., Champion E., Lambert S., Lauhoff G., and Desai M. (2011). *Advanced Double Exchange-Break PMR Media Structures,* Digests of TMRC (A2, San Diego).
51. Honda N., Yamakawa K., and Ouchi K. (2007). Simulation study of factors that determine write margins in patterned media, *IEICE Trans. Electron.*, **E90-C**(8), pp. 1594–1598.
52. Honda N. (2013). Analysis of magnetic switching of 2 to 4 layered exchange coupled composite structures, *J. Magn. Soc. Jpn.*, **37**, pp. 126–131.
53. Tahk Y. W., et al. (2004). Transition noise characteristics of PMR media with the synthetic antiferromagnetic coupled soft underlayer, *IEEE Trans. Magn.* **40**(4), p. 2564.
54. Jiang W., Velu E. M. T., Malhotra S., Jung H.S., Kwok C. K., Bertero G., and Wachenschwanz D. (2005). Recording performance characteristics of granular perpendicular media, *IEEE Trans. Magn.,* **41**(2), p. 587.
55. Aziz M. M. (1999). *Signal and Noise Properties of Longitudinal Thin-Film Disk Media,* PhD thesis, Faculty of Science and Engineering, University of Manchester.
56. Mallary M., and Benakli M. (2008). Perpendicular magnetic recording write head design, *Int. J. Prod. Dev.,* **5**(3/4), pp. 226–237.
57. Okada T., et al. (2005). Newly developed wraparound shielded head for perpendicular recording, *IEEE Trans. Magn.,* **41**(10), p. 2899.
58. Piramanayagam S. N., and Chong T. C. eds. (2011). *Developments in Data Storage: Materials Perspective* (John Wiley and Sons).
59. Kryder Mark H. (2011). An introduction to magnetic recording heads, magnetic storage systems beyond 2000, *NATO Sci. Ser.,* **41**, 2001, pp. 449–466.
60. Fedeli J. M. (1993). Magnetic heads for high density recording, magnetic recording sreries, *NATO ASI Ser.,* **229**, pp. 251–279.
61. Jeffers F. (1986). High-density magnetic recording heads, *IEEE Trans. Magn.,* **74**, p. 1540.
62. Freitas P. P., Ferreira R., Cardoso S., and Cardoso F. (2007). Magnetoresistive sensors, *J. Phys.: Condens. Matter,* **19**, 165221 (21pp).
63. Fert A., Nguyen Van Dau F., Petroff F., Etienne P., Creuzet G., Friederich A., and Chazelas J. (1988). Giant magnetoresistance of (001)Fe/(001)Cr magnetic superlattices, *Phys. Rev. Lett.,* **61**(21), pp. 2472–2475.

64. Binasch G., Grünberg P., Saurenbach F., and Zinn W. (1989). Enhanced magnetoresistance in layered magnetic structures with antiferromagnetic interlayer exchange, *Phys. Rev. B*, **39**(7), pp. 4828–4830.
65. Camley R. E., and Barnaś J. (1989). Theory of giant magnetoresistance effects in magnetic layered structures with antiferromagnetic coupling, *Phys. Rev. Lett.*, **63**(6), pp. 664–667.
66. Valet T., and Fert A. (1993). Theory of the perpendicular magnetoresistance in magnetic multilayers, *Phys. Rev. B*, **48**(10), pp. 7099–7113.
67. Baibich M. N., et al. (1989). Giant magnetoresistance of (001)Fe/(001)Cr magnetic superlattices, *Phys. Rev. Lett.*, **61**(21), pp. 2472–2475.
68. Julliere M. (1975). Tunneling between ferromagnetic films, *Phys. Lett.*, **54A**, pp. 225–226.
69. Marchon B., Pitchford T., Hsia Y.-T., and Gangopadhyay S. (2013). The head-disk interface roadmap to an areal density of 4 Tbit/in_2, *Adv. Tribol.*, **2013**, Article ID 521086.
70. Klaassen K. B. (2006). Electrical low-frequency noise in tunneling magnetoresitive heads: phenomena and origins, *TMRC 2006*, C-5.
71. Smith N., and Arnett P. (2001). White-noise magnetization fluctuations in magnetoresistive heads, *Appl. Phys. Lett.*, **78**(10), p. 1448.
72. Safonov V. L., and Bertram H. N. (2002). Non-uniform thermal magnetization noise in thin films: application to GMR heads, *J. Appl. Phys.*, **91**(10), p. 7279.
73. Delgado F., Lopez K., Ferreira R., and Fernandez-Rossier J. (2013). Intrinsic spin noise in MgO magnetic tunnel junctions, *Appl. Phys. Lett.*, **102**, p. 063102.
74. Guan Y., Sun J. Z., Jiang X., Moriya R., Gao L., and Parkin S. S. P. (2009). Thermal-magnetic noise measurement of spin-torque effects on ferromagnetic resonance in MgO-based magnetic tunnel junctions, *Appl. Phys. Lett.*, **95**, p. 082506.
75. Heinonen O., and Cho H. S. (2004). Thermal magnetic noise in tunneling readers, *IEEE Trans. Magn.*, **40**(4), p. 2227.
76. Smith N., Katine J. A., Childress J. R., and Carey M. J. (2005). Angular dependence of spin torque critical currents for CPP-GMR read heads, *IEEE Trans. Magn.*, **41**(10), p. 2935.
77. Chandra Sekhar M., et al. (2015). Enhanced stability against spin torque noise in current perpendicular to the plane self-biased differential dual spin valves, *J. Magn. Magn. Mater.*, **374**, pp. 740–743.

78. Yuan L., Shen J. X., and Pant B. B. (2004). Locating magnetic noise sources in TMR and GMR recording heads using scaning probe microscopy, *IEEE Trans. Magn.*, **40**(4), p. 2233.

79. Bogy D.B., Gupta V., and Juang J.-Y. (2004). Nanoscale stability and active control in hard disk drives, in *Proceedings of the International Symposium on Micro-Nanomechatronics and Human Science*, pp. 7–13.

80. Matsuoka K., Obata S., Kita H., and Toujou F. (2001). Development of FDB spindle motors for HDD use. *IEEE Trans. Magn.*, **37**(2), pp. 783–788.

81. Bi C., Jiang Q., Lin S., Low T. S., and Mamun A. A. (2003). Reduction of acoustic noise in FDB spindle motors by using drive technology, *IEEE Trans. Magn.*, **39**(2), pp. 800–805.

82. Hredzak B., Guo G., and Zhang J. (2003). Investigation of the effect of hard disk drive unbalance on repeatable and non-repeatable runout, in *Proceedings of the 2003 IEEE International Conference on Power Electronics and Drive Systems*, pp. 1359–1363.

83. McAllister J. (1996). The effect of disk platter resonances on track misregistration in 3.5 inch disk drives, *IEEE Trans. Magn.*, **32**(3), pp. 1762–1766.

84. Guo L., and Chen Y. (2001). Disk flutter and its impact on HDD servo performance, *IEEE Trans. Magn.*, **37**(2), pp. 866–870.

85. Srikrishna P., and Kasetty K. (2000). Predicting track misregistration (TMR) from disk vibration of alternate substrate materials, *IEEE Trans. Magn.*, **36**(1), pp. 171–176.

86. Kang S.-W., Han Y.-S., Hwang T.-Y., and Koo J.-C. (2000). Noise source identification of hard disk drive using sound intensity and its control, *Asia-Pacific Magnetic Recording Conference 2000*, APMRC.

87. Deeyiengyang S., and Ono K. (2001). Suppression of resonance amplitude of disk vibrations by squeezing air bearing plate, *IEEE Trans. Magn.*, **37**(2), pp. 820–825.

88. Beghi, Oboe R., Capretta P., and Soldavini F. (2001). Loop shaping issues in hard disk drive servo system design, in *Proceedings of the IEEE/ASME International Conference on Advanced Intelligent Mechatronics*, Como, Italy, pp. 828–832.

89. Chang J., and Ho H. (1999). LQR/LTR frequency loop shaping to improve TMR budget, *IEEE Trans. Magn.*, **35**(5), pp. 2280–2282.

90. Gahinet P., and Apkarian P. (1993). An LMI-based parametrization of all H∞ controllers with applications, in *Proceedings of the 32th Conference on Decision and Control*, San Antonio, TX, pp. 656–661.

91. Kobayashi M., Nakagawa S., and Nakamura S. (2002). A phase-stabilized servo controller for dual-stage actuators in hard disk drives, in *Digest of Asia-Percific Magnetic Recording Conference*, Singapore, p. BA4.
92. Li Y. F., and Horowitz R. (2000). Analysis and self-tuning control of dual stage servo with MEMS micro-actuator, in *Proceedings of the American Control Conf*erence, Chicago, IL, pp. 2627–2632.
93. Yi L., and Tomizuka M. (1999). Two-degree-of-freedom control with robust feedback control for hard disk servo systems, *IEEE/ASME Trans. Mechatron.*, **4**(1), pp. 17–24.
94. Fujimoto H., Hori Y., and Kawamura A. (2001). Perfect tracking control based on multirate feedforward control with generalized sampling periods, *IEEE Ind. Electron.*, **48**(3), pp. 636–644.
95. Kobayashi M., Yamaguchi T., Yoshida T., and Hirai H. (1998). Multi-sensing servo with carriage-acceleration feedback for magnetic disk drives, in *Proceedings of the Conference on Decision and Control*, Philadelphia, PA, pp. 3038–3042.
96. Yamada H., Sasaki M., and Nam Y. (2004). Control of a microactuator for harddisk drives using self-sensing, 8th IEEE International Workshop on Advanced Motion Control.
97. Li Y., Horowitz R., and Evans R. (2003). Vibration control of a PZT actuated suspension dual-stage servo system using a PZT sensor, *IEEE Trans. Magn.*, **39**(2), Pt. 1, pp. 932- 937.
98. Tan Y. Z., Pang C. K., Hong F., Won S., and Lee T. H. (2011). Hysteresis compensation of piezoelectric actuators in dual-stage hard disk drives, in *Proceedings of the 2011 8th Asian Control Conference (ASCC)*, Kaohsiung, Taiwan, pp. 1024 –1029.
99. Franklin G., Powell J., and Workman M. (1990). *Digital Control of Dynamic Systems* (Addsion-Wesley, Reading, MA).
100. Mamun A. A., Guo G., and Bi C. (2007). *Hard Disk Drive: Mechatronics and Control* (CRC Press, Boca Raton, FL).
101. Chen B. M., Lee T. H., and Venkatakrishnan V. (2002). *Hard Disk Drive Servo System* (Spinger-Verlag, London).
102. Tomizuka M. (1997). Mechatronics for computer data storage devices, in *Proceedings of the IEEE International Conference on Intelligent Engineering Systems*, pp. 25–30.
103. Abramovitch D., and Franklin G. (2002). A brief history of disk drive control, *IEEE Control Syst. Mag.*, pp. 28–42.

104. Abramovitch D., Hurst T., and Henze D. (1997). The PES pareto method: uncovering the strata of position error signals in disk drives, in *Proceedings of the American Control Conference*, Albuquerque, NM, pp. 2888–2895.

105. Ehrlich R., and Curran D. (1999). Major HDD TMR sources and projected scaling with TPI, *IEEE Trans. Magn.*, **35**(2), pp. 885–891.

106. Heath J. (2000). Boosting servo bandwidth, in *Digest of Asia-Pacific Magnetic Recording Conference*, p. MP20.

107. Kim Y.-H., Kang C.-I., Lee H. S., and Tomizuka M. (2005). Adaptive and optimal rejection of disk mode-induced non-repeatable PES in computer hard disk drives, in *Proceedings of the IEEE/ASME International Conference on Advanced Intelligent Mechatronics*, pp. 1–6.

108. Gu Y., and Tomizuka M. (2000). Digital redesign and multi-rate control for motion control-a general approach and application to hard disk drive servo system, in *Proceedings of the 6th International Workshop on Advanced Motion Control,* Nagaya, Japan, pp. 246–251.

109. Du, Xie L., Teoh J. N., and Guo G. (2008). Control for head positioning in axial and radial dimensions for self-servo track writing, *IEEE Trans. Control Syst. Technol.*, **16**(1), pp. 177–181.

110. Nie J., and Horowitz R. (2011). Control design of hard disk drive concentric self-servo track writing via H2 and H∞ synthesis, *IEEE Trans. Magn.*, **47**(7), pp. 1951–1957.

111. Yada, H. (1996). Clock jitter in a servo-derived clocking scheme for magnetic disk drives, *IEEE Trans. Magn.*, **32**, pp. 3283–3290.

112. Sacks A. H., Messner W. C., and Bodson M. (1996). MR head effects on PES generation: simulation and experiment, *IEEE Trans. Magn.*, **32**(3), Pt. 2, pp. 1773–1778.

113. Nagashima M., and Nakagawa S. (2008). Identification of microactuator resonance at frequencies beyond the Nyquist rate in dual-stage actuator hard disk drives, *SICE Annual Conference*, pp. 2549–2554.

114. Atsumi T., Okuyama A., S. Nakagawa. (2008). Vibration control above the nyquist frequency in hard disk drives, *IEEE Trans. Ind. Electron.*, **55(**10), pp. 3751–3757.

115. Yamaguchi T., Shishida K., Tohyama S., and Hirai H. (1996). Mode switching control design with initial value compensation and its application to head positioning control on magnetic disk drives, *IEEE Trans. Ind. Electron.*, **43**(1), pp. 65– 73.

116. Schroeck S., and Messner W. C. (1999). On compensator design for linear timeinvariant dual-input/single-output systems, in *Proceedings of the American Control Conference*, pp. 4122–4126.

117. Messner W., and Bodson M. (1994). Design of adaptive feedforward controllers using internal model equivalence, in *Proceedings of the American Control Conference*, Baltimore, MD, pp. 1619–1623.

118. Kempf C., Messner W., Tomizuka M., and Horowitz R. (1993). Comparison of four discrete-time repetitive control algorithms, *IEEE Control Syst. Mag.*, **13**(6), pp. 48–54.

119. Sri-Jayantha S.M., Dang H., Sharma A., Yoneda I., Kitazaki N, and Yamamoto S. (2001). TrueTrackTM servo technology for high TPI disk drives, *IEEE Trans. Magn.*, **37**(2), pp. 871–876.

120. Kim Y.-H., Kang C.-I., and Lee H. J. (2006). Written-in RRO correction based on disturbance observer, in *ASME MIPE/ISPS 2006*, Santa Clara, CA, pp. 21–23.

121. Anderson B., and Moore J. (1990). *Optimal Control-Linear Quadratic Method* (Prentice Hall, Englewood Cliffs, NJ, USA).

122. Seron M., Braaslavsky J., and Goodwin G. (1997). *Fundamental Limitations in Filtering and Control* (Springer, London).

123. White M., and Lu W. (1999). Hard disk drive bandwidth limitations due to sampling frequency and computational delay, in *Proceedings of the IEEE/ASME International Conference on Advanced Intelligent Mechatronics*, pp. 120–125.

124. Tang W., Temesvary V., Miller R., Desai A., Tai Y., and Miu D. (1995). Silicon micromachined electromagnetic microactuators for rigid disk drives, *IEEE Trans. Magn.*, **31**(6), pp. 2964–2966.

125. Fan L., Ottesen H., Reiley T., and Wood R. (1995). Magnetic recording head positioning at very high track densities using a microactuator-based two-stage servo system, *IEEE Trans. Ind. Electron.*, **42**(3), pp. 222–233.

126. Fujita H., Suzuki K., Ataka M., and Nakamura S. (1999). A microactuator for head positioning system of hard disk drives, *IEEE Trans. Magn.*, **35**(2), pp. 1006–1010.

127. Hirano T., Fan L., Lee W., Hong J., Imaino W., Pattanaik S., Chan S., Webb P., Horowitz R., Aggarwal S., and Horsley D. (1998). High-bandwidth high-accuracy rotary microactuator for magnetic hard disk drive tracking servos, *IEEE/ASME Trans. Mechatron.*, **3**(3), pp. 156–165.

128. Imamura T., Katayama M., Ikegawa Y., Ohwe T., Koishi R., and Koshikawa T. (1998). MEMS-based integrated head/actuator/slider for hard disk drives, *IEEE/ASME Trans. Mechatron.*, **3**(3), pp. 166–174.

129. de Callafon R., Harper D., Skelton R., and Talke F. (1999). Experimental modeling and feedback control of a piezobased milliactuator, *J. Inf. Storage Proc. Syst.*, **2**(1), pp. 217–224.

130. Guo L., Martin D., and Brunnett D. (1999). Dual-stage actuator servo control for high density disk drives, in *Proceedings of the IEEE/ASME International Conference on Advanced Intelligent Mechatronics*, Atlanta, GA, pp. 132–137.

131. Rapley P. R., and Messner W.C. (2001). Designing controllers for two stage disk drive actuator systems using the PQ method and the sbode plot, *IEEE Trans. Magn.*, **37**(2), pp. 944–948.

132. Oboe R., Beghi A., and Murari B. (1999). Modeling and control of a dual stage actuator hard disk drive with piezoelectric secondary actuator, in *Proceedings of the IEEE/ASME International Conference on Advanced Intelligent Mechatronics*, Atlanta, GA, pp. 138–143.

133. Du C., Tan P., and Yang J. Three-stage control for high servo bandwidth and small skew actuation, *IEEE Trans. Magn.*, **51**(1), pp. 1–7.

134. Herrmann G., Turner M. C., Postlethwaite I., and Guo G. (2004). Practical implementation of a novel anti-windup scheme in a HDD-dual-stage servo-system, *IEEE/ASME Trans. Mechatron.*, **9**(3), pp. 580–592.

135. Evans R. B., and Karaman M. (2000). Closed loop testing of a suspension based piezoelectric microactuator, *Digest of Asia-Pacific Magnetic Recording Conference 2000*, pp. 6–8.

136. Vasic B., and Kurtas E., eds. (2005). *Coding and Signal Processing for Magnetic Recording* (CRC Press).

137. Hwang E., et al. (2014). Skew-dependent analysis performance evaluation for array-reader based magnetic recording with dual reader, in *Proceedings of the Magnetic Recording Conference 2014*.

138. Reed I. S., and Solomon G. (2004). Polynomial codes over certain finite fields, *J. Soc. Ind. Appl. Math.*, **1960**.

139. Lin S., and Costello D. (1983). *Error Control Coding: Fundamentals and Applications*, (Prentice-Hall).

140. Gallager R. (1962). Low-density parity-check codes, *IRE Trans. Inf. Theory*, **8**(1), pp. 21–28.

141. Tanner R. M. (1981). A recursive approach to low complexity codes, *IEEE Trans. Inf. Theory*.
142. MacKay D., and Neal R. (1995). Good codes based on very sparse matrices, in *Cryptography and Coding, 5th IMA Conference*.
143. Hagenauer J., and Hoeher P. (1989). A Viterbi algorithm with soft decision outputs and its applications, in *Proceedings of GLOBECOM 89*, Dallas, pp. 47.1.1–47.1.7.
144. Quantum Corporation. (1995). *PRML Read Channels: Bringing Higher Densities and Performance to New-Generation Hard Drives*.
145. Marcus B. H., Siegel P. H., and Wolf J. K. (1992). Finite state modulation codes for data storage, *IEEE J. Sel. Areas Commun.*, **10**(1).
146. Caroselli J., and Wolf J. K. (1996). Application of a new simulation model for media noise limited magnetic recording channels, *IEEE Trans. Magn.*, **32**.
147. Krishnan A. R., et al., (2009) 2D magnetic recording: read channel modeling and detection, *IEEE Trans. Magn.*, **45**(10).
148. Lender A. (1964). Correlative digital communication techniques, *IEEE Trans. Commun. Tech.*, **COM-12**.
149. Kobayashi H., and Tang D. T. (1970). Application of partial response channel coding to magnetic recording systems, *IBM J. Res. Dev.*, **15**.
150. Wood R. W., and Peterson D. A. (1986). Viterbi detection of Class IV partial response on a magnetic recording channel, *IEEE Trans. Commun.*, **COM-34**.
151. Forney G. D. (1972). Maximum-likelihood sequence estimation of digital sequences in the presence of intersymbol interference, *IEEE Trans. Inf. Theory*, **IT-18**.
152. Thapar H. K., and Patel A. M. (1987). A class of partial response systems for increasing storage density in magnetic recording, *IEEE Trans. Magn.* **23**(5).
153. Moon J., and Zeng W. (1995). Equalization for maximum likelihood detectors, *IEEE Trans. Magn.*, **31**(2).
154. Eleftheriou E., and Hirt W. (1996). Noise predictive maximum likelihood (NPML) detection for the magnetic recording channel, in *Proceedings of the IEEE International Conference on Communication*, pp. 556–560.
155. Caroselli J., Altekar S. A., McEwen P., and Wolf J. K. (1997). Improved detection for magnetic recording systems with media noise, *IEEE Trans. Magn.*, **33**(5), pp. 2779–2781.

156. Kavcic A., and Moura J. M. (2000). The Viterbi algorithm and Markov noise memory, *IEEE Trans. Inf. Theory*, **46**(1), pp. 291–301.

157. Moon J., and Park J. (2001). Pattern-dependent noise prediction in signal-dependent noise, *IEEE J. Sel. Areas Commun.*, **19**(4), pp. 730–743.

158. Hassner M., and Grochowski E. (2015). *4K Byte-Sector HDD-Data Forma Standard*, www.idema.org/wp-content/downloads/1259.do.

159. Hocevar D. E. (2004). A reduced complexity decoder architecture via layered decoding of LDPC codes, in *IEEE Workshop on Signal Processing Systems*, pp. 12–15.

160. Ryan W. E. (2004). An introduction to LDPC codes, in *CRC Handbook for Coding and Signal Processing for Recording Systems*, ed., Vasic B. (CRC Press).

161. Richardson T., Shokrollahi A., and Urbanke R. (2001). Design of capacity-approaching irregular low-density parity-check codes, *IEEE Trans. Inf. Theory*, **47**.

162. Chung S. Y., Forney G. D., Richardson T. J., and Urbanke R. (2001). On the design of low-density parity0check codes within 0.0045 dB of the Shannon limit, *IEEE Commun. Lett.*, **5**.

163. Siegel P. H. (2005). Applications of the Viterbi algorithm in data storage technology, in *Viterbi Conference: Advancing Technology through Communications Sciences*, University of Southern California.

164. Rottmayer R. E., et al. (2006). Heat-assisted magnetic recording, *IEEE Trans. Magn.*, **42**(10), pp. 2417–2421.

165. Wood R., Williams M., Kavcic A., and Miles J. (2009). The feasibility of magnetic recording at 10 terabits per square inch on conventional media, *IEEE Trans. Magn.*, **45**(2), Pt. 2, pp. 917–923.

166. Terris B., Thomson T., and Hu G. (2006). Patterned media for future magnetic data storage, *Microsyst. Technol.*, **13**(2), pp. 189–196.

Chapter 3

Conventional Perpendicular Magnetic Recording Media

Hong-Sik Jung
Seagate Technology PLC, RMO, 47010 Kato Road, Fremont, CA 94538, USA
hong-sik.jung@seagate.com

Perpendicular magnetic recording (PMR) in hard disk drives (HDDs) for computer and consumer electronic applications has been a long-awaited technology. It was first proposed by S. Iwasaki in 1975 [1] but commercialized in 2005 by Toshiba [2] with 40 GB/1.8-inch platter (133 Gbits/in^2). Implementation of PMR technology has almost taken three decades due to many challenges and making it competitive over well-established longitudinal magnetic recording (LMR) technology. In LMR, a continuous reduction of grain size in metallic CoCrPt-based alloys is the major contribution for improving areal densities. A further increase in the recording density of $\geq$100 Gbits/in^2 needed a grain size smaller than 9 nm and a narrower grain size distribution; this grain size was approaching the fundamental superparamagnetic limit [3], which leads to thermal instability in the recording medium. A combination of oriented LMR media with antiferromagnetically coupled longitudinal magnetic layers [4] allowed improving thermal stability caused by $M_r t$ (a product of remanent magnetization and thickness) cancellation,

Ultrahigh-Density Magnetic Recording: Storage Materials and Media Designs
Edited by Gaspare Varvaro and Francesca Casoli

ISBN 978-981-4669-58-0 (Hardcover), 978-981-4669-59-7 (eBook)
www.panstanford.com

thus partially extending the limit of LMR. A significant increase of areal density was obtained by introducing PMR technology with a thicker magnetic recording layer, which provides sufficient thermal stability and high M_rt as well as high head field strength and gradient. The possibility of PMR to achieve 1 Tbit/in^2 was speculated around 2000 [5, 6]. PMR has made remarkable progress in the past one decade and the areal density is now approaching ~920 Gbits/in^2. After analyzing different generations of PMR media, significant reduction of magnetic cluster size is highlighted, while keeping similar structural grain size [7]. Shingled magnetic recording (SMR) [8] is considered a possible extension of current PMR technology for areal densities >1 Tbit/in^2. SMR media with track density benefit are developing with similar PMR materials but advanced head and media design concepts for further improving areal densities.

3.1 Introduction

Implementation of perpendicular magnetic recording (PMR) technology into hard disk drives (HDDs) has almost taken three decades since it was proposed by S. Iwasaki in 1975 [1]. The areal density of longitudinal magnetic recording (LMR) media with the magnetization in the recorded bits parallel to the disk surface was improved continuously, reducing the physical grain size. In 1997 Charap et al. [3] pointed out that thermal instability can occur when the magnetic energy becomes comparable with thermal energy. Materials with a uniaxial magnetocrystalline anisotropy constant (K_u) higher than current Co-based alloys are highly attractive [9]. PMR media with magnetization in the recorded bits perpendicular to the disk surface have a low demagnetization field (H_{demag}) between recorded bits compared to LMR media, which allows more recorded bits at the same area. These two main benefits are accelerating the transition from LMR to PMR, even when facing many challenges.

Initial PMR integration [10] uniquely used a monopole write head and perpendicularly oriented CoCr media with a magnetic soft underlayer (SUL). An SUL is considered part of the write head as magnetic flux return paths. It enhances the magnetic write field strength and its field gradient with the same head materials as longitudinal heads did, which allows for the use of high coercivity (H_c) media. PMR media require magnetically perpendicular orientation with proper SUL thickness and anisotropic permeability (μ) to match write pole properties and saturation magnetization ($4\pi M_s$). However, these simple requirements created many challenges for practical implementation of PMR [11] in 1996: low nucleation field (H_n), high noise in the SUL, high stray field sensitivity, and head-induced erasure issues of adjacent track erasure (ATE), side track erasure (STE), and return field–induced partial erasure (RFPE). After extensive works to solve these issues, new technical approaches were reported around 2000: high H_n media [12, 13], trailing-shielded (TS) pole write head combined with reverse-flying longitudinal head benefits [14], and tunnel magnetoresistance (TMR) reader with crystalline MgO insulator barrier [15–17]. Mallary et al. [6] with high-field-gradient shielded-pole writer design and high H_c media provided the technical feasibility of 1 Tbit/in^2. The trade-off between writability and field gradient remains a major challenge for head and media integration.

Media signal instability after writing in early PMR media became one of the big issues. Typical sheared Kerr hysteresis loops with positive H_n were observed in early CoCr-based PMR media [18]. This sheared loop strongly depends on H_{demag} caused by film thickness and poor grain isolation, which is in good agreement with the simulation results [19]. Reversed grains in a bit due to high H_{demag} reduce the signal and become the source of high direct current (DC) media noise. H_{demag} is controlled by proper material selection, degree of grain isolation, and crystallographic c axis orientation in media. CoCr media isolated by Cr-rich boundaries achieve better crystalline orientation, less H_{demag}, and better signal-to-noise ratio (SNR) than conventional CoCr media [20]. However, a Cr-rich CoCr alloy with lower K_u and M_s is not sufficient. Negative-H_n media are essential to stabilize magnetization against disturbance fields during the recording process and to improve media SNR. It is reported [12, 13] that CoCrPtO single-layered media with amorphous oxygen-rich grain boundaries show high H_c with a steep hysteresis loop and a high squareness ratio. Both high Ar pressure and oxygen-reactive sputtering process are key parameters to control both physical and chemical isolations among magnetic grains. Addition of SiO_2 to the CoPtCr alloy also provides a similar well-isolated fine-grain microstructure without disturbing the heteroepitaxial growth of CoPtCr grains on top of the Ru underlayer [21]. This significant improvement of H_n is achieved by improvement of c axis orientation and a well-isolated microstructure in higher-K_u CoPtCr alloys on top of a proper Ru underlayer. Media performance has been continuously improved by implementing new magnetic design concepts of dual magnetic layers with a coupled granular/continuous (CGC) [22] or capping layer [23], static tilted magnetic switching media [24], and dynamic tilted magnetic switching media for exchange-coupled composite (ECC) media [25] with a single-breaking-layer (BL), exchange-spring (ES) media [26, 27] with K_u gradients, and magnetic switching block–controlled (or segmented) media [28, 29] with multiple BLs.

The SUL as part of the write head needs to follow write pole properties of dimension and $4\pi M_s$. On the basis of modeling [30] with a monopole write head, the minimum SUL thickness is simply estimated under the conservation of magnetic flux for a write pole. However, Cain et al. [11] point out the critical role of an SUL and its related problems. A strong relationship between SUL and media noise was present in the early stage. A thick SUL with high μ creates high noises due to magnetic domain instability, even at small stray magnetic fields: spike noise by formation of 180° domain walls and high DC noise by formation of stripe domains. Simulation results [31] show that the sensitivity to stray fields is greatly reduced, with little effect on recording performance when $\mu < 100$. To avoid domain noise of an SUL, it is asserted that a reversible and nonhysteretic field response is required. When the SUL is too thin, this saturation effect of the SUL reduces the write field and field gradient and causes the recording performance to degrade [32]. New technical ways of single-domain SUL [33, 34] and synthetic antiferromagnetic (SAF)-coupled SUL [35–37] are approached for improving stray field robustness and better manufacturability.

Typical configuration of TS pole write/TMR read heads and CoCrPtO-based hard/soft stacked composite media with SAF-coupled SULs has been used for PMR drive integration. The first generation of PMR drives provided a capacity of 80 GB/2.5-inch

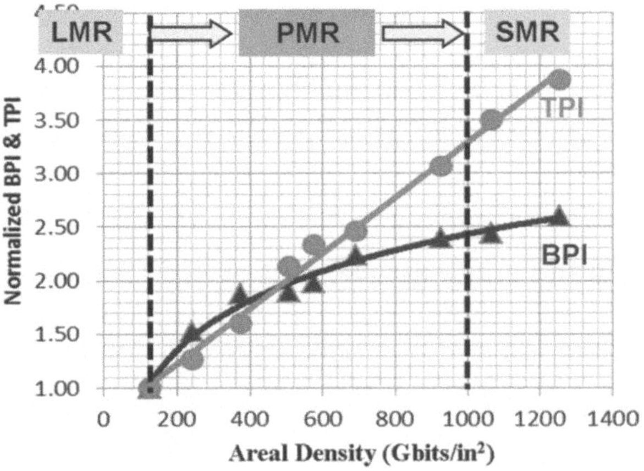

Figure 3.1 The trends of linear density (LD) and track density (TD) normalized to the value used for 130 Gbits/in^2 in 2005 as a function of areal density.

platters, equivalent to a 130 Gbits/in^2 areal density with 1000 kBPI for linear density (LD) and 130 kTPI for track density (TD), where LD = 1/bit length and TD = 1/track pitch. The increase in areal densities has been achieved by improving LD and/or TD but it strongly depends on the write/read head type and media design. For a given write/read head, both the magnetic design and the sputter process are tuned by considering media writability, amplitude, and grain isolation. Head/media mismatch becomes one of the hot issues with increasing areal density. Owing to remarkable progress in media, head, channel, and system optimization, the HDD industry is now making the product with areal densities of ~920 Gbits/in^2. Figure 3.1 shows the trends of LD and TD normalized to the value on the first-generation PMR products in 2005 as a function of areal density. Both LD and TD were improved together up to 500 Gbits/in^2, while a significant increase in TD is observed for areal densities larger than 500 Gbits/in^2. The intense activities aimed to significantly increase TD create many critical issues: lower amplitude, weaker writability, less margin of ATE and track misregistration, higher sensitivity of head-to-media spacing (HMS), and larger variation of population and performance in writer pole dimension. The most effective way to overcome worse writability and lower amplitude is to reduce HMS but limited by mechanical issues. Therefore, the role of media with a significant fast turnaround time from thin-film deposition to measurement is getting more important in order to achieve higher areal densities.

3.2 Evolution and Improvement of PMR Media

3.2.1 *Definition and Functions of Layer Structures*

PMR media consist of an adhesion layer (ADL), an SUL, an intermediate layer (IL), a magnetic recording layer (MAG), a carbon overcoating (COC), and a lubricant layer (lube), as shown in Fig. 3.2. AlMg substrates for desktop and most server products and glass substrates for mobile and partial server products are typically used. Substrate smoothness and contamination are highly critical for head flyability and media defects when HMS is getting lower

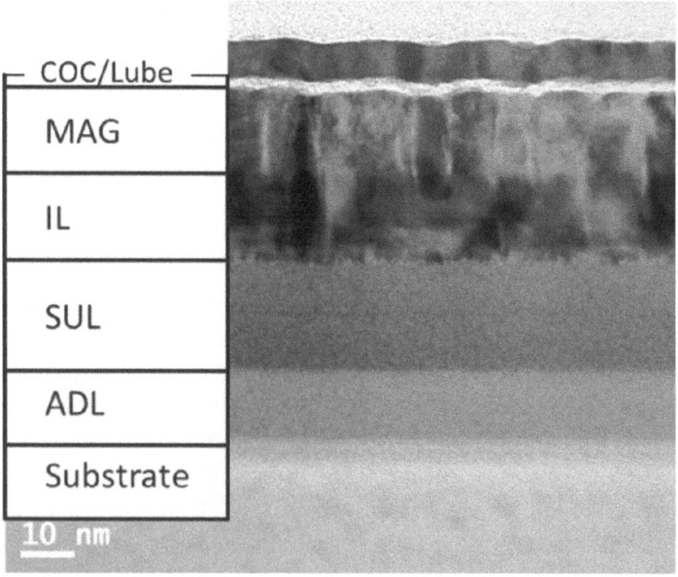

Figure 3.2 Definition of layer structure on a cross-sectional TEM image of typical perpendicular media.

at higher areal densities [38]. The mechanical texturing process for glass substrates is still used to remove asperities to make the surface smoother and of better quality. When a metallic SUL with less substrate affinity and weak corrosion robustness is directly deposited on the substrate, the films are highly stressed and easily delaminated or peeled off at the early PMR stage. Typical amorphous alloys of AlTi, CrTi, or NiTa are used, especially on a glass substrate. This amorphous ADL, including the elements of Ti, Cr, Ta, etc., and having high affinity with oxide surface, significantly improves corrosion robustness and also enhances the substrate heating effect for better growth of the IL. The SUL as part of the writer for magnetic flux returns is deposited on top of the ADL. The IL, consisting of a seed layer and an interlayer, acts as an exchange BL between the SUL and the MAG and as template of grain growth and grain isolation for the MAG, and it consists of a seed layer and an interlayer. The MAG structure is complicated, with many magnetic layers deposited on top of the IL, but it divides into two parts of CoPtCr-based oxide-containing layers and metallic CoCrPtB-based layers. The COC deposited on top of the MAG consists of diamond-like carbon and thin CN_x to protect magnetic layers from corrosion. Diamond-like carbon is deposited by ion beam deposition (IBD) or plasma-enhanced chemical vapor deposition (PECVD) and CN_x deposited by either sputtering carbon with reactive N_2 gas or the nitrogen implementation process. Finally, 9–14 Å thin lube is coated in order to reduce the friction between head and disk and to protect the surface from oxidation and mechanical damage. Higher bonding to the carbon surface and smoother roughness are needed for less lube pickup and better clearance. The total thickness of the COC and lube needs to be minimized due to HMS loss if corrosion and mechanical performance are satisfied. The roles and evolution on each layer structure in typical PMR media are described in detail in this section.

3.2.2 *Magnetic Soft Underlayer*

3.2.2.1 Roles of an SUL

One of uniqueness in PMR is to use an SUL in the medium as part of the writer for magnetic flux return paths. The SUL

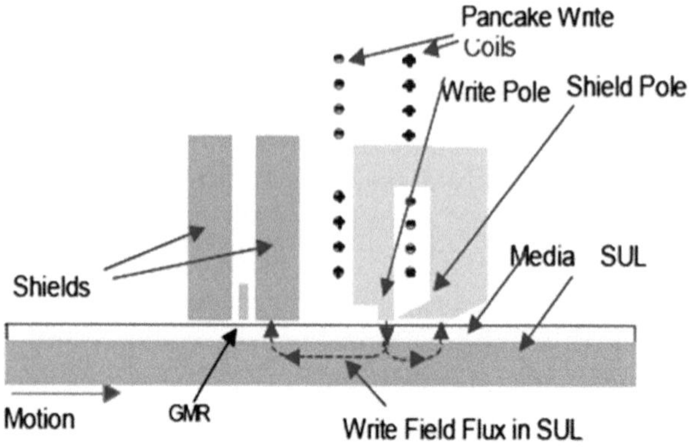

Figure 3.3 Typical PMR configuration with trailing-shielded pole writer/giant magnetoresistance (GMR) reader/perpendicular media showing uniaxial magnetic flux return paths. © [2002] IEEE. Reprinted, with permission, from Ref. [6].

requirement is that $4\pi M_s$ and thickness should match the record pole properties to optimize the record field and gradient. The other important parameter is the permeability (μ), indicating the degree of magnetization that a magnetic material obtains in response to an applied external magnetic field. The product of $4\pi M_s$ and thickness, $4\pi M_s t$, and μ in the SUL are a measure of the amplitude and speed of magnetic return fluxes, respectively.

By assuming SUL fluxes in the down-track and cross-track directions, the minimum SUL thickness is simply estimated by Eq. 3.1 [39]:

$$t_{\text{SUL}} \approx \frac{1}{2} \frac{(T \times W)}{(T + W)} \frac{M_{\text{s Pole}}}{M_{\text{s SUL}}} \tag{3.1}$$

where M_{sPole} and M_{sSUL} are the saturation magnetization of the write pole and the SUL, respectively, and T and W are the thickness and width of the monopole writer, respectively. Figure 3.3 shows a typical PMR configuration with TS pole writer/giant magnetoresistance (GMR) reader/perpendicular media showing uniaxial magnetic flux return paths [6]. This GMR reader was replaced with a TMR reader. The PMR configuration with a TMR reader was used in the first-generation PMR products in 2005. The calculated effective switching-field strengths between mono- and TS poles with two different SUL thicknesses of 40 and 160 nm

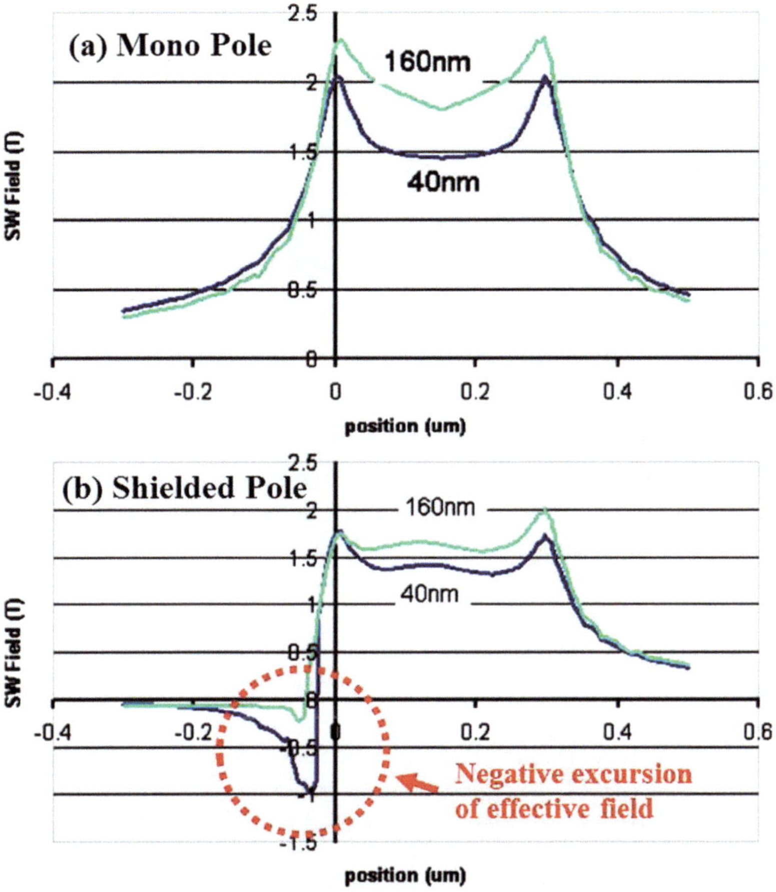

Figure 3.4 Comparison of effective switching (SW) fields between monopole and trailing-shielded pole with two different SUL thicknesses of 40 and 160 nm. Reprinted with permission from Ref. [40]. Copyright 2005, AIP Publishing LLC.

are directly compared in Fig. 3.4 [40]. The TS pole writer has a much lower effective switching-field strength than the monopole but a significantly higher field gradient. Uniquely, the TS pole writer shows the existence of negative excursion of the effective field under the trailing shield and less dependence of SUL thickness. This negative excursion of the effective field induces reversed magnetic

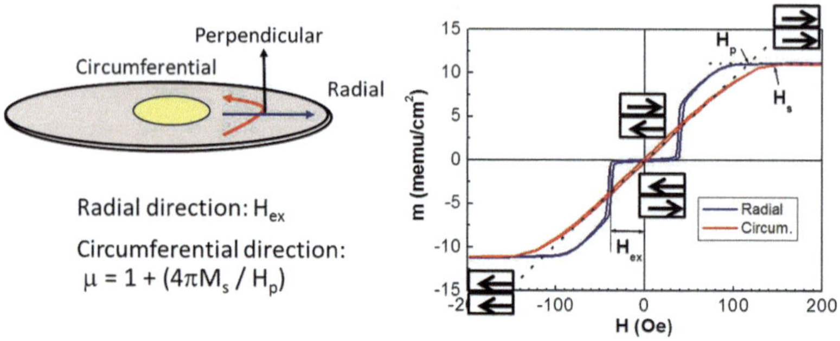

Figure 3.5 Typical M–H loop and important parameters of SAF-coupled CoTaZr/Ru/CoTaZr SUL.

grains in a bit, called RFPE [41]. Either a thinner SUL or a higher write current at the TS pole writer makes the negative excursion field higher due to the dominant effect of the longitudinal head field at the write gap, but it enhances field gradient.

The SUL saturation effect is observed when a thin SUL is used at the TS pole writer. It provides a smaller writing bubble with relatively less perpendicular magnetic fluxes and more longitudinal magnetic fluxes. This initial saturation effect induces a worse reverse overwrite (ROW), a higher TD, and a better SNR by the enhanced field gradient. However, the SNR is degraded when further reduction of SUL thickness cannot provide sufficient write field strength. Higher-H_n media are needed to overcome the negative excursion field. However, a thinner SUL at the monopole writer degrades both the field gradient and the effective field strength.

Figure 3.5 shows a typical M–H loop and important parameters of an SAF-coupled CoTaZr/Ru/CoTaZr SUL. Values of μ along the circumferential direction are predicted by the formulas of Eq. 3.2:

$$\mu = \frac{B}{H} = 1 + \left(\frac{4\pi M_s}{H_p}\right) \tag{3.2}$$

where H_p is the saturation field linearly-extrapolated at zero field along the circumferential direction, $4\pi M_s$ is the intrinsic property in a given ferromagnetic material. H_{ex} is an indicator of stray field robustness. The loop shift of the M–H loop along the radial direction is determined by SUL materials and thickness: H_{ex} is higher for systems with a higher M_s and thinner thickness. μ is considered the magnetic field sensitivity and controlled by H_p, which is determined by magnetic softness, degree of magnetic radial orientation, and exchange-coupling strength interacted with a pinning layer. A high μ is good for magnetic flux returns, but it increases both media noise and ATE and STE. Therefore, it is important to select proper SUL materials and thickness and to control anisotropic μ. Crystalline soft magnetic FeCo, CoNiFe, or NiFe used as write head materials [42] are highly considered as SUL materials during the early development of PMR media. However, the SUL has additional roles, not emphasized at an early stage. It also provides a smoother surface and enhances grain texture growth and crystallinity of the interlayer on top of the SUL. Glass substrates at an early stage have a rougher surface, so a certain thickness of the amorphous SUL is needed to cover the rougher surface of the glass substrate. This amorphous surface initially provides better aligned grain formation of the seed layer. The substrate heating effect during the sputtering of a thick SUL improves crystallinity of the seed layer and interlayer as a grain template of the MAG. The amorphous microstructure additionally limits material selection of the SUL. Amorphous Fe-Co-(B, Ta, Zr)-based alloys are being used instead of crystalline write head materials.

3.2.2.2 Noise sources and solutions

A very thick SUL from 200 nm to 10 μm with the range of μ from 200 to 2000 is used in an early PMR stage [43]. Typical noises of a NiFe SUL are the spike noise occurring at a very low frequency, less than 40 kHz, and medium noise in the frequency range from 1 to 10 MHz. Spike noise is observed due to 180° domain walls for an SUL with a high μ but it is reduced or removed with a high pass filter. Medium noise less than 10 MHz is exhibited in an SUL with a low μ, explained by the formation of stripe domains [44] first reported in NiFe films. Stripe domains are formed due to perpendicular magnetocrystalline and/or magnetostrictive anisotropies. The critical stripe domain width (δ) [45] is calculated by Eq. 3.3:

$$\delta = \pi \sqrt{\frac{A}{K_{\text{perp}}}} \tag{3.3}$$

where A is the exchange stiffness of the magnetic material and K_{perp} is the effective uniaxial perpendicular anisotropy constant.

The stripe domain onset thickness depends on stress and texture in sputtered $Co_{65}Fe_{35}$ and FeAlN films [46]. The formation of stripe domains deteriorates soft magnetic properties in SULs.

Another serious media noise comes from high sensitivity of a stray magnetic field caused by domain instability in SULs with a high μ. This head-induced erasure appears right after writing in the down-track direction and data are also erased along the cross-track direction. This selective erasure occurs on certain locations on a disk. Numazawa et al. [47] reported the magnetic orientation effect in SULs: the SNR significantly improved on an oriented SUL medium over a randomly oriented SUL medium. Hard-axis orientation in SULs along the circumferential direction of the disk is needed but the difficulty of uniformly anisotropic orientation in the disk at all locations is addressed. In a typical $Co_{90}Ta_5Zr_5$ SUL, Ta and Zr elements with a large atomic size make it amorphous and induce a directional pair ordering mechanism along the magnetic field direction [48], resulting in formation of a magnetic easy axis. Co atoms are aligned along the radial magnetic field direction from the magnet backpack during the deposition. A magnet backpack design with uniform radial magnetic fields is essential. When a soft SUL is coupled with a magnetic hard layer, SUL dynamic properties are also changed dramatically: slower time response of the perpendicular component of the magnetization in an SUL under the presence of stray fields [49]. High external stray field sensitivity on a monopole head becomes one of the hot issues compared to a longitudinal ring head [11]: The applied external field of 5 Oe is magnified to 1.6 kOe in the medium for a monopole head but only 45 Oe for a ring head. This high sensitivity of the external stray field is dramatically improved by reducing μ in an SUL. On the basis of magnetic simulation [31], the reduction of μ from 2000 to 100 remarkably suppresses the stray field–induced head field by 70% in the recording layer under the external stray field of 10 Oe, while writing and reading sensitivities are not significantly affected. A higher H_c in an SUL is not a good way to reduces μ due to additional source of hysteresis loss.

Using two NdFeB permanent magnets placed in the vicinity of the media, it is demonstrated [30] that a magnetically biased SUL with a pseudo-single-domain state remarkably reduces the media noise. Magnetic biasing is necessary to control the magnetic domain structure on SULs. To avoid magnetic domain noises and improve stray field robustness in SULs, a single domain or few domains with an anisotropic μ of $\sim$100 along the down-track direction are needed. The pinning concept to stabilize domains has been proposed to use various SULs coupled with a CoSm hard magnetic layer [50], an FeMn or IrMn antiferromagnetic layer [33, 51], an SAF configuration [35–37], or a CoCrTa hard magnet–pinned SAF configuration [34, 52]. Figure 3.6 shows magnetic domains measured by a Candela optical surface analyzer (OSA) and typical *M*–*H* loops on three different SUL structures of (a) a single SUL (CoTaZr), (b) an SAF-coupled SUL (CoTaZr/Ru/CoTaZr), and (c) a hard magnet–pinned SAF-coupled SUL (CoCrTa/Ru/CoTaZr). A single SUL with an H_c lower than the magnetostatic H_{demag} along the edge of the disk has many 180° domain walls in Fig. 3.6a but directional domain wall distribution is observed due to the existence of a stray field of 1–2 Oe within the Candela OSA tester, which is higher than H_c <1 Oe of single CoTaZr. An SAF-coupled SUL clearly reduces

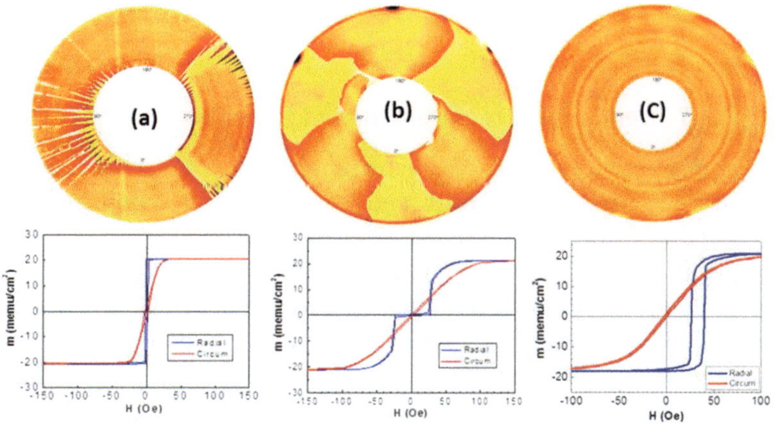

Figure 3.6 Comparison of magnetic domains measured by a Candela optical surface analyzer (OSA) and typical M-H loops on three different SUL structures: (a) a single SUL (CoTaZr), (b) an SAF-coupled SUL (CoTaZr/Ru/CoTaZr), and (c) a hard magnet-pinned SAF-coupled SUL (CoCrTa/Ru/CoTaZr). Reprinted with permission from Ref. [34]. Copyright 2006, AIP Publishing LLC.

magnetic domains to a few domains (Fig. 3.6b), and a hard magnet–pinned SAF SUL eventually achieves a single domain on a disk (Fig. 3.6c). Ring patterns in Fig. 3.6c are caused by a polished NiP/AlMg substrate. However, strong stray fields from the CoCrTa hard magnetic layer create a strong magnetic ripple structure at a 60 nm thick CoTaZr SUL, even with a single domain [52]. This ripple structure clearly contributes to DC-erase noise especially on the hard magnet–pinned layer so a thin SUL of $<$60 nm can't be used. An SAF-coupled SUL by indirect coupling between two ferromagnetic layers through a nonmagnetic metallic spacer [53] is very attractive to manufacturability due to the simple structure as well as SUL domain stability. This SUL structure is used in most PMR media. Jiang et al. [54] confirmed that applying an in-plane stray field close to 20 Oe induces domain instability in CoTaZr/Ru/CoTaZr films with $4\pi M_s \approx$ 13kG, $H_{ex} \approx$ 20 Oe, and $H_p \approx$ 50 Oe. The estimated μ from $4\pi M_s$ and H_p is 260. The media with SAF-coupled SULs show no remarkable spike noise throughout the disk, and STE is effectively suppressed [55]. Lower media noise is originated from the suppressed vortex formation in the SUL and the reduced magnetically perturbed region [56].

3.2.2.3 Trend of SUL requirement and materials

The aggressive push of TD for increasing areal densities requires a narrower write pole width and less ATE, while minimizing the loss of write field strength. To meet the requirements, write head types are getting more complicated, as shown in Fig. 3.7: mono- and TS pole writers with a thick return pole in Fig. 3.7a to a side-shielded wraparound pole with a thin return pole in Fig. 3.7b and a fully wraparound pole without a return pole in Fig. 3.7c. The bottom images show a magnified write pole area in the red-colored circle. The trapezoidal pole shape in a write head is specially designed to avoid data erasure in neighboring tracks when the head is operated at a skew angle with respect to the inner or outer diameter of the disk. Magnetic flux return paths need to be diverse from uniaxial to multiple directions. It is important to select proper SUL materials and thickness and to control anisotropic μ.

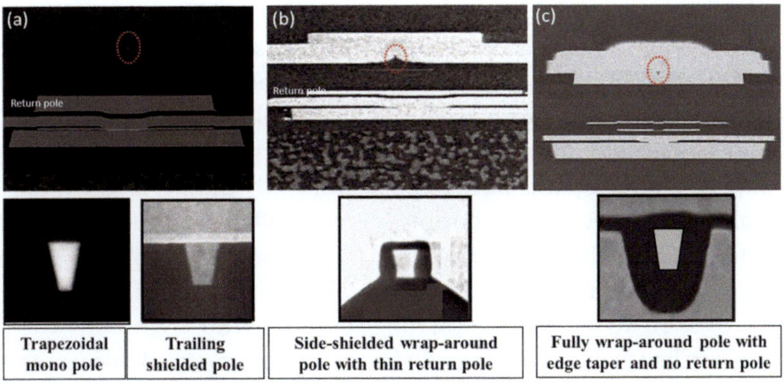

Figure 3.7 Air-bearing surface view of three different PMR heads: (a) monopole and trailing-shielded (TS) pole writer/reader, (b) side-shielded wraparound pole writer/reader, and (c) fully wraparound pole writer/reader. The bottom images show a magnified write pole area in the red-colored circle.

The TS pole writer in Fig. 3.7a is used in the first-generation PMR products. This TS pole writer [14] enhances the head field gradient and changes the write field angle into the media by creating an adequate longitudinal field. It makes the magnetic switching field lower but this effect is compensated with the reduced write field strength. Magnetic flux return paths are divided into two directions of trailing shield and leading shield along the down-track direction. For the TS pole head in the first-generation PMR products, a ~200 nm thick SAF-coupled CoTaZr-based SUL with a high $4\pi M_s$ of 12–15 kG and a low μ of 30–50 is used and considered as the first-generation SUL. SUL thickness is considerably reduced owing to the use of a narrower write pole width and trailing shield, while maintaining high anisotropic μ between radial and circumferential directions of a disk. High μ for down-track (circumferential) direction but low μ for cross-track (radial) direction, this anisotropic μ controls the priority of magnetic flux return paths and its efficiency.

For improving TD by enhancing the ATE margin, a side-shielded wraparound write head with a thinner return pole (Fig. 3.7b) is used, even with a significantly reduced write field strength compared to the TS pole. Simply increasing the write current to enhance the write field strength creates another problem of ATE and RFPE. Thus, improvement of media writability requires a new design concept of MAGs and SULs. This side-shielded write head needs multiple return paths of magnetic fluxes from the write pole to the trailing shield, side shield, and return pole. Second-generation SULs for supporting multiple return paths have to be significantly thinner and with less anisotropic μ. A less anisotropic μ is controlled by reducing both M_s and H_k in the SUL and decreasing H_{ex} in the SAF-coupled SUL. The $4\pi M_s$ in an SAF-coupled CoFeTaZr-based SUL reduces from 12–15 kG to 8–10 kG and the typical range of μ is 50–80.

Further reduction of the write pole width and enhancement of the ATE margin for areal densities of >500 Gbit/in^2, either deeper side-shielded or fully wraparound poles (Fig. 3.7c), are used. For more multiple flux return paths, $4\pi M_s$ in an SAF-coupled FeCoTa-based SUL further reduces from 8–10 kG to 6–8 kG and the typical range of μ is 80–120. It shows much less anisotropic μ for better supporting multiple flux return paths compared to first- and second-generation SULs.

3.2.3 *Intermediate Layer: Seed Layer and Interlayer*

The magnetic switching field in uniaxial materials is sensitive to the angle between the applied field and the magnetic easy axis. The crystallographic perpendicular orientation and alignment of media grains are highly considered to achieve narrow switching-field distribution (SFD) [57]. However, crystallographic *c* axis orientation does not represent magnetic perpendicular easy-axis orientation. There are controversial results: narrower [58] or significantly wider [59] magnetic easy-axis dispersion than crystallographic *c* axis orientation, which is complicated by the coexistence of lateral intergranular exchange interactions and H_{demag} in the recording layers. Both understanding the roles of the IL between the MAG and the SUL and designing the IL structure are important to achieve narrow distribution of crystallographic and magnetic perpendicular orientation.

The IL as an exchange BL between the MAG and the SUL and a template of grain growth for the MAG consists of a seed layer and an interlayer. The crystalline seed layer and interlayer significantly affect both the microstructure and magnetic properties. The control of IL thickness (t_{IL}) is very limited so it is difficult to check the effect of intrinsic t_{IL} on recording performance. A thinner IL typically provides smaller magnetic grains with higher lateral intergranular exchange coupling strength (A_{ex}) and worse crystallinity, leading to large magnetic cluster size even with physical smaller grains [60–62]. It is not easy to experimentally prove the effect of intrinsic t_{IL} on recording performance. It is still questionable that a thin IL is essential for improving areal densities. However, this reduction of t_{IL} provides better mechanical performance—smoother surface and better corrosion robustness. Ideally, a highly crystallized, thin IL with a well-isolated grain template is needed but is difficult to achieve together.

3.2.3.1 Intrinsic IL roles as an exchange breaking layer

The effect of intrinsic t_{IL} on read/write processes in perpendicular recording is investigated by using magnetic simulation due to the difficulty of experimental proof [63]. Figure 3.8 shows a comparison of write field strength on two different types of perpendicular media as a function of t_{IL}. As t_{IL} is reduced, a higher perpendicular field (H_{perp}) and lower in-plane field ($H_{in\text{-}plane}$) are observed along the down-track direction at the center of the track in Fig. 3.8a. Improvement of intrinsic media writability at a thinner IL is anticipated. Interestingly, effective field strength ($H_{effective}$) strongly depends on magnetic design with different magnetic switching behaviors, as shown in Fig. 3.8b. Static or dynamic tilted switching media with incoherent magnetic switching behavior significantly enhances both $H_{effective}$ and the field gradient compared to conventional Stoner–Wohlfarth media with coherent magnetic switching behavior. Combination of a thin IL with static or dynamic tilted switching media will maximize media performance. However, the conventional media with coherent switching behavior don't show the benefits of media writability at a thinner IL and no dependence of magnetic write width (MWW) and transition curvature on t_{IL}. A thin IL may not be required for the conventional perpendicular media.

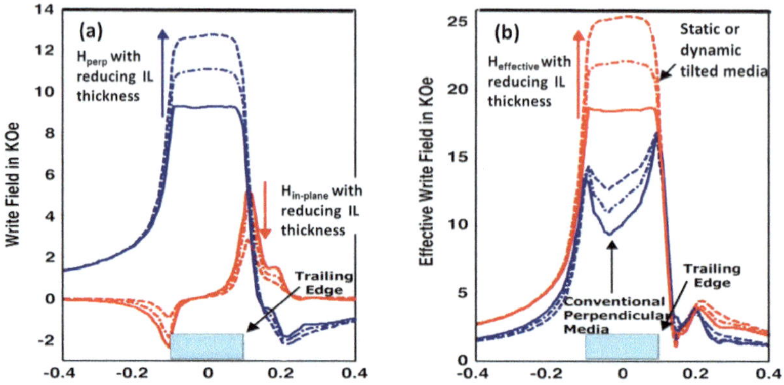

Figure 3.8 Comparison of write field strength on two different perpendicular media as a function of intermediate layer thickness (t_{IL}). (a) Perpendicular (H_{perp}) and in-plane write field ($H_{in\text{-}plane}$) along the down-track direction at the center of the track. (b) Effective field strength versus t_{IL} for static or dynamic tilted media with incoherent switching behavior and conventional Stoner-Wohlfarth media with coherent switching behavior. © [2008] IEEE. Reprinted, with permission, from Ref. [63].

The roles of the intrinsic IL as an exchange BL between the MAG and the SUL are investigated experimentally by using an amorphous and nonmagnetic CrTa filling layer between an SUL and a seed layer [64, 65]. This amorphous CrTa thickness is widely controlled in the IL structure of CrTa(0–25)/NiW(6)/Ru(14 nm) on top of an amorphous SUL. A minor substrate heating effect is observed at a thick CrTa but it does not change magnetic switching behaviors relatively close to coherent switching behavior. The effect of t_{IL} on the recording performance of ROW, signal resolution (RES), magnetic core width (MCW), and signal-to-noise ratio at $2T$ S_oNR_{2T} is checked in the t_{IL} range of 20–45 nm. $SoNR_2T$ is measured by a root-mean-square (rms) signal at $12T$ and the integrated media noise at $2T$ square wave, where T is typically the minimum transition spacing in the drive operation. ROW is measured by overwriting $10T$ patterns over $1.3T$ patterns and RES is determined by root-mean-square (rms) signal ratio of $2T$ and $12T$. MCW including MWW and erasure band (EB) is measured at $10T$. S_oNR_{2T} is measured by the rms signal at $12T$ and the integrated media noise at $2T$. As t_{IL} increases from 20 to 45 nm, ROW decreases

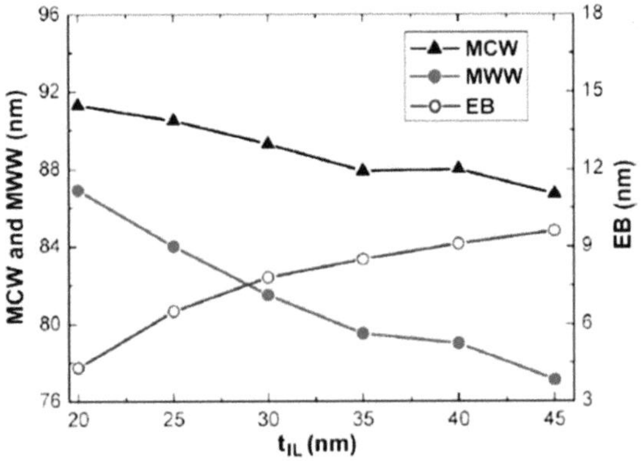

Figure 3.9 Effect of intermediate layer thickness (t_{IL}) on magnetic core width (MCW), magnetic write width (MWW), and erasure band (EB). Note MCW = MWW + EB. © [2010] IEEE. Reprinted, with permission, from Ref. [64].

from 42 to 34 dB, while RES values increases from 33% to 39%. Interestingly, MCW shows less dependence of t_{IL} in Fig. 3.9 because of the compensation effect between narrower MWW and wider EB at a thicker IL. Improvement of S_oNR_{2T} and degradation of RES at a thinner IL are mostly due to the enhancement of amplitude at $12T$ by 13%, considerably higher than the values at high frequencies of $2T$ and $1.3T$, as shown in Fig. 3.10. The enhancement of ROW at a thinner IL can be explained by the increase in field strength beneath the center of the writer pole, as shown in the simulation plot with conventional perpendicular media of Fig. 3.8b, which provides stronger and more field penetration into the media. However, the maximum located near the gap changes very little in comparison. This similar transition quality supports no improvement of the SNR at the middle and high frequencies. Reduction of the EB at a thinner IL is understood by a higher effective field gradient along the cross-track direction. All the recording parameters with increasing t_{IL} are qualitatively consistent with simulation results.

Reduction of t_{IL} clearly improves ROW and reduces the EB. However, it only enhances the SNR at low frequency. The effects of t_{IL} on the SNR are strongly impacted by the writing process. Optimal t_{IL} for the best SNR depends on the writing condition. A thin IL is not es-

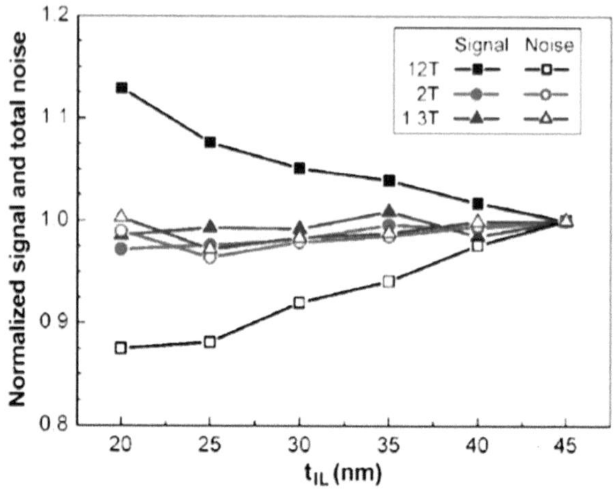

Figure 3.10 Effect of intermediate layer thickness (t_{IL}) on rms signal and total noise normalized by the values at $t_{IL} = 45$ nm at three different frequencies of $12T$, $2T$, and $1.3T$. © [2010] IEEE. Reprinted, with permission, from Ref. [64].

sential for conventional perpendicular media except for the cases of ROW shortage and extra smoothness of magnetic surface roughness.

3.2.3.2 Template for grain growth and grain isolation

The hard magnetic layers with well-isolated grains and higher K_u on top of the dome-shaped Ru interlayer are essential for better thermal stability and higher TD. However, the existence of a highly exchange-coupled initial thin magnetic layer on top of the Ru layer creates undesirable side effects of significant incoherent switching behavior and high magnetic activation volume [66]. It corresponds to a high-noise microstructure, as observed in electroplated CoNiP films [67]. Figure 3.11 shows cross-sectional microstructures and schematic models of (a) low-noise media and (b) high-noise media with and without a grain isolation template on top of the IL, respectively. Final columnar structures look similar but the initial nucleation and growth mechanism is very different. The columnar structure of high-noise media without a grain isolation template has a considerably larger grain size at an early stage in the growth and is followed by formation of subgrains with grain isolation.

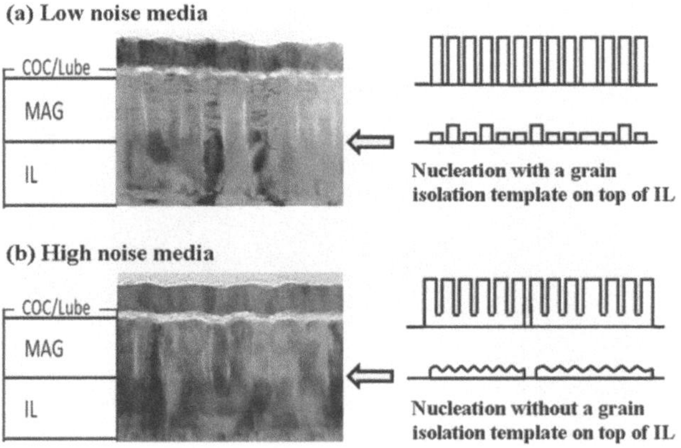

Figure 3.11 Cross-sectional microstructure and schematic models of (a) low-noise media and (b) high-noise media with and without a grain isolation template on top of the IL, respectively.

The highly exchange-coupled initial layer, defined as a transition layer caused by a nonuniform microstructure, increases lateral A_{ex} and provides soft magnetic properties. It creates more incoherent switching behavior and higher magnetic activation volume (V_{act}). It should be removed ideally but minimized at least for reducing media transition noise. Further improvement of media performance can be achieved by using a good template of the IL for grain growth and grain isolation. Ideally, a highly crystallized thin IL template with well-isolated grains is needed for achieving higher areal densities.

Magnetic properties and the microstructure in the bottom hard magnetic layers are particularly sensitive to the microstructure of the IL. A thinner IL typically provides high media transition noise caused by less crystallinity, worse grain isolation, and broader crystallographic *c* axis orientation ($\Delta\theta_{50}$) in given magnetic layers. There are many efforts to achieve a proper IL microstructure as a template for the magnetic recording layer. The conventional dual Ru_1/Ru_2 layer structure deposited with low and high Ar pressures, respectively, has been used for obtaining both good $\Delta\theta_{50}$ and a rougher surface [68, 69]. This unique Ru layer matches well with the CoCrPt-based oxide alloy so Co magnetic grains are heteroepitaxially grown on top of the rough Ru interface deposited with high Ar pressure, as shown in Fig. 3.12. Yellow- and red-

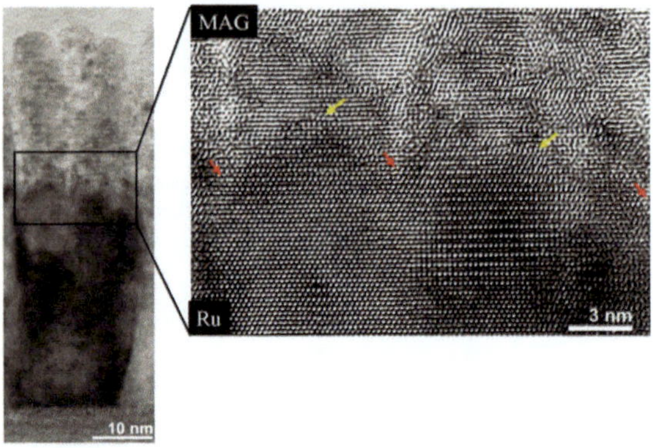

Figure 3.12 High-resolution cross-sectional TEM images showing heteroepitaxial growth of Co magnetic grains on top of a rough Ru interface deposited with high Ar pressure. Yellow- and red-colored arrows with a solid line indicate hills and valleys of the Ru grains, respectively. Amorphous grain boundaries are observed on the valleys. Courtesy of Unoh Kwon.

colored arrows with solid line indicate hills and valleys of the Ru grains, respectively. Amorphous grain boundaries are observed on the valleys. This epitaxial growth supports good $\Delta\theta_{50}$ and bulk magnetic properties. A dual Ru_1/Ru_2 interlayer enhances grain isolation in the magnetic layer, but it doesn't reduce grain size. Addition of Cr into Ru up to 40% provides a better lattice match with a CoCrPtB film heteroepitaxially grown on top of Ta/RuCr [70]. Either an additional Ru oxide layer [71, 72] on top of Ru or insertion of a thin carbon synthetic nucleation layer [73] between the MAG and the IL effectively reduced grain size of the recording layers. Typical reduction of grain size induced by the IL structure makes H_c lower due to worse crystallinity caused by interfacial stress from lattice mismatch and/or oxide formation at grain boundaries. An alternative way is to use a face-centered cubic (fcc) nonmagnetic IL with stacking faults induced by adding Cr to fcc Pt or Pd elements [74, 75], but it is not used in real media due to not-better performance and high material cost.

Here are two examples to understand the correlation between ILs and recording performance. The first case is to replace Ru with RuCr for improving crystallinity [70]. It is known [76] that a higher ratio of substrate temperature and melting point (T_{mp}) in a target

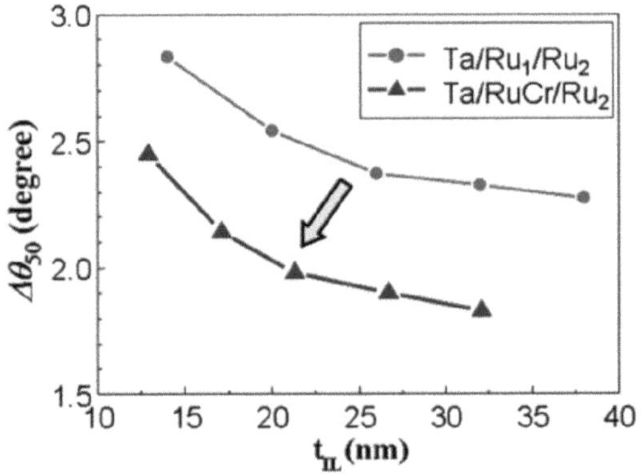

Figure 3.13 Effect of intermediate layer thickness (t_{IL}) on crystallographic c axis orientation ($\Delta\theta_{50}$), full width at half maximum of Ru (0002) peak in two different IL types. © [2008] IEEE. Reprinted, with permission, from Ref. [61].

material provides more surface and bulk diffusions in growing films, resulting in denser columnar grains with fewer defects. In a given sputtering condition, higher atomic mobility can be achieved by reducing T_{mp} in a target material. Addition of Cr into Ru reduces T_{mp} from 2330°C at Ru to 1800°C at $Ru_{0.6}Cr_{0.4}$. Magnetic switching behavior in CoCrPtSi–O single-layer media on two IL types of conventional Ru_1/Ru_2 and $RuCr/Ru_2$ is compared [61]. Replacement of Ru with RuCr provides narrower $\Delta\theta_{50}$ than conventional Ru_1/Ru_2 and achieves superior $\Delta\theta_{50}$ <2.5°, as shown in Fig. 3.13. It also improves crystallinity at relatively thin IL thickness. $RuCr/Ru_2$ enhances coherent switching behaviors in CoCrPtSi–O films but it induces considerably higher V_{act} than Ru_1/Ru_2 (Fig. 3.14). Higher V_{act} for the media with $RuCr/Ru_2$ can be attributed by higher lateral A_{ex} and/or larger grain size. The use of RuCr improves both H_c and coherent switching behavior at thinner t_{MAG}. This improvement is explained by contributions of narrower $\Delta\theta_{50}$, thinner transition layer thickness showing incoherent switching behavior, and higher V_{act}. The increased V_{act} in the media with RuCr can be understood by the increase in physical grain size. Reduction of V_{act} by using a proper IL while maintaining sufficient thermal stability factor (K_uV/kT) in the bottom hard layer can further reduce media transition noise and improve the SNR.

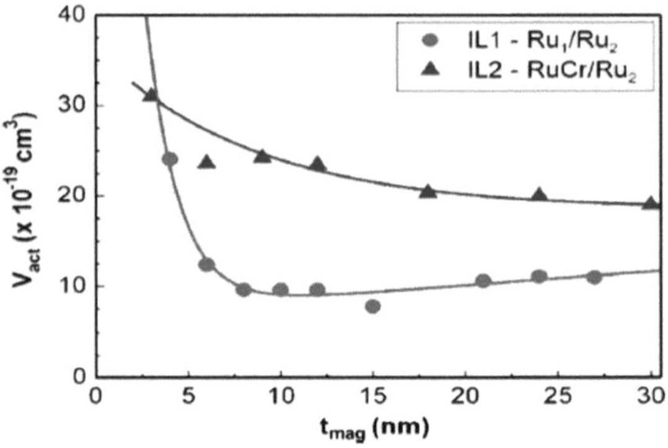

Figure 3.14 Effect of magnetic layer thickness (t_{MAG}) on magnetic activation volume (V_{act}) in CoCrPtSi–O films with two different IL types. © [2008] IEEE. Reprinted, with permission, from Ref. [61].

The second case is to additionally use a thin Ru oxide layer as a well-defined granular template on top of Ru_2. A 5 nm thick Ru–SiO_2 layer on top of Ar–ion etched Ru effectively reduces grain size up to ~4 nm but it significantly decreased H_c [77]. The grain size dependence on K_u could be understood by the higher amount of stacking faults in smaller magnetic grains [78]. However, the use of a 2 nm thick Ru oxide layer deposited on top of physically segregated Ru_2 grains is simple and effective to enhance grain isolation without losing H_c [71]. In Fig. 3.15a, the cross-sectional transmission electron microscopy (TEM) image shows the well-defined columnar structure of the magnetic layer and Ru/Ru oxide interlayer. A good heteroepitaxial relationship is maintained throughout the Ru/Ru oxide and Ru oxide/magnetic layer interfaces. These plan-view TEM images of Fig. 3.15b (1–3) clearly shows the difference in grain isolation of Ru grains at roughly bottom (1), middle (2), and top (3) regions in Fig. 3.15 (a), respectively. Compared to the regions (1 and 2) closer to the Ta seed layer, the 2 nm thick Ru oxide layer significantly enhances grain isolation at the top of Ru grains with uniform size distribution without disturbing the epitaxial relationship between the magnetic layer and the interlayer. This grain template effect considerably enhances H_c by ~2 kOe in CoCrPt oxide layers as shown in Fig. 3.15c. This is unique to achieving good grain isolation and crystallinity together.

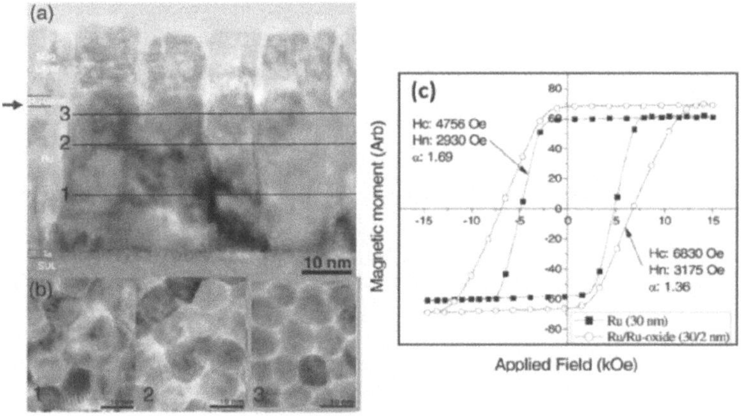

Figure 3.15 (a) Cross-sectional TEM image of perpendicular media with the layer structure of CoCrPtO(15)/Ru oxide(2)/Ru(26)/Ta(2 nm)/SAF-SUL, (b) plan-view TEM images the media with Ru oxide(2)/Ru(26)/Ta(2 nm)/SAF-SUL, and (c) hysteresis Kerr loops of CoCrPtO media with and without 2 nm thick Ru oxide. © [2005] IEEE. Reprinted, with permission, from Ref. [71].

Magnetic clustering, thermal stability, and recording performance on advanced media with multilayered H_k gradient CoPtCr oxide/capping layers with various Ru oxide layer thicknesses ($t_{\text{Ru oxide}}$) on top of Ru_2 are studied [79]. With increasing $t_{\text{Ru oxide}}$ from 0 to 1.3nm, H_c and H_s are clearly enhanced but H_n is reduced. Magnetic correlation length (D_n) extracted from a set of major and minor loops in Fig. 3.16, an indicator of magnetic cluster size significantly decreases but intrinsic SFD remains unaffected. A short-time switching field proportional to H_k increases linearly, while K_uV/kT remains unchanged. No change in K_uV/kT is explained by compensation of the reduced D_n with the enhanced K_u induced by a thin Ru oxide. However, thermal decay rate degrades from 0.06% to 0.32%/decade, which correlates well with lower H_n and smaller D_n. Similar values of initial minor loop slopes indicate no change in magnetic switching behavior. A 1 nm thick Ru oxide layer as a well-defined granular template significantly improves intrinsic recording performance (Fig. 3.17): a higher SNR_{2T} by 1.4 dB and a narrower MCW at $10T$ by 8 nm are observed, even at lower ROW by 8 dB, compared to the media without the Ru oxide layer.

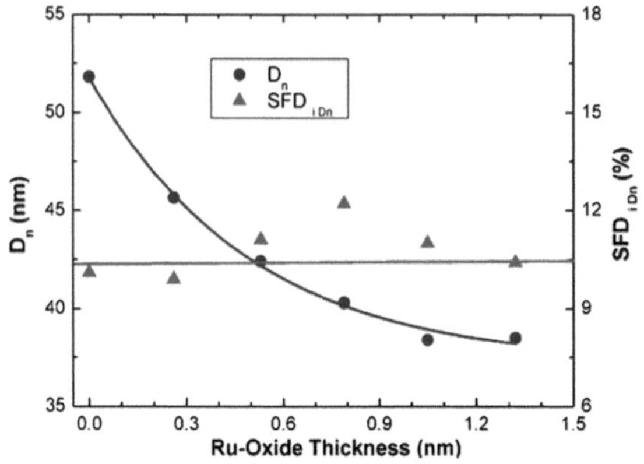

Figure 3.16 Effect of Ru oxide thickness on magnetic correlation length (D_n) and intrinsic switching-field distribution ($\mathrm{SFD}_{i\,D\mathrm{n}}$) from a set of major and minor loops of the full-stacked media. Reprinted with permission from Ref. [79]. Copyright 2012, AIP Publishing LLC.

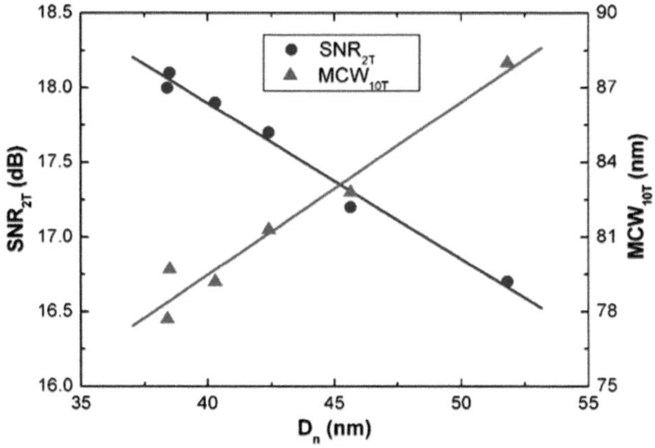

Figure 3.17 Change in the signal-to-noise ratio (SNR) at 2T and magnetic core width (MCW) at 10T as a function of magnetic correlation length (D_n) on the full stacked media with various Ru oxide thicknesses. Reprinted with permission from Ref. [79]. Copyright 2012, AIP Publishing LLC.

An alternate way is to use a nonmagnetic or superparamagnetic CoCr-based high-oxide-containing layer [80] on top of Ru_2 instead of a thin Ru oxide layer, called the grain isolation magnetic anisotropy layer (GIMAL).

This GIMAL with $M_s < 200$ emu/cm^3 has similar lattice parameters as the bottom hard layer and it acts as a buffer layer for reducing interface strains caused by lattice mismatch and a template for enhancing grain isolation. A 1 nm thick GIMAL enhances H_c by 0.6 kOe and shows similar trends of recording performance except for higher M_rt and K_uV/kT as a thin Ru oxide layer does: a higher SNR and a narrower MCW but worse ROW.

3.2.3.3 Trend of IL materials

An amorphous CoZr [81] or Ta [82] seed layer on top of a nanocrystalline NiFeMo SUL is initially used to provide a smoother surface and better aligned grain formation. However, the purpose of the seed layer is changed to directly make an initial crystallized phase with hexagonal close-packed (hcp) (0001) preferred orientation when an amorphous SUL is used. As a seed layer, Ru or RuX-based alloys for hcp (0001) orientation are used at an early PMR stage. The Ru element is a unique material for CoCrPt-based alloys. Ru has hcp (0001) preferred orientation and a high T_{mp} of 2334°C. This Ru provides a template for crystallographic grain growth and c axis orientation of Co with the same hcp structure. Surface roughness as a source of physical grain isolation is easily controlled by the sputter pressure profile. A unique dual Ru_1/Ru_2 IL deposited with low and high Ar pressure, respectively, is developed. Ru_1 provides good crystallinity and c axis orientation and Ru_2 with a physical rougher surface induces good grain isolation in magnetic recording layers. Combination of Ru_1/Ru_2 with CoCrPt oxide layers leads to significantly improved H_n and recording performance in PMR media.

The use of Ru for PMR media made the price soar from $35/oz. in 2003 to $800/oz. in 2007 [83]. As an alternative material of Ru_1, Ni-(X,Y)-based alloys are chosen to release this price pressure. Ni has an fcc phase and a low T_{mp} of 1453°C, as well as a significantly low material cost. As one of the examples, substitution of Ni with W up to 12% maintains the fcc phase and also expands the lattice constant to match with Ru. Compared to Ru_1, NiW alloys provide higher H_c due to better crystallinity with more freedom to grow fcc (111) orientation. It leads to a much rougher surface than Ru_1 at the same thickness so it can't replace Ru_1 completely. To improve corrosion robustness, either addition of Cr into NiW or simple reduction of NiW thickness is used. Another approach is to add Cr or Mn into Ru to enhance crystallinity, while maintaining an hcp structure. A typical IL structure consists of Ni-(X,Y)/Ru_1 or Ru-(X,Y)/Ru_2.

A highly crystallized thin IL template with well-isolated grains has been developed continuously for achieving higher areal densities. A template for enhancing both grain isolation and H_k highly attracts interest in designing PMR media with thermally stable isolated small grains. The enhancement of both grain isolation and H_k is reported when the CoPtCr oxide recording layer is exchange-coupled with antiferromagnetic IrMn [84] or MnRu [85]. However, the enhancement of K_u is insignificant and its origin is not clear. High content of Mn also limits corrosion robustness. However, a 2 nm thin Ru oxide layer on top of Ru_2 clearly enhances grain isolation and significantly increases H_c by $\sim$2 kOe in a 15 nm thick CoPtCr oxide single layer [71]. A plausible cause of the enhanced H_c with increasing Ru oxide thickness is chemical grain isolation induced by solid-state oxidation reaction between the Ru–O layer and metallic Cr of the magnetic oxide top layer, as has been observed at the interfaces between Co–O and Al and between CoFe–O and Ta [86, 87]. A 1 nm thick Ru oxide layer is very effective to reduce magnetic cluster size as a well-defined granular template and to increase H_c. It significantly improves both SNR and MCW, even at worse ROW. Using a thin nonmagnetic or superparamagnetic CoCr-based high-oxide-containing layer on top of Ru_2 also shows similar effects on bulk magnetic properties and recording performance, except for high $M_r t$ and $K_u V/kT$ as a thin Ru oxide layer does [80]. An advanced IL structure is getting complicated to meet multiple purposes.

As areal densities increase continuously, head writability is getting worse due to narrower write pole width and limitation of highest $4\pi M_s$ material of FeCo. Further improvement of ROW in media is needed. As one of activities, the Ni-(X,Y)-based nonmagnetic alloy is replaced with CoNiFe [88] or NiFe-(X,Y)-based magnetic soft alloys for boosting ROW by reducing head-to-SUL spacing. The head writability becomes stronger at lower head-to-SUL spacing but the influence of side fringing fields on the nearest adjacent track highly depends on the write head type. However, this complicated IL structure doesn't provide a smaller grain template, even though magnetic cluster size is reduced significantly. To achieve a highly crystallized thin IL template with well-isolated small grains still remains as one of hot challenges.

3.2.4 *Magnetic Recording Layers*

PMR media have been developing with different magnetic design concepts for increasing areal densities. Figure 3.18 shows a schematic view of different magnetic design concepts for PMR media: (a) early single magnetic oxide media [12, 13], (b) hard/soft stacked composite media with coupled granular/continuous (CGC) [22] or capping layer [23], dynamic tilted switching media for (c) exchange-coupled composite (ECC) media with a single BL [25], (d) anisotropy-graded exchange spring (ES) media [26, 27], and (e) magnetic switching block–controlled (or segmented) media with multiple BLs [28, 29]. In this section, magnetic design concepts from early to latest research works are described. Fundamental grain isolation mechanism and cons and pro on each magnetic design are discussed.

3.2.4.1 Initial PMR media with CoCrPt-based oxide-containing layers only

Early PMR media used a CoCr-based metallic alloy with lower K_u and high lateral A_{ex}, as well as poor *c* axis orientation compared to current PMR media. A typical sheared Kerr hysteresis loop with positive H_n on a CoCr-based alloy creates high DC noise caused by instability of amplitude against magnetostatic H_{demag} and disturbance fields during the recording process. This high DC media noise became one of the big challenges to implementing PMR technology. Two major breakthroughs occurred around 2000. One is to use CoCrPtO single-layered media with high K_u in core grains and amorphous thick oxygen-rich boundaries [12, 13], which demonstrates high H_c with a steep hysteresis loop and a high

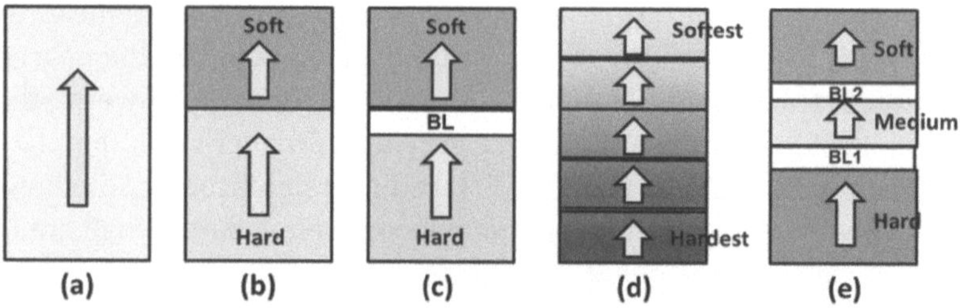

Figure 3.18 Schematic view of different magnetic design concepts for PMR media: (a) early single magnetic oxide media [12, 13], (b) hard/soft stacked composite media with coupled granular/continuous (CGC) [22] or capping layer [23], dynamic tilted magnetic switching media for (c) exchange-coupled composite (ECC) media with a single breaking layer [25], (d) anisotropy-graded exchange spring (ES) media [26, 27], and (e) magnetic switching block–controlled (or segmented) media with multiple breaking layers [28, 29].

squareness ratio. Enhancement of physical and chemical grain isolation is achieved by high Ar pressure and the oxygen reactive sputtering process. Similar effect is obtained later by adding SiO_2 to the CoPtCr alloy on top of the Ru interlayer [21]. The other is to use a dual Ru_1/Ru_2 layer structure [68, 69] deposited with low and high Ar pressures, which is described in the previous IL section. This Ru IL uniquely provides good *c* axis orientation and rougher surface together. The combination of CoCrPt oxide alloys with dual Ru ILs significantly improves both H_c and H_n compared to early CoCr-based alloys but still a lower negative H_n for practical use. It is not solved until either the magnetic capping layer or the CGC layer on top of the well-isolated CoCrPt oxide granular layer is used.

Both the fundamental grain isolation mechanism and an understanding of the correlation between microstructure and recording performance are highly important to improve media performance. The media noise depends on oxygen incorporation that forms amorphous oxide grain boundaries, which are highly sensitive to process conditions in single CoCrPtSi–O media [89–91]. Zheng et al. [90] pointed out the existence of two different zones, as shown in Fig. 3.19: Zone I with higher H_c and constant M_s up to oxygen content (OC) = 15% owing to the enhancement of grain isolation but Zone II with lower H_c and M_s at OC >15% due to the overoxidation

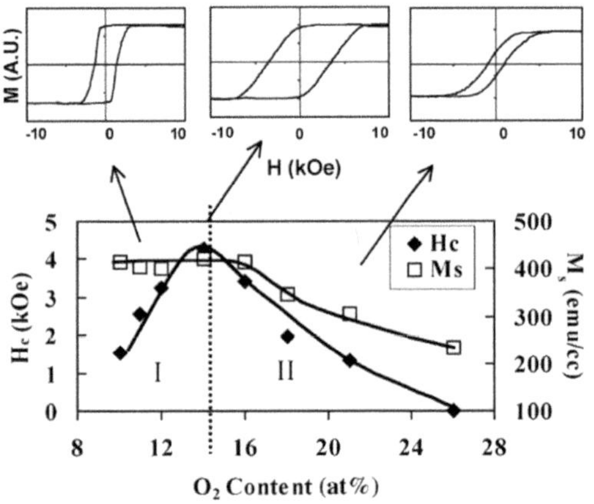

Figure 3.19 The change in H_c and M_s as a function of oxygen content in CoCrPtSi–O perpendicular media. © [2004] IEEE. Reprinted, with permission, from Ref. [90].

of Co in core grains. In CoCrPtSi–O films [92], M_s of the grains gradually increased with increasing SiO_2 content, suggesting the possibility of a reduction of Cr content in CoCrPt grains. Oxygen incorporation controls lateral A_{ex} as well as compositional change in core grains. To understand the grain isolation mechanism caused by oxygen incorporation, the effect of oxygen incorporation on the microstructure and media performance in CoCrPtSi–O films with various OCs within Zone I is studied systematically [66, 93–95].

Figure 3.20 shows the trends of (a) H_c and (b) coercivity squareness (S^*) as a function of OC and magnetic thickness (t_{MAG}) in CoCrPtSi–O perpendicular media [94]. H_c increases with increasing OC from 3% to 10% and it reaches 6.6 kOe at $t_{MAG} = 18$ nm with OC = 10%. The peak of H_c is obtained at a lower t_{MAG} for higher OC due to overoxidation. Plan-view TEM images in Fig. 3.21 clearly show the big change in the microstructure of 18 nm thick CoCrPtSi–O media with different OCs. Smaller grains and thicker amorphous grain boundaries are observed with increasing OC. Values of M_s increase from 320 emu/cm^3 at OC = 3% to 450 emu/cm^3 at OC = 10%. The increase in H_c and M_s with increasing OC indicates the compositional change in core grains. The change in compositions of core grains at $t_{MAG} = 18$ nm with different OCs is confirmed by TEM

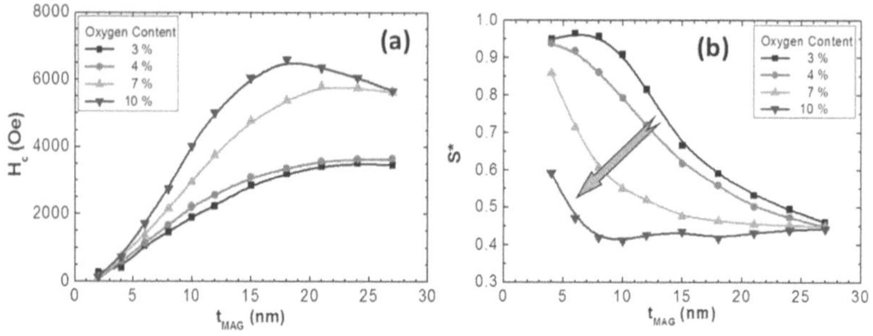

Figure 3.20 Trends of (a) H_c and (b) coercivity squareness (S^*) as a function of oxygen content and magnetic thickness (t_{MAG}) in CoCrPtSi–O perpendicular media. © [2007] IEEE. Reprinted, with permission, from Ref. [94].

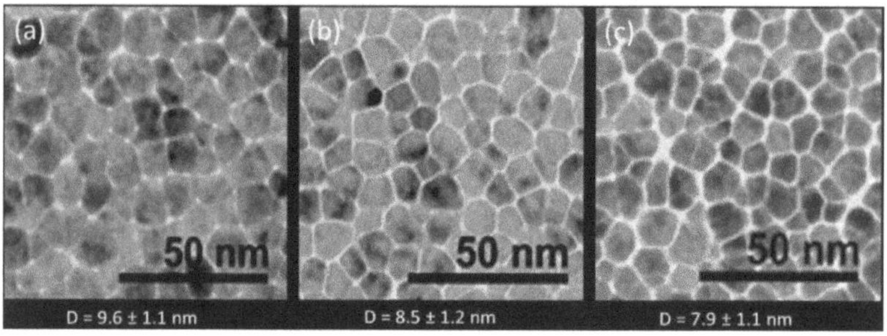

Figure 3.21 Plan-view TEM images of the 18 nm thick CoCrPtSi–O media with different oxygen contents of (a) 3%, (b) 7%, and (c) 10%. © [2007] IEEE. Reprinted, with permission, from Ref. [94].

energy-dispersive X-ray spectroscopy (EDS) with an electron probe size that is less than 2 nm. The Cr content decreases from 19% to 13% with increasing OC, while the Pt content increases from 24% to 30% (Fig. 3.22). This result explains higher H_c and M_s in films with increasing OC.

Although S^* is typically complicated by the intrinsic SFD and H_{demag}, S^* is used as an indicator of lateral A_{ex}. The value of S^* decreases with increasing t_{MAG} (Fig. 3.20b). A higher OC significantly reduces S^*, especially for thinner layers. However, S^* considerably increases for $t_{MAG} < 8$ nm even when OC = 10%. It indicates the existence of possible sources of high lateral A_{ex} at a thinner thickness. Plan-view TEM images of CoCrPtSi–O media with OC = 10%

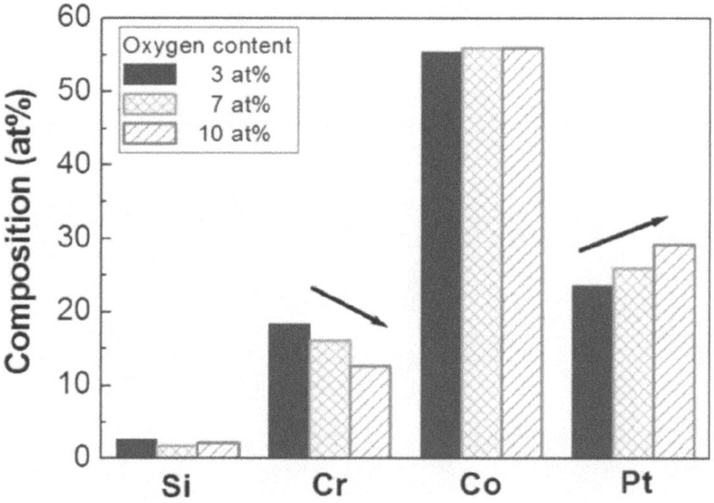

Figure 3.22 Comparison of composition in core grains in the 18 nm thick CoCrPtSi–O layer with different oxygen contents measured by TEM-EDS with an electron probe size that is less than 2 nm. © [2007] IEEE. Reprinted, with permission, from Ref. [94].

at various t_{MAG} of (a) 4 nm, (b) 10 nm, (c) 18 nm, and (d) 27 nm in Fig. 3.23 reveal the existence of nonuniform microstructure along the film thickness of the CoCrPtSi–O layer. It explains the discrepancy between H_c and the degree of grain isolation: the highest H_c at $t_{MAG} = 18$ nm but the best microstructure with good grain isolation at $t_{MAG} = 10$ nm. Grain isolation is incomplete and a large fraction of Co grains are physically connected to adjacent grains at $t_{MAG} = 4$ nm. It also explains the higher $S*$ at an initial thin magnetic layer, indicating more lateral A_{ex}. At $t_{MAG} = 10$ nm, thicker amorphous oxide grain boundaries in the magnetic layer are formed along the grain boundaries of the Ru interlayer, resulting in a significant decrease in S^*. However, the Co grains grown on the same Ru grain are further divided into two or three subgrains at $t_{MAG} = 18$ nm, while maintaining a well-isolated grain structure. The formation of these subgrains is more clearly visible at $t_{MAG} = 27$ nm and the grain isolation becomes worse. Later, high-angle dark-field TEM images reveal a cavity in the center of many grains in the range of 5–8 nm, called *tooth growth mechanism* [96]. As metallic CoPt on top of Ru

Figure 3.23 Plan-view TEM images of CoCrPtSi–O media with 10% oxygen content at various magnetic layer thicknesses of (a) 4 nm, (b) 10 nm, (c) 18 nm, and (d) 27 nm. Reprinted with permission from Ref. [66]. Copyright 2007, AIP Publishing LLC.

domes nucleates preferably first and grows faster from the sides, voids start to be created. This void at the center acts as a subgrain boundary and it will be filled out with amorphous oxides of Ti, Si, Cr, and O. The reduction in H_c at t_{MAG} >18 nm can be understood by lower K_u caused by the formation of subgrains and higher stacking fault density and less grain isolation.

The effect of a nonuniform microstructure as a function of t_{MAG} on magnetic switching behaviors is checked by using normalized angular dependence of remanent H_c ($H_{cr45^\circ}/H_{cr0^\circ}$), as shown in Fig. 3.24. The data with $1/\cos\theta$ and S-W indicate domain wall switching and coherent switching in Stoner–Wohlfarth particles, respectively. Surprisingly, the 4 nm thick film with OC = 7% clearly exhibits a high degree of incoherent switching behavior. $H_{cr45^\circ}/H_{cr0^\circ}$ continuously increases with the increasing angle between the easy

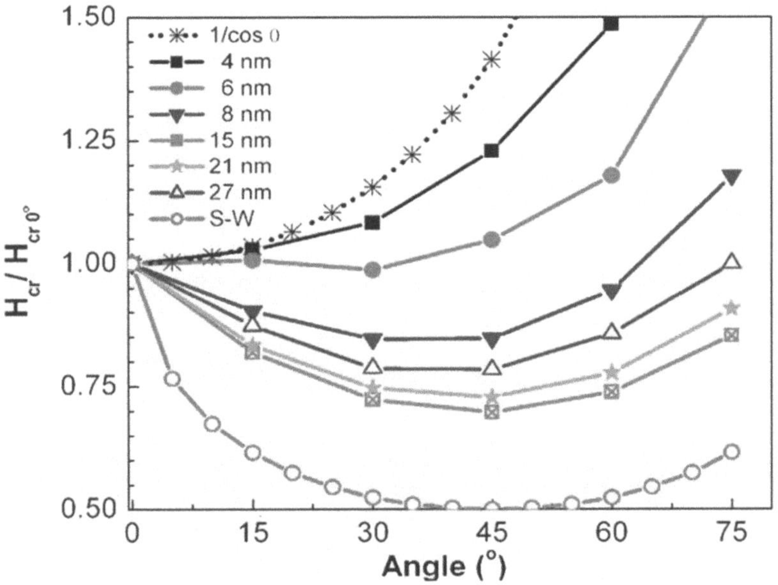

Figure 3.24 Angular dependence of remanent coercivity (H_{cr}) normalized by H_{cr} at 0° in CoCrPtSi–O media with 7% oxygen content at various magnetic layer thickness. The data with $1/\cos\theta$ and S-W indicate domain wall switching and coherent switching in Stoner–Wohlfarth particles, respectively. Reprinted with permission from Ref. [66]. Copyright 2007, AIP Publishing LLC.

axis and the applied field direction, close to domain wall switching behavior. Values of $H_{cr45°}/H_{cr\ 0°}$ decrease with increasing t_{MAG}, reach a minimum at $t_{MAG} = 15$ nm, and then increase at $t_{MAG} >15$ nm. The minimum value of 0.7 at 45° on the medium with $t_{MAG} = 15$ nm is significantly higher than 0.5 in the S-W particles.

For relative comparison of magnetic cluster volume, activation volume of magnetic reversal (V_{act}) is estimated by using the waiting-time method [97] independent of H_{demag} based on coherent switching, which is more reliable and accurate than the magnetic viscosity method [98]. A linear relationship between magnetic switching volume and t_{MAG} is expected for uniform grain isolation along the thickness direction. However, values of V_{act} are considerably high at all thin layers, as shown in Fig. 3.25, and they rapidly decrease with

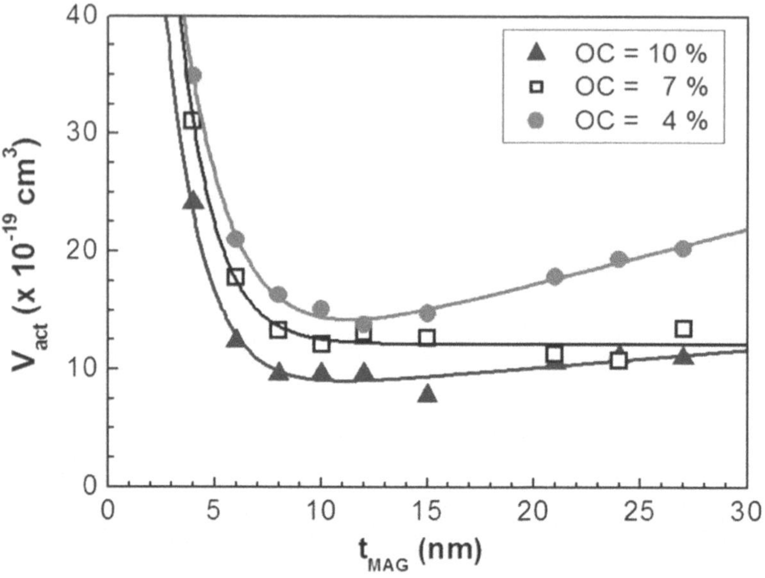

Figure 3.25 Magnetic activation volume (V_{act}) in CoCrPtSi–O media with various oxygen contents as a function of magnetic layer thickness (t_{MAG}). Reprinted with permission from Ref. [66]. Copyright 2007, AIP Publishing LLC.

increasing t_{MAG}. Overall higher OC reduces V_{act} but large V_{act} exists at $t_{MAG} \leq 6$ nm regardless of OC in the films. The X-ray photoelectron spectroscopy (XPS) depth profile in Fig. 3.26 clearly supports these phenomena: the magnetic layer on top of Ru initially has a CoPt-rich phase and later is incorporated with Cr and oxide phase. The formation of an initial CoPt alloy will provide better lattice match with Ru, which can be understood as a stress relief [99]. The nonuniform microstructure at initial highly-exchange-coupled thin CoPt layer significantly degrades recording performance due to the main source of large magnetic cluster formation. Reduction or removal of the highly-exchange-coupled initial layer is achieved by inserting either a thin Ru oxide layer [71, 79] or a nonmagnetic or superparamagnetic CoCr-based high-oxide-containing layer [80] between the bottom hard layer and Ru_2, as previously described in the IL section.

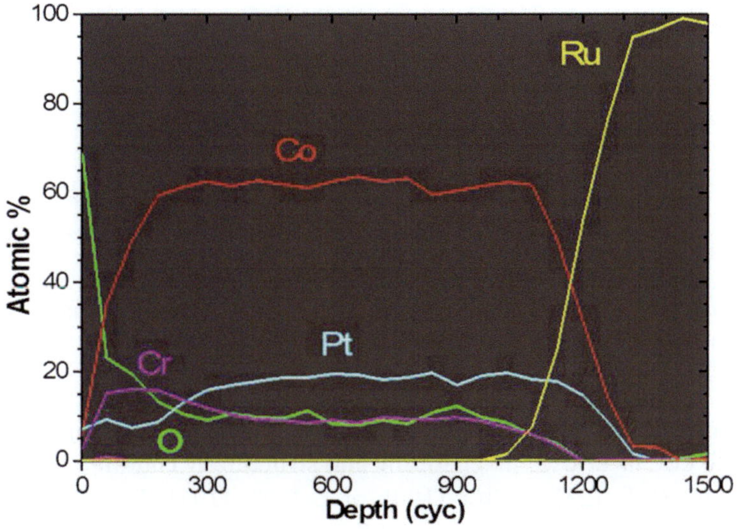

Figure 3.26 X-ray photoelectron spectroscopy (XPS) depth profile in the 18 nm thick CoCrPtSi-O medium with 10% oxygen content. Reprinted with permission from Ref. [95]. Copyright 2008, AIP Publishing LLC.

3.2.4.2 Hard/soft stacked composite media: coupled granular/continuous layer and capping layer

PMR media with high H_n are required for improving robustness against magnetostatic H_{demag} and head-leakage field. Enhancement of grain isolation in CoCrPt oxide media provides smaller grains with thicker grain boundaries, while maintaining high H_c. However, it simultaneously induces undesirable side effects of nonuniform grain isolation, larger composition variation in core grains, broader magnetic anisotropy dispersion, and higher temperature coefficient of H_c. Thus, a broad SFD is inevitable in single-layer media with a well-isolated grain structure. To improve SFD and ROW, hard/soft stacked composite PMR media by using soft magnetic layers directly exchange-coupled with a well-isolated high-K_u CoCrPt oxide layer are proposed.

This soft magnetic layer is called as either a CGC layer [22, 100–102] or a capping layer [23, 103–105], depending on the physical concept. Figure 3.27 shows a schematic view of hard/soft stacked

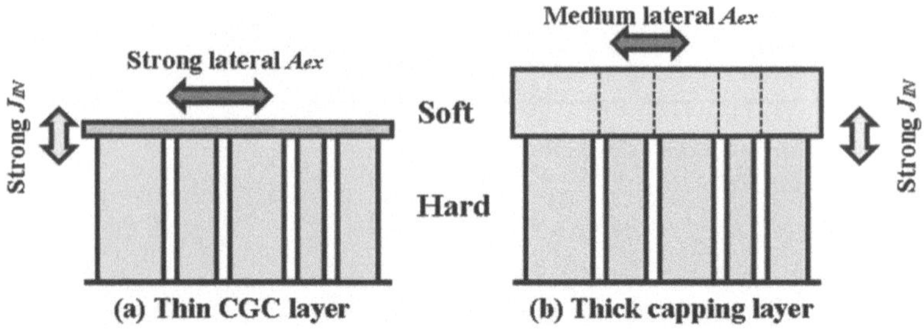

Figure 3.27 Schematic view of hard/soft stacked composite media with (a) thin coupled granular/continuous (CGC) layer and (b) thick capping layer.

composite media with (a) a CGC layer and (b) a capping layer. The CGC layer uses magnetically and structurally continuous layers of $[Co/Pt]_n$ [22], CoPtCr [100], or CoPrTb [101]. This CGC layer provides strong interlayer exchange-coupling strength (J_{IL}) directly with bottom grains and indirect grain-to-grain interaction via the CGC layer with strong lateral A_{ex}.

However, the capping layer uses a magnetically continuous but structurally discontinuous metallic CoCrPtB-based alloy layer [23, 103–105] so it has medium lateral A_{ex} less than the CGC layer. A soft top layer typically has high M_s, low K_u, and larger lattice mismatch between hard and soft layers. Values of J_{IL} are particularly sensitive to the degree of heteroepitaxial growth between hard and soft layers, which can affect SFD and angular dependence of the switching field. Strong J_{IL} and low A_{ex} can maximize benefits for hard/soft stacked media. High M_s and A_{ex} and weak J_{IL} in a soft top layer will provide strong H_{demag}, resulting in low H_n. This causes an increase in DC noise and adjacent track erasure (ATE). It is reported [25] that a soft layer with in-plane K_u provides much smaller thermal energy barrier than a soft layer with out-of-plane K_u. Figure 3.28 compares recoiled Kerr loops of hard/soft stacked composite media with different soft layers of (a) amorphous CoTaZr–SiO_2 for the case of low A_{ex}, in-plane low K_u, and weak J_{IL}; (b) low Pt-containing CoCrPtSi–O for the case of low A_{ex}, out-of-plane low K_u, and medium J_{IL}; and (c) metallic CoCrPtB as a conventional capping layer for the case of high A_{ex}, out-of-plane medium/high K_u, and strong J_{IL} [105]. When H_{demag} is greater than

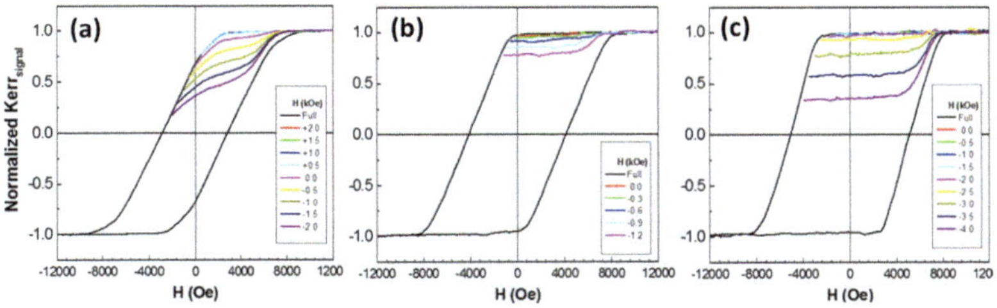

Figure 3.28 Recoiled Kerr loops of hard/soft stacked composite media with different soft layers of (a) amorphous CoTaZr–SiO_2, (b) low Pt-containing CoCrPtSi–O, and (c) metallic CoCrPtB as a conventional capping layer. Reprinted from Ref. [105]. Copyright (2008), with permission from Elsevier.

H_n, magnetizations in a soft layer can be tilted statically and exhibit reversible minor hysteresis loops. The medium with a 10 nm thick CoTaZr–SiO_2 shows positive H_n and reversible minor loops, while the medium with a 12 nm thick CoCrPtSi–O exhibits small negative H_n and irreversible minor loops. This suggests that CoCrPtO has less H_{demag} than CoTaZr–SiO_2. However, values of negative H_n are still low compared to the conventional capping layer media (Fig. 3.28c).

The use of a soft top layer in hard/soft stacked media provides both benefits and undesirable side effects. Basically, a highly-exchange-coupled soft magnetic layer on top of a well-isolated CoCrPt oxide layer enhances A_{ex} locally, especially on subgrains or smaller grains. This leads to higher H_n and lower H_s, resulting in a narrow SFD. This hard/soft composite media concept significantly improves media writability, amplitude, and thermal stability over the single-oxide-layer media. The capping layer relatively provides a thicker soft magnetic layer sputtered with low Ar pressure compared to the hard magnetic layer. A thick capping layer helps reduce surface roughness and provides better carbon quality with less OC. Corrosion robustness is significantly improved over the single-oxide-layer media. However, the CGC layer has lateral A_{ex} considerably higher than vertical J_{IL} so its thickness is limited. Corrosion in PMR media nucleates in the porosity of grain boundaries with amorphous oxides and creates single-needle growths along individual intergranular sites [106]. Thus, a thin CGC layer has corrosion weakness, while improving recording performance as the capping layer media do. This thin CGC layer reduces the total magnetic thickness. A higher effective write field strength and field gradient as the HMS effect are benefits at thin CGC media. The capping layer media are used in the first-generation PMR products.

Understanding the roles of soft magnetic layer with high lateral A_{ex} and strong J_{IL} in hard/soft stacked composite media is critical to designing magnetic layer structure. According to modeling results of CGC media [107], hysteresis loop squareness is approaching unity when a 3–6 nm CGC layer is coupled with a CoCrPt granular layer. The domain wall pinning effect and transition smoothing due to interlayer exchange coupling between the two layers improve both the SNR and thermal stability. In CoPtCrO/CoCrPtB capping media [23, 103–105], the CoCrPtB capping layer stabilizes thermally unstable small Co grains through lateral A_{ex}, resulting in the reduction of DC noise and improvement of H_n. The SNR in the capping layer media is strongly dependent on the magnetization of the capping layer. High H_n prevents small grains from magnetization reversals so ATE and return field–induced partial erasure (RFPE) are considerably reduced. These capping or CGC media have much less sensitivity of grain size distribution and H_k dispersion compared to the single-oxide-layer media.

In CoPtCrO/CoCrPtB capping media, higher M_s in the capping layer than M_s in the bottom hard layer is needed for obtaining a higher SNR [103]. When the soft CoCrPtB capping layer with low $H_c \approx 1.2$ kOe is deposited on the bottom hard layer with $H_c \approx 3.7$ kOe, lowering H_c is anticipated but higher H_c and H_n are achieved [104]. This unique phenomenon relating to exchange coupling effect between hard and soft magnetic layers is highly interesting. The effect of capping layer thickness (t_{Cap}) on media properties in CoPtCrO-based granular composite media is studied [108]. Total magnetic thickness (t_{MAG}) consists of an 11.3 nm thick bottom magnetic layer and variable t_{Cap}. Figure 3.29 shows two distinct zones when t_{Cap} increases. H_c increases from 4.9 to 5.9 kOe with increasing t_{Cap} from 0 to 3.8 nm (Zone I) and then decreases to 4.1 kOe at $t_{Cap} = 8.3$ nm (Zone II). The considerably increase in H_n is observed at Zone I but the significant decrease in H_s is exhibited at Zone II. In the case of a thicker bottom CoPtCrO layer only, H_c is actually decreased due to nonuniform microstructure, formation

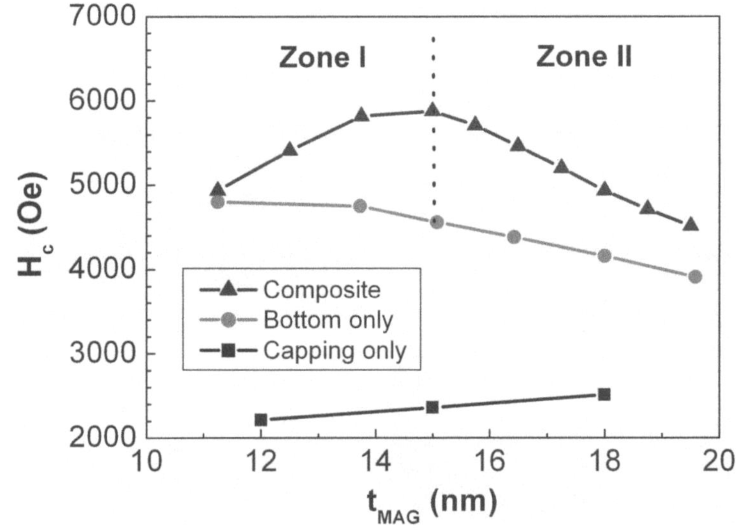

Figure 3.29 Effect of the recording layer thickness (t_{MAG}) on coercivity (H_c) of the media with a bottom layer only, a capping layer only, and a bottom/capping composite media. Values of t_{MAG} in composite media consist of thicknesses of an 11.3 nm bottom layer and a variable capping layer. © [2007] IEEE. Reprinted, with permission, from Ref. [108].

of subgrains, and more stacking faults along the thickness direction [93]. Figure 3.30 shows plan-view TEM images in hard/soft stacked composite media with different t_{Cap} of (a) 0.0 nm, (b) 3.8 nm, and (c) 8.3 nm. Slightly bigger grain size and less grain boundary width are observed in Fig. 3.30b but larger grains and much less grain boundary widths are exhibited in Fig. 3.30c. Zone I exhibits a relatively constant loop slope (α) at H_c and V_{act}, while Zone II shows an increase in α and V_{act}, which is consistent with plan-view TEM results with increasing t_{Cap}. Angular dependence of H_c in Zone II shows less coherent switching behavior compared to Zone I. $K_u V/k_B T$ continuously improves from 59 to 107 with increasing t_{Cap}. The increase in $K_u V/k_B T$ in Zone I is explained by the increase in K_u, while the increase in $K_u V/k_B T$ in Zone II is understood by the increase in V. Enhancement of K_u and V in Zone 1 can be plausibly explained by the solid-state oxidation mechanism between Co–O in the bottom hard layer and Cr in the capping layer, as described at

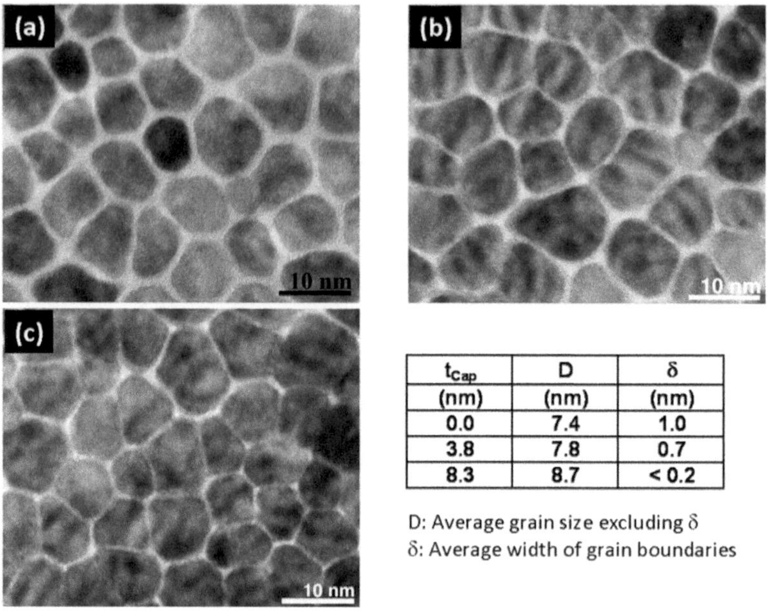

t_{Cap} (nm)	D (nm)	δ (nm)
0.0	7.4	1.0
3.8	7.8	0.7
8.3	8.7	< 0.2

D: Average grain size excluding δ
δ: Average width of grain boundaries

Figure 3.30 Plan-view TEM images in hard/soft stacked composite media with different capping layer thickness (t_{Cap}) of (a) 0.0 nm, (b) 3.8 nm, and (c) 8.3 nm. © [2007] IEEE. Reprinted, with permission, from Ref. [108].

the interfaces between Ru oxide and Cr in the magnetic hard layer [79]. This solid-state oxidation will extend oxide grain boundaries into the capping layer, as shown in Fig. 3.31. It will increase the shape anisotropy constant but it only contributes to an increase of 3% in H_c based on the demagnetizing factors of cylinders [109]. Magnetic simulation reveals that the capping layer with a higher K_u provides more increase in H_c. The increase of 20% in H_c in Zone I can be explained by the enhancement of K_u in a thin t_{Cap} induced by the solid-state oxidation mechanism but this assumption requires further experimental proof. The decrease in H_c at Zone II is understood by the increase in lateral A_{ex} at a thick t_{Cap}. Significant enhancement of amplitude and ROW occurs at Zone II, while MCW is getting wider. Typical t_{Cap} is optimized at Zone II.

Tight control of average lateral A_{ex} and its variation is critical to improve the media SNR. It strongly depends on thickness variation of the amorphous oxide grain boundary and lateral A_{ex} of a capping

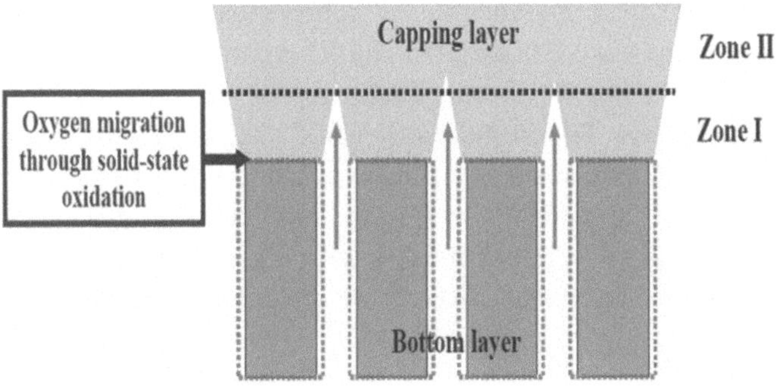

Figure 3.31 Schematic view of hard/soft stacked CoPtCrO/CoCrPtB composite media with grain-isolated capping layer at Zone I through the solid-state oxidation process.

or a CGC layer in perpendicular media. Both variations induce random A_{ex} and degrade the media SNR caused by the broadening magnetic cluster size distribution. The magnetic clusters are not a limiting factor in recording performance if these are breakable in a given head with a sufficiently high field gradient [110]. Grain size reduction is essential to improve intrinsic media performance but most experimental results do not show intrinsic SNR gain on the media with smaller grain sizes. This can be explained due to the dominant effect of random A_{ex}. The magnetic simulation result [111] shows a better SNR on the media with a 9 nm grain size and no lateral A_{ex} compared to media with a 6 nm grain size and random A_{ex}. Thus, A_{ex} needs to be optimized in given heads while minimizing random A_{ex}.

3.2.4.3 Static and dynamic tilted magnetic switching media

To improve media writability and thermal stability as well as high sensitivity of easy-axis dispersion in conventional perpendicular media, static tilted magnetic switching media with the magnetic easy axis tilted at 45° from the perpendicular direction was proposed by Bertram and Gao [24] in 2002 for areal densities beyond 1 Tbit/in^2. This new concept utilizes that a normalized switching field reduces to 1/2 at 45° on the basis of the Stoner–Wohlfarth model with coherent magnetic switching [112]. The switching field is almost independent of angle variation when the external field is applied into 45° to the easy axis of the media. This can allow us to use smaller grains with higher K_u and to provide less sensitivity of SFD on easy-axis dispersion. The simulation result [24, 113] with a tapered-neck monopole head supports improvement of the media writability and SNR. Three different ways are tried to fabricate static tilted media: to make CoCrPt thin films with a tilted columnar structure and magnetic easy axis directly deposited by collimated oblique sputtering [114], to make a tilted magnetic easy axis induced by using seed layers deposited by oblique physical vapor deposition [115], and to make a composite grain of in-plane anisotropy material CoIr with negative K_u and perpendicular anisotropy material CoPt with positive K_u [116]. However, there is a difficulty in fabricating uniformly easy-axis-tilted media on a disk.

To solve the fabrication issue of static tilted media, the idea of ECC media was introduced by Victora and Shen [25] in 2005. ECC media consist of magnetically hard and soft regions within each grain and a nonmagnetic exchange layer between two regions, as shown in Fig. 3.18c. The unique role of an exchange BL moderates the interaction between hard and soft regions. An insertion of a thin BL between hard and soft regions significantly changes the angle of effective fields into the hard region. This can mimic static tilted switching media during the recording process only while holding perpendicular magnetization, called dynamic tilted switching media. It provides freedom to fabricate uniformly tilted media on a disk.

More realistic models [117, 118] of ECC media with full magnetostatic interactions are performed. In a hysteresis loop, the insertion of a BL shows significantly reduced H_s but little effect on H_n, which is dominated by the coherent rotation of magnetizations within the soft region. It confirms that effective field strength and its field gradient are significantly improved by the distinct angular dependence of the switching field of ECC media. The write field profile for ECC media is also narrower in the cross-track direction than that for conventional perpendicular media. This will provide a significant reduction in ATE. However, this ATE improvement is compensated with the inhomogeneity of ECC to enhance the thermal attempt frequency. This concept was tested experimentally by Wang et al. [119] in $[Co/PdSi]_{14}/PdSi/FeSiO$ as a hard/BL/soft ECC layer structure. ECC media with incoherent switching behavior caused by reducing J_{IL} show both significantly reduced H_c and good thermal stability in bulk magnetic properties. As a more practical ECC structure, CoCrPt–SiO_2 (high Pt)/Pt/CoCrPt–SiO_2 (low Pt) media demonstrated the possible benefits of ECC media in spin-stand level [120]: great enhancement of writability and thermal stability as well as comparable SNR over the best capping media. However, on-track improvement in the SNR and ROW is frequently achieved for easy-to-write media so it often doesn't improve areal densities due to TD loss [121]. Therefore, the media with higher-K_u hard layer are needed to get a narrower write width and better ATE in ECC media.

The effect of J_{IL} in ECC media with the layer structure of CoPtCrSi–O/BL/CoPtCrB on magnetic switching behavior and recording characteristics was studied [122]. J_{IL} is controlled by adjusting M_s and thickness of the BL. Coherent switching behavior is typically shown in the granular CoPtCrSi–O layer only while incoherent switching behavior is exhibited when either the BL with weaker J_{IL} or the capping layer with high lateral A_{ex} is exchange-coupled with the granular oxide layer. Insertion of BL thicknesses (t_{BL}) reduces H_s but the further increase in t_{BL} eventually creates the kink of polar Kerr loops showing two distinct switching stages, which indicates independently magnetic switching of CoPtCrSi–O hard and CoPtCrB capping layers. Figure 3.32 shows the normalized saturation field $H_s/H_s(\text{ox})$ as a function of t_{BL} with various M_s, where $H_s(\text{ox})$ means H_s at the granular oxide layer only. $H_s/H_s(\text{ox})$ decreases continuously with increasing t_{BL}. When t_{BL} is too thick, the switching torque from the capping layer is less delivered to the granular oxide layer, resulting in higher H_s. Critical BL thickness showing the minimum of H_s/H_s (ox), $t_{\text{BL critical}}$, depends on M_s in BL: thicker $t_{\text{BL critical}}$ at higher M_s in the BL. A thicker BL generates a more nonuniform switching reversal process between the granular oxide layer and the capping layer. When $t_{\text{BL}} \gg t_{\text{BL min.}}\ H_s$, domain wall motion in the capping layer is dominated. Both ROW and MWW are maximized near $t_{\text{BL min.}}\ H_s$, while both values are dramatically decreased at $t_{\text{BL}} \gg t_{\text{BL min.}}\ H_s$ due to the high switching field of the granular oxide.

Figure 3.33 shows magnetic force microscopy (MFM) patterns of the ECC media with different t_{BL}: (a) $t_{\text{BL}} = t_{\text{BL min.}}\ H_s$, (b) $t_{\text{BL}} >$

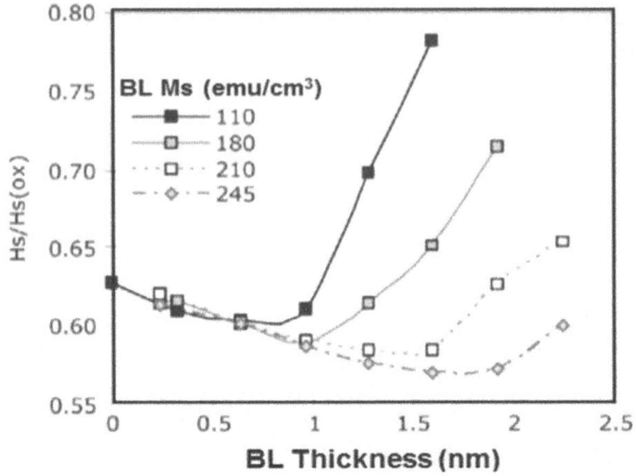

Figure 3.32 Normalized saturation field H_s/H_s(ox) at polar Kerr loops of the ECC media as a function of BL thickness with various M_s. Note H_s(ox) means H_s at the granular oxide layer only. © [2009] IEEE. Reprinted, with permission, from Ref. [122].

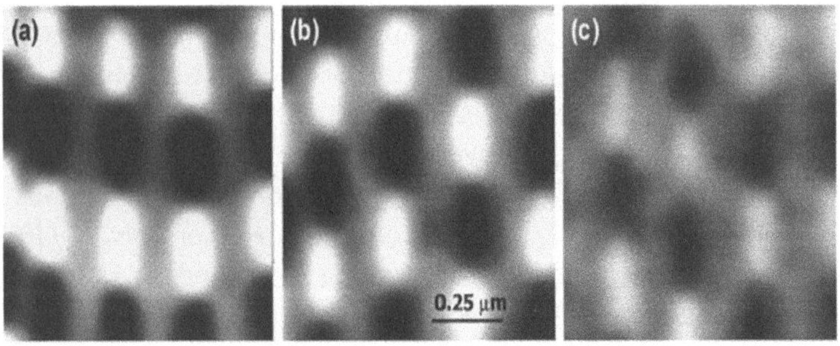

Figure 3.33 Magnetic force microscopy (MFM) patterns of the ECC media with different BL thickness (t_{BL}): (a) $t_{BL} = t_{BLmin.}$ H_s, (b) $t_{BL} > t_{BL\,min.}$ H_s, and (c) $t_{BL} \gg t_{BL\,min.}$ H_s. Note $t_{BL\,min.}$ H_s means critical BL thickness showing the minimum of H_s/H_s(ox). © [2009] IEEE. Reprinted, with permission, from Ref. [122].

$t_{BL\,min.}$ H_s, and (c) $t_{BL} \gg t_{BL\,min.}$ H_s. In Fig. 3.33c with very weak J_{IL}, significantly narrower track width and irregular bit shapes indicate that poor writability and high media noise are dominated by granular oxides and the capping layer with high lateral A_{ex}, respectively. ECC media with optimal J_{IL} improve both writability and ATE without losing the intrinsic SNR.

Another design approach for dynamic tilted media, ES media, was proposed by Suess et al. [26] in 2005 and extended to anisotropy-graded media (Fig. 3.18d) [27]. This ES concept in hard/soft stacked media is to use domain wall switching in the soft layer along the perpendicular direction. The soft layer is fully exchange-coupled with the hard layer and a magnetic domain wall is created next to the hard/soft interface. It acts as a magnetic spring that initiates the reversal of the hard layer, as observed in composite permanent magnets [123]. ES media need strong J_{IL} and domain wall switching, which are differentiated from ECC media with weak J_{IL} and incoherent switching. On the basis of simulation results [26, 124], it significantly reduces H_c by a factor of 6 but still shows thermal stability two times higher than single-oxide media. For a given easy-axis dispersion, ES media with less angular dependence of H_c can anticipate SNR gain by a factor of 2.5. This concept is extended from two layers to N layers with continuous variation of the anisotropy, called anisotropy graded media [27]. H_c decreases with $1/N$, while thermal stability is only controlled by the magnetic properties of the hardest layer. Although ES media show benefits similar to ECC media, there exists the difficulty in implementing ES media into real products. The limitations are strong J_{IL} and high K_u of the hard layer. A strong J_{IL} needs very good epitaxial growth between hard and soft regimes. However, it is not easy to make J_{IL} sufficient to provide domain wall switching along the perpendicular direction due to different alloy compositions and sputter conditions. A thicker soft layer than the domain wall width of the hard layer is needed so that total magnetic layer thickness is determined by K_u of the hard layer. K_u of the hardest layer should be too high to get reasonable soft-layer thickness. Current CoCrPt-based alloys cannot reach the K_u level for ES media. This limits material selection. Even at the difficulty of practical fabrication, this physical concept of ES media helps further improve media performance by combining with the concept of ECC media.

In practical PMR media with ECC and K_u gradient in multilayered oxides [125], magnetic switching behavior and writability are studied. The magnetic switching behavior strongly depends on magnetic volume of laterally exchange-coupled capping layer and J_{IN} between oxide and cap layers. The magnetic switching field is significantly reduced due to incoherent switching mode. Higher-K_u−graded oxide media further enhance writability and improve resolution as well as the SNR.

3.2.4.4 Magnetic switching block–controlled (or segmented) media

Improvement of intrinsic media writability and the SNR is essential in magnetic media design for continuous improvement of areal densities above 500 Gbits/in^2. New media design is getting a more complicated structure of multilayered granular oxide and BL layers by using mixed physical concepts of hard/soft stacked and dynamic tilted switching media. The HDD industry considers additional BLs into ECC media for improving intrinsic media writability under the narrow heads. Bertero et al. [29] presented experimental results of advanced double BL media in Fig. 3.18e and micromagnetic simulation results were reported later by Zhu and Wang [28]. To use dual BLs into the media structure means three different switching blocks of bottom, middle, and top layers, called switching block–controlled or segmented media. Switching blocks have an anisotropy gradient along the perpendicular direction: the bottom layer on top of the IL has the highest K_u and the middle and top layers have medium and low K_u, respectively. Basically, this structure provides more freedom of magnetic design but the switching mechanism is more complicated. J_{IL} between blocks is mostly controlled by both M_s and thickness of the BL. A weaker J_{IL} provides a lower switching field and less sensitivity to a high K_u in the bottom layer. It leads to significantly improved SFD and ROW but it degrades ATE due to lower H_n.

The latest realistic experimental results were reported by Choe and Ikeda [126]. This paper investigates switching behavior for granular oxide media with different lateral and vertical exchange-coupling strengths. A nonmagnetic or weak magnetic BL controls the J_{IL} between switching blocks: The BL1 layer between the bottom and middle magnetic oxide layers and the BL2 layer between the middle magnetic oxide layer and top metallic magnetic layer. The

H_s value in the Kerr loop correlates well with media writability. BL1 typically is more effective in reducing H_s than BL2. J_{IL} highly depends on M_s and thickness of the BL and thermal stability of the switching segment. When the lateral A_{ex} between oxide grains gets strong, the effect of the BL on switching behavior diminishes. An optimum media design is likely achievable by tuning the switching mode through this BL concept. The multiple BL design concept is extending from dual- to three-BL layers [127]. However, the MCW is getting wider in easy-to-write magnetic design. One of the big challenges is to maximize ECCness without losing TD. This magnetic design approach is promising to provide intrinsic gain of media writability and SNR.

3.3 Challenges of Conventional PMR Media

The aggressive push of TD is mainly driving areal densities of >700 Gbits/in^2, as shown in Fig. 3.1. A further reduced track width creates undesirable side effects. The decrease in physical dimensions of the writer pole and the reader width for a narrower track width significantly reduces both write field strength and amplitude. It also induces more interference between tracks due to high track misregistration caused by a higher repeatable run-out on servo tracks. Further improvement of ATE needs a narrower side-shielded gap and a deeper or fully wraparound shield, as shown in Fig. 3.7b,c. A higher field gradient is essential to improve LD but it is obtained by reducing the TS gap. The reduction of the side-shield gap and the TS gap significantly reduces write field strength. However, the FeCo alloy with the highest $4\pi M_s$ of 24 kG is being used as write head material. The enhancement of write field strength needs to be achieved by other ways such as modification of 3D write pole geometry, control of writer protrusion, or reduction of HMS. Either a higher rate of writer protrusion during the writing or improved clearance of the flying height boosts write field strength, but it brings more mechanical concerns of head burnish, head smear and scratch, operation shock, etc. Another concerning part is a wider head population of write width when the write head dimension is getting smaller. More variation of head writability and amplitude is anticipated. A typically weaker ROW penalty at narrower heads determines head usage of the wafer and drive yield. This fundamental shortage of writability in narrow heads should be overcome by media magnetic design. In a reader head, its sensitivity needs to be improved to detect a weaker signal from narrower tracks but reader sensor technology is also limited. New technology beyond the TMR reader sensor has not appeared yet. At the drive level, reducing HMS is the most effective way to enhance both ROW and the signal, but it will have the high risk of mechanical and reliability issues. HMS can be reduced by a thinner COC, thinner lube, and more clearance by reducing interfacial roughness. Media have significant fast turnaround time from thin-film deposition to measurement compared to heads. Therefore, the role of media to achieve higher areal densities is getting more important. Ideally easier-to-write media with better ATE, higher amplitude, and narrower MCW are needed. These requirements are controversial in magnetic design and add more complexity in the magnetic design concept by considering mechanical and reliability concerns. Both head/media match and control of tiny clearance are critical for achieving higher areal densities.

Upcoming technology is shingled magnetic recording (SMR) [8] for >1 Tbit/in^2, considered as the extension of current PMR technology. Shingled write tracks sequentially overlapped in one direction can significantly increase the areal density margin by $\sim$30% with the current head and media technology [128]. This SMR provides many benefits of higher writability, less skew sensitivity, less restriction of writer pole and its width tolerance, and less concern of ATE and STE, but HDD performance is slowed down. New drive firmware for maintaining the data rate and emulating random access is more challenging than ADC achievement. Degradation of the SNR in SMR is caused by higher lateral A_{ex}, higher head skew, and narrower track pitch [129]. TD capability highly depends on a read width around the 500 kTPI region. Signal loss and more side reading at a narrower track pitch can be emphasized. Kagami et al. [130] successfully demonstrated areal densities of 1.26 Tbits/in^2 in 2011 by using an optimized fully wraparound shielded writer with an extended write gap and an improved reader with less side reading. Tighter control of track misregistration, less interference between tracks, and a new reader design concept will help further improve ADC in the near future. SMR media with smaller grain templates for reducing magnetic cluster size and its distribution are continuously developing with advanced magnetic design concepts.

References

1. Iwasaki S., and Takemura K. (1975). An analysis for the circular mode of magnetization in short wavelength recording, *IEEE Trans. Magn.*, **11**, pp. 1173–1175.
2. Toshiba 40GB/1.8″ platter, http://storage.toshiba.com/techdocs/PressRelease57-40GB1.8-InchHDDCaterstoMobileStorageDemands.pdf.
3. Charap, S. H., Lu, P. L., and He Y. (1997). Thermal stability of recorded information at high densities, *IEEE Trans. Magn.*, **33**, pp. 978–983.
4. Fullerton E. E., Margulies D. T., Schabes M. E., Carey M., Gurney B., Moser A., Best M., Zeltzer G., Rubin K., Rosen H., and Doerner M. (2000). Antiferromagnetically coupled magnetic media layers for thermally stable high-density recording, *Appl. Phys. Lett.*, **77**, pp. 3806–3808.
5. Wood R. (2000). The feasibility of magnetic recording at Terabit per square inch, *IEEE Trans. Magn.*, **36**, pp. 36–42.
6. Mallary M., Torabi A., and Benakli M. (2002). One terabit per square inch perpendicular recording conceptual design, *IEEE Trans. Magn.*, **38**, pp. 1719–1724.
7. Wang T., Mehta V., Ikeda Y., Do H., Takano K., Florez S., Terris B. D., Wu B., Graves C., Shu M., Rick R., Scherz A., Stohr J., and Hellwig O. (2013). Magnetic design evolution in perpendicular magnetic recording media as revealed by resonant small angle X-ray scattering, *Appl. Phys. Lett.*, **103**, p. 112403.
8. Wood R., Williams M., Kavic A., and Miles J. (2009). The feasibility of magnetic recording at 10 terabits per square inch on conventional media, *IEEE Trans. Magn.*, **45**, pp. 917–923.
9. Weller D., Moser A., Folks L., Best M. E., Lee W., Toney M. F., Schwickert M., Thiele J.-U., and Doerner M. F. (2000). High K_u materials approach to 100 Gbits/in^2, *IEEE Trans. Magn.*, **36**, pp. 10–15.
10. Iwasaki S., and Nakamura Y. (1977). An analysis of the magnetization mode for high density magnetic recording, *IEEE Trans. Magn.*, **13**, pp. 1272–1277.

11. Cain W., Payne A., Baldwinson M., and Hempstead R. (1996). Challenges in the practical implementation of perpendicular recording media, *IEEE Trans. Magn.*, **32**, pp. 97–102.
12. Takeo A., Oikawa S., Hikosaka T., and Tanaka Y. (2000). A new design to suppress recording demagnetization for perpendicular recording, *IEEE Trans. Magn.*, **36**, pp. 2378–2380.
13. Oikawa S., Tanaka A., Hikosaka T., and Tanaka Y. (2000). High performance CoCrPtO single layered perpendicular media with no recording demagnetization, *IEEE Trans. Magn.*, **36**, pp. 2393–2395.
14. Mallary M. (1987). *Vertical Magnetic Recording Arrangement*, US Patent 4656546.
15. Butler W. H., Zhang X.-G., Schulthess T. C., and MacLaren J. M. (2001). Spin-dependent tunneling conductance of Fe/MgO/Fe sandwiches, *Phys. Rev. B*, **63**, p. 054416.
16. Mathon J., and Umerski A. (2001). Theory of tunneling magnetoresistance of an epitaxial Fe/MgO/Fe (001) junction, *Phys. Rev. B*, **63**, p. 220403.
17. Bowen M., Cros, V., Petroff F., Fert, A., Martınez Boubeta C., Costa-Krämer J. L., Anguita J. V., Cebollada A., Briones F., de Teresa J. M., Morellón L., Ibarra, M. R., Güell F., Peiró F., and Cornet A. (2001). Large magnetoresistance in Fe/MgO/FeCo(001) epitaxial tunnel junctions on GaAs(001), *Appl. Phys. Lett.*, **79**, pp. 1655–1657.
18. Iwasaki S., and Ouchi K. (1978). Co-Cr recording films with perpendicular magnetic anisotropy, *IEEE Trans. Magn.*, **14**, pp. 849–851.
19. Chang C.-R., and Fredkin D. R. (1987). The shearing of perpendicular loop with columnar structure, *IEEE Trans. Magn.*, **23**, pp. 2052–2054.
20. Ikeda Y., Sonobe Y., Uchida H., and Toyooka T. (1996). Noise and $\Delta M_{\perp}$ analysis of a Cr-rich Co-Cr perpendicular medium for an MR head, *IEEE Trans. Magn.*, **32**, pp. 3828–3830.
21. Oikawa T., Nakamura M., Uwazumi H., Shimatsu T., Muraoka H., and Nakamura Y. (2002). Microstructure and magnetic properties of CoPtCr-SiO_2 perpendicular recording media, *IEEE Trans. Magn.*, **38**, pp. 1976–1978.
22. Sonobe Y., Weller D., Ikeda Y., Takano K., Schabes M. E., Zeltzer G., Yen B. K., and Best M. E. (2001). Coupled granular/continuous medium for thermally stable perpendicular magnetic recording, *J. Magn. Magn. Mater.*, **235**, pp. 424–428.
23. Bertero G. A., Wachenschwanz D., Malhotra S., Velu S., Bian B., Stafford D., Wu Y., Yamashita T., and Wang S. X. (2002). Optimization of granular

double-layer perpendicular media, *IEEE Trans. Magn.*, **38**, pp. 1627–1631.

24. Bertram H. N., and Gao K. Z. (2002). Magnetic recording configuration for densities beyond 1 Tb/in^2 and data rates beyond 1 Gb/s, *IEEE Trans. Magn.*, **38**, pp. 3675–3683.
25. Victora R. H., and Shen X. (2005). Composite media for perpendicular magnetic recording, *IEEE Trans. Magn.*, **41**, pp. 537–542.
26. Suess D., Schrefl T., Fahler S., Kirschner M., Hrkac G., Dorfbauer F., and Fiedler J. (2005). Exchange spring media for perpendicular recording, *Appl. Phys. Lett.*, **87**, p. 012504.
27. Suess D. (2006). Multilayer exchange spring media for magnetic recording, *Appl. Phys. Lett.*, **89**, p. 113105.
28. Zhu J., and Wang Y. (2011). SNR enhancement in segmented perpendicular media, *IEEE Trans. Magn.*, **47**, pp. 4066–4072.
29. Bertero G., Acharya R., Malhotra S., Srinivasan K., Champion E., Lauhoff G., and Desai M. (2010). Double exchange-break PMR media structures, in *Digest of the 21st Magnetic Recording Conference (TMRC 2010)*, p. A2.
30. Litvinov D., Kryder M. H., and Khizroev S. (2001). Recording physics of perpendicular media: soft underlayers, *J. Magn. Magn. Mater.*, **232**, pp. 84–90.
31. Muraoka H., Sato K., Sugita Y., and Nakamura Y. (1999). Low inductance and high efficiency single-pole writing head for perpendicular double layer recording media, *IEEE Trans. Magn.*, **35**, pp. 643–648.
32. Schabes M. E., Lengsfield B., and Schrefl T. (2002). Micromagnetic modeling of soft underlayer magnetization processes and fields in perpendicular magnetic recording, *IEEE Trans. Magn.*, **38**, pp. 1670–1675.
33. Jung H. S., and Doyle W. D. (2001). FeTaN/IrMn exchange-coupled multilayers as soft underlayers for perpendicular media, *IEEE Trans. Magn.*, **37**, pp. 2294–2297.
34. Jung H. S., Velu E. M. T., Avenell M. R., Malhotra S. S., and Bertero G. (2006). Magnetic domain-free hard magnet-biased soft magnetic underlayers for perpendicular media, *J. Appl. Phys.*, **99**, p. 08Q901.
35. Shukh A. M., Singleton E. W., Khizroev S., and Litvinov D. (2002). *Perpendicular Recording Medium with Antiferromagnetic Exchange Coupling in Soft Magnetic Underlayer*, US Patent US2002/0028357 A1.
36. Byeon S. C., Misra A., and Doyle W. D. (2004). Synthetic antiferrormagnetic soft underlayers for perpendicular recording media, *IEEE Trans. Magn.*, **40**, pp. 2386–2388.

37. Acharya B. R., Zhou J. N., Zheng M., Choe G., Abarra E. N., and Johnson K. E. (2004). Anti-parallel coupled soft under layers for high-density perpendicular recording,*IEEE Trans. Magn.*, **40**, pp. 2383–2385.
38. Sung S. H., and Tan A. H. (2011). Investigation of substrate surface morphology effects on media performance, *IEEE Trans. Magn.*, **47**, pp. 528–531.
39. Xiao M., Wilson B., Takano K., Do H., Ikeda Y., and Rosen H. (2005). Soft underlayer noise in perpendicular recording and its impact on error rate, *IEEE Trans. Magn.*, **41**, pp. 3154–3156.
40. Benakli M., Mallary M. L., Marshall S. B., and Torabi A. F. (2005). Write performance of shielded pole heads on media with thin soft underlayers, *J. Appl. Phys.*, **97**, p. 10P103.
41. Bai D. Z., Luo P., Torabi A., Terrill D., Wang J., Stoev K., Liu F., Moneck M., Tang Y., and Zhu, J.-G. (2007). Return field-induced partial erasure in perpendicular recording using trailing-edge shielded writers, *IEEE Trans. Magn.*, **43**, pp. 600–604.
42. Kief M. T., Inturi V., Benakli M., Tabakovic I., Sun M., Heinonen O., Riemer S., and Vas'ko V. (2008). High magnetic saturation poles for advanced perpendicular writers, *IEEE Trans. Magn.*, **44**, pp. 113–118.
43. Sato K., Wakamatsu H., and Shinohara M. (1993). Noise characteristics of CoCrTa/NiFe perpendicular recording media, *IEEE Trans. Magn.*, **29**, pp. 3730–3732.
44. Saito N., Fujiwara H., and Sugita Y. (1964). A new type of magnetic domain structure in negative magnetostriction Ni-Fe films, *J. Phys. Soc. Jpn.*, **19**, pp. 1116–1125.
45. Murayama Y. (1966). Micromagnetics on stripe domain films, *J. Phys. Soc. Jpn.*, **21**, pp. 2253–2265.
46. Zou P., Yu W., and Bain J. A. (2002). Effect of stress on stripe domain onset in sputtered FeAlN and CoFe films, *J. Appl. Phys.*, **91**, pp. 7830–7832.
47. Numazawa J., Yoneda Y., and Yokoyama K. (1986). Underlayer induced noise and carrier to noise ratio in a double-layer medium, *J. Magn. Magn. Mater.*, **54**, pp. 1588–1590.
48. Chikazumi S., and Oomura T. (1955). On the origin of magnetic anisotropy induced by magnetic annealing, *J. Phys. Soc. Jpn.*, **10**, pp. 842–849.
49. Litvinov D., Lyberatos A., Wolfson J., Bain J., and Khizroev S. (2002). Recording layer influence on the dynamics of a soft underlayer, *IEEE Trans. Magn.*, **38**, pp. 1994–1996.

50. Ando T., and Nishihara T. (1997). Triple-layer perpendicular recording media for high SN ratio and signal stability, *IEEE Trans. Magn.*, **33**, pp. 2983–2985.

51. Takahashi S., Yamakawa K., and Ouchi K. (1999). A design of soft magnetic backlayer for double-layered perpendicular magnetic recording medium, *J. Mag. Soc. Jpn.*, **23**(S2), pp. 63–66.

52. Jung H. S., Velu E. M. T., Malhotra S. S., Jiang W., and Bertero G. (2006). Recording performance of CoCrPtO perpendicular media with single-domain hard magnet-biased soft magnetic underlayers, *J. Appl. Phys.*, **99**, p. 08E702.

53. Grünberg P., Schrieber R., Pang Y., Brodsky M., and Sowers H. (1986). Layered magnetic structures: evidence of antiferromagnetic coupling of Fe layers across Cr interlayers, *Phys. Rev. Lett*, **57**, pp. 2442–2445.

54. Jiang W., Velu E. M. T., Malhotra S., Jung H. S., Kwok C. K., Bertero G., and Wachenschwanz D. (2005). Recording performance characteristics of granular perpendicular media, *IEEE Trans. Magn.*, **41**, pp. 587–592.

55. Tanahashi K., Arai R., and Hosoe Y. (2005). Exchange-biased soft underlayers for perpendicular recording, *IEEE Trans. Magn.*, **41**, pp. 577–580.

56. Tahk Y. W., Hong S. Y., Hong D.-H., Lee H. J., and Lee T. D. (2004). Transition noise characteristics of PMR media with the synthetic antiferromagnetic coupled soft underlayer, *IEEE Trans. Magn.*, **40**, pp. 2564–2566.

57. Berger A., Marinero E., Doerner M., Bian X., Tang K., and Polcyn A. (2005). New technique to determine the magnetic easy axis orientation in recording media, *J. Appl. Phys.*, **97**, p. 10N115.

58. Wu X. W., Zhou H., van de Veerdonk R. J. M., Klemmer T. J., and Weller D. (2004). Measurement of the magnetic easy axis dispersion in perpendicular media, *J. Magn. Magn. Mater.*, **278**, pp. 285–288.

59. Voznuk V. G., Misra A., Doyle W. D., Visscher P. B., and Chantrell R. W. (2005). Angular dispersion of the anisotropy in perpendicular media, *J. Appl. Phys.*, **97**, p. 10N508.

60. Supper N., Margulies D. T., Olson T., Ikeda Y., Moser A., Lengsfield B., and Berger A. (2006). Interlayer thickness, transition parameter, and intergranular exchange in perpendicular oxide media, *IEEE Trans. Magn.*, **42**, pp. 2333–2335.

61. Jung H. S., Kuo M., Malhotra S. S., Bertero G., and Torabi A. F. (2008). Reduction of transition layer thickness showing incoherent switching

behavior in single CoCrPt-SiO perpendicular media, *IEEE Trans. Magn.*, **44**, pp. 3484–3487.

62. Yonemura S., Saito S., Hashimoto A., and Takahashi M. (2008). Topological bumped surface induced by the lattice extension of the RuCr intermediate layer for granular-type perpendicular recording media, *J. Magn. Magn. Mater.*, **320**, pp. 3053–3056.
63. Gao K.-Z., Sapozhnikov V., Zheng A., and Heinonen O. (2008). Effect of interlayer on read write processes in perpendicular recording, *IEEE Trans. Magn.*, **44**, pp. 3400–3403.
64. Jung H. S., Choe G., Zhang K., Ghaderi A., Olson T., and Lengsfield B. (2010). Comparison of experiment and simulation results of interlayer thickness effect in perpendicular recording media, *IEEE Trans. Magn.*, **46**, pp. 1825–1828.
65. Zhang K., Jung H. S., and Choe G. (2010). An experimental study of effects of interlayer thickness on recording performance under various writing and reading clearances, *IEEE Trans. Magn.*, **46**, pp. 1722–1725.
66. Jung H. S., Kuo M., Malhotra S. S., and Bertero G. (2007). Origin of incoherent magnetic switching behavior in CoCrPt–SiO_2 perpendicular magnetic recording media, *Appl. Phys. Lett.*, **91**, p. 212502.
67. Rauch G. C., Byun C., Johns E. R. C., Messinger C., and Greggi, J., Jr. (1992). Effect of cluster size on media noise in Co-Ni-P thin films, *IEEE Trans. Magn.*, **28**, pp. 3105–3107.
68. Mukai R., Uzumaki T., and Tanaka A. (2005). Microstructure improvement of thin Ru underlayer for CoCrPt–SiO granular perpendicular media, *IEEE Trans. Magn.*, **41**, pp. 3169–3171.
69. Park S. H., Kim S. O., Lee T. D., Oh H. S., Kim Y. S., Park N. Y., and Hong D. H. (2006). Effect of top Ru deposition pressure on magnetic and microstructural properties of CoCrPt–SiO media in two-step Ru layer, *J. Appl. Phys.*, **99**, p. 08E701.
70. Wierman K. W., Klemmer T. J., Lu B., Ju G., Howard K. J., Roy A. G., and Laughlin D. E. (2002). Ru Cr/Ta underlayer for Co-alloy perpendicular magnetic recording, *J. Appl. Phys.*, **91**, pp. 8031–8033.
71. Kwon U., Sinclair R., Velu E. M. T., Malhotra S., and Bertero G. (2005). Ru/Ru-oxide interlayer for CoCrPtO perpendicular recording media, *IEEE Trans. Magn.*, **41**, pp. 3193–3195.
72. Yuan H., Laughlin D. E., Zhu X., and Lu B. (2008). Ru + oxide interlayer for perpendicular magnetic recording layer, *J. Appl. Phys.*, **103**, p. 07F513.

73. Piramanayagam S. N., and Srinivasan K. (2007). Sub-6-nm grain size control in polycrystalline thin films using synthetic nucleation layer, *Appl. Phys. Lett.*, **91**, p. 142508.

74. Hashimoto A., Saito S., Itagaki N., and Takahashi M. (2006). Improvement of magnetic properties of granular perpendicular recording media by using a fcc nonmagnetic intermediate layer with stacking faults, *Appl. Phys. Lett.*, **89**, p. 262508.

75. Nemoto H., Araki R., and Hosoe Y. (2006). Pt-Cr alloy intermediate layer for CoCrPt-SiO_2 granular perpendicular recording media, *IEEE Trans. Magn.*, **42**, pp. 2336–2338.

76. Thornton J. A. (1974). Influence of apparatus geometry and deposition conditions on the structure topography of thick sputtered coatings, *J. Vac. Sci. Technol.*, **11**, pp. 666–670.

77. Yuan H., and Laughlin D. E. (2008). Microstructure and magnetic properties of perpendicular media with reduced grain size, *Appl. Phys. Lett.*, **93**, p. 102511.

78. Yuan H., and Laughlin D. E. (2009). Stacking faults in smaller grain size perpendicular media, *IEEE Trans. Magn.*, **45**, pp. 3569–3571.

79. Jung H. S., Ikeda Y., Choe G., and Shi Z. (2012). Recording performance and thermal stability in perpendicular media with enhancement of grain isolation as well as magnetic anisotropy field, *J. Appl. Phys.*, **111**, p. 07B705.

80. Jung H. S., Stafford D., Acharya B. R., Malhotra S. S., and Bertero G. A. (2013). *Perpendicular Magnetic Recording Medium with Grain Isolation and Magnetic Anisotropy Layer*, US Patent 8488276 B1.

81. Lee I. S., Ryu H., Lee H. J., and Lee T. D. (1999). Role of a paramagnetic amorphous CoZr seed layer in CoCrPt/Ti perpendicular recording media, *J. Appl. Phys.*, **85**, pp. 6133–6135.

82. Zheng M., Choe G., Johnson K. E., Gao L., and Liou, S-H. (2002). Seedlayer and preheating effects on crystallography and recording performance of CoCrPtB perpendicular media, *IEEE Trans. Magn.*, **38**, pp. 1979–1981.

83. Creamer Media. (2007). *Ruthenium Price Soaring to Great Heights, Say Marketers,* http://www.miningweekly.com/article/ruthenium-price-soaring-to-great-heights-say-marketers-2007-03-02.

84. Srinivasan K., Piramanayagam S. N., and Sbiaa R. (2008). Antiferromagnetic iridium manganese based intermediate layers for perpendicular magnetic recording media, *Appl. Phys. Lett.*, **93**, p. 072503.

85. Liao J.-W., Dumas R. K., Hou H.-C., Huang Y.-C., Tsai W.-C., Wang L.-W., Wang D.-S., Lin M.-S., Wu Y.-C., Chen R.-Z., Chiu C.-H., Lau J. W., Liu K., and Lai C.-H. (2010). Simultaneous enhancement of anisotropy and grain isolation in CoPtCr-SiO_2 perpendicular recording media by a MnRu intermediate layer, *Phys. Rev. B,* **82**, p. 014423.

86. Kottler V., Gillies M. F., and Kuiper A. E. T. (2001). An in situ X-ray photoelectron spectroscopy study of AlOx spin tunnel barrier formation, *J. Appl. Phys.*, **89**, pp. 3301–3306.

87. Shen F., Xu Q. Y., Yu G. H., Lai W. Y., Zhang Z., Lu Z. Q., Pan G., and Al-Jibouri A. (2002). A specular spin valve with discontinuous nano-oxide layers, *Appl. Phys. Lett.*, **80**, pp. 4410–4412.

88. Choe G., Minardi M., Zhang K., and Mirzamaani M. (2008). Head-to-SUL spacing reduction with a magnetic seed layer and the effect on perpendicular recording characteristics, *IEEE Trans. Magn.*, **44**, pp. 3480–3483.

89. Ikeda Y., Supper N., Do H., Takano K., and Lengsfield B. (2003). Exchange coupling optimization on CoPtCrO perpendicular media, *IEEE Trans. Magn.*, **39**, pp. 2341–2343.

90. Zheng M., Acharya B. R., Choe G., Zhou J. N., Yang Z. D., Abarra E. N., and Johnson K. E. (2004). Role of oxygen incorporation in Co-Cr-Pt-Si-O perpendicular magnetic recording media, *IEEE Trans. Magn.*, **40**, pp. 2498–2500.

91. Yamane H., Watanabe S., Ariake J., Honda N., Ouchi K., and Iwasaki S. (2005). Effect of oxygen on magnetic properties of Co-Pt-Cr-SiO thin films for perpendicular recording media, *J. Magn. Magn. Mater.*, **287**, pp. 153–158.

92. Shimatsu T., Sato H., Oikawa T., Inaba Y., Kitakami O., Okamoto S., Aoi H., Muraoka H., and Nakamura Y. (2005). High-potential magnetic anisotropy of CoCrPt-SiO perpendicular recording media, *IEEE Trans. Magn.*, **41**, pp. 566–571.

93. Kwon U., Jung H. S., Kuo M., Velu E. M. T., Malhotra S. S., Jiang W., Bertero G., and Sinclair R. (2006). Effect of magnetic recording layer thickness on media performance in CoCrPt-oxide perpendicular media, *IEEE Trans. Magn.*, **42**, pp. 2330–2332.

94. Jung H. S., Kwon U., Kuo M., Velu E. M. T., Malhotra S. S., Jiang W., and Bertero G. (2007). Effect of oxygen incorporation on media performance and microstructure in CoCrPt-SiO_2 perpendicular magnetic recording media, *IEEE Trans. Magn.*, **43**, pp. 615–620.

95. Jung H. S., Kuo M., Malhotra S. S., and Bertero G. (2008). Effect of non-uniform microstructure on magnetic switching behavior in CoCrPt-SiO_2 perpendicular recording media, *J. Appl. Phys.*, **103**, p. 07F515.
96. Laughlin D. E., Nuhfer N. T., Park S., Yuan H., and Zhu, J.-G. (2009). Topology and elemental distribution in Co alloy: oxide perpendicular media, *J. Appl. Phys.*, **105**, p. 07B739.
97. El-Hilo M., O'Grady K., and Chantrell R. W. (2002). Fluctuation fields and reversal mechanisms in granular magnetic systems, *J. Magn. Magn. Mater.*, **248**, pp. 360–373.
98. Greaves S., and Muraoka H. (2005). Fluctuation fields in perpendicular media, *IEEE Trans. Magn.*, **41**, pp. 3124–3126.
99. Thayer G. E., Ozolins V., Schmid A. K., Bartelt N. C., Asta M., Hoyt J. J., Chiang S., and Hwang R. Q. (2001). Role of stress in thin film alloy thermodynamics: competition between alloying and dislocation formation, *Phys. Rev. Lett.*, **86**, pp. 660–663.
100. Sonobe Y., Muraoka H., Miura K., Nakamura Y., Takano K., Moser A., Do H., Yen B. K., Ikeda Y., Supper N., and Weresin W. (2002). Thermally stable CGC perpendicular recording media with Pt-Rich CoPtCr and thin Pt layers, *IEEE Trans. Magn.*, **38**, pp. 2006–2011.
101. Shimatsu T., Muraoka H., Nakamura Y., and Sonobe Y. (2002). High thermal stability in CoPrTb/CoCrTa composite perpendicular media, *J. Appl. Phys.*, **91**, pp. 8061–8063.
102. Sonobe Y., Tham K. K., Wu L., Umezawa T., Takasu C., Dumaya J. A. H., Onoue T., Leo P., and Liau M. (2006). CGC perpendicular recording media with CoCrPt-SiO_2 alloy as granular layer, *IEEE Trans. Magn.*, **42**, pp. 2351–2353.
103. Choe G., Zheng M., Acharya B. R., Abarra, E. N., and Zhou J. N. (2005). Perpendicular recording CoPtCrO composite media with performance enhancement capping layer, *IEEE Trans. Magn.*, **41**, pp. 3172–3174.
104. Acharya B. R., Zheng M., Choe G., Yu M., Gill P., and Abarra E. N. (2005). Anisotropy enhanced dual magnetic layer media design for high-density perpendicular recording, *IEEE Trans. Magn.*, **41**, pp. 3145–3147.
105. Jung H. S., Velu E. M. T., Malhotra S. S., Bertero G., and Kwon U. (2008). Comparison of media properties between hard/soft stacked composite and capping layer perpendicular recording media, *J. Magn. Magn. Mater.*, **320**, pp. 3151–3156.

106. Wu Y., and Liu H. K. (2009). Quantifying corrosion growths on perpendicular magnetic recording media, *IEEE Trans. Magn.*, **45**, pp. 3012–3016.
107. Muraoka H., Sonobe Y., Miura K., Goodman A. M., and Nakamura Y. (2002). Analysis on magnetization transition of CGC perpendicular media, *IEEE Trans. Magn.*, **38**, pp. 1632–1636.
108. Jung H. S., Velu E. M. T., Malhotra S. S., Kwon U., Suess D., and Bertero G. (2007). CoCrPtO-based granular composite perpendicular recording media, *IEEE Trans. Magn.*, **43**, pp. 2088–2090.
109. Chen D. X., Brug J. A., and Goldfarb R. B. (1991). Demagnetizing factors for cylinders, *IEEE Trans. Magn.*, **27**, pp. 3601–3619.
110. Greaves S. J., Goodman A. M., Muraoka H., and Nakamura Y. (2005). Exchange coupling and grain-size distributions in perpendicular recording media, *J. Magn. Magn. Mater.*, **287**, pp. 66–71.
111. Wang Y., and Zhu, J.-G. (2010). Understanding noise mechanism in small grain size perpendicular thin film media, *IEEE Trans. Magn.*, **46**, pp. 2391–2393.
112. Stoner E. C., and Wohlfarth E. P. (1948). A mechanism of magnetic hysteresis in heterogeneous alloys, *Trans. R. Soc. London, A*, **240**, pp. 599–642.
113. Wang J. P., Zou Y. Y., Hee C. H., Chong T. C., and Zheng Y. F. (2003). Approaches to tilted magnetic recording for extremely high areal density, *IEEE Trans. Magn.*, **39**, pp. 1930–1935.
114. Zheng Y. F., Wang J. P., and Ng V. (2002). Control of the tilted orientation of CoCrPt/Ti thin film media by collimated sputtering, *J. Appl. Phys.*, **91**, pp. 8007–8009.
115. Klemmer T. J., and Pelhos K. (2005). Seed layer control for tilted magnetic recording media, *Appl. Phys. Lett.*, **88**, p. 162507.
116. Park S., Zhu J.-G., and Laughlin D. E. (2009). A novel crystalline soft magnetic intermediate layer for perpendicular recording media, *J. Appl. Phys.*, **105**, p. 07B723.
117. Victora R. H., and Shen X. (2005). Exchange coupled composite media for perpendicular magnetic recording, *IEEE Trans. Magn.*, **41**, pp. 2828–2833.
118. Shen X., Kapoor M., Field R., and Victora R. H. (2007). Issues in recording exchange coupled composite media, *IEEE Trans. Magn.*, **43**, pp. 676–681.
119. Wang J., Shen W. K., Bai J. M., Victora R. H., Judy J. H., and Song W. L. (2005). Composite media (dynamic tilted media) for magnetic recording, *Appl. Phys. Lett.*, **86**, p. 142504.

120. Wang J. P., Shen W., and Hong, S.Y. (2007). Fabrication and characterization of exchange coupled composite media, *IEEE Trans. Magn.* **43**, pp. 682–686.

121. Berger A., Supper N., Ikeda Y., Lengsfield B., Moser A., and Fullerton E. E. (2008). Improved media performance in optimally coupled exchange spring layer media, *Appl. Phys. Lett.*, **93**, p. 122502.

122. Choe, G, Ikeda Y., Zhang K., Tang K., and Mirzamaani M. (2009). Control of exchange coupling between granular oxide and highly exchange coupled cap layers and the effect on perpendicular magnetic switching and recording characteristics, *IEEE Trans. Magn.*, **45**, pp. 2694–2700.

123. Kneller E. F., and Hawig R. (1991). The exchange-spring magnet: a new material principle for permanent magnets, *IEEE Trans. Magn.*, **27**, pp. 3588–3600.

124. Suess D., Schrefl T., Kirschner M., Hrkac G., Dorfbauer F., Ertl O., and Fidler J. (2005). Optimization of exchange spring perpendicular recording media, *IEEE Trans. Magn.*, **41**, pp. 3166–3168.

125. Choe G., Park J., Ikeda Y., Lengsfield B., Olson T., Zhang K., Florez S., and Ghaderi A. (2011). Writeability enhancement in perpendicular magnetic multilayered oxide media for high areal density recording, *IEEE Trans. Magn.*, **47**, pp. 55–62.

126. Choe G., and Ikeda Y. (2014). Magnetic switching behavior of granular perpendicular magnetic media with different lateral and vertical exchange coupling, *IEEE Trans. Magn.*, **50**, p. 3202604.

127. Lengsfield B., Olson T., Park J., and Torabi A. F. (2014). A general energy barrier model for switching of exchange spring media in an external field, *IEEE Trans. Magn.*, **50**, p. 3200506.

128. Guarisco D., Xing X., Moser A., Stoev K., Gider S., and Mauri D. (2011). Magnetic recording at extreme track densities, in *IEEE International Magnetics Conference (INTERMAG 2011)*, p. FA-02.

129. Greaves S., Kanai Y., and Muraoka H. (2009). Shingled recording for 2–3 Tbit/in^2, *IEEE Trans. Magn.*, **45**, pp. 3823–3829.

130. Kagami T., Kiyono H., Nakada O., Mori T., Hirata K., Anagawa K., Nishio N., Uesugi T., Machita T., Degawa N., and Oike T. (2011). An areal-density capability study of SMR by using improved write and read heads, in *IEEE International Magnetics Conference (INTERMAG 2011)*, p. FA-01.

Chapter 4

Energy-Assisted Magnetic Recording

Edward Gage,[a] Kai-Zhong Gao,[a] and Jian-Gang (Jimmy) Zhu[b]

[a] *Seagate Technology LLC, 1280 Disk Drive, Shakopee, MN 55379, USA*
[b] *Data Storage Systems Center, Carnegie Mellon University, Pittsburgh, PA 15213-3890, USA*
edward.gage@seagate.com

4.1 Introduction

Energy-assisted magnetic recording (EAMR) is a promising technology to extend the impressive areal density growth of rotating magnetic data storage. Two energy-assisted recording technologies are active areas of research: Microwave-assisted magnetic recording (MAMR) uses a high-frequency oscillating magnetic field to assist the recording process, while heat-assisted magnetic recording (HAMR) or thermally assisted magnetic recording (TAMR) uses thermal energy to modify the magnetic recording process [1, 2].

From 1955 to 2015, many developments have been responsible for the capacity growth of hard disk drives (HDDs). Over the past two decades, the areal data storage density in an HDD has increased by almost three orders of magnitude, from 1 Gbits/in^2 to nearly 1 Tbit/in^2. The scaling of three primary factors, grain

Ultrahigh-Density Magnetic Recording: Storage Materials and Media Designs
Edited by Gaspare Varvaro and Francesca Casoli

ISBN 978-981-4669-58-0 (Hardcover), 978-981-4669-59-7 (eBook)
www.panstanford.com

size in the recording media, the write field, and the effective down-track and cross-track write field gradient, have continued to provide the ability to increase the storage areal density for a given recording system signal-to-noise ratio (SNR) requirement [3]. In addition, significant progress in advanced channel and coding has enabled a reduced SNR requirement for a given reliable HDD system bit error rate (BER) criteria. With current recording media materials, superparamagnetic effects prevent further reduction in media grain size [4]. To date, lack of a higher magnetic moment head field material and the poor magnetic field confinement as we scale down the write pole dimensions prevent the improvement in the write field gradients [3]. As result, the medium grain size cannot be further reduced without weakening thermal stability, and the areal density growth in HDDs has slowed to less than 20% per year in 2014. There is an urgent need for a technology breakthrough to renew the areal density growth that has propelled hard drives as the world's dominant data storage technology. Energy-assisted recording modifies the media's magnetic properties or magnetization switching dynamics to allow recording on novel media designs with smaller grain size and larger down-track and cross-track effective write field gradients.

Figure 4.1 shows the 2014 technology roadmap from the Advanced Storage Technology Consortium (ASTC). The expectation from the ASTC is that HAMR technology will likely appear in HDD products after 2018 and allow areal density growth up to 4 $Tbits/in^2$ [5, 6]. As we will discuss further, HAMR still has a number of reliability and manufacturability challenges. Although not shown on the roadmap, MAMR is also an active area of research, targeting products even earlier than HAMR.

4.2 Microwave-Assisted Magnetic Recording

MAMR provides an additional alternative to conventional perpendicular recording for significantly extending the areal density capability (ADC) of HDDs [1, 7, 8]. The technology employs a spin torque oscillator (STO), embedded inside the write head gap, for generating an alternating magnetic field in the microwave-frequency regime in

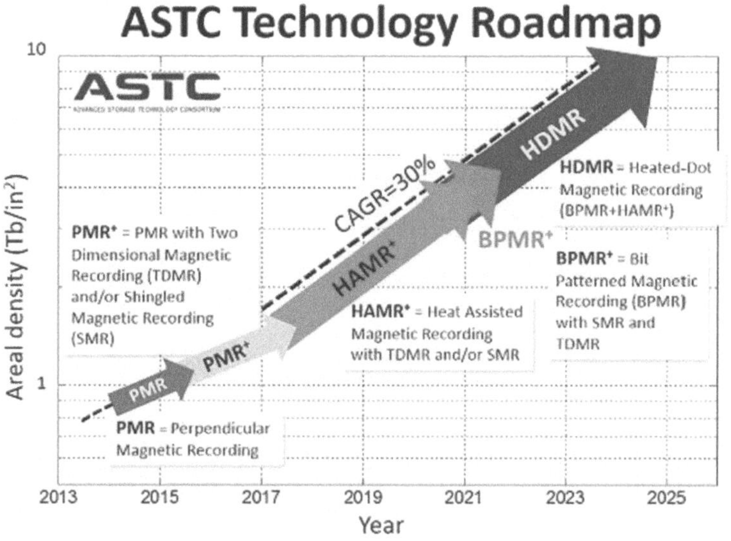

Figure 4.1 The Advanced Storage Technology Consortiums Spring 2014 technology roadmap.

the thin-film media beneath the recording head [1, 7, 9, 10]. The application of this additional STO field can significantly enhance writability and, more importantly, the effective write field gradient in the media. Consequently, recording media with smaller grain sizes and higher magnetic anisotropy can be used for substantially higher recording ADC than that of the conventional perpendicular magnetic recording.

When an alternating current (AC) magnetic field is applied roughly transverse to the easy axis of a magnetic grain of uniaxial anisotropy, the energy is coupled into the magnetic spins if the frequency of the AC field is within a certain region. Figure 4.2 shows the magnetic switching with the assist of a linear AC field for a magnetic grain with anisotropy field $H_K = 2K_u/M_s$. The magnitude of the AC field is 0.1 H_K. An applied direct current (DC) field of magnitude $0.5H_K$ is applied opposite to the initial magnetization. Without the AC field, the magnetization would not reverse to the direction of the applied magnetic field due to the insufficient magnitude of the applied field. In the presence of the AC

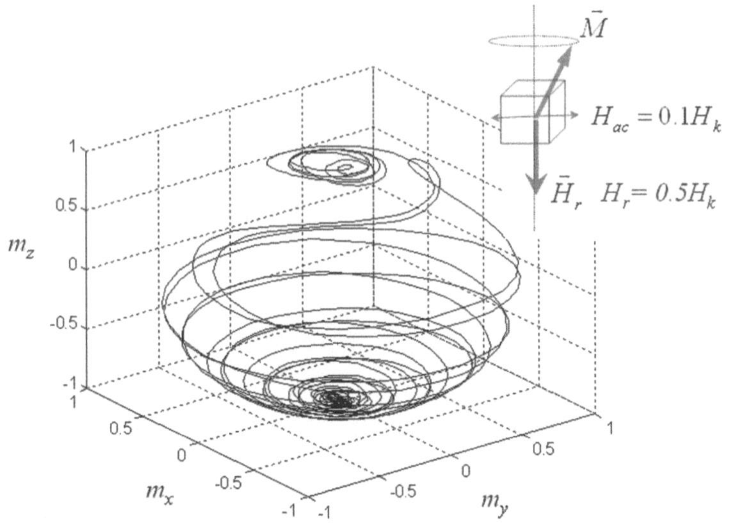

Figure 4.2 Calculated trajectory of volume-averaged magnetization of a magnetic grain with uniaxial anisotropy in the presence of a transverse AC field. The reversing field H_r is $0.5H_K$. The AC field is at an angular frequency of $0.45(\gamma/2\pi)H_K$ with an amplitude of $0.1H_K$. The trajectory is over 2 ns duration, starting from the rise of the reversing field pulse.

field, the magnetization will precess around the easy axis. Then the reversing field causes the angle of the magnetization with respect to the easy axis to quickly increase.

As the magnetization precesses across the equator of the magnetization sphere, the anisotropy field reverses its sign and precession chirality reverses. Consequently, the magnetization irreversibly reverses with respect to its initial orientation. Note the frequency of the AC field is $f = 0.45(\gamma/2\pi)H_K$, where γ is the gyromagnetic ratio with $\gamma/2\pi \approx 28$ GHz/T.

The underlying basic principle of the AC field–assisted switching described above is related to ferromagnetic resonance (FMR), except that FMR usually only considers magnetization precession in a linear regime, while reversal is a nonlinear process. The instantaneous frequency of magnetization precession in the absence of the AC field in the above setting is determined by the instantaneous anisotropy

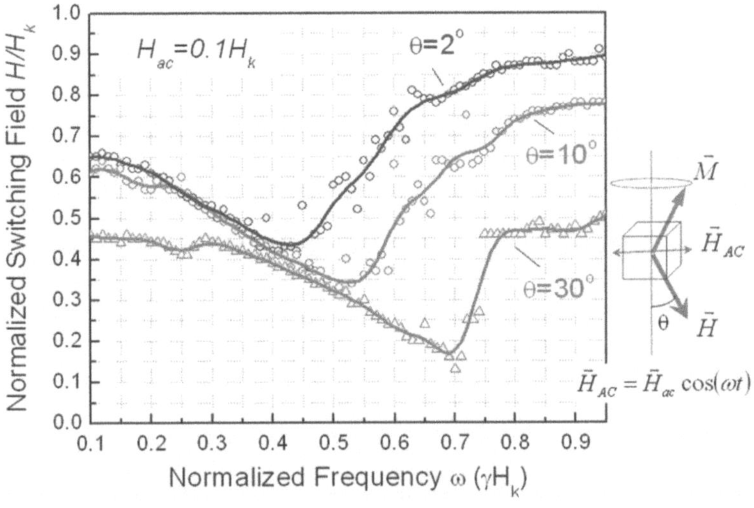

Figure 4.3 Calculated switching-field threshold as a function of AC field frequency for three different angles of the applied reversing field with respect to the easy axis. The pulsed reversing field has duration of 1 ns. The AC field amplitude is $0.1H_K$. The solid line is a fitting to the simulation results (individual marks).

field:

$$\omega_p = \gamma \cdot H = \gamma \cdot \left(H_k \cos\beta - H_{\text{applied}}\right) \tag{4.1}$$

where β is the angle between the magnetization and the easy axis, referred to as the precession angle. An increase of precession angle during a reversal would yield a decrease of the precessional frequency. This difference from the conventional FMR significantly broadens the frequency range over which the AC field can assist the switching process [11].

Figure 4.3 shows the calculated magnetic grain switching field as a function of AC field frequency [1, 12]. The AC field is linear with an amplitude of $0.1H_K$. The three different curves correspond to three different angles between the applied magnetic field and the magnetization easy axis. The switching field is defined as the minimum applied field value under which the magnetization can be irreversibly switched within a 1 ns time interval.

It would have been very difficult to implement the MAMR technology without the utilization of the perpendicular spin torque

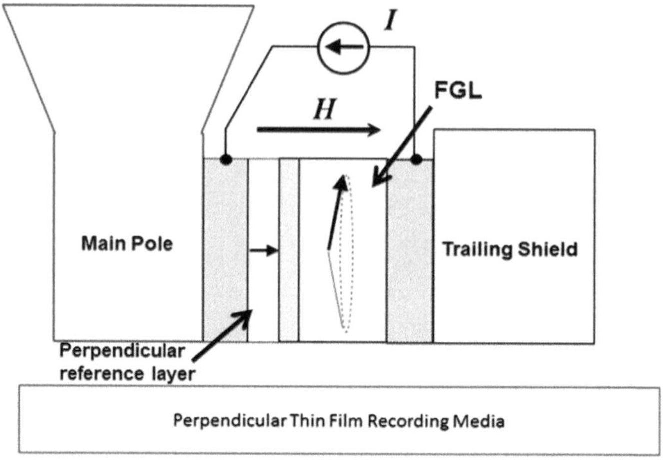

Figure 4.4 Schematic drawing of a perpendicular STO embedded in the write gap of a recording head. The magnetization precession in the oscillating layer will generate a high-frequency AC field in the thin-film recording media. The stray field strength inside the write gap is around $1.2T$. If the required AC field frequency is below 33 GHz, the perpendicular layer in the STO can be omitted.

oscillator (PSTO) [8, 9]. The PSTO is a two-terminal magnetic multilayer device comprising a spin polarization layer and a magnetization precessing layer or an oscillating layer. By embedding the PSTO in the write gap of the recording head between the main write pole and trailing shield, we can generate the needed AC magnetic field in the recording media. Within the gap, there is a relatively strong magnetic field more or less parallel to the air-bearing surface (ABS) of the head, of a magnitude about 1.2 T or larger. Figure 4.4 shows a diagram of a recording head with a PSTO embedded in the trailing shield gap. The PSTO shown in the figure illustrates the most basic PSTO design, which consists of a magnetic spin polarization layer (left layer) and a relatively thick magnetic layer (right layer) for generating the AC field, referred to as the field-generating layer (FGL). The two magnetic layers are separated by a noble metal layer. In operation, a DC current is injected into the PSTO, as shown in the figure, and the corresponding electron flow within the FGL becomes spin polarized due to the spin polarization layer. The spin angular momentum transfer from the spin polarized current to the local magnetization within the FGL effectively generates a torque on the magnetization of FGL [13–16]. If the current direction is as shown in the figure, the torque pulls the magnetization away from the field direction into the film plane, creating a nonequilibrium state for the magnetization of FGL and facilitating a steady magnetization precession with a constant angle between the magnetization and the field [10].

The generated AC magnetic field in the medium would provide substantial assistance to the recording head field, enabling high-fidelity recording in the medium with effective anisotropy substantially greater than the recording head field. To generate the AC field of significant amplitude in the recording media, the oscillating layer, also referred to as FGL, needs to be relatively thick, around 10 nm to 15 nm, and have a large saturation magnetization, greater than 2 T. If the technology is applied to next HDDs targeted for product in about 2018, the intended track width will be around 40 nm. At such a narrow track width, the magnetic poles on the side edges of the FGL when the magnetization of the FGL is cross-track also produce significant field in the media. Consequently, the generated AC field on the recording track in the media becomes circularly polarized, as illustrated in Fig. 4.5. The circles in the figure indicate the trajectory of the STO-generated AC field vectors, hence the polarization of the AC field.

By applying the AC field, saturation recording can be achieved in a medium with significantly higher anisotropy for a given recording head field. Figure 4.6 gives a quantitative description of the switching-field reduction in the presence of either linear or circular AC fields. Here, the recording field is a pulsed field with 1 ns pulse duration and 0.1 ns rise time and is applied 30° off the easy axis of the grain, opposite to the initial magnetization. The calculated switching field is normalized to the anisotropy field of the grain, H_K and the AC field is continuous and its frequency is normalized to γH_K.

As shown in the figure, the circular AC field is much more effective in terms of switching-field reduction with respect to the grain anisotropy field than the linear AC field. However, for the circular AC field, the chirality has to match the chirality of the grain

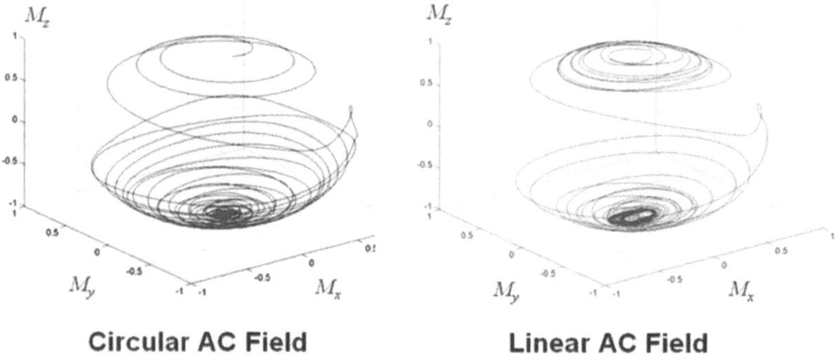

Figure 4.7 Simulated magnetization trajectory of a grain during magnetization switching in the presence of a circular AC field (left) and a linear AC field (right). The magnitude of the two AC fields is the same. The irreversible points during switching, where the effective magnetic field of the grain reverses its direction, occur when the chirality of the magnetization precession reverses.

energy to overcome the energy barrier, as shown in Fig. 4.8. The intrinsic damping of the grain magnetic energy is important since it describes the rate of energy "leaking" during the assisted switching process. As shown in the left graph of Fig. 4.9, increasing the Gilbert damping constant will make the AC field assist less effective with smaller switching-field reduction. However, the rise of the switching field in the circular AC field case is much smaller than that in the linear AC field case [7].

More in-depth understanding on the effect of Gilbert damping constant is important. Figure 4.10 shows the calculated switching field as a function of AC field frequency for a segmented grain with two exchange break layers (EBLs) [19–22]. In the figure, three cases with three different values of the Gilbert damping constant for the grain are shown. Again, the recording field is a 1 ns pulse, applied in the direction 30° off the anisotropy easy axis, while a circular AC field is applied continuously. The red cell in the map indicates a magnetization reversal, while the blue region indicates a reversal did not occur. The switching field of the grain without the AC field is around 1.2 T. Note that a higher value of the damping constant mainly shifts the high-frequency boundary region of the switching

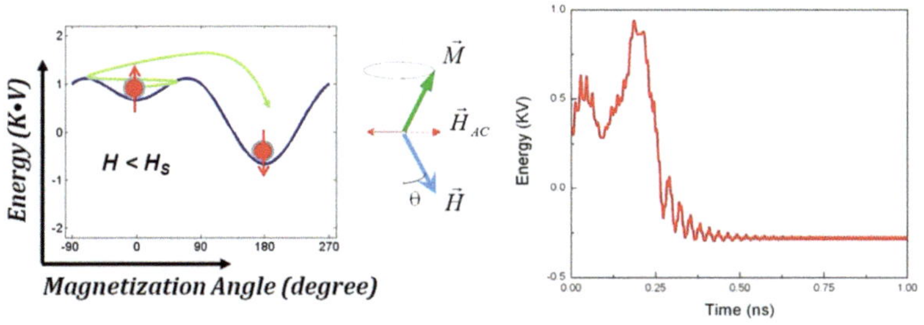

Figure 4.8 Illustration of the mechanism for switching-field reduction in the presence of an AC field. (Left) Schematic drawing of grain magnetic energy as a function of magnetization angle in the presence of a recording field H. (Right) The magnetic energy of a grain during switching in the presence of a linear AC field.

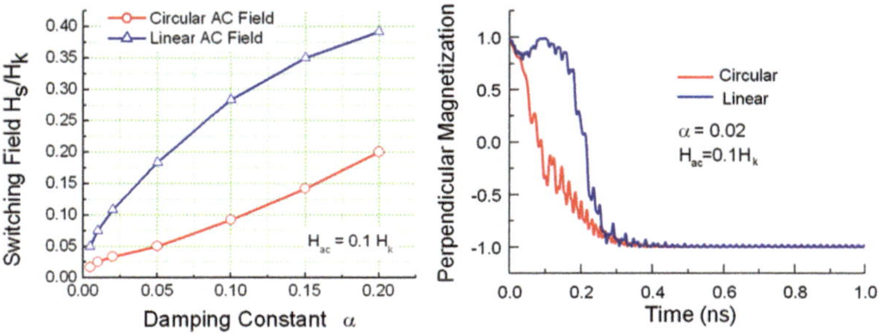

Figure 4.9 (Left) Calculated normalized switching field in the presence of a circular AC field (red) and a linear AC field (blue) as a function of the Gilbert damping constant of the grain. (Right) Transient process of the easy-axis magnetization component during switching for the two cases. The recording field is a 1 ns pulse at the direction 30° off the easy axis.

map toward lower frequencies, while the switching map at lower frequencies is virtually unchanged. By operating at the lower range of the effective AC field frequencies, we can eliminate or significantly alleviate the impact of damping constant dispersion among the grains in the media. In practice, one should avoid operating near the high-AC-field-frequency boundary of the switching map where the switching is marginal and sensitive to intrinsic parameter variations.

Recording processes on perpendicular granular thin-film media with segmented grains have been simulated by superimposing a

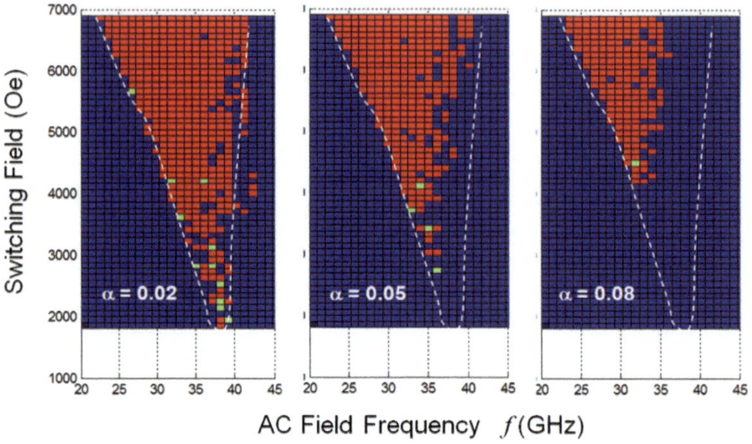

Figure 4.10 Calculated switching maps for a segmented grain with two exchange break layers inserted for three damping constants of the magnetic particles. The reversal field is a 1 ns pulse at the direction 30° off the grain anisotropy easy axis, while a steady circular AC field is applied in the horizontal plane. A red cell indicates irreversible magnetization switching, and blue otherwise. The switching field of the grain without the AC field is 1.2 T.

practical recording head field and the AC field generated by the PSTO embedded in the write gap of the recording head. Figure 4.11 shows an example of the simulation in which case each grain has three 4 nm thick segments in the perpendicular (along grain height) direction with optimized exchange coupling between adjacent segments. It is interesting to note that the recording footprint is no longer the footprint of the recording head write pole but rather a small region near the trailing edge of the write pole where the AC field is significant. Actually, the location of the FGL inside the write gap becomes important and critical for yielding optimal recording performance, that is, the optimal SNR of the resulted recording data pattern.

The segmentation of medium grain is important for MAMR [23]. Since the PSTO-generated AC field decays significantly faster through the depth of the media away from the surface than that of the recording head field, the segmentation with adequately reduced exchange coupling between the surface segment and the segment

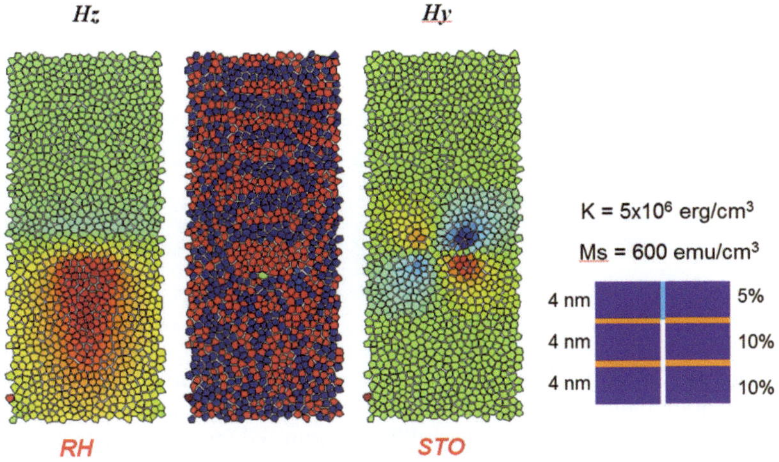

Figure 4.11 Map of the perpendicular (left) and down-track (right) components of the recording field, along with the resulting transition pattern (middle). The head moves down relative to the medium. *x* is in the cross-track direction, *y* the down-track direction, and *z* the perpendicular direction. The figure on the far right shows an illustration of the layered grain structure, in cross-sectional view, used in the MAMR study. Each layer is with a grain thickness of 4 nm. The anisotropy distributions of the three layers are 5%, 10%, and 10% respectively, as shown in the figure.

below will enable one to fully take advantage of the AC field near the medium surface. The anisotropy field of the top segment also should be significantly higher than that of present conventional perpendicular recording media because the effective gradient of the combined head field and PSTO AC field is substantially higher than that of the recording head field alone. It has been shown that the effective field gradient for an optimally located FGL with a conventional perpendicular recording head can be over 3000 Oe/nm, at least six times greater than that of an optimized conventional recording head without a PSTO. This factor is of great importance since a significantly enhanced write field gradient is also essential for enabling ADC gain, in addition to the writability gain [7].

To reduce grain size, the anisotropy energy constant of the grain needs to be sufficiently high to ensure sufficient thermal stability. Figure 4.12 shows segmented grain stack designs for different grain

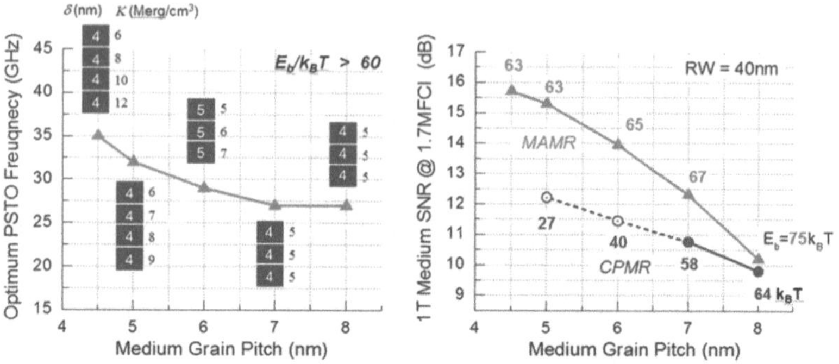

Figure 4.12 (Left) Segmented grain stack designs for different grain sizes. The grain diameter is assumed to be 95% of the grain pitch. The vertical exchange coupling through each EBL is optimized for obtaining maximum SNR. (Right) SNR at 1.7 million flux changes per inch (MFCI) as a function of mean grain pitch for the designs shown on the left (MAMR) vs. conventional perpendicular magnetic recording (CPMR). The read track width is kept at 40 nm and the written track width is approximately 50 nm.

pitch, thereby different grain size. In each case, the anisotropy strength of the segments is chosen such that the energy barrier E_b of an isolated grain always exceeds $60k_BT$. The actual energy barrier should be higher at film level due to the small intergranular exchange coupling among the top segments of the adjacent grains. The through EBL exchange coupling is optimized between the segments for each grain stack design to obtain the maximum SNR. The insertion of the EBL does yield a slight reduction of the energy barrier compared to the addition of energy barriers of the segments within a grain [24]. However, this reduction becomes smaller for higher-anisotropy designs at smaller grain sizes, since the vertical coupling through EBL needs to be stronger.

Note that the anisotropy energy constant of the top segment is 6×10^6 erg/cm^3 or lower because above this anisotropy recording becomes difficult, even with the assistance of the STO-generated AC field. The segmentation by inserting EBLs enables grain switching of lower segments with higher anisotropy strength, similar to today's medium design, to maintain the grain energy barrier as grain size reduces. With the segmented design, increasing the energy barrier does not require an equal amount of writability enhancement. As a result, the optimum AC field frequency only increases slightly as grain size reduces. Only when grain pitches become smaller than 5 nm does the STO design need to have a perpendicular anisotropy layer exchange-coupled to the FGL to produce an oscillation frequency higher than that supported by the in-gap field. Otherwise, the perpendicular anisotropy layer can be omitted.

With the AC field produced by the PSTO at the corresponding frequencies, excellent recording can be obtained for all the medium designs shown in Fig. 4.12. The right graph shows the calculated SNR for all the designs shown in the left graph. The SNR performance of the optimized PMR grain stack design is also plotted in the figure, even though the energy barrier is too low to be viable at small grain pitches. The number shown at each data point indicates the corresponding energy barrier in units of $k_B T$, with $T = 300$ K. Again, these energy barrier values are for individual grains. In the case of MAMR, the SNR increases with decreasing grain pitch, and at small grain pitches, the SNR is substantially higher than that of the PMR case. At grain pitches of 8 nm and beyond, present PMR as well as MAMR essentially create grain pitch–limited transitions. At smaller grain pitches, transitions created in the MAMR case becomes sharper than that in the PMR case, yielding higher signal and lower jitter noise due to the significantly higher, effective write field gradient. It is important to note that enhanced writability alone is not sufficient for ADC gain, which relies on the ability to produce grain pitch–limited transitions at small grain pitches. Both modeling studies and experimental investigation have shown that MAMR at optimal conditions significantly enhances the effective write field gradient, enabling the creation of grain pitch–limited transitions, even at very small grain pitches [7, 25–27].

4.3 Heat-Assisted Magnetic Recording

4.3.1 *Introduction*

As discussed throughout this book, magnetic hard drives have enjoyed significant areal density growth from 1956 to the present (2015). Superparamagnetic effects have imposed a trade-off in

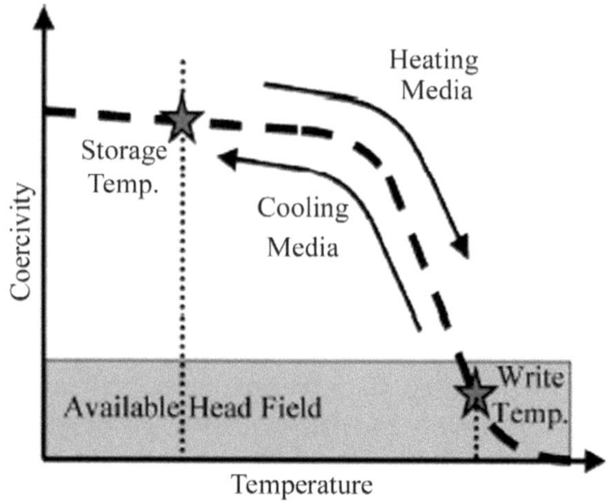

Figure 4.13 The coercivity versus temperature and the associated available effective head magnetic field illustrating the HAMR concept.

the SNR, thermal stability, and writability that has prevented the HDD industry from scaling storage areal densities at the historical rate of ~40% per year [3, 4]. HAMR uses a temporary heating of the recording layer to break the required trade-off imposed by superparamagnetism [2].

Figure 4.13 illustrates the HAMR concept. The recording medium has a very large anisotropy and coercivity, so it is thermally stable even at grain sizes much smaller than today's PMR media. The smaller grain size allows a reduction of the data bit area on the disk with a large enough number of grains per bit to maintain a sufficient SNR. The coercivity of the recording layer at room temperature greatly exceeds the available head field, making the medium unwritable with a conventional PMR head.

When the medium is heated to near the Curie temperature (T_C), the magnetization will be aligned to the head field direction. Rapid cooling and tight confinement of the thermal energy prevent thermal agitation from deteriorating the recording quality. In a similar writing process, magneto-optical recording heats a much larger region [28]. In another difference, HAMR uses a magnetic resistive sensor to read the data, while magneto-optical systems use an optical readout based on Kerr rotation.

HAMR research started in the late 1990s as a candidate to extend the capacity and performance of removable magneto-optical drives and as a response to the difficulty of scaling longitudinal magnetic recording due to superparamagnetic effects. Companies such as Terastor [29] and QUINTA Corporation [30] tried to put a magneto-optical recording system into the HDD architecture. Unfortunately, their targeted densities, $\sim$10 Gb/in^2 were soon surpassed by all magnetic recording as longitudinal magnetic recording was successfully scaled to over 100 Gb/in^2 and the subsequent recording technology [31], perpendicular magnetic recording is being scaled to beyond 1 Tb/in^2.

When HAMR work was started, there were a number of large technical challenges [32]. How should the thermal energy be confined to the sizes required for 1 Tb/in^2 and beyond? What type of high-anisotropy recording layer could be produced that meets all of the HDD requirements and has the right grain size and thermal properties? What would be the source of the thermal energy, and how would it be integrated into the HDD architecture? What materials would be required to produce a robust head–disk interface (HDI) and recording subsystem (RSS) as the medium is heated to near the Curie temperature (T_c)? HAMR development has answered some of these challenges, and some are still active areas of research.

An example of an early HAMR recording architecture with a free space light delivery is illustrated in Fig. 4.14 (top). The recording head includes a magnetoresistive read sensor, an electromagnetic field source, a light path with an integrated waveguide, and a near-field transducer (NFT). The light generated by an external laser source is grating-coupled into the waveguide and is directed to the NFT to generate surface plasmons. The high-intensity field generated from the NFT heats the media as the media surface passes under the flying recording head. The media coercivity is reduced with temperature, as shown in Fig. 4.13. The magnetization can be switched when the media coercivity drops below the recording head field. For an efficient HAMR system, this requires HAMR media with a proper heat sink and a soft underlayer (SUL) to optimize the thermal and magnetic properties for the recording process.

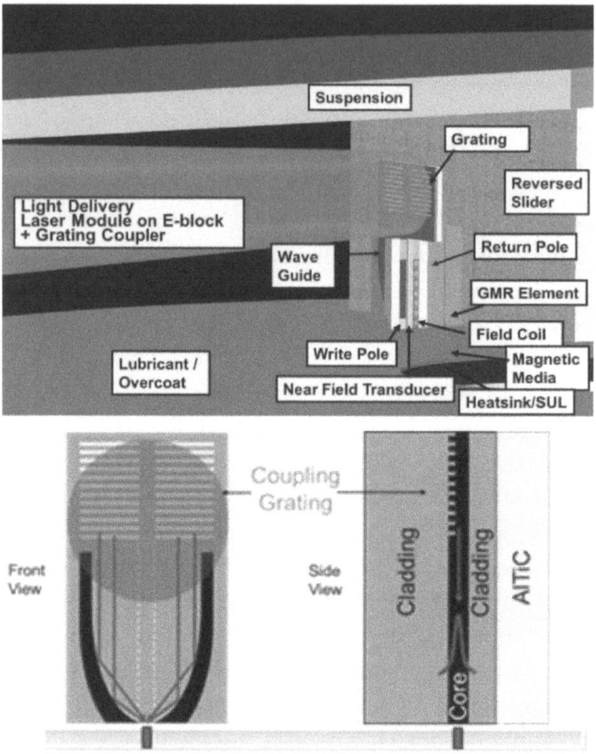

Figure 4.14 (Top) An illustration of the HAMR recording system showing a cutaway of the HAMR head. The HAMR head consists of a read and write element, a light path, and an NFT flying over the surface of HAMR media with a heat sink and a soft underlayer (SUL). (Bottom) An optical design is shown that uses grating coupling, a waveguide, and a solid immersion mirror to focus light toward the NFT.

The optical elements in the slider from Fig. 4.14 (top) are further illustrated in Fig. 4.14 (bottom) [33].

4.3.2 *Recording Head*

4.3.2.1 Near-field transducer

A number of designs were considered to locally heat the recording layer, such as a local tunneling current or a resistive heater at the ABS of the slider. By far the preferred architecture is a tightly confined

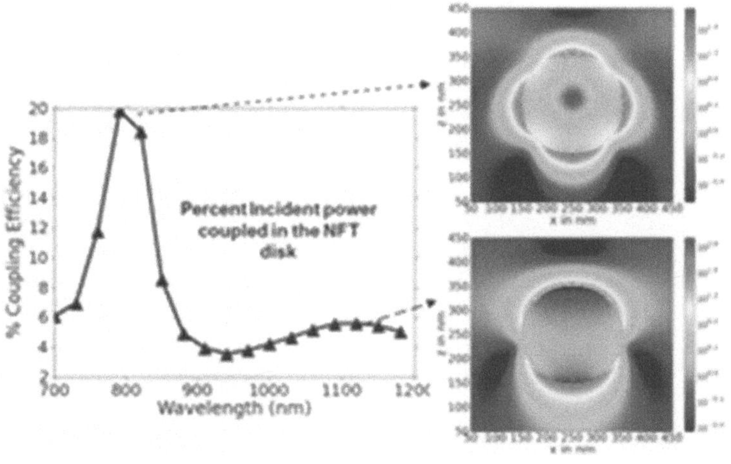

Figure 4.15 (Left) Coupling efficiency vs. wavelength for an NFT disk. (Right) Illustration of the field intensity distribution at the disk surface when the focused light hit the gold disk at two different wavelengths.

optical spot. The optical spot creates a thermally heated region whose cross-track dimension is on the order of the track pitch. At densities near 1 Tb/in^2 and a bit aspect ratio of approximately 4, an optical spot with full width at half maximum (FWHM) of 50 nm or less is required. This is well beyond diffraction-limited optics with visible light, even with solid immersion lenses or mirrors. For example a Blu-Ray drive uses a blue laser diode to create an optical spot of 280 nm, while the Terastor system used a red or near-infrared laser diode and a solid immersion lens to create an optical spot of about 200 nm. HAMR uses a plasmonic NFT to create spot sizes that can achieve an FWHM of 50 nm or less.

Figure 4.15 illustrates the electron density optical field resonance that occurs when light is incident on a gold disk, with the optical configuration where the focused light has a π phase shift between left and right reflections from a solid immersion mirror (SIM) [34]. The electron–light interaction causes regions of high field or high electron density to have dimensions that are much smaller than the wavelength of the incident light. To further enhance the field confinement, a small antenna or peg is added to the gold disk to form the lollipop-shaped NFT. When the recording medium is

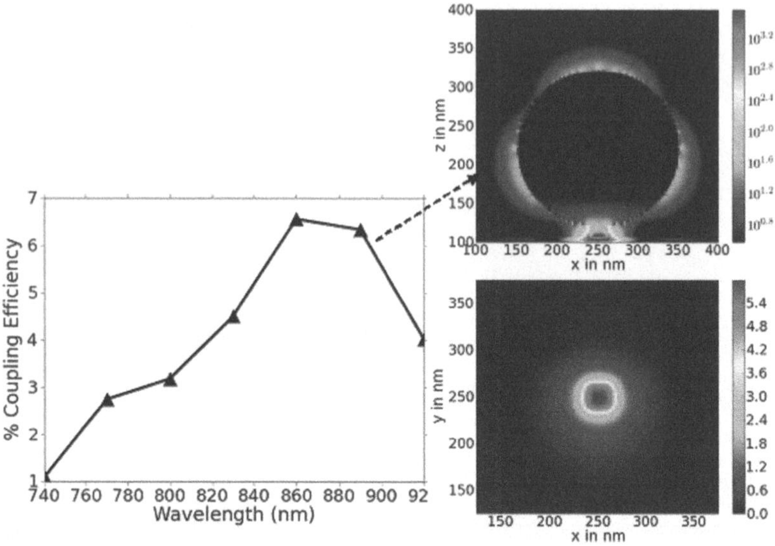

Figure 4.16 (Left) The coupling efficiency vs. incident light wavelength. (Right) An example of field intensity distribution at the NFT surface (top right) and the storage layer (bottom right) for a lollipop type of NFT design.

brought into proximity of the end of the peg, the electric field is coupled in to the recording medium.

Figure 4.16 illustrates the field in the transducer and the storage medium for a fixed-head media separation with the geometry given by Fig. 4.14 (bottom) [34]. The energy is confined in both the cross-track and down-track dimensions. The heating efficiency very much depends on the media design but typically the NFT conversion efficiency, defined by the total energy absorbed in the recording media over total optical energy impinging on the NFT surface, is 1% to 10%. This is efficient enough to allow media heating with inexpensive laser diodes but the wasted energy results in heating of the NFT, the head, and the HDI, which requires robust materials to provide a reliable HAMR system.

4.3.2.2 Light delivery path

A number of designs were proposed for mechanically and optically integrating a laser diode into the recording head. In the most

Figure 4.17 An illustration of the laser-on-slider approach. Courtesy of Seagate Technology.

common design from the academic or patent literature, the laser diode is mounted on a submount that is attached to the top surface of the slider. This laser-on-slider design is shown in Fig. 4.17. The laser emission is coupled into a single-mode waveguide in the recording head.

The couplers are typically on the order of 50% efficient and require alignment tolerances that are on the order of 100 nm to a micron. This often necessitates active alignment of the laser diode submount subassembly to the slider. This active alignment requires sophisticated assembly equipment and will add to the cost of HAMR heads.

The laser light is propagated from the coupler to the NFT using either a channel waveguide or a series of integrated mirrors and a planar waveguide as illustrated in Fig. 4.18 and Fig. 4.13 [34, 35]. For the lollipop design, the efficiency is optimized with two incident focused beams that are phase shifted by 180°. This

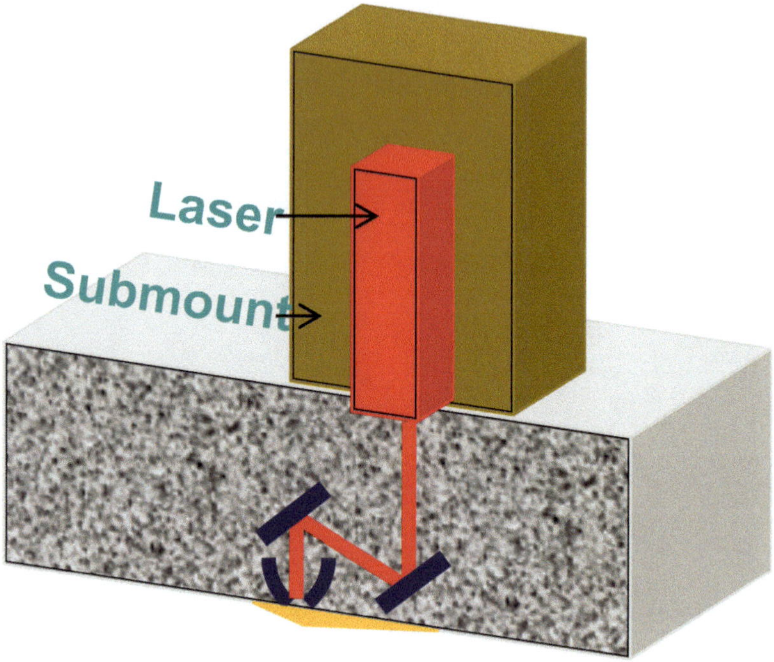

Figure 4.18 A laser-on-slider light delivery path is illustrated. The laser path is shown as a red solid line and the blue lines are mirrors. Courtesy of Seagate Technology.

can be accomplished by shifting the two segments of the planar parabolic SIM. The waveguides can use tantala (index of refraction $n \approx 2.1$) for the core and alumina for the cladding ($n \approx 1.6$). The integrated optical elements, coupler, waveguides, and mirrors can all be fabricated in the heads during wafer processing. This ensures very accurate placement of the optical elements and minimizes the cost impact to the head.

Figure 4.19 shows an example of an NFT with the lollipop design [36]. The disk portion of the NFT is on the order of 200 nm and a small peg exposed at the ABS helps to define a very small thermal spot in the recording media on the order of 50 nm or less. The NFT materials are typically made of good plasmonic materials and alloys to reduce absorption, and improve efficiency.

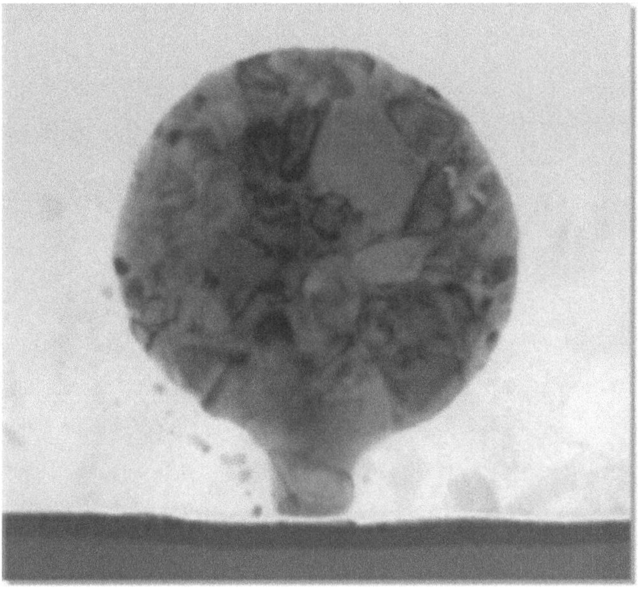

Figure 4.19 An SEM image is shown of a lollipop NFT as utilized for recording beyond 1 Tb/in^2.

The thermal spot in the media cools as it moves away from the NFT. In this region the applied magnetic field causes the recording layer to orient or freeze in the direction of the magnetic field as the media cools. The applied head field is provided by a magnetic writer for perpendicular magnetic recording that was describer earlier. A writer design with the optical elements in the gap of a magnetic "ring head" is shown in Fig. 4.20. To illustrate each component in the head, the image shows a cross section view of a recording head cut from the center line. The optimization of the recording process depends on write field and the NFT-to-write-pole spacing.

The HAMR head also needs to contain the elements found in the PMR head in today's HDD products. This would include a magnetoresistive element to read the recorded magnetic data, a contact detection sensor to define a minimum head-to-media spacing (HMS), and one or two heater elements in the head to use thermal expansion to control HMS in both read and write mode. A photodetector can also be included in the laser submount to

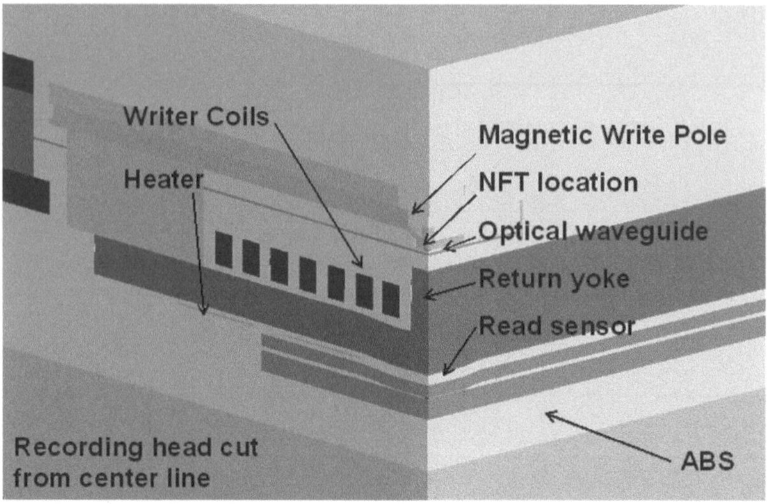

Figure 4.20 A cross-sectional view is shown of the HAMR recording head, where the light path and NFT are integrated within the magnetic "ring head" gap.

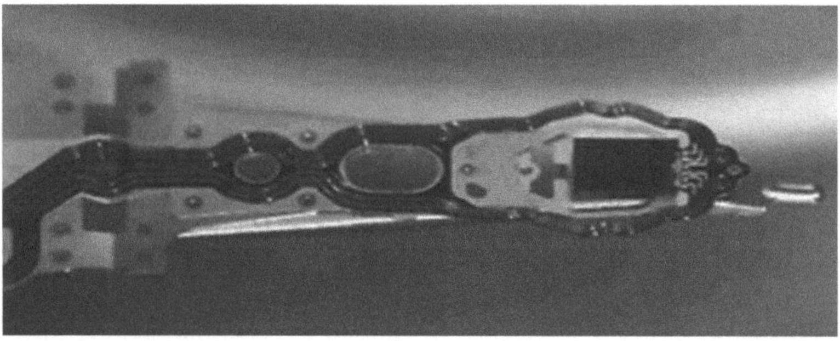

Figure 4.21 Illustration of an HAMR head gimbal assembly. Courtesy of TDK Corporation.

monitor laser output and correct laser power due to laser aging or temperature changes. All of these electrical components must be connected to the controlling circuits through the trace gimbal assembly. A complete HAMR head gimbal assembly is shown in Fig. 4.21.

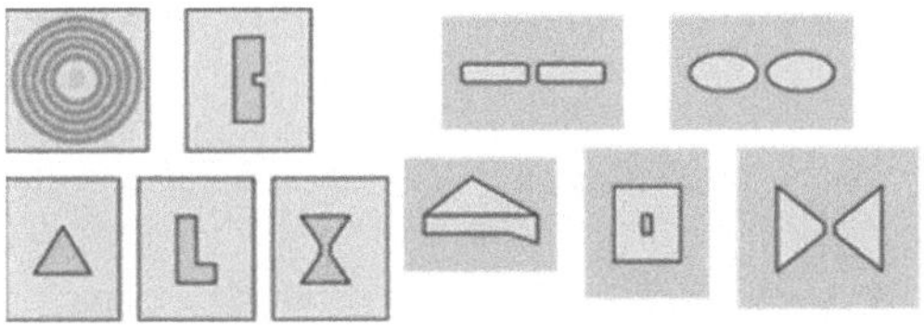

Figure 4.22 Illustration of NFT designs studied for HAMR over the past.

The description above is based on one of the potential HAMR head designs. There are several alternative NFT designs, and currently it is still a very active area of research. Figure 4.22 illustrates some of the examples, including the apertures type of designs such as triangle, C aperture, waveguides, and the antenna type of designs such as bow tie, beaked triangle, nanorods etc. [36–41]. Light delivery options include the laser in slider, integrating surface emitting laser, or even more integrated designs where the NFT is integrated directly with the laser structure [42, 43].

To fully understand and optimize the HAMR process, sophisticated models are needed for the underlying electromagnetic, thermal, and magnetic physics. Finite-difference time domain simulations were used to model the media heating to optimize the lollipop NFT in Fig. 4.19 [36]. Today powerful optimization methods lead to very sophisticated NFT designs that hold the promise for greater efficiency [41].

4.3.3 *HAMR Media*

HAMR's promise has always been that high-anisotropy magnetic materials could be thermally stable at smaller grain sizes but could then be recorded via heating during the recording process. Thus the material selection for the recording layer focuses on high-anisotropy materials, with a reasonable Curie temperature, T_c, and all of the traditional attributes of recording media such as low corrosion rate in air, low surface roughness, etc. Figure 4.23 is a plot of anisotropy of candidate materials [44, 45]. Progress over the last 10 years [46–49] has led to the development of FePt as the overwhelming choice for the HAMR media recording layer. With the proper segregant selection, underlayers, and deposition methods, it has been shown

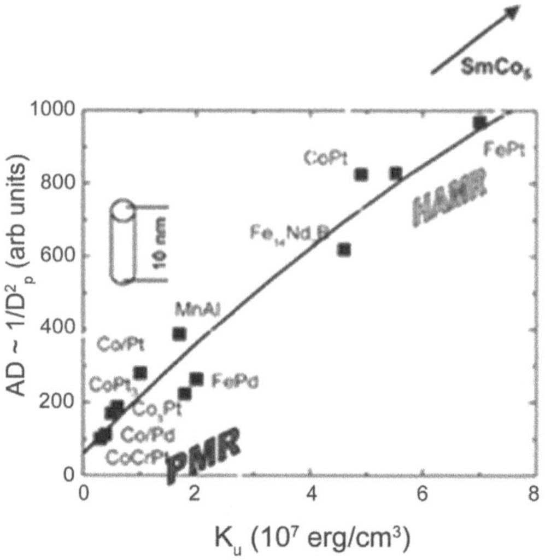

Figure 4.23 The anisotropy of different materials suitable for PMR and HAMR are shown with the grain size–limited areal density.

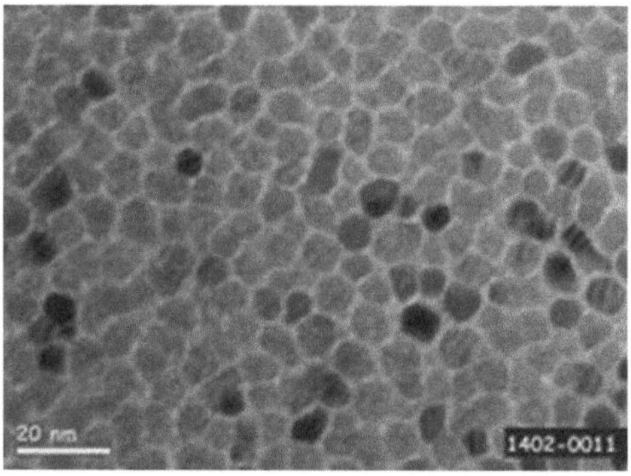

Figure 4.24 A TEM planar view is shown of today's HAMR media based on FePt–C alloys. From Ref. [47]. Courtesy of Ramamurthy Acharya.

that FePt media can produce high magnetic anisotropy and good orientation, with a microstructure similar to today's PMR media [48].

A transmission electron microscopy (TEM) planar view image of an FePt–C medium is shown in Fig. 4.24 [47]. This medium has a

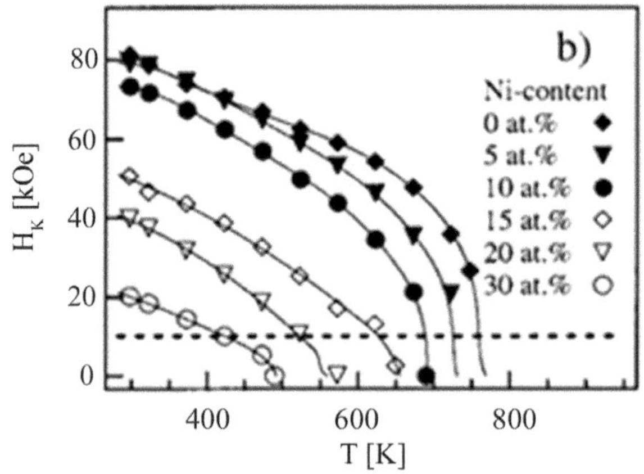

Figure 4.25 FePt media T_C and H_K reduction are plotted as a function of percentage Ni doping. Reprinted with permission from Ref. [50]. Copyright 2002, AIP Publishing LLC.

grain-to-grain spacing on the order of 10 nm with an anisotropy field of larger than 50 kOe.

As stated earlier, it is important that the media cool quickly to prevent thermal destabilization. This requires a new element in the media thin-film stack—a heat sink to quickly conduct the heat from the recording layer after the media reaches the desired temperature. The media will also require a protective overcoat and lubricant that can withstand the peak temperatures in the recording process, which will be near T_c. As shown in Fig. 4.25, we can reduce the T_c of FePt with doping but this will reduce the anisotropy and ultimately the achievable areal density [50]. The typical T_c is ~750 K for today's FePt media [50].

The deposition of $L1_0$–FePt requires a chemical ordering that in today's sputtering methods requires high temperatures. This necessitates a disk substrate that is chemically and mechanically robust at sputtering temperatures up to 500°C. These have been developed and are readily available for HAMR media manufacturing. A commercially available HAMR disk and associated structure are shown in Fig. 4.26.

As shown in Fig. 4.27, to achieve high areal density, recording media grain size, and grain size distribution have steadily reduced

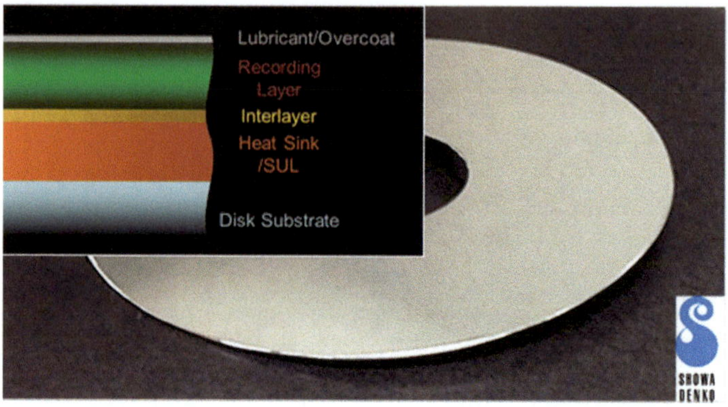

Figure 4.26 An example of a HAMR disk and associated layer structure. Disk courtesy of Showa Denko Corporation.

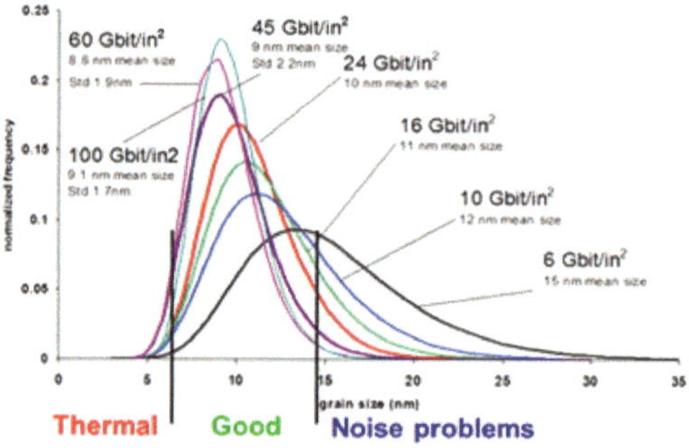

Figure 4.27 Media grain size and grain size distribution over the past two decades. Reprinted from Ref. [48].

over the past two decades [48]. The primary reasons to drive this reduction of grain size as well as grain size distribution are due to media SNR requirement (for smaller grain size) and thermal stability requirement (for smaller grain size distribution) [3]. Over the past decade, PMR has been limited by the media switching-field distribution (SFD), primarily determined by anisotropy (H_K) and grain size distributions ($\sigma H_K/H_K$ and σ_D/D) [48, 51].

In HAMR media, in addition to the above distributions, the SFD also comes from the media T_c distribution (due to the materials' composition and lattice structure variation), H_K distribution due to

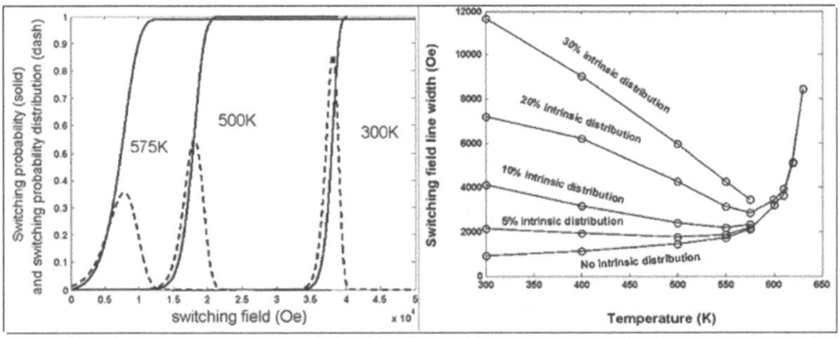

Figure 4.28 (Left) Media switching field reduction and switching-field distribution (dashed line) increasing with temperature for individual media grain. (Right) The switching line width (i.e., FWHM of the dashed line in the left figure) vs. temperature for media with different anisotropy (H_K) distributions.

grain-to-grain order parameters change, and thermal effects during the recording process [51, 52]. The T_c distribution can cause the SFD to be increased at the elevated temperature. Thermal fluctuations also lead to the increased SFD with temperature as the switching field is reduced, as illustrated in Fig. 4.28. Combined with grain size and anisotropy distribution-induced SFD, at the writing position, the total SFD as a function of temperature reaches an optimized condition as the temperature drops below T_c [52].

Furthermore, to achieve recording density beyond 1 Tb/in^2, ultimately the media grain size needs to be reduced. As has been shown before, with a finite grain boundary between the polycrystalline grain structures, the measured media switching field (H_c) will be reduced for smaller grain size, which will be an ongoing challenge for areal density beyond 2 Tb/in^2 [53].

4.3.4 *HAMR Recording System*

4.3.4.1 Define/achieve HAMR transition

A typical HAMR recording process is illustrated in Figs. 4.29–4.31 using a simplified model.

As the media moves away from the recording head, it absorbs optical energy from the NFT peg along the center of the track as shown in Fig. 4.29. In practice it is difficult to measure the actual

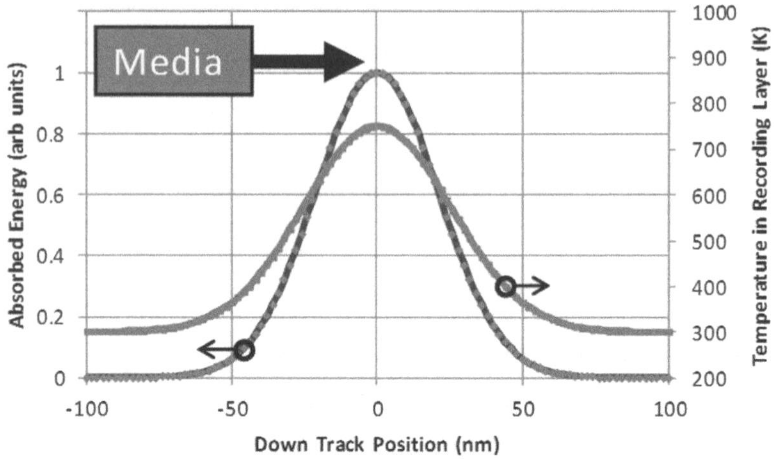

Figure 4.29 The normalized absorbed energy density and temperature in the recording layer in the HAMR write process.

absorbed power in the recording layer. To first-order approximation, the optical intensity for a given design as shown in Fig. 4.19 has a bell-shaped profile. For illustration, one can assume the optical profile has a Gaussian distribution in the media and, in this example, a down-track FWHM of 50 nm. This absorbed energy results in a temperature distribution in the media that depends on the optical energy as well as the specifics of the media design. We note that in this simplified model we neglect the optical impact of other features in the head, such as the write pole, which can lead to an asymmetric optical and thermal spot in the recording layer.

The efficiency of the NFT will also depend on HMS. To first order, the peak temperature in the recording layer can be scaled by adjusting the laser diode operating current. In Fig. 4.29 we have adjusted the optical power to produce a peak temperature of 750 K from the simplified model. Due to thermal spread, the thermal spot has an FWHM of 60 nm and we have assumed an ambient temperature of 300 K. Note that the coercivity in Fig. 4.30 drops to zero where the temperature is above the Curie point (T_c). In this model the region of zero coercivity is narrower than the FWHM.

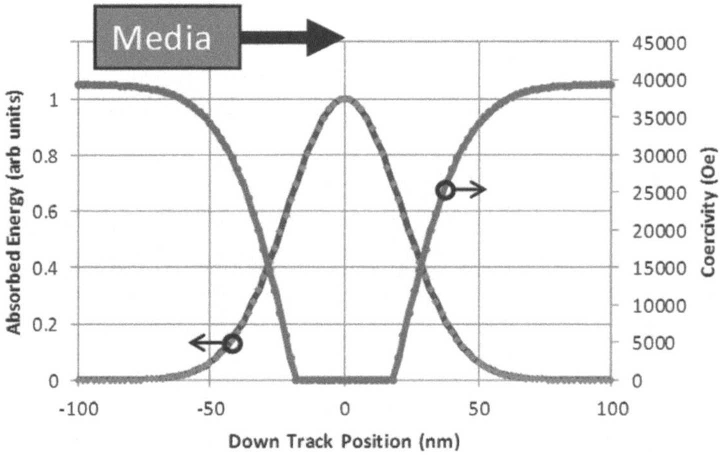

Figure 4.30 The normalized absorbed energy density and media coercivity in the recording layer during the HAMR write process.

If we assume the temperature dependent coercivity of the FePt is as shown in Fig. 4.25, then

$$H_c(T) = H_c(0)[1 - (T/T_c)]^n \tag{4.3}$$

where in Fig. 4.30, $H_c(0\text{ K}) = 50$ kOe, $n = 2$, and $T_c = 650$ K. This T_c corresponds to the 15% Ni doping and the n value in Eq. 4.3 is measured as shown in Ref. [50]. Under the NFT, the medium is heated to above the T_c and then regains its anisotropy as it quickly cools with a typical time scale of a few nanoseconds. The magnetic write pole produces fields in the media that orient the magnetization as the media cools. This freezing near the Curie temperature takes place in a very short period of time and the exact nature of the effective field is still an area of active research [54–56]. The transitions are written when the magnetic writer polarity is changed at the freezing point.

In Fig. 4.31, we show the effective switching field due to write field angle given by the Stoner–Wohlfarth relation [57] from the magnetic write pole shown in Fig. 4.20 along the center of the track, assuming the leading edge of the write pole is about 25 nm down-track from the center of the NFT. At the last point where the media coercivity is less than the applied field, the magnetic state is frozen and the information is recorded. In this case, this occurs right under

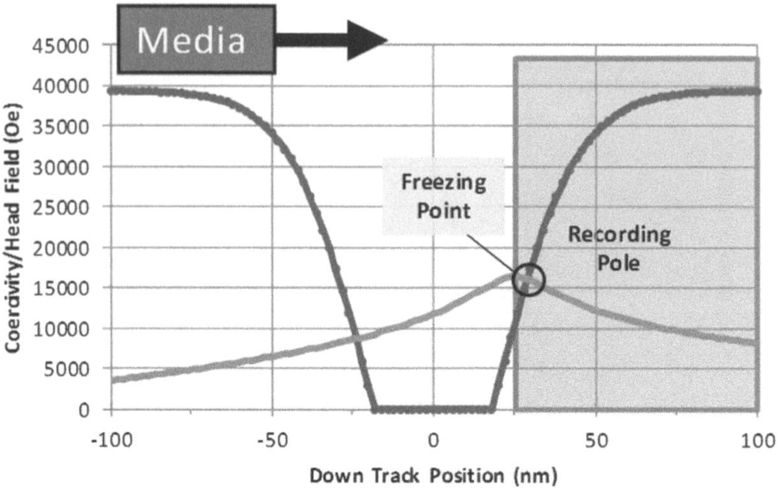

Figure 4.31 Effective magnetic field from the writer and the media coercivity as a function of down-track position in the HAMR write process (media moving from left to right).

the leading edge of the write pole. Although this simple model used a simplified write field and temperature distributions and did not include thermal fluctuations, it illustrates many important attributes of the HAMR process. For example, although the medium was heated to 100° above the T_c, the recording takes place well below the T_c, at approximately 550 K in this case, where the medium H_c is equal to the applied head field. The magnetic state is frozen in during the cooling process in alignment with the applied field that is greater than 10 kOe. We could increase the gradient and decrease the required peak temperature if we could put the NFT closer to the magnetic pole, but in practice the optically lossy metal in the write pole disturbs the NFT resonance and the electromagnetic energy heats up the write pole [58].

4.3.4.2 HAMR performance improvement

An important parameter in optimizing the performance of magnetic recording is to achieve a high effective write field gradient for the recording process. A high gradient enables narrower transitions and larger tolerance to media SFDs [51, 52, 59, 60]. The typical transition width, as defined by the William–Comstock formula, is

$$\frac{dM}{dx} = \frac{dM}{dH}\frac{dH}{dx} \tag{4.4}$$

In HAMR, the field gradient needs to include temperature variation from recording media [61]

$$\frac{dH}{dx} = \frac{dH_{\mathrm{h}}}{dx} - \frac{dH_{\mathrm{c}}}{dT}\frac{dT}{dx} \tag{4.5}$$

where H_{h} is the head field and H_{c} is the temperature-dependent media coercivity.

From Figs. 4.30 and 4.31 we can estimate the effective write field gradient. At the freezing point the effective thermal gradient is about 10 K/nm and the effective write field gradient is about 1500 Oe/nm that is about 5 times higher than in PMR recording [62, 63]. Note that at the place where HAMR defines the transition, the media SFD is higher due to media T_{c} distribution and thermal fluctuations, in addition to anisotropy field (H_{K}) and grain size distributions. Therefore, a much higher effective field gradient for HAMR is a requirement to achieve a high SNR or ADC. As seen from the previous discussion there are a number of inputs to the HAMR recording system that determine the recording quality.

One of the key metrics is the thermal gradient in the media. Recently a number of groups have reported the ability to measure the thermal gradient by modulating the laser diode current and measuring the corresponding bit shift. This requires a means to estimate the temperature rise per milliampere of laser current. Proposals include assuming the recording occurs at a fixed offset from the T_{c} or by modulating the writer current and extracting the freezing temperature [63, 64]. Given the large number of input variables (NFT-to-pole spacing, NFT-to-media spacing, media heat sink thickness, etc.), it is important to have metrics such as thermal gradient to optimize the system. Since in a typical HAMR system the recording head magnetic field profile is much broader than the NFT-defined thermal spot, the write width measurement under different laser powers can be used to estimate the cross-track thermal gradient [65].

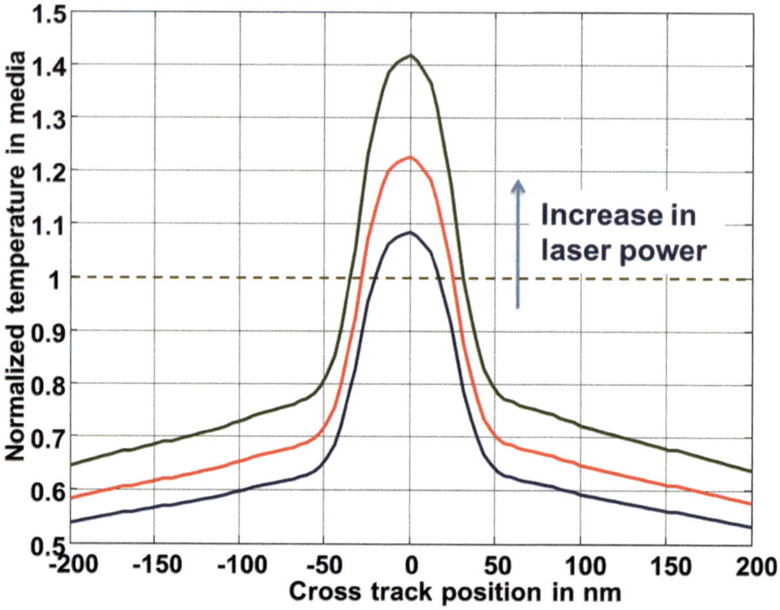

Figure 4.32 The HAMR thermal spot change for different laser powers. The temperature is normalized to the transition temperature in the recording process.

As shown in Fig. 4.32, for a given transition temperature, defined by the media properties and the write field ($H_w = H_c(T)$), the write width can be varied arbitrarily by change of the laser power. Due to high effective field gradient in HAMR, in principle, a narrow track pitch can be obtained. This is different from perpendicular recording where sub-50 nm track cannot be easily directly achieved. With an optimized media and NFT, a narrow track pitch down to 30 nm becomes possible [66]. However, we note that at high track density, other factors may impact HAMR system performance, such as the impact of curvature as well as the limitation due to the read width and playback process [65]. In addition, using lower laser powers, the thermal gradient in down- and cross-track directions will be reduced.

Figure 4.33 shows the magnetic force microscopy (MFM) image of an HAMR-recorded pattern based on FePt media. The results show a small track pitch (50 nm) can be defined accurately using HAMR.

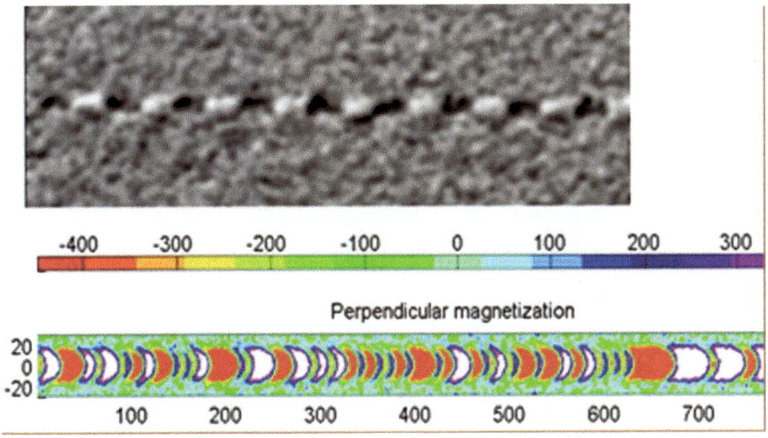

Figure 4.33 (Top) MFM image, an HAMR NFT recording pattern defined at 500 kTPI. (Bottom) Micromagnetic simulation result of average magnetization of HAMR recording at close to 3 Tb/in^2.

The lower chart in Fig. 4.33 illustrates micromagnetic simulation results for HAMR operation at close to 3 Tb/in^2. The results show that in theory it is possible to achieve such areal density without introducing other technology such as bit-patterned media (BPM).

4.3.4.3 HAMR recording model, design, and understanding of recording process

To fully understand the HAMR process, research has focused on understanding the magnetization reversal process at elevated temperatures. At a temperature close to T_c, the governing equation is based on Landau–Lifshitz–Bloch (LLB) formalism with renormalization of the magnetic properties of the recording media [54, 55]. As the media cools down to the temperature where the transition is defined, both Landau–Lifshitz–Gilbert (LLG) and LLB methods are used to understand the formation of the transition. At a temperature sufficiently below T_c, the difference between LLB and LLG methods is reduced.

Note that the detailed interaction terms and how the reordering of the magnetization occurs before the transition is formed is an area of active research. These models are challenged by the fact

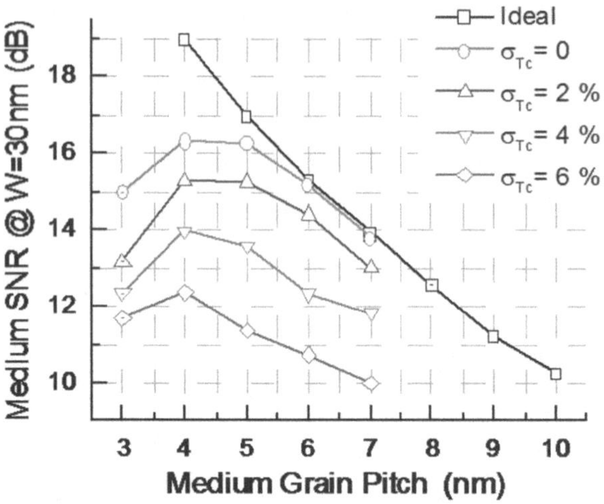

Figure 4.34 The micromagnetic HAMR system model results for the SNR plotted vs. grain size for different T_C distributions.

that the magnetic reordering takes place as the media cools below T_c at a very rapid rate, which for our 10 K/nm thermal gradient means a cooling rate of 200 K/ns at 20 m/s disk velocity. There are several groups working together to get an improved HAMR model for system study [54, 55, 67–71]. The inputs to a HAMR system model requires an accurate picture of the media temperature profile, the head field profile, and the media magnetic properties at different temperatures. In addition, interactions between different components are crucial for an optimized HAMR design.

Figure 4.34 shows one of the HAMR system model analyses on the impact of T_c distribution and medium grain size [68]. The results clearly show that the significant penalty will occur if the T_c distribution, one of the leading contributing factors to SFD, is not well controlled.

4.3.5 *Head–Disk Interface*

The head–disk interface (HDI) in HAMR has all of the challenges of today's recording systems with the addition of high temperatures and new materials in the head and media. In our recording example,

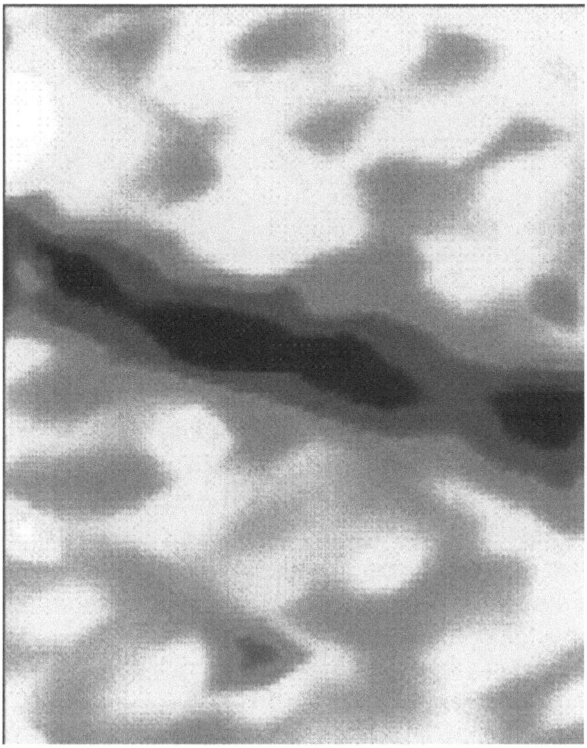

Figure 4.35 XPS image of the media surface showing the lubricant desorption along the recorded track.

the peak media temperature was 100° above the T_c. With pure FePt that would mean temperatures in the recording layer of over 800 K. At this temperature, neither the conventional lubricant nor the media overcoat is expected to be stable. The head will also be heated from the optical losses in the optical path, the laser diode, and the media back-heating in addition to the usual heat sources, the magnetic writer and the heater.

It was observed early in the optically assisted recording that the lubricant could be damaged from the optical spot. Figure 4.35 shows an X-ray photoelectron spectroscopy (XPS) image of the disk surface after recording in the QUINTA Corporation optically assisted drive [30]. This led to HDI failure. In HAMR the much shorter

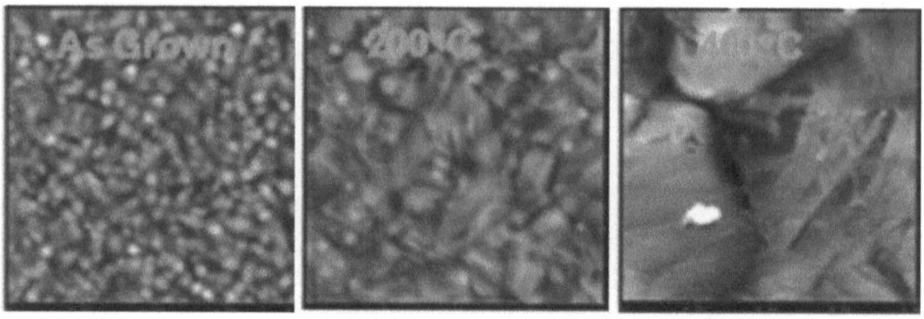

Figure 4.36 The gold microstructure seen in the as-deposited state and after heating to 250°C and 400°C. Courtesy of Mike Kautzky.

heating time does reduce the lubricant loss [72] but conventional perfluoropolyether (PFPE) lubricants such as Z-Dol 4000 or Z-Tetrol, which are used to coat current magnetic recording media surfaces, cannot withstand HAMR recording. Thermogravimetric analysis is often used to compare the thermal stability of different lubricants. Increased-molecular-weight species are known to be more stable [73]. New lubricants have been developed that are much more stable at higher temperatures [74, 75]. These new materials hold the promise of meeting all of the HDD requirements as well as being stable during HAMR recording.

The NFT needs to be near the ABS to couple energy to the recording medium efficiently. The primary ingredient in the NFT is typically gold. During the recording the NFT temperatures can reach 250°C–400°C, where the gold is not mechanically stable [76, 77]. Changes in the gold microstructure impact its density and shape that change the media heating efficiency and eventually degrade the optical confinement and the ability to heat the media to the required recording temperature. Figure 4.36 shows the changes in the gold microstructure caused by heating to 250°C and 400°C [78]. The large increase in grain size will impact the NFT behavior. Current research is focused on NFT and media designs that limit head temperature increases and material and seed layer modifications that improve NFT robustness.

In addition to the above HDI concerns, media and head overcoats will need to provide the usual corrosion and mechanical protection but also be stable at elevated temperatures. Also the dynamic temperature changes during the transition to writing will lead to head deformation that will make controlling the HMS a challenge. Progress in HAMR HDI is a key to commercialization.

4.3.6 *Status and Challenges*

A number of significant announcements of HAMR progress toward commercialization have been made since 2010. In the fall of 2011, TDK Corporation announced achieving 1 Tb/in^2 with an on-track error rate of 10^{-2} [79]. With high-coercivity HAMR media (FePt based), 1100 kBPI by 915 kTPI was achieved that could not be recorded with a conventional head. Then in March 2012, Seagate announced 1 Tb/in^2 at the same error rate but at a commercializable aspect ratio, 1975 kBPI by 510 kTPI [80, 81]. The key was the improved thermal confinement from the NFT and the FePt media with an optimized heat sink. These demonstrations showed parity with PMR technology and established the areal density promise of HAMR.

A number of fully integrated HAMR drives have been publicly demonstrated. In August 2012, Seagate used a 7200 rpm external HAMR drive to present a paper on HAMR technology [82] at The Magnetic Recording Conference (TMRC). A photo of the drive is shown in Fig. 4.37. In October 2013, TDK received the Grand Prix in the Key Technology Category of the CEATEC AWARDS 2013 for the demonstration of a 10,000 rpm HAMR head in a Seagate drive [83]. The drives had clear covers, as shown in Fig. 4.38, and were running various video streams. Also in 2013, Western Digital presented a conference presentation at the 2013 China (Ningbo) International Forum on Advanced Materials and Commercialization from a PC with an internal HAMR 2.5" drive [84]. These drives demonstrate the ability to integrate HAMR technology into an HDD.

As discussed in the previous section, HAMR faces a number of challenges in NFT and HDI reliability. In early HAMR demonstrations the head could only write for very short times (second to minutes). At TMRC in August 2014, Seagate announced the successful completion of a 1000-write-power-on-hour test for a 10,000 rpm enterprise drive [85], as shown in Fig. 4.39. This is an important demonstration of HAMR durability. Of course, for HAMR to become

Figure 4.37 The Seagate 7200 rpm HAMR drive from TMRC 2012.

the mainstream HDD technology, the drives need to hit common mean time between failures of 300,000 and 1,200,000 hours of operation (read and write hours).

Recent HAMR demonstrations in areal density, drive integration, and reliability are the reason that HAMR appears as the next technology on the ASTC roadmap. To achieve productization, further developments will be needed and these are active areas of research at all of the HDD companies and suppliers.

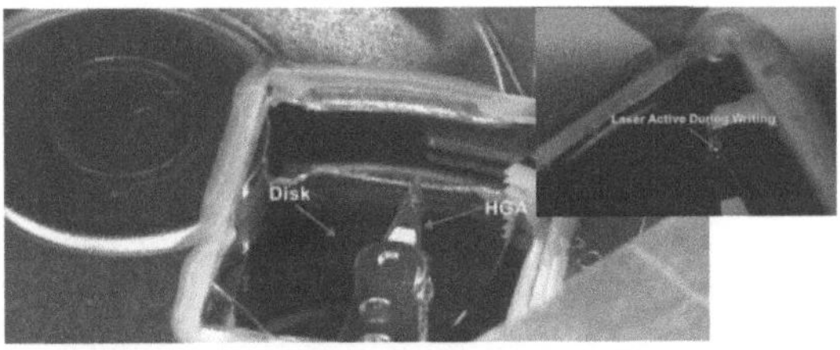

Figure 4.38 An HAMR drive with a clear cover from the TDK/Seagate Ceatec demo. The laser is lasing in the inset.

Figure 4.39 Demonstration of Seagate's HAMR drive with more than 1000 write-hour single-head life test. The test was conducted at 10,500 rpm on a 3400-track test band, equivalent to 2 full drive-writes per day for 5 years.

4.4 Conclusions

The unabated exponential demand for data storage has been responsible for and enabled by the continued areal density and capacity growth of HDDs. It is seen in current HDD products that the cumulative annual growth rate (CAGR) has fallen below historical averages. This threatens our ability to economically store the world's expanding data volume or for HDDs to lose their dominant position to other technologies most notably flash storage. This desire to continue the areal density growth has resulted in large investments in future HDD technologies.

In the current view, energy-assisted magnetic recording (EAMR) is the most likely new technology to restore the historical areal density growth. The ability to add microwave or thermal energy to the recording process allows us to envision a head and media that can reach much higher areal densities than today's PMR systems. This additional energy does bring new challenges in reliability and manufacturability that must be overcome. With the recent progress in performance and understanding and the large investments by the HDD industry, we expect to see commercially available EAMR HDDs in this decade. As shown in the ASTC roadmap, Fig. 4.1, EAMR is expected to enable the 1 to 4 Tb/in^2 HDD products. Beyond 4–5 Tb/in^2, we need to further reduce the number of grains per bit with technologies such as BPM.

Acknowledgments

For the MAMR topic, the authors would like to thank Drs. Xiaochun Zhu, Yuhui Tang, Yiming Wang, and Matt Moneck for their contribution to this work while at Carnegie Mellon University. The research has been supported in part by the Data Storage Systems Center at Carnegie Mellon and some of its industrial sponsors and by the National Science Foundation MRSEC through Johns Hopkins University.

For the HAMR topic, the authors would like to thank Drs. Chubing Peng, Hua Zhou, Chris Rea, Jie Zou, Nils Gokemeijer, Yimin Niu, Lien Lee, Yongjun Zhao, Lijuan Zou, James Kiely, Amit Itagi, William Challener, Peter Czoschke, Heidi Olson, Mark Ostrowski, Darren Karns, Xiaobin Wang, Mike Seigler, Cynthia Hipwell, Amit Sharma, Alexander Wu, Timothy Klemmer, Yingguo Peng, Xiaobin Zhu, Yinfeng Ding, Eric KC Chang, Yukiko Kubota, Hassib Amini, Thomas Rosecamp, Jan-Ulrich Thiele, Ganping Ju, Thomas Chang, Steve Hwang, Tim Rausch, Jon D. Trantham, Alfredo S. Chu, Housan Dakroub, Jason W. Riddering, Charles P. Henry, Deepak Prakash, Ying Qi, Bruce Hunstad, Kerry Mahoney, Qisuo Chen, Michael Wissbaum, Kevin Heim, Cal Hardie, Pu-Ling Lu, MinJie Ma, John W. Dykes, and

Mark Re for their contribution to HAMR work in Seagate. In addition, the authors would like to thank Showa Denko, TDK/SAE/Headway, and the ASTC for their genuine support for HAMR.

We especially thank Sangita Kalarickal, Lynn Li, and Pu-Ling Lu for critically reading this chapter.

References

1. Zhu J. G., Zhu X., and Tang Y. (2008). Microwave assisted magnetic recording, *IEEE Trans. Magn.*, **44**, pp. 125–131.
2. Kryder M. H., Gage E. C., McDaniel T. W., Challener W. A., Rottmayer R. E., Ju G., Hsia Y. T., and Erden M. F. (2008). Heat assisted magnetic recording, *Proc. IEEE*, **96**, pp. 1810–1835.
3. Gao K. Z., Heinonen O., and Chen Y. (2009). Read and write processes, and head technology for perpendicular recording, current perspectives: perpendicular recording, *J. Magn. Magn. Mater.*, **321**(6), pp. 495–507.
4. Charap S. H., Lu P., and He Y. (1997). Thermal stability of recorded information at high densities, *IEEE Trans. Magn.*, **33**, pp. 978–983.
5. Wang X., Gao K. Z., Zhou H., Itagi A., Seigler M., and Gage E. (2013). HAMR recording limitations and extendibility, *IEEE Trans. Magn.* **49**, p. 128.
6. IDEMA. (2014). *ASTC Technology Roadmap 2014*. The Advanced Storage Technology Committee of the International Disk Equipment and Materials Association, San Jose, CA, Sept. 2, 2014, http://www.idema.org/?page_id=5868
7. Zhu J., and Wang Y. (2010). Microwave assisted magnetic recording utilizing perpendicular spin torque oscillator with switchable perpendicular electrodes, *IEEE Trans. Magn.*, **46**, pp. 751–757.
8. Shiroishi Y., Fukuda K., Tagawa I., Iwasaki H., Takenoiri S., Tanaka H., Mutoh H., and Yoshikawa N. (2009). Future options for HDD storage, *IEEE Trans. Magn.*, **45**, pp. 3816–3822.
9. Xi H., Gao K. Z., and Shi Y. (2004). Microwave generation by a direct current spin-polarized current in nanoscale square magnets, *Appl. Phys. Lett.*, **84**, p. 4977.
10. Zhu X., and Zhu J. G. (2006). Bias-field-free microwave oscillator driven by perpendicularly polarized spin current, *IEEE Trans. Magn.*, **42**, pp. 2670–2672.
11. Thirion C., Wernsdorfer W., and Mailly D. (2003). Switching of magnetization by nonlinear resonance studied in single nanoparticles, *Nat. Mater.*, **2**, pp. 524–527.

12. Scholz W., and Batra S. (2008). Micromagnetic modeling of ferromagnetic resonance assisted switching, *J. Appl. Phys.*, **103**, p. 07F539.
13. Berger L. (1996). Emission of spin waves by a magnetic multilayer traversed by a current, *Phys. Rev. B*, **54**, pp. 9353–9358.
14. Berger L. (2001). Effect of interfaces on Gilbert damping and ferromagnetic resonance linewidth in magnetic multilayers, *J. Appl. Phys.*, **90**, p. 4632.
15. Slonczewski J. C. (1996). Current-driven excitation of magnetic multilayers, *J. Magn. Magn. Mater.*, **159(**1–2), pp. L1–L7.
16. Kiselev S. I., Sankey J. C., Krivorotov I. N., Emley N. C., Schoelkopf R. J., Buhrman R. A., and Ralph D. C. (2003). Microwave oscillations of a nanomagnet driven by a spin-polarized current, *Nature*, **425(**6956), pp. 380–383.
17. Wang Y., Tang Y., and Zhu J.-G. (2009). Media damping constant and performance characteristics in microwave assisted magnetic recording with circular AC field, *J. Appl. Phys.*, **105**, p. 07B902.
18. Igarashi M., Suzuki Y., Miyamoto H., Maruyama Y., and Shiroishi Y. (2009). Effect of elliptical high-frequency field on, *IEEE Trans. Magn.*, **45**, pp. 3711–3713.
19. Victora R. H., and Shen X. (2005). Exchange coupled composite media for perpendicular magnetic recording, *IEEE Trans. Magn.*, **41**, pp. 2828–2833.
20. Wang J., Shen W., and Bai J. (2005). Exchange couple composite media for perpendicular magnetic recording, *IEEE Trans. Magn.*, **41**, pp. 3181–3186.
21. Tang K., Bian X., Choe G., Takano K., Mirzamaani M., Wang G., Zhang J., Xiao Q.-F., Ikeda Y., Risner-Jamtgaard J., and Xu X. (2009). Design consideration and practical solution of high-performance perpendicular magnetic recording media, *IEEE Trans. Magn.*, **45**, pp. 786–792.
22. Suess D., Schrefl T., Fähler S., Kirschner M., Hrkac G., Dorfbauer F., and Fidler J. (2005). Exchange spring media for perpendicular recording, *Appl. Phys. Lett.*, **87**, p. 012504.
23. Zhu J. (2014). SNR and areal density gain in MAMR with segmented media, *IEEE Trans. Magn.* **50**, p. 3200809.
24. Zhu, J. G., and Wang Y. (2011). SNR enhancement in segmented perpendicular media, *IEEE Trans. Magn.*, **47**, pp. 4066–4072.
25. Nozaki Y., Ishida N., Soeno Y., and Sekiguchi K. (2012). Room temperature microwave-assisted recording on 500-Gbpsi-class perpendicular medium, *J. Appl. Phys.*, **112**, p. 083912.

26. Matsubara M., Shiimoto M., Nagasaka K., Sato Y., Udo Y., Sugiura K., Hattori M., Igarashi M., Nishida Y., Hoshiya H., Nakamoto K., and Tagawa I. (2011). Experimental feasibility of spin-torque oscillator with synthetic field generation layer for microwave assisted magnetic recording, *J. Appl. Phys.*, **109**, p. 07B741.

27. Shiimoto M., Igarashi M., Sugiyama M., Nishida Y., and Tagawa I. (2013). Effect of effective field distribution on recording performance in microwave assisted magnetic recording, *IEEE Trans. Magn.*, **49**, pp. 3636–3639.

28. McDaniel T. W., and Victora R. (1996). *Handbook of Magneto-Optic Data Recording, Materials, Subsystems, Techniques* (Noyes, Salem, MA) ISBN: 978-0-8155-1391-9.

29. Knight G. (1998). Near-field recording: TeraStor's mass storage benchmark technology for the next decade, in *Proceedings of the THIC Meeting*, San Mateo, CA.

30. Heanue J. F. (1999). Micro-optical requirements for optically-assisted Winchester recording heads, *Proc. SPIE*, **3778**, pp. 98–103.

31. Abarra E. N., Acharya B. R., Inomata A., and Okamoto I. (2001). Synthetic ferrimagnetic media, *IEEE Trans. Magn.*, **37**, pp. 1426–1431.

32. McDaniel T. W., Challener W. A., and Sendur K. (2003). Issues in heat-assisted perpendicular recording, in *Digest of Joint NAPMRC 2003*, DOI:10.1109/NAPMRC.2003.1177070

33. Peng C., and Challener W. A. (2004). Input-grating couplers for narrow Gaussian beam: influence of groove depth, *Opt. Exp.*, **12**, 6481.

34. Challener W. A., Gage E., Itagi A., and Peng C. (2006). Optical transducers for near field recording, *J. Appl. Phys.*, **45**(pt.1), p. 6632.

35. Rausch T., Bain A., Stancil D. D., and Schlesinger T. E. (2004). *IEEE Trans. Magn*, **40**, p. 137.

36. Challener W. A., Peng C., Itagi A. V., Karns D., Peng W., Peng Y., Yang X., Zhu X., Gokemeijer N. J., Hsia Y. T., Ju G., Rottmayer R., Seigler M., and Gage E. C. (2009). Heat-assisted magnetic recording by a near-field transducer with efficient optical energy transfer, *Nat. Photon.*, **3**, p. 220.

37. Stipe B. C., Strand T. C., Poon C. C., Balamane H., Boone T. D., Katine J. A., Li J.-L., Rawat V., et al, (2010). Magnetic recording at 1.5 PB/m^2 using an integrated plasmonic antenna, *Nat. Photon.*, **4**, pp. 484–488.

38. Mizuno T., Tsuchiya Y., Shimazawa K., and Chou T. (2009). *Near-Field Light Transducer Comprising Propagation Edge with Predetermined Curvature Radius*, US Patent 8169731 B2.

39. Zhou, N, Kinzel E. C., and Xu X. (2011). Nanoscale ridge aperture as near-field transducer for heat-assisted magnetic recording, *Appl. Opt.*, **50**, pp. G42–G46.

40. Jin E. X., and Xu X. (2005). Obtaining super resolution light spot using surface plasmon assisted sharp ridge nano-aperture, *Appl. Phys. Lett.*, **86**, p. 111106.

41. Bhargava S., and Yablonovitch E. (2014). Multi-objective inverse design of sub-wavelength optical focusing structures for heat assisted magnetic recording, *Proc. SPIE*, **9201**, p. 92010M.

42. Bell B. W. (1992). *Flying Magnetooptical Read/Write Head Employing an Optical Integrated Circuit Waveguide*, US Patent 5199090; Peng C., and Gage E. (2012). *Light Delivery*, US Patent 8750081.

43. Goulakov A., Gao K. Z., and Jin X. (2010). *Direct Waveguide Light Delivery to NFT for Heat Assisted Magnetic Recording*, US Patent 8649245. The laser in the slider was presented by T. E. Schlesinger in an INSIC ATP project in 2002 but was not officially published.

44. Weller D., Moser A., Folks L., Best M. E., Lee W., Toney M. F., Schwickert M., Thiele, J. U., and Doerner M. F. (2000). High K_u materials approach to 100Gbits/in^2, *IEEE Trans. Magn.*, **36**, pp. 10–15.

45. Klemmer T., Hoydick D., Okumura H., Zhang B., and Soffa W. A. (1995). Magnetic hardening and coercivity mechanisms in L10 ordered FePd ferromagnets, *Scr. Metall. Mater.*, **33**, pp. 1793–1805.

46. Weller D., Parker G., Mosendz O., Champion E., Stipe B., Wang X., Klemmer T., Ju G., and Ajan A. (2014). A HAMR media technology roadmap to an areal density of 4Tb/in^2, *IEEE Trans. Magn.*, **50**, p. 3100108.

47. Acharya R. (2014). Challenges and recent developments in heat assisted magnetic recording media, in *TMRC Conference*, Berkeley, August 2014.

48. Richter H. J. (2007). The transition from longitudinal to perpendicular recording, *J. Phys. D: Appl. Phys.*, **40**, pp. R149–R177.

49. Rausch T., Trantham J. D., Chu A. S., Dakroub H., Riddering J. W., Henry C. P., Kiely J. D., Gage E., and Dykes, J. W. (2013). HAMR drive performance and integration challenges, *IEEE Trans. Magn.*, **49**, pp. 730–733.

50. Thiele J. U., Coffey K. R., Toney M. F., Hedstrom J. A., Kellock A. J. (2002). Temperature dependent magnetic properties of highly chemically ordered $Fe_{55-x}Ni_xPt_{45}L_{10}$ films, *J. Appl. Phys.*, **91**, p. 6595.

51. Gao K. Z., and Bertram H. N. (2003). Transition jitter estimates in tilted and conventional perpendicular recording media at 1Tb/in^2, *IEEE Trans. Magn.*, **39**, p. 704.

52. Wang X., Gao K. Z., Hohlfeld J., and Seigler M. (2010). Switching field distribution and transition width in energy assisted magnetic recording, *Appl. Phys. Lett.*, **97**, p. 102502.
53. Gao K. Z., Fernandez-de-Castro J., and Bertram H. N. (2005). Micromagnetic study of the switching fields in polycrystalline magnetic thin-film media, *IEEE Trans. Magn.*, **41**, p. 4236.
54. Evans R. F. L., Hinzke D., Atxitia U., Nowak U., Chantrell R. W., and Chubykalo-Fesenko O. (2012). Stochastic form of the Landau-Lifshitz-Bloch equation, *Phys. Rev B.*, **85**, p. 014433.
55. Victora R. H., and Huang P. W.(2013). Simulation of heat-assisted magnetic recording using renormalized media cells, *IEEE Trans. Magn.*, **49**, p. 751.
56. Wang X., Gao K. Z., and Seigler M. (2011). Magnetization switching variations and transition width limit in heat assisted magnetic recording, *IEEE Trans. Magn.*, **47**, p. 2371.
57. Stoner E. C., and Wohlfarth E. P. (1948). A mechanism of magnetic hysteresis in heterogeneous alloys, *Philos. Trans. R. Soc. London, A*, **240**, p. 74.
58. Xu B. X., Liu Z. J., Ji R., Toh Y. T., Hu J. F., Li J. M., Zhang J., Ye K. D., and Chia C. W. (2012). Thermal issues and their effects on heat assisted magnetic recording system, *J. Appl. Phys.*, **111**, p. 07B701.
59. Wang X., and Bertram H. N, (2003). Simple transition parameter expression including grain size and intergranular exchange, *J. Appl. Phys.*, **93**, p. 7005.
60. Wang X., Valcu B., and Yeh N. (2009). Transition width limit in magnetic recording, *Appl. Phys. Lett.*, **94**, p. 202508.
61. Rausch T., Bain J. A., Stancil D. D., Schlesinger T. E. (2004). Thermal Williams-Comstock model for predicting transition lengths in a heat assisted magnetic recording system, *IEEE Trans. Magn.*, **40**, p. 137.
62. Kanai Y., Matsubara R., Watanabe H., Muraoka H., and Nakamura Y. (2003). Recording field analysis of narrow track SPT head with side shields, tapered main pole, and tapered return path for 1Tb/in^2, *IEEE Trans. Magn.*, **39**, p. 1955.
63. Hohlfeld J., Zheng X., and Benakli M. (2015). Measuring temperature and field profiles in heat assisted magnetic recording, to be published in *J. Appl. Phys.* 2015
64. Richter H., Parker G., Staffaroni M., Grobis M., Stipe B. C. (2014). Heat assisted magnetic recording with laser pulsing, *IEEE Trans. Magn.*, **50**, p. 3001707.

65. Gao K. Z. (2014). Impact of transition curvature for heat assisted magnetic recording, *IEEE Magn. Lett*, to be published.
66. TDK. Next-generation thermal assisted magnetic recording using a near field light, *TDK Tech. Mag.*, 2, http://www.global.tdk.com/techmag/inductive/vol2/index3.htm; Shiloy A. (2012). *TDK to Start Making Heads for TAMR-Based Hard Drives in 2014: TDK Demos Magnetic Head for Thermal-Assisted Recording*, http://www.xbitlabs.com/news/storage/display/20121002220536_TDK_Expects_TAMR_Based_Hard_Drives_in_2014.html
67. Greaves S., Kanai Y., and Muraoka H. (2012). Magnetization switching in energy assisted recording, *IEEE Trans. Magn.*, **48**, p. 1794.
68. Li H., and Zhu J. G. (2014). Understanding the impact of T_C and H_K variation on signal-to-noise ratio in heat assisted magnetic recording, *J. Appl. Phys.*, **115**, p. 17B744.
69. Chang R., Fu S., Lubarda M., Menarini M., and Lomakin V. (2014). Numerical modeling of heat assisted magnetic recording system, in *ASTC Research Report and Radio Science Meeting (Joint with AP-S Symposium)*, 2014 USNC-URSI.
70. Champion E. (2014). Micromagnetic modeling in HAMR head-media systems, in *Intermag 2014* Dresden, p. AD-03.
71. Li Z., Wei D., and Wei F. (2008). Micromagnetic modeling for heat assisted magnetic recording, *J. Magn. Magn. Mater.*, **320(**22), p. 3108.
72. Ma Y., Gonzaga L., An C., and Liu B. (2011). Effect of laser heating duration on lubricant depletion in heat assisted magnetic recording, *IEEE Trans. Magn.*, **47**, p. 3445.
73. Li L., Jones, P. M., and Hsia Y.-T. (2004). Effect of chemical structure and molecular weight on high-temperature stability of some fomblin Z-type lubricants, *Tribol. Lett.*, **16(**1–2), pp. 21–27.
74. Zhang J., Ji R., Xu J. W., Ng J. K. P., Xu B. X., Hu S. B., Yuan H. X., and Piramanayagam S. N. (2006). Lubrication for heat-assisted magnetic recording media, *IEEE Trans. Magn.*, **42**, pp. 2546–2548.
75. Gauvin M., Zheng H., Suen B., Lee J., Kang H. J., and Talke F. E. (2011). Enhanced photo-thermal stability of modified PFPE lubricants under laser beam exposure, *IEEE Trans. Magn.*, **47**, pp. 1849–1854.
76. Zhou N., Xu X., Hammack A. T., Stipe B. C., Gao K. Z., Scholz W., and Gage E. C. (2014). Plasmonic near-field transducer for heat-assisted magnetic recording, *Nanophotonics*, **3**(3), pp. 141–155.
77. Estrine E. C., Tabakovic I., Seaton N. C. A., and Stadler B. J. H. (2014). Deposition and characterization of gold films and nanostructures for use in heat assisted magnetic recording, in *Intermag 2014* Dresden.

78. Karim S., Toimil-Molares M. E., Balogh A. G., Ensinger W., Cornelius T. W., Khan E. U., and Neumann R. (2006). Morphological evolution of Au nanowires controlled by Rayleigh instability, *Nanotechnology* **17** p. 5954; Karim S., Toimil-Molares M. E., Ensinger W., Balogh A. G., Cornelius T. W., Khan E. U., and Neumann R. (2007). Influence of crystallinity on the Rayleigh instability of gold nanowires, *J. Phys. D: Appl Phys*, **40**, p. 3767.
79. http://techon.nikkeibp.co.jp/article/NEWS/20111003/198905/
80. Seagate Press Release. (2012). *Seagate Reaches 1 Terabit per Square Inch Milestone in Hard Drive Storage with New Technology Demonstration*, March 19, 2012 http://www.seagate.com/newsroom/press-releases/terabit-milestone-storage-seagate-master-pr/?paramChannelName=newsroom
81. Wu A. Q., Kubota Y., Klemmer T., Rausch T., Peng C., Peng Y., Karns D., Zhu X., Ding Y., Chang E. K. C., Zhao Y., Zhou H., Gao K. Z., Thiele J.-U., Seigler M., Ju G., and Gage E. (2013). HAMR areal density demonstration of 1+ Tbpsi on spinstand, *IEEE Trans. Magn.*, **49**, pp. 779–782.
82. http://ieeexplore.ieee.org/stamp/stamp.jsp?arnumber=6417002
83. http://www.ceatec.com/2013/news/en-webmagazine/e023
84. Western Digital Press Release. (2013). *WD® Demonstrates Heat Assisted Magnetic Recording Hard Drive Technology at the 2013 China (Ningbo) International Forum on Advanced Materials and Commercialization*, November 13, 2013, http://wdc.com/en/company/pressroom/releases/?release=dc8e1c07-6a5b-48ce-b931-e090e566da29
85. Rausch T., Chu A. S., Lu P. L., Puranam S., Nagulapally D., Dykes J., Lammers T., and Gage E. (2015). Recording performance of a pulsed HAMR architecture, presentation B5, in *TMRC*, August 11, 2014 (to be published in 2015).

Chapter 5

$L1_0$–FePt Granular Films for Heat-Assisted Magnetic Recording Media

Kazuhiro Hono and Yukiko K. Takahashi
Magnetic Materials Unit, National Institute for Materials Science, 1-2-1 Tsukuba, 305-0047, Japan
kazuhiro.hono@nims.go.jp

FePt–C nanogranular films are coming close to the actual application as recording media for the next-generation high-density recording system, heat-assisted magnetic recording (HAMR). To optimize the $L1_0$–FePt nanogranular structure suitable for high-areal-density recording, various combinations of segregants and seed layer materials have been explored. In this chapter, we review recent investigations aiming at achieving the ideal nanogranular structure for HAMR media.

5.1 Introduction

Hard disk drives (HDDs) have been a major storage device in computers for the last 50 years and more recently in home video-recording systems and cloud computing due to their large

Ultrahigh-Density Magnetic Recording: Storage Materials and Media Designs
Edited by Gaspare Varvaro and Francesca Casoli

ISBN 978-981-4669-58-0 (Hardcover), 978-981-4669-59-7 (eBook)
www.panstanford.com

storage capacity, low bit cost, and nonvolatility. Although solid-state devices (SSDs) are now replacing the storage devices for notebook computers, HDDs will continue to be a major player in the data storage industry. The number of HDDs sold in 2012 was approximately 800 million, nearly one-tenth of the world population! To meet the demand for rapidly increasing digital data and to keep the competitiveness of HDDs with other emerging storage devices, the areal density of HDDs must keep increasing.

The areal density of the current HDD is approximately 800 Gbit/in^2. In the past, it had shown an increase of 100% every year, but the recent growth rate has slowed down to 30%, reaching saturation [1]. Although various technological improvements have been made in both heads and media in the current perpendicular magnetic recording (PMR) system, ~1 Tbit/in^2 is considered to be the limit for the PMR method. Due to the recent improvements in signal processing using shingled magnetic recording, the areal density of PMR is expected to increase marginally above 1 Tbit/in^2. However, a transition to a new magnetic recording method must be realized in the near future to sustain the growth of the recording industry.

Heat-assisted magnetic recording (HAMR) is a technology close to commercialization. However, a new medium that is totally different from the current one must be developed using a high-magnetocrystalline-anisotropy ferromagnetic material. The current PMR recording media are CoCrPt–SiO_2 nanogranular films, in which nanosized CoCrPt columnar grains with the hexagonal close-packed (hcp) structure are dispersed uniformly with a strong [0001] fiber texture, as shown in Fig. 5.1. Such films are deposited on glass substrates with an amorphous CoTaZr (a-CoTaZr) soft magnetic underlayer and a Ru seed layer that optimizes the grain size and crystallographic orientation of the CoCrPt grains. Since the easy axis of magnetization of the CoCrPt alloy is [0001], this crystallographic texture gives rise to a strong perpendicular magnetic anisotropy (PMA). Each grain of approximately 8 nm in diameter is magnetically isolated, and 1 bit contains multiple CoCrPt grains that are magnetized in the same direction, as shown in Fig. 5.2 [2]. The bit size for the areal density of 1 Tbit/in^2 will be ~25 × 25 nm^2, in which more than 10 ferromagnetic grains must be contained to attain a

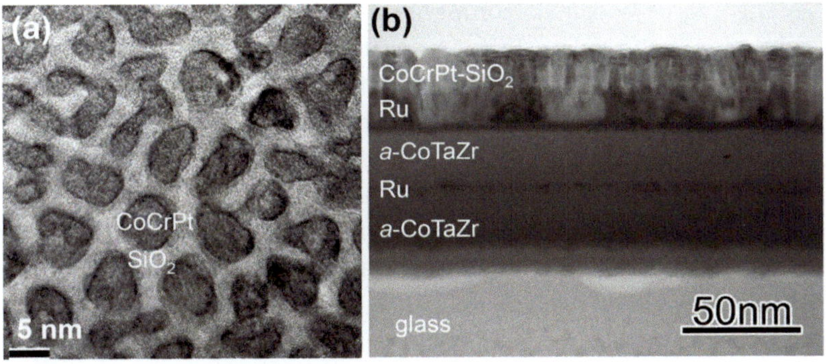

Figure 5.1 (a) Plan-view TEM bright-field images and (b) cross-sectional TEM image of the currently used perpendicular recording media. CoCrPt ferromagnetic grains of ~8 nm are dispersed in a SiO_2 matrix. On the glass substrate, an amorphous-CoTaZr soft magnetic underlayer and a Ru interlayer that align the [0001] axis of the CoCrPt alloy in the perpendicular direction to the film plane are grown.

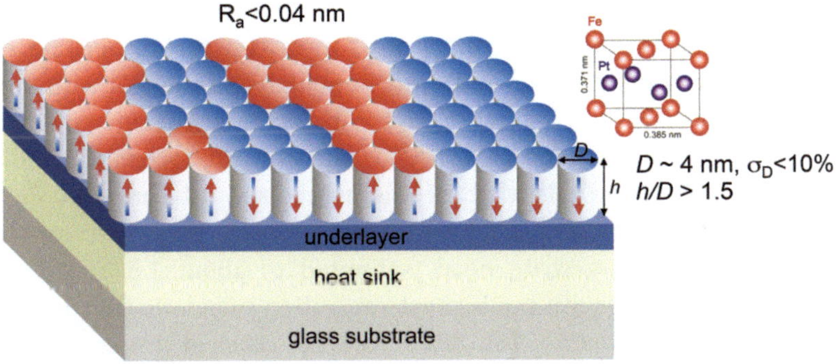

Figure 5.2 The ideal structure of an HAMR medium for 4 $Tbit/in^2$ areal density. Red and blue colors show the direction of magnetic poles. The $L1_0$–FePt crystal (upper-right figure) must grow columnarly separated by the nonferromagentic segregant. The column-height-to-grain-diameter ratio h/D must be higher than 1.5 and the diameter of the columnar grains is desired to be less than 4 nm.

sufficiently high signal-to-noise ratio (SNR). This suggests that the ferromagnetic grains must be refined further to ~4 nm for an areal density above 1 $Tbit/in^2$. This requirement makes the ferromagnetic grains thermally unstable as the magnetocrystalline anisotropy energy K_uV becomes comparable to the thermal energy k_BT, where K_u is the uniaxial magnetocrystalline anisotropy constant, V is the volume of the grain, and k_B is the Boltzmann constant. To keep the recorded information for longer than 10 years, K_uV/k_BT must be larger than 60. This means that ferromagnetic materials with high magnetocrystalline anisotropy such as $L1_0$–FePt, $L1_0$–CoPt, $Nd_2Fe_{14}B$, and $SmCo_5$ must be used for high-density recording, where V becomes as small as $\sim 4 \times 10^{-26}$ m^3 [3, 4]. As the media require long-term endurance, the rare-earth compounds are unlikely to be selected as they are very susceptible to oxidation and corrosion. As a result, the $L1_0$–FePt phase is currently considered to be the most promising material for HAMR media.

While K_u of the CoCrPt alloy is ~0.3 MJ/m^3, $L1_0$-ordered FePt has an order of magnitude larger magnetocrystalline anisotropy of 6.6 MJ/m^3, which makes the minimum size of stable ferromagnetic grains ~4 nm for spheres and 2.4 nm for cylinders [5]. In addition, the magnetic polarization of the $L1_0$–FePt phase is $\mu_0 M_s \approx 1.4$ T, which is large enough to gain a sufficient SNR with a film thickness of less than 8 nm. However, if hard magnetic grains become so small, the magnetic field required to switch the magnetization of the grain (H_N) or coercivity (H_c) will increase enormously. The typical switching field, or $\mu_0 H_c$, for the current recording media is about 0.8 T, while that of the nanosized $L1_0$-ordered FePt will be higher than 3 T. The highest magnetic field that can be generated using a write head is limited to about 1.5 T using a soft ferromagnet with the highest saturation magnetic flux density of 2.35 T for $Fe_{65}Co_{35}$ [6]. Since the reduction of grains of hard magnetic materials to a nanosized dimension leads to substantial increase of coercivity larger than 1.5 T, these grains will not be writable with the head field. This is known as "trilemma" of magnetic recording. Ultrahigh-density magnetic recording requires nanosized ferromagnetic grains, which become thermally unstable below certain sizes. To overcome this problem, a high-K_u material must be used; then the switching field of the grains becomes too large to write.

To overcome the trilemma, the magnetization switching must be assisted by some external energy such as heat or microwave. HAMR technology uses heat to assist the magnetization switching of high-

K_u grains using a well-focused laser beam [7, 8]. As temperature increases, the coercivity of the grains decreases because of reduced anisotropy field. Writing is done at the temperature slightly lower than Curie temperature, where the coercive field becomes smaller than the head field of about 1 T. When temperature decreases rapidly after writing, the high-K_u nanosized grain has sufficient thermal stability for permanent recording. To realize HAMR, new technologies are required in both the write head and the medium. The HAMR head has been demonstrated recently using a plasmonic antenna [7]. The tolerance of near-field optical transducers is the most critical issue for improving the reliability of HAMR HDDs to a practical level. As to the media, the FePt–C system is the only viable system for HAMR media at this moment. However, the grain size and in-plane magnetic component are not reduced to the desired level yet. Although the first HAMR HDD is anticipated to use the FePt–C-based media with a recording density slightly higher than the upper limit of PMR, a further increase in the areal density toward 4 Tbit/in^2 depends on how we can optimize the nanogranular structure of the FePt–C-based media on glass disks through various buffer layers for heat sinking and crystal orientation alignments. In this paper, we review the history of the development and optimization of the $L1_0$-FePt-based nanogranular films based on the published literature[a].

5.2 Ideal Structure for HAMR Media

The desirable HAMR media should be comprised of densely dispersed ferromagnetic grains exhibiting high magnetocrystalline anisotropy of larger than 5 MJ/m^3 with a uniform grain size of $\sim$4 nm with a center-to-center distance of 5 nm. The size distribution of the grains must be lower than 10% for attaining sufficiently low switching-field distribution (SFD). For a sufficient SNR, the grain shape should be columnar with the grain aspect ratio $h/D > 1.5$, where h is the height and D is the diameter of

[a] Many investigations carried out in the industry are kept confidential. It is beyond the authors' knowledge to make a comprehensive review including the efforts carried out in industrial laboratories. Hence, the chapter is based on the published literature only.

Table 5.1 HAMR media requirements for 2 and 4 Tbit/in^2 released from the Advanced Storage Technology Consortium (ASTC)

Areal density (Tbit/in^2)	2	4
Grain size, D (nm)	6	4.3
Center-to-center distance, D_P (nm)	7	5.1
Size distribution, σ (%)	10–15	10–15
$\mu_0 M_s$ of ferromagnetic layer (T)	0.88	1
$\mu_0 M_s$ of ferromagnetic grain (T)	1.1	1.26
K_u (MJ/m^3)	3.5	5
$\mu_0 H_K$ (T)	8	10
$\sigma H_K/H_K$	5–10	5–10
T_c (°C)	480	430
$\sigma T_c/T_c$ (%)	2	2
Angular distribution, σ_θ (°)	2	0.8
Media thickness, t (nm)	9	8.2
Grain aspect ratio (h/D)	1.29	1.6
Grains/bit	6.7	6.2

FePt grains. Such an ideal film structure is schematically shown in Fig. 5.2 [2], and the proposed HAMR media requirements for 2 and 4 Tbit/in^2 released from the Advanced Storage Technology Consortium (ASTC) are shown in Table 5.1 [9]. The film must exhibit high coercivity above 3 T, more than four times larger than that of the current recording media. In addition, the microstructure should be realized on commercially viable substrates such as glass. The key factors to control the microstructure of HAMR media are (i) ordering the A1–FePt phase to the $L1_0$ structure by annealing without grain coarsening, (ii) obtaining a strong PMA with a low in-plane component, (iii) attaining a nanometer diameter (ideally 4 nm) with a narrow size distribution of less than 10%, and (iv) attaining low surface roughness (R_a<0.4 nm) because of narrow head–media distance for high-density recording. Fulfilling all of the above requirements has been challenging, and many investigations on the optimization of FePt-based nanogranular structure have been made since 2000. Seagate has recently demonstrated HAMR over the areal density of 1 Tbit/in^2 using supposedly FePt-based granular media [10]. However, further microstructure optimization is essential to extend the areal density to 4 Tbit/in^2.

5.2.1 $A1 \rightarrow L1_0$ Ordering

Although the $L1_0$-FePt phase is thermally stable below 1300°C, the FePt films sputter-deposited below 400°C have the disordered A1 structure because of a kinetic constraint. Since the melting temperature of the equiatomic FePt alloy is about 1830 K, the volume diffusion is very sluggish below 917 K (~640°C). This means that the sputtered A1 film has to be annealed above 640°C to make it transform to the $L1_0$ structure by volume diffusion [11] or must be deposited on heated substrates [12]. Hence, lowering the kinetic ordering temperature from the A1 to the $L1_0$ structure was considered to be essential, and the effect of various alloying elements to enhance the kinetics of $L1_0$ ordering was investigated. Among various elements, Cu and Ag were reported to be effective to reduce the kinetic ordering temperatures in continuous polycrystalline films [12–14]. By alloying FePt with Cu, the ordering temperature decreases to 400°C, while binary A1 FePt does not order below 600°C [12]. The Cu alloying, however, leads to the reduction of the magnetocrystalline anisotropy energy [15]. Note that most of these investigations on the effect of additives on ordering were carried out on continuous polycrystalline films, in which ordering was accompanied by recrystallization of nanocrystalline films [16]. So the elements that are effective in reducing the ordering temperature of continuous films do not necessarily work on the FePt–X granular films, where FePt grains are isolated by segregants [17]. It should be also noted that there is a size dependence of $L1_0$ ordering when the grain size becomes smaller than 4 nm [18–20]. This is because the surface energy becomes dominant compared to the volume free energy, where the A1 phase becomes more stable than the $L1_0$ phase. This size dependence of order may become significant when the grain size is successfully reduced to less than 4 nm for a 4 Tbit/in^2 areal density in the future.

5.2.2 [001] Texture

To apply the $L1_0$-ordered FePt thin film to HAMR media, the magnetic easy axis must be aligned normal to the film plane. However, the FePt thin films deposited onto amorphous substrates tend to develop a [111] texture because the (111) plane of the A1 phase has the lowest surface energy. In early studies of FePt thin films, (001) oriented MgO single-crystalline substrates were used to grow PMA films [21–27]. MgO has the NaCl structure with a lattice parameter of 0.421 nm, while the lattice parameters of $L1_0$–FePt are a = 0.386 nm and c = 0.373 nm in the facecentered tetragonal (fct) lattice or $a = 0.273$ nm and $c = 0.373$ nm in the bodycentered tetragonal (bct) lattice. The (001) FePt grows on (001) MgO with the cube-to-cube orientation relationship with a lattice mismatch of about 8.3%. Because of this large misfit, FePt is subject to receive expansion stress to the a and b axes directions, thereby growing only one variant for A1$\rightarrow$$L1_0$ ordering on a heated MgO substrate. This gives rise to a strong PMA to $L1_0$-ordered FePt thin films.

On the basis of detailed transmission electron microscopy (TEM) observations, Shima et al. showed the $L1_0$-ordered FePt film grows with the Volmer–Weber (VW) mode on (001) single-crystalline MgO at 700°C [26]. The coercivity of these films was as high as 4 T, which decreases abruptly when the isolated grains are percolated for nominal film thickness of larger than 40 nm. The growth mode changes from the VW mode to the Frank–Van der Merwe (FV) mode when the substrate temperature is decreased below 350°C [26], as shown in Fig. 5.3. This nature of the VW growth on the MgO substrate is essential for the formation of well-isolated FePt grains in recording media. The degree of order increases above 450°C, about 200°C lower than $T_m/2$, where T_m is the melting temperature of FePt. This is because $L1_0$ ordering progresses in situ on a heated substrate by surface diffusion. The coercivity of the films depends on both the morphology of the FePt grains and the degree of $L1_0$ order. Even if the degree of the $L1_0$ order is high, H_c can be low if FePt becomes continuous, as the magnetization switching can occur by the domain wall motion without any pinning after a nucleation. Single-domain grains of about 50 nm show a very large coercivity of 7 T at room temperature, which is nearly 50% of the anisotropy field H_A [28]. This level of coercivity with respect to the anisotropy field has never been observed in other hard magnetic systems.

One may wonder if the large mismatch between MgO and FePt may influence the intrinsic magnetic properties of the $L1_0$-

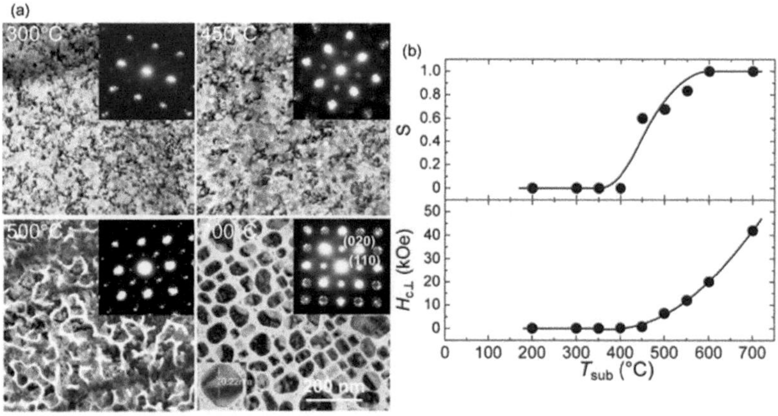

Figure 5.3 (a) Plan-view TEM bright-field images of FePt films deposited on MgO (001) single-crystalline substrates at different temperatures and (b) the degree of the $L1_0$ order and coercivity of the films as a function of substrate temperature (T_{sub}).

ordered FePt films, such as magnetization M_s and K_u. Hotta et al. investigated the thickness dependence of the magnetic properties of FePt films deposited on heated MgO, $MgAl_2O_4$, and $SrTiO_3$ (001) single-crystalline substrates that induce lattice mismatches of ~9%, 5%, and 1%, respectively [29]. On the basis of the results, they concluded that the magnitude of the lattice mismatch does not give any influence on M_s, K_u, and the long-range order parameter S. On the other hand, Ding et al. investigated the K_u of an FePt film sputter-deposited on MgO(001) substrates through various buffer layers, such as Pt, Cr, and CrMo, and concluded that the critical lattice mismatch around 6% is the most suitable for improving the $L1_0$ ordering of FePt [30]. However, the results need to be carefully interpreted because of the possibility of interdiffusion of Cr into the FePt layer.

In recording applications, single-crystalline substrates cannot be used because of their high cost. While FePt thin films were prepared on a heated single-crystalline (001) MgO substrate in early fundamental studies, a practical seed layer must be deposited on either glass or amorphous substrates. Jeon et al. reported the growth of PMA FePt polycrystalline films on a (001)-textured

MgO seed layer that was deposited on a thermally oxidized Si substrate [31]. Thereafter, Xu et al. reported the successful growth of (001)-textured FePt polycrystalline films ($t = 50$ nm) on a $Cr_{91}Ru_9$ substrate with PMA [32]. However, since Cr diffuses into FePt at elevated substrate temperature, the growth temperature was limited to 400°C. Hence, only a low perpendicular coercivity of ~0.4 T was reported. Although the (110) Cr plane is the most close-packed plane, a [001] texture develops in a Cr buffer layer at an optimized condition, which led to the [001] texture of FePt on CrRu. The lattice mismatch of 5.8% with FePt (001) causes the expansion of the in-plane FePt (100) lattice. Since the c/a ratio of the $L1_0$–FePt phase is 0.964, the slightly higher lattice mismatch between Cr (110) and FePt (100) assists the transformation from the A1 phase to the $L1_0$ phase. By adding Ru into Cr thin films, the lattice mismatch can be tuned as the atomic radius of Ru is larger than that of Cr. Xu et al. showed that 10 at% Ru optimized the lattice mismatch for $L1_0$ ordering [32]. With in situ annealing, perpendicular FePt [001] texture with a good $L1_0$ order was obtained at a relatively lower temperature of 350°C, whereas a temperature over 600°C was required to achieve a high $L1_0$ order by postdeposition annealing. This means that a reduction of the ordering temperature by 200°C was achieved by the in situ annealing using the CrRu underlayer. Although a Cr-based underlayer reduces the ordering temperature, Cr diffusion into FePt deteriorates the magnetocrystalline anisotropy. Chen et al. inserted a thin MgO layer as a diffusion barrier of Cr and reported the growth of (001)-textured $L1_0$–FePt polycrystalline films with strong PMA on glass substrates [33]. In their film stack, the orientation relationships of FePt(001)[100]//MgO(001)[100]//CrRu(200)[110] were confirmed. The coercivity was enhanced to 1.2 T in this film stack.

Later, Perumal et al. reported [001]-textured FePt particulate films grown on a [001]-textured MgO seed layer that was predeposited on thermally oxidized Si [34]. This work has shown that the intermediate layer such as CrRu is not essential to control the texture and the microstructure of FePt films. However, there is one drawback in the MgO seed layer. Since it is an insulator, the MgO seed layer must be grown using radio frequency (rf) sputtering, which causes slow sputtering rates and some undesired grains. Ideally, the seed layer could be sputtered with a direct current (DC) power, so electrically conductive materials are more desirable as a seed layer. Varaprasad et al. reported a [001]-textured FePt film grows on a $(Mg_{0.2}Ti_{0.8})O$ seed layer. $(Mg_{1-x}Ti_x)O$ (MTO) is electrical conductive having the NaCl structure for a wide range of x [35]. This seed layer shows strong [001] texture on amorphous substrates and can induce the growth of a [001] FePt ferromagnetic layer. However, the surface energy appears to be lower than that of MgO(001). Hence, the wettability of FePt on MTO is good, hindering the particulate growth of FePt layers using the FePt–C system. MTO has the identical rock salt structure for a wide range of x, that is, $0 < x < 0.8$, and the surface energy of Mg-rich MTO becomes close to that of MgO. So MTO has potential to be a seed layer to replace MgO, if the nanogranular structure of the FePt layer can be optimized.

5.2.3 *FePt–C Granular Films*

To use thin films as a recording medium, exchange-decoupled nanosized ferromagnetic grains must be densely dispersed in two dimensions. Early investigations on FePt thin films were mainly done to obtain high coercivity in magnetically coupled polycrystalline films [10, 12–14], in which ordering occurs accompanied by recrystallization [13], but the ordering process in completely isolated grains must progress continuously without grain boundary migration [17]. One of the earliest works on the FePt–Al–O granular film by Watanabe et al. [36, 37] reported randomly oriented grains of about 5 nm in diameter in the as-sputtered condition. In this work, FePt was co-sputtered with aluminum and oxygen on a thermally oxidized Si substrate at room temperature. As a result, the structure of the FePt grains in the as-sputtered condition was A1 without any crystallographic texture. High-temperature annealing was necessary to order the A1 phase to the $L1_0$ structure, but this caused a coarsening and coalescence of the $L1_0$–FePt grains. Luo et al. [38] reported perpendicular anisotropy films fabricated by annealing FePt/B_2O_3 multilayers deposited on glass substrates. However, the microstructure reported in the work is far from the ideal microstructure that can be used as a recording medium. Peng also investigated the multilayer deposition of $[FePt/MgO]_n$ on a

[001]-textured MgO seed layer to form FePt–MgO composite films, but the FePt grains were not well isolated and no PMA was observed [39].

In 2003, Ko et al. [40] and Perumal et al. [41, 42] reported the FePt–C granular structure with PMA having an average grain size of 4 nm. They co-sputtered FePt and C on MgO (001) single-crystalline substrates at 400°C. Since C has no solubility with Fe and Pt, C was rejected from FePt nanograins, making the FePt grains magnetically isolated through intergranular carbon. $L1_0$ ordering occurs during the film deposition with a strong [001] texture. Following this work, Chen et al. have grown FePt–C films by co-sputtering FePt and C targets on glass substrates through $Cr_{90}Ru_{10}$/Pt buffer layers at 350°C [43]. They obtained a very fine uniform granular structure of about 6 nm grain size with a perpendicular coercivity of about 0.37 T. Because of the usage of the CrRu underlayer, the substrate temperature was limited to 350°C, which was not sufficient to give a high degree of the $L1_0$ order. Subsequently, Chen et al. [44–47] deposited FePt–C films on a glass substrate through a CrRu buffer layer and MgO seed layer and demonstrated a [001]-textured FePt–C granular film with a higher coercivity of ~1.4 T. From the cross-sectional image of the FePt–C magnetic layer, they have shown the formation of a second layer in the FePt–C magnetic layer. Also, they reported that columnar grains can be grown by increasing the sputtering rate. However, the in-plane TEM observation showed the FePt grains were interconnected.

In 2008, Perumal et al. reported FePt–C-based PMA granular films with narrow size distribution of isolated $L1_0$–FePt grains [48]. The average grain size of the $L1_0$-ordered FePt was ~5.5 nm with 2.3 nm size dispersion. The films were grown by co-sputtering Fe, Pt, and C on a MgO seed layer that was predeposited on thermally oxidized Si substrate. This is the first media-like structure obtained from FePt-based films. The key to the success was to sputter-deposit FePt and C on a heated substrate so that the ordering to the $L1_0$ structure can progress during the film growth at a much lower temperature than that required for the postdeposition annealing (~640°C). During the growth of the films at elevated temperature, atomic transport occurs by surface diffusion, thereby enhancing the kinetics for $L1_0$ ordering at the substrate temperature. While $L1_0$

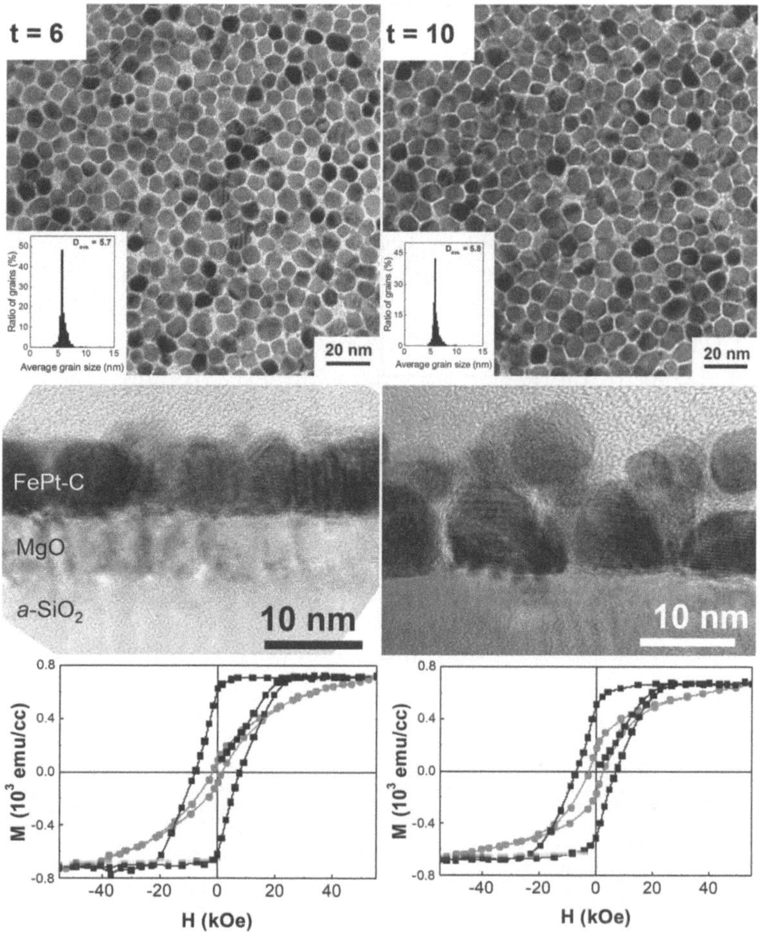

Figure 5.4 Plan-view and cross-sectional TEM bright-field images and the magnetization curves of FePt–C granular films with a nominal thickness, *t*, of 6 nm and 10 nm. Up to $t = 6$ nm, the film grows with a single-layer granular structure, but a second granular layer appears for $t > 6$ nm.

ordering progresses during the growth, carbon is rejected from FePt, forming a thin channel of amorphous carbon, as shown in Fig. 5.4 [48]. However, the coercivity was only in the order of 0.8–1.5 T due to the low degree of the $L1_0$ order in the FePt grains, and a higher coercivity had to be achieved for the HAMR media.

5.2.4 $(FePt)_{0.9}Ag_{0.1}$–C Granular Films

The excellent FePt–C nanogranular structure demonstrated by Perumal et al. convinced the community that FePt–C could be an HAMR medium. However, the coercivity of the film was still too low for the HAMR media requirement [48, 49]. In early studies of FePt continuous films, Platt et al. showed that the kinetic ordering temperature of FePt can be reduced by the addition of Ag [50]. Following this concept, Zhang et al. succeeded in processing a nanograin-dispersed PMA thin film with a mean grain diameter of 6.1 nm, size dispersion of 1.8 nm, and coercivity of 3.7 T by adding Ag to the FePt–C granular films [51]. To align the [001] crystal orientation perpendicular to the film, a thin layer of MgO was grown on a Si substrate, followed by sputtering a (FePt)Ag–C magnetic layer. The $(FePt)_{0.9}Ag_{0.1}$–40vol.%C film had the highest grade of grain dispersion and the [001] texture of $L1_0$–FePt. The in-plane TEM image in Fig. 5.5a shows the uniform microstructure with an average grain size of 6.1 nm and the size distribution of 1.8 nm [52]. The pitch distance (center to center) was about 9.6 nm. The selected area (electron) diffraction (SAED) pattern shows the FePt grains have a strong c axis texture. Figure 5.5b shows the magnetization curves of out-of-plane and in-plane directions [52]. The film has a strong PMA. The perpendicular $\mu_0 H_c$ is about 3.5 T and the anisotropy field estimated by the cross point of the in-plane and out-of-plane curves is about 6.5 T. The estimated K_u of the FePt grains is 4.3×10^6 J/m^3, which is half of that reported for a bulk $L1_0$–FePt. The high K_u and $\mu_0 H_c$ result from the high degree of the $L1_0$ order, 0.9. Because of the high anisotropy FePt grains, the thermal stability $(K_u V/k_B T)$ estimated by the time-dependent H_c measurement was about 200, indicating that the stored information in the media is thermally stable for more than 10 years [53]. Subsequent HAMR static tester results demonstrated the areal density of 550 Gbit/in^2 on this medium, as shown in Fig. 5.5c [52]. This recording density was the highest one achieved by HAMR at that time and was comparable to that of the conventional PMR method.

To understand the mechanism of the enhancement of H_c and the anisotropy of the FePt–C granular film by the addition of Ag, a

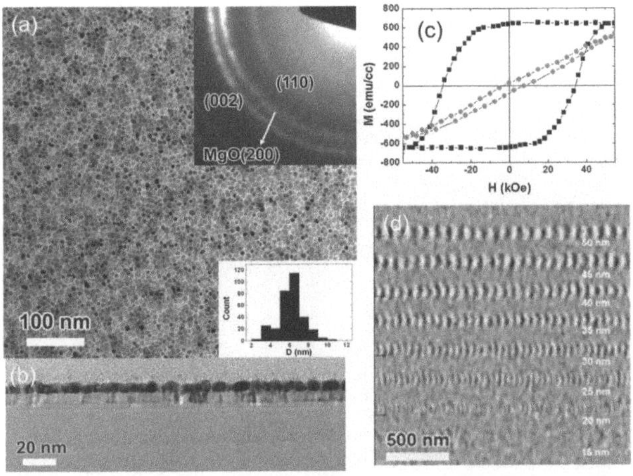

Figure 5.5 (a) Plan-view and (b) cross-sectional TEM bright-field images of the $(FePt)_{0.9}Ag_{0.1}$–40vol.%C granular film, (c) magnetization curves, and (d) patterns of recording bits by the static HAMR head. The mean grain diameter of the FePt grains is 6.1 nm, and their size dispersion is 1.8 nm. Coercivity is 3.7 T, which is five times larger than that of conventional magnetic recording media. In the recording pattern with the static HAMR head, 15 nm bits were observed with a bit width of 92 nm. Converting to a recording density, this is equivalent to 550 Gbpsi.

detailed chemical analysis was carried out using aberration-corrected scanning transmission electron microscopy (STEM). Figure 5.6 shows STEM energy-dispersive X-ray spectroscopy (STEM-EDS) elemental maps of Fe and Ag in the $(FePt)_{0.9}Ag_{0.1}$–28vol%C film. It shows that the FePt grains are enveloped by Ag-rich shells [54]. In addition, X-ray absorption fine structure (XAFS) measurements revealed that Ag of the shell had a face-centered cubic (fcc)-like structure. Ag was confirmed to be dissolved in FePt grains when the $(FePt)_{0.9}Ag_{0.1}$–C was deposited at room temperature, and the formation of the Ag-rich shell can be interpreted as a result of Ag rejection from the FePt grains during the deposition process. The Ag rejection is possible only with the diffusion of Ag through vacancy motion, suggesting that the Ag alloying enhances the diffusion in FePt crystals, thereby enhancing the kinetics of $A1 \rightarrow L1_0$ ordering.

Since the (FePt)Ag–C nanogranular films can be deposited easily by sputtering on the MgO seed layer, where crystal orientations

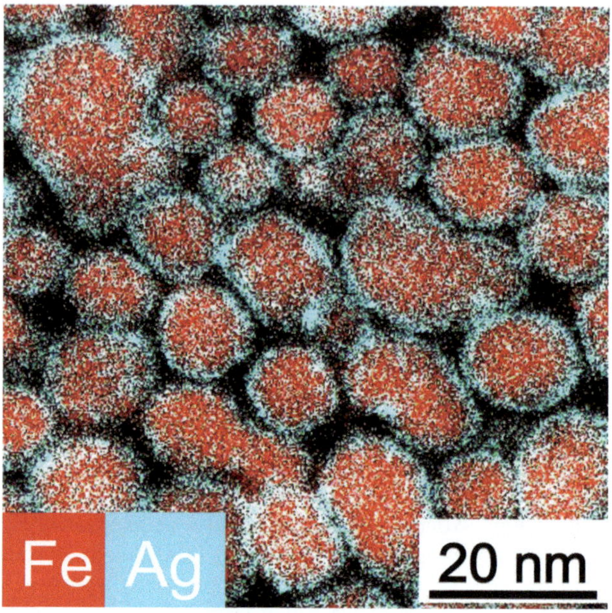

Figure 5.6 Scanning transmission electron microscope–energy dispersive X-ray spectroscopy (STEM-EDS) elemental maps of Fe and Ag in the $(FePt)_{0.9}Ag_{0.1}$–28vol%C granular film.

are naturally aligned to the [001] direction during film growth, they can be fabricated on substrates other than thermally oxidized Si, including glass substrates and amorphous metals. From this viewpoint, the (FePt)Ag–C system has high potential to be extended to industrially viable production lines. Using a semi-industrial sputtering machine with a triode target, Hitachi Global Storage Technologies (HGST) reported a similar microstructure on glass substrates with a NiTa amorphous heat sink layer [55, 56].

Although the microstructure and magnetic properties attained in the (FePt)Ag–C were so close to those required for practical use, there have been two critical issues to be resolved: one is to obtain a small roughness on the surface and the other is to grow FePt grains in a columnar shape with an aspect ratio $h/D > 1.5$, as shown in Fig. 5.2. These two problems stem from the strong driving force for the phase separation between FePt and C. The reason why the FePt–C system shows a well-isolated planar structure is that the

FePt and C are entirely immiscible with a strong repulsive nature. Hence, while the plan view shows nice grain separation, the cross-sectional image shows the formation of a second layer on top of the FePt grains that were grown on a MgO seed layer epitaxially, as shown in Fig. 5.4 [48]. Since the second layer has no orientation relationship with the underlayer being separated by amorphous carbon, the perpendicular anisotropy is substantially degraded. This is because C has too strong a driving force for phase separation. Although strong phase separation makes the grain isolation better in the lateral direction, this in turn interrupts the columnar growth in the growth direction and causes another grain to nucleate on an underlying grain. To overcome this issue, various segregants other than C have been explored.

5.2.5 *Exploration for a New Segregant for FePt–X Media*

Before the FePt–C system was found to be suitable for achieving the granular structure for an HAMR medium, various FePt–X systems, including FePt–SiO_x, were thoroughly investigated. Seki et al. reported the formation of a fine FePt–SiO_2 granular structure [57]. However, coercivity becomes very low when the substrate temperature is decreased to achieve a grain size of less than 6 nm. Yang et al. also reported columnar grain growth in the cross-sectional TEM image of FePt–SiO_2 films, but the plan-view TEM image showed the grains were interconnected in the planar direction [58]. Various other systems were investigated, but when a columnar structure is observed in a cross-sectional TEM image, in most of the cases the grains are interconnected in the planar direction [59, 60]. Figure 5.7 shows typical granular microstructure observed in the FePt–TiO_x system [60]. The FePt grains are interconnected in the planar direction, while a uniform columnar structure with a smooth surface is obtained in the cross-sectional view. However, the film showed soft magnetic properties because of the suppression of $L1_0$ ordering due to the dissolution of Ti into FePt grains. This means that the driving force for the phase separation between FePt and TiO_x as well as SiO_2 is too small to obtain well-isolated grains. From this standpoint, Varaprasad et al. have grown FePt–(C,SiO_2) and FePt–(C,TiO_x) granular films and confirmed the

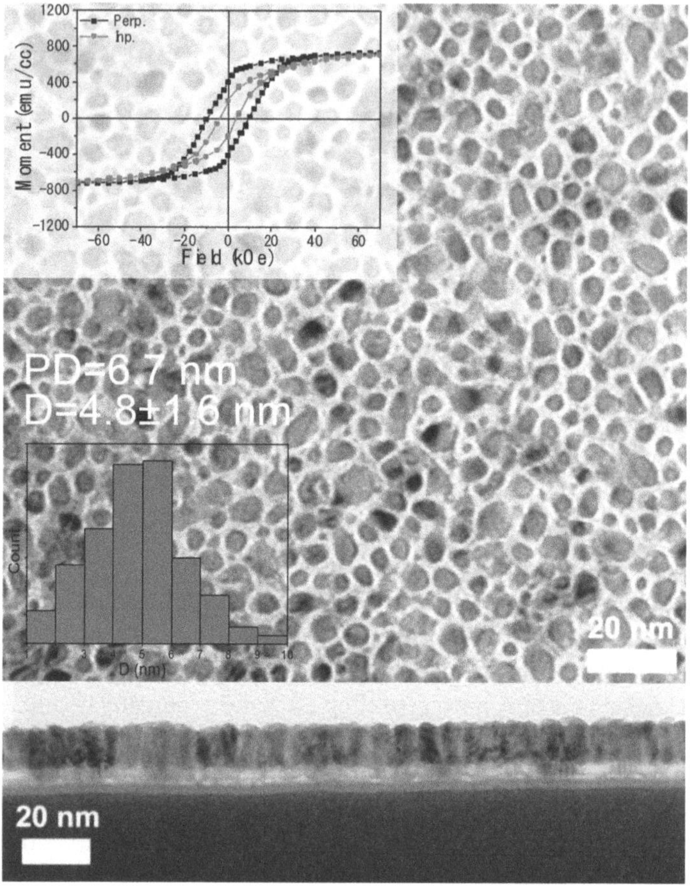

Figure 5.9 In-plane and cross-sectional TEM bright-field images of the FePt–40vol.%Cr_2O_3 film deposited on a glass substrate with a MgO seed layer.

FePt–C on a MgO seed layer and the coercivity is only 10 kOe with a large in-plane component. The average grain size is 4.8 nm, very fine as compared to FePt–C. However, the size distribution is about 33%, which is too large for a recording medium. Recent investigations on FePt–ZnO_2 [63], FePt/AlN [64], and FePt–B–Ag [65] all suggest that no systems can match the FePt–C system in view of planar isolation, narrow size distribution, and high coercivity. Thereafter, research has been focused on how to optimize the granular microstructure in the FePt–C system.

5.3 Optimization of Nanostructure of the FePt–C System

The remaining issues for the FePt–C system are summarized as follows:

- How to grow $L1_0$-FePt grains columnarly in the perpendicular direction to the film without the formation of a second ferromagnetic layer
- How to reduce the average grain size to achieve a pitch distance of 7 nm or less, while keeping the grain size distribution less than 15%
- How to remove the in-plane component of magnetization

5.3.1 *Columnar Grain Growth*

To grow FePt grains columnarly for a thickness larger than 6 nm, suppression of the phase separation between FePt and C in the perpendicular direction to the film plane is necessary. When the thickness of the FePt ferromagnetic layer becomes more than 6 nm, carbon appears to segregate on the surface of FePt grains by phase separation in the growth direction, which triggers the formation of a second FePt layer. This problem appears to be overcome by modifying the composition of the segregant and the sputtering conditions from the fact that aspect ratios exceeding 1.5 were reported in recent unpublished reports from industrial laboratories [66, 67]. However, no details on the materials and processing conditions for achieving the columnar grain structure have been revealed. Therefore, there is a need for getting a better understanding of the condition of columnar growth of FePt grains.

Although the process is totally different from the industrial one, Varaprasad et al. reported a way to achieve columnar growth of FePt grains by using a concentration-graded co-sputtering method [68]. Figure 5.10a,b shows the plan-view and cross-sectional TEM images of the FePt–30vol.%C films deposited on

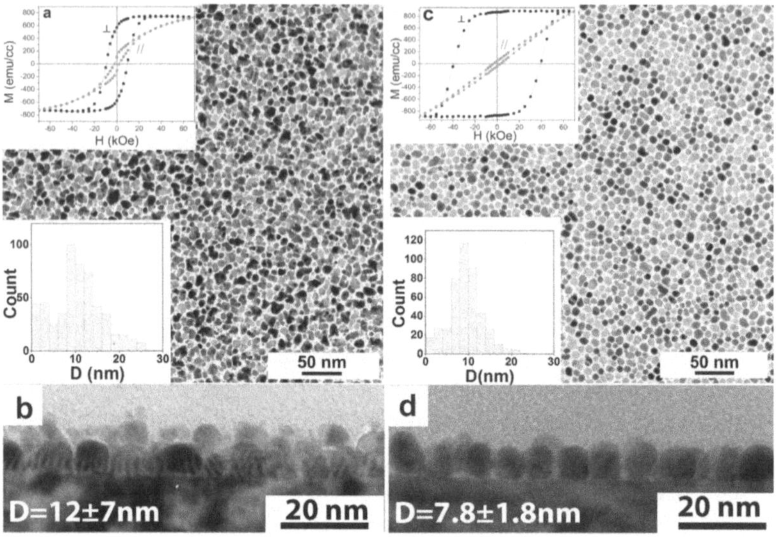

Figure 5.10 Plan-view (a) and cross-sectional (b) TEM bright-field images of FePt–30vol.%C films deposited by the standard co-sputtering method and the compositionally graded process (c, d). The inset figures show the magnetization curves and size distributions of the grains. The compositionally graded process means that the films were grown with a sequence of FePt–25vol%C(2)/FePt–30vol%C(2)/FePt–35vol.%C(2)/FePt–25vol%C(2)/FePt–30vol%C(2), where the numbers in the parentheses are the thickness of each layer in nanometers.

MgO (001) single crystals by the standard co-sputtering method and Fig. 5.10c,d of the compositionally graded process [68]. The compositionally graded process means that the films were grown with a sequence of FePt–25vol%C(2)/FePt–30vol%C(2)/FePt–35vol.%C(2)/FePt–25vol%C(2)/FePt–30vol%C(2), where the numbers in the parentheses are the thickness of each layer in nanometers. The total thickness of these films is 10 nm, 1.7 times thicker than the critical thickness of the single layer grown by co-sputtering (Fig. 5.3 [26]). In the cross-sectional TEM image of the co-sputtered sample, Fig. 5.10b, the film shows the second layer with randomly oriented FePt grains. The plan-view TEM image shows FePt grains are overlapped with other grains and are interconnected, resulting in a large average grain size of 12 nm. Because of this second layer formation, PMA of the film is very poor, as shown in Fig. 5.10a. On the other hand, the film fabricated by the compositionally graded process shows a uniform microstructure with an average grain size of 7.8 nm and an aspect ratio of 1.5. The co-sputtered film has small perpendicular H_c and large in-plane hysteresis due to the relatively large grain size and second-layer grains. On the other hand, the film grown with the compositionally graded process shows a large perpendicular H_c of 3.9 T without in-plane hysteresis. This suggests that if we control the supply of an excess amount of carbon after the nucleation of FePt grains, the formation of the second FePt layer will be surpressed by the continuous columnar growth of the grains.

5.3.2 *Reduction of In-Plane Magnetic Component*

SFD has a close correlation with the SNR of magnetic recording and is influenced by the distributions of size, σ_D, distribution of the anisotropy field of individual grains because of some heterogeneities in chemical composition and the degree of order, and the easy-axis distribution. Pisana et al. concluded that the H_k distribution is the major reason for SFD, and the grain size distribution and angular distribution of the *c* axis of FePt grains are the secondary reasons [69]. They also point out that the in-plane and randomly oriented grains and the grain size distribution do not make much contribution to SFD. On the other hand, Acharya reported that the SNR and jitter noise receive significant influence from the in-plane component of magnetization when the grain size of FePt becomes smaller than 7 nm [66]. In the prototype HAMR media, a MgO seed layer is used to induce strong [001] texture in the FePt–C magnetic layer [69, 70]. However, all FePt–C granular media grown on MgO seed layers contain substantial in-plane components and the origin of the misaligned grains has been a subject of recent TEM studies [71–73].

To clarify the influence of grain boundary and textures of the MgO seed layer, Wang et al. investigated the easy-axis distribution in FePt–C media deposited on two kinds of MgO [73], that is, one is a single-crystalline substrate and the other is a polycrystalline MgO (poly-MgO) underlayer deposited on glass substrates through an

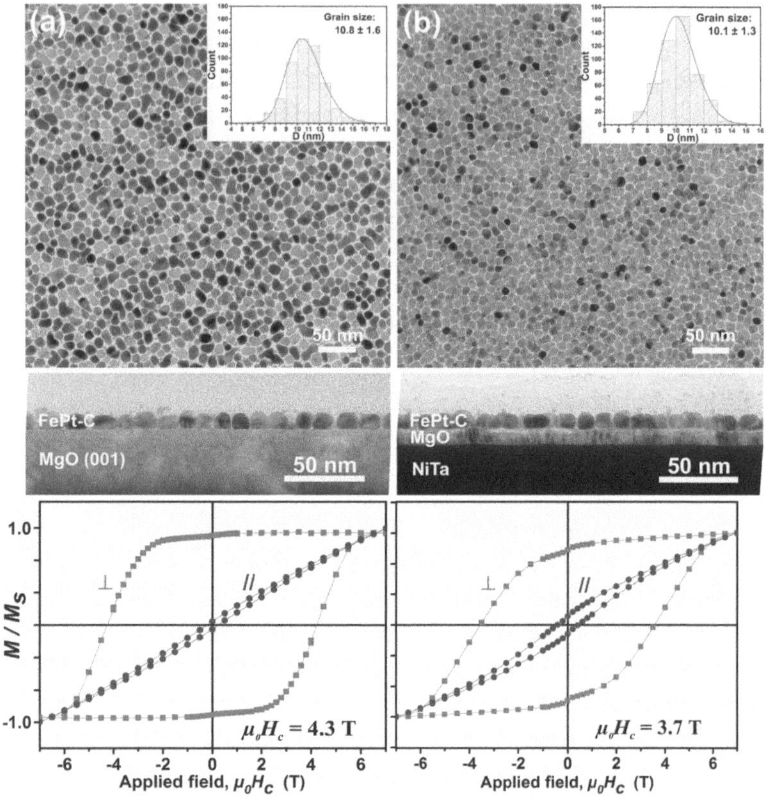

Figure 5.11 Plan-view and cross-sectional TEM bright-field images of an FePt–C granular film on (a) a MgO single crystal and (b) a polycrystalline MgO underlayer. (Inset: FePt grain size distribution, nm.) The FePt–C magnetic layer was deposited with the concentration-graded process of [FePt–48vol.%C (0.25 nm)/FePt (0.15 nm)]$_{25}$ to obtain the nominal thickness of the FePt–C layer as 10 nm. In-plane and out-of-plane magnetization curves of the FePt–C granular film on (a) a MgO single crystal and (b) a polycrystalline MgO underlayer.

a-NiTa buffer layer. Figure 5.11 shows in-plane and cross-sectional TEM images and their magnetization curves of the FePt–C film deposited on (a) a MgO single-crystalline substrate and (b) a poly-MgO seed layer [73]. Both films show a similar microstructure with an average grain size of about 10 nm and a grain distribution of 1.4

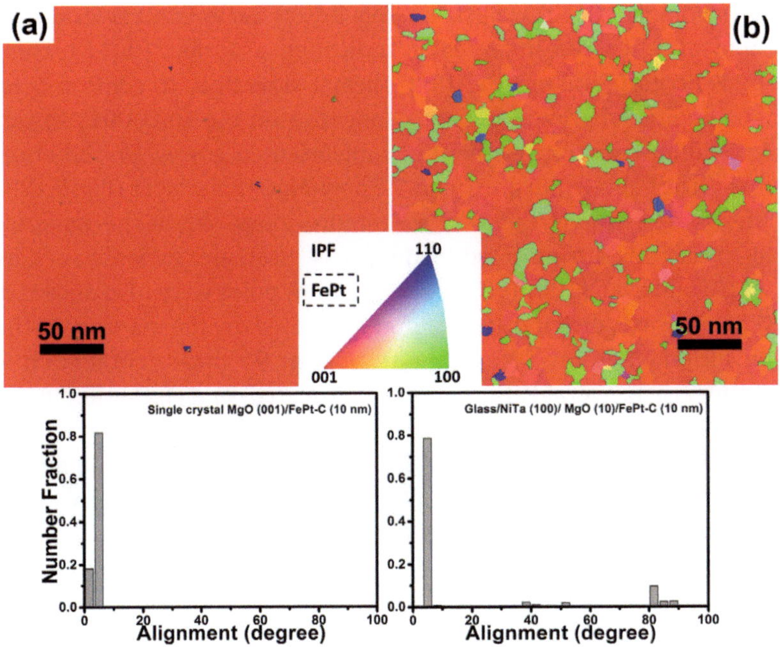

Figure 5.12 Orientation maps of FePt–C granular films deposited on (a) a MgO single-crystalline substrate and (b) a polycrystalline MgO underlayer. The orientation of each grain is given by the reference triangle.

nm. The nominal thickness of the FePt–C layer is 10 nm. Although the films grown on polycrystalline MgO show a strong PMA and H_c, as seen from the magnetization curves in Fig. 5.11, the squareness and gradient of the demagnetization curve of the film deposited on the MgO single-crystalline substrate is far better than that on the poly-MgO seed layer. This indicates that either the angular distribution of the MgO grains or the presence of grain boundaries in the polycrystalline seed layer causes in-plane component of the magnetization in the FePt–C layer. Figure 5.12 shows the orientation maps of the FePt–C granular films grown on a MgO(001) single-crystalline substrate and a poly-MgO underlayer obtained from the plan-view specimens, together with the easy-axis distribution [73]. The out-of-plane orientation map for the film grown on the single-crystalline MgO(001) substrate (Fig. 5.12a) demonstrates an almost homogeneous red index color, indicating a perfect [001] texture of FePt grains along the film normal direction. It confirms the good epitaxial growth of FePt nanograins on the MgO(001) single-crystalline substrate without any misoriented grains. On the other hand, the FePt–C film grown on the poly-MgO underlayer (Fig. 5.12b) shows a distinctly different orientation map. While the majority of the grains are oriented in the [001] direction, variants such as [110] and even [100] (90° misorientation) poles were also detected. The easy-axis distribution histograms also show the majority of the misoriented grains have either 48°–52° or 90° misorientation. It is known that several FePt grains are nucleated on a MgO grain in the seed layer [73], so the easy-axis distribution within the seed layer grain should be very low. This suggests that if the misalignment of the [001] texture in the MgO seed layer is improved, the easy-axis distribution will be substantially improved. However, even if a MgO(001) single-crystalline substrate is used, a strong texture does not always develop [68], suggesting misaligned particles may be nucleating from atomic steps on the MgO (001) surface. On the basis of high-resolution TEM observation, Wicht et al. [74] reported that the misalignment of the MgO crystal is smaller than that of FePt grains. Their work suggests highly textured FePt grains do not develop even if the angular distribution of MgO is small. They conclude that FePt and MgO are slightly misaligned from the perfect cube-to-cube orientation relationship because of the nucleation of FePt from step edges of the MgO seed layer. As the misfit between MgO and FePt on the (001) plane is large, some of the faceted planes in FePt show epitaxy with the step edge of MgO, which causes misalignment of FePt on the [001]-textured MgO seed layer. On the other hand, Ho et al. [72] reported the FePt grains crossing the grain boundaries of the MgO underlayer cause the formation of the $L1_0$ lattice with 90° misalignment. These recent investigations suggest the controlled surface roughness, grain size, and texture of the MgO seed layer are essential to reduce the easy-axis distribution of FePt–C media. In addition, search of an alternative seed layer is also important. Some of the promising seed layers are MTO [35] and TiN, both of which can be grown with DC sputtering.

5.4 Concluding Remarks

Recent progress made on the development of FePt-based granular films for HAMR media has been reviewed. After many investigations of various FePt–X systems, only the FePt–C and (FePt)Ag–C systems remain as promising candidates that can fulfill the requirements for 2 Tbit/in^2 HAMR media. However, there are still several issues to be overcome: reduction of grain size, columnar growth of grains, and reduction of easy-axis distribution. To resolve these issues, fundamental studies on the growth mechanisms of FePt–C ferromagnetic layers on various seed layers are necessary. Some approaches to these issues are addressed in this chapter. Recent industrial demonstrations of HAMR recording over 1 Tbit/in^2 and its improving endurance test result are encouraging, and the commercial implementation of HAMR appears to be right around the corner. However, further improvement of the microstructure to optimize the read and write properties as well as heat control is essential for achieving higher areal density beyond 2 Tbit/in^2 in the future.

Acknowledgments

Financial support from IDEMA-ASTC for our HAMR media work is acknowledged. We thank A. Perumal for valuable comments.

References

1. Shiroishi Y., Fukuda K., Tagawa I., Iwasaki H., Takenoiri S., Tanaka H., Mutoh H., and N. Yoshikawa N. (2009). Future options for HDD storage, *IEEE Trans. Magn.*, **45**, p. 3816.
2. Varaprasad B. S. D. Ch. S., Takahashi Y. K., and Hono K. (2013). Microstructure control of L1-ordered FePt granular film for heat-assisted magnetic recording (HAMR) media, *JOM*, **65**, pp. 853–861.
3. Klemmer K., Hoydick D., Okumura H., Zhang B., and W. A. Soffa W. A. (1995). Magnetic hardening and coercivity mechanisms in L1$_0$ ordered FePd ferromagnets, *Scripta Metall. Mater.*, **33**, p. 1793.

4. Weller D., and Moser A. (1999). Thermal effect limits in ultrahigh-density magnetic recording, *IEEE Trans. Magn.,* **35**, p. 4423.
5. Rottmayer R. E., Batra S., Buechel D., Challener W. A., Hohlfeld J., Kubota Y., Lu B., Michalcea C., Mountfield K., Pelhos K., Peng C., Rausch T., Seigler M. A., Weller D., and Yang X. M. (2006). Heat-assisted magnetic recording, *IEEE Trans. Magn.,* **42**, p. 2417.
6. Guarisco D., and Nguy H. (2003). High linear density in perpendicular recording, *J. Appl. Phys.,* **93**, p. 6745.
7. Kryder M. H., Gage E. C., McDaniel T. W., Challener W. A., Rottmayer R. E., Ju G., Hsia Y. T., and Erden M. F. (2008). Heat assisted magnetic recording, *Proc. IEEE,* **96**, p. 1810.
8. Stipe B. C., Strand T. C., Poon C. C., Balamane H., Boone T. D., Katine J. A., Li J. L., Rawat V., Nemoto H., Hirotsune A., Hellwig O. Ruiz R., Dobisz E., Kercher D. S., Robertson N., Albrecht T. R., and Terris B. D. (2010). Magnetic recording at 1.5 Pb m^{-2} using an integrated plasmonic antenna, *Nat. Photon.,* **4**, p. 484.
9. Weller D., Parker G., Mosendz O., Champion E., Stipe B., Wang X., Klemmer T., Ju G., and Ajan A. (2014). A HAMR media technology roadmap to an areal density of 4 Tb/in^2, *IEEE Trans. Magn.,* **50**, p. 3100108.
10. Wu A.Q. Kubota Y., Klemmer T., Rausch T., Peng C., Peng Y., Karns D., Zhu X., Yinfeng Ding, Chang E. K. C., Zhao Y., Zhou H., Gao K., Thiele J-H., Seigler M., Ju G., and Edward Gage E. (2013). HAMR areal density demonstration of 1+ Tbpsi on Spinstand, *IEEE Trans. Magn.* **49**, p. 779.
11. Ristau R. A., Barmak K., Lewis L. H., Coffey K. R., and Howard J. K. (1999). On the relationship of high coercivity and L10 ordered phase in CoPt and FePt thin films, *J. Appl. Phys.,* **86**, pp. 4527–4533.
12. Visokay M. R., and Sinclair R. (1995). Direct formation of ordered CoPt and FePt compound thin films by sputtering, *Appl. Phys. Lett.,* **66**, pp. 1692–1694.
13. Maeda T., Kikitsu A., Kai T., Nagase T., and Akiyama J. (2002). Reduction of ordering temperature of an FePt-ordered alloy by addition of Cu, *Appl. Phys. Lett.,* **80**, p. 2147.
14. Takahashi Y. K., Ohnuma M., and Hono K. (2002). Effect of Cu on the structure and magnetic properties of FePt sputtered film, *J. Magn. Magn. Mater.,* **246**, p. 259.
15. Platt C. L., Wierman K. W., Svedberg E. B., Veerdonk R., Howard J. K., Roy A. G., and Laughlin D. E. (2002). L1 ordering and microstructure of FePt thin films with Cu, Ag, and Au additive, *J. Appl. Phys.,* **92**, p. 6104.

16. Maret, M., Brombacher C., Matthes P., Makarav D., Boudet N., and Albercht M. (2012). Anomalous x-ray diffraction measurements of long-range order in (001)-textured L1 FePtCu thin films, *Phys. Rev. B*, **86**, p. 024204.

17. Takahashi Y. K., Ohnuma M., and Hono K. (2003). Ordering process of sputtered FePt films, *J. Appl. Phys.*, **93**, p. 7580.

18. Takahashi Y. K., and Hono K. (2005). On low temperature ordering of FePt films, *Scripta Mater.*, **53**, pp. 403–409.

19. Takahashi Y. K., Ohkubo T., Ohnuma M., and Hono K. (2003). Size effect on the ordering of FePt granular films, *J. Appl. Phys.*, **93**, pp. 7166–7168.

20. Takahashi Y. K., Koyama T., Ohnuma M., Ohkubo T., and Hono K. (2004). Size dependence of ordering in FePt nanograins, *J. Appl. Phys.*, **95**, pp. 2690–2696.

21. Miyazaki T., Kitakami O., Okamoto S., Shimada Y., Akase Z., Murakami Y., Shindo D., Takahashi Y. K., and Hono K. (2005). Size effect on the ordering of L10 FePt nanograins, *Phys. Rev. B.* **72**, p. 144419.

22. Watanabe, M. Takanashi, K., and Fujimori, H. (1991). Perpendicular magnetic anisotropy and magneto-optial Kerr effect of FePt/Pt multi-layer films, *J. Magn. Magn. Mater.*, **113**, pp. 110–114.

23. Risokay, M. R., and Sinclair R. (1995). Direct formation of ordered CoPt and FePt compound thin films by sputtering, *Appl. Phys. Lett.*, **66**, pp. 1692–1694.

24. Farrow R. F. C., Weller D., Marks R. F., Toney M. F., Cebollada A., and Harp G. R., (1996). Control of the axis of chemical ordering and magnetic anisotropy in epitaxial FePt films, *J. Appl. Phys.*, **79**, pp. 5967–5969.

25. Thiele J. U., Folks L., Toney M. F., and Weller D. (1998). Perpendicular magnetic anisotropy and magnetic domain structure in sputtered epitaxial FePt (001) L1 films*J. Appl. Phys.*, **84**, pp. 5686–5692.

26. Shima T. Takanashi K. Takahashi Y. K. Hono K. (2002). Preparation and magnetic properties of highly coercive FePt films, *Appl. Phys. Lett.*, **81**, pp. 1050–1052.

27. Takahashi Y. K., Hono K., Shima T., Takanashi K. (2003). Microstructure and magnetic properties of FePt thin films epitaxially grown on MgO (001) substrates, *J. Magn. Magn. Mater.*, **267**, pp. 248–255.

28. Shima T., Takanashi K., Takahashi Y. K., and Hono K. (2004). Coercivity exceeding 100 kOe in epitaxially grown FePt sputtered films, *Appl. Phys. Lett.*, **85**, pp. 2571–2573.

29. Hotta A., Ono T., Hatayama M., Tsumura K., Kikuchi N., Okamoto S., Kitakami O. and Shimatsu T. (2014). Magnetic anisotropy and order

structure of L1-FePt(001) single-crystal films grown epitaxially on (001) planes of MgO, $SrTiO_3$, and $MgAl_2O_4$ substrates, *J. Appl. Phys.*, **115**, p. 17B712.

30. Ding Y. F., Chen J. S., Liu E., Sun C. J., and Chow G. M. (2005). Effect of lattice mismatch on chemical ordering of epitaxial L1 FePt films, *J. Appl. Phys.*, **97**, p. 10H303.
31. Jeong S., McHenry M. E., and Laughlin D. E. (2001). Growth and characterization of L1 FePt and CoPt 001 textured polycrystalline thin films*IEEE Trans. Magn.*, **37**, p. 1309.
32. Xu, Y. F., Chen, J. S., and Wang J. P. (2002). In situ ordering of FePt thin films with face-centered-tetragonal (001) texture on Cr100ÀxRux underlayer at low substrate temperature, *Appl. Phys. Lett.*, **80**, p. 3325.
33. Chen, J. S. Lim, B. C. Hu, J. F. Lim Y. K., Liu, B., and Chow. G. M. (2007). High coercivity L1 FePt films with perpendicular anisotropy deposited on glass substrate at reduced temperature, *Appl. Phys. Lett.*, **90**, p. 042508.
34. Perumal A. Takahashi Y. K. Seki T. O., and Hono K. (2008). Particulate structure of L10 ordered ultrathin FePt films for perpendicular recording, *Appl. Phys. Lett.*, **92**, p. 132508.
35. Varaprasad B. S. D. Ch. S., Takahashi Y. K. Ajan A., and Hono K. (2013). Electrically conductive $(Mg_{0.2}Ti_{0.8})O$ underlayer to grow FePt-based perpendicular recording media on glass substrates, *J. Appl. Phys.*, **113**, p. 203907.
36. Watanabe M, Masumoto T, Ping D. H., and Hono K. (2000). Microstructure and magnetic properties of FePt–Al–O granular thin films, *J. Appl. Phys.*, **76**, pp. 3971–3973.
37. Ping D. H., Ohnuma M., Hono K., Watanabe M., Iwasa T., and Masumoto T. (2001). Microstructures of FePt–Al–O and FePt–Ag nanogranular thin films and their magnetic properties, *J. Appl. Phys.*, **90**, pp. 4708–4716.
38. Luo C.P. Liou S. H. Gao L. Liu Y., and Sellmyer D. J. (2000). Nanostructured FePt:B_2O_3 thin films with perpendicular magnetic anisotropy, *Appl. Phys. Lett.*, **77**, pp. 2225–2227.
39. Peng Y, Zhu J. G., and Laughlin D. E. (2006). L1 FePt–MgO perpendicular thin film deposited by alternating sputtering at elevated temperature, *J. Appl. Phys.*, **99**, p. 08F907.
40. Ko H. S., Perumal A., and Shin S-C., (2003). Fine control of L10 ordering and grain growth kinetics by C doping in FePt films, *Appl. Phys. Lett.*, **82**, pp. 2311–2313.

41. Perumal A., Ko H. S., and Shin S. C. (2003). Perpendicular thin films of carbon-doped FePt for ultrahigh-density magnetic recording media, *IEEE Trans. Magn.*, **89**, pp. 2320–2322.
42. Perumal A., Ko H. S., and Shin S. C. (2003). Magnetic properties of carbon-doped FePt nanogranular films, *Appl. Phys. Lett.*, **83**, pp. 3326–3328.
43. Chen J. S., Ding Y. F., Lim B. C., and Liu E. J. (2006). Nanogranular L1 FePt:C composite films for perpendicular recording, *IEEE Trans. Magn.*, **42**, pp. 2363–2365.
44. Chen, J. S. Lim B. C. Hu, J. F. Liu B. Chow G. M., and Ju G. (2007). Low temperature deposited L1 FePt–C (001) films with high coercivity and small grain size, *Appl. Phys. Lett.*, **91**, p. 132506.
45. Chen J. S., Hu J. F., Lim B. C., Y. K. Lim Y. K., Liu B., Chow G. M., and Ju G. (2008). High coercive L1 FePt–C (001) nanocomopsite films with small grain size for perpendicular recording media, *J. Appl. Phys.*, **103**, p. 07F517.
46. Chen J. S., Lim B. C., Hu J. F., Ding Y. F., Chow G. M., and Ju G. (2008). Microstructural and magnetic properties of L10 FePt–C (0 0 1) textured nanocomposite films grown on different intermediate layers, *J. Phys. D: Appl. Phys.*, **41**, p. 205001.
47. Li, H., Hu H., Ju J. F., Chow G. M., and Chen J. S. (2011). Effects of CrRu–SiOx underlayer with MgO intermediate layer on the microstructure and magnetic properties of FePt–C thin film, *J. Appl. Phys.*, **109**, p. 07A736.
48. Perumal A., Takahashi Y. K., and Hono K. (2008). L1 FePt-C nanogranular perpendicular anisotropy films with narrow size distribution, *Appl. Phys. Express* **1**, p. 101301.
49. Perumal, A. Takahashi Y. K., and Hono K. (2009). FePt-C nanogranular films for perpendicular magnetic recording, *J. Appl. Phys.*, **105**, p. 07B732.
50. Platt C. L., Wierman K. W., Svedberg E. B., van de Veerdonk R., Howard J. K., Roy A. G., and Laughlin D. E. (2002). L1 ordering and microstructure of FePt thin films with Cu, Ag, and Au additive, *J. Appl. Phys.*, **92**, p. 6104.
51. Zhang L., Takahashi Y. K., Perumal A., and Hono K. (2010). L1-ordered high coercivity (FePt)Ag–C granular thin films for perpendicular recording, *J. Magn. Magn. Mater.*, **322**, pp. 2658–2664.
52. Zhang, L. Takahashi, Y. K. Hono, K. Stipe, B. C. Juang, J. Y., Grobis, M. (2011). L1-ordered FePtAg-C granular thin film for thermally assisted magnetic recording media, *J. Appl. Phys.*, **109**, p. 07B703.

53. Zhang L., Takahashi Y. K., Hono K., Stipe B. C., Juang J. Y., and Grobis M. (2011). FePtAg-C nanogranular film as thermally assisted magnetic recording (TAR) media, *IEEE Trans. Magn.*, **47**, p. 4062.
54. Varaprasad, B. S. D. C. S., Takahashi YK., Wang J., Ina T., Nakamura T., Ueno W., Nitta K., Uruga T., and Hono K. (2014). Mechanism of coercivity enhancement by Ag addition in FePt-C granular films for heat assisted magnetic recording media, *Appl. Phys. Lett.*, **104**, p. 222403.
55. Mosendz O., Pisana S., Reiner J. W., Stipe B., and Weller D. (2012). Ultra-high coercivity small-grain FePt media for thermally assisted recording, *J. Appl. Phys.*, **111**, p. 07B729.
56. Pisana S., Mosendz O., Parker G. J., Reiner J. W., Santos T. S., McCallum A. T., Richter H. J., and Weller D. (2013). Effects of grain microstructure on magnetic properties in FePtAg-C media for heat assisted magnetic recording, *J. Appl. Phys.*, **113**, p. 043910.
57. Seki T. O. Takahashi Y. K., and Hono K. (2008). Microstructure and magnetic properties of FePt-SiO_2 granular films with Ag addition, *J. Appl. Phys.*, **103**, p. 023910.
58. Yang E., and Laughlin D. E. (2008). L1 FePt-oxide columnar perpendicular media with high coercivity and small grain size, *J. Appl. Phys.*, **104**, p. 023904.
59. Chen J. S., Lim B. C., Ding Y. F., Hu J. F., Chow G. M., and Ju, G. (2009). Granular L1 FePt–X (X=C, TiO2, Ta2O5) nanocomposite films with small grain size for high density magnetic recording, *J. Appl. Phys.*, **105**, p. 07B702.
60. Varaprasad B. S. D. Ch. S., Chen M., Takahashi Y. K., and Hono K. (2013). L10-ordered FePt-based perpendicular magnetic recording media for heat-assisted magnetic recording, *IEEE Trans. Magn.*, **49**, pp. 718–722.
61. Shiroyama T., Abe T., Takahashi Y. K., and Hono K. (2013). Microstructure and magnetic properties of FePt-MO granular films, *IEEE Trans. Magn.*, **49**, p. 3616.
62. Shiroyama T., Varaprasad B. S. D. C. S., Takahashi Y. K., and Hono K. (2014). Microstructure and magnetic properties of FePt-Cr_2O_3 films, *IEEE Trans Magn.*, **50**, p. 3202404.1.
63. Dong K. F., Li H. H., Peng Y. G., Ju G., Chow G. M., and Chen J. S. (2014). L1 FePt-ZrO_2 (001) nanostructured films with high aspect ratio columnar grains, *Appl. Phys. Lett.*, **104**, p. 192404.
64. Feng C., Zhang E., Xu C. C., Li N., Jiang Y., Yu G. H., and Li B. H. (2011). Magnetic properties and microstructure of L1-FePt/AlN perpendicular nanocomposite films, *J. Appl. Phys.*, **110**, p. 063910.

65. Granz S. D., Barmak K., and Kryder M. K. (2012). Granular L1 FePt-B and FePt-B-Ag (001) thin films for heat assisted magnetic recording, *J. Appl. Phys.*, **111**, p. 07B709.
66. Acharya A. (2014). Challenges and recent developments in heat assisted magnetic recording media, *TMRC 2014*, Berkley, August 11–13, 2014.
67. Hellwig O. (2015). Challenges in developing FePt L1 granular thin film media for heat assisted magnetic recording (HAMR), *Intermag 2015*, p. AA-05, Beijing, May 11–15, 2015.
68. Varaprasad B. S. D. Ch. S., Wang J., Shiroyama T., Takahashi Y. K., and Hono K. (2015). Columnar structure in FePt–C granular media for heat-assisted magnetic recording, *IEEE Trans. Magn.*, DOI:10.1109/TMAG.2015.2435027.
69. Pisana S., Mosendz O., Parker G. J., Reiner J. W., Santos T. S., McCallum A. T., Richter H. J., and Weller D. (2013). Effects of grain microstructure on magnetic properties in FePtAg-C media for heat assisted magnetic recording, *J. Appl. Phys.*, **113**, p. 043910.
70. Weller D., Mosendz O., Gregory G., Pisana S., and Santos T. S. (2013). FePtX–Y media for heat-assisted magnetic recording, *Phys. Stat. Sol.*, **A210**, pp. 1245–1260.
71. Wicht S., Neu V., Schultz L., Weller D., Mosendz O., Parker G., Pisana S., and Rellinghaus B. (2013). Atomic resolution structure–property relation in highly anisotropic granular FePt-C films with near-Stoner-Wohlfarth behaviour, *J. Appl. Phys.*, **114**, p. 063906.
72. Ho H., Zhu J., Kulovits A., Laughlin D. E., and Zhu J. G. (2014). Quantitative transmission electron microscopy analysis of multi-variant grains in present L1-FePt based heat assisted magnetic recording media, *J. Appl. Phys.*, **116**, p. 193510.
73. Wang J., Hata S., Takahashi Y. K., Sepehri-Amin H., Varaprasad B. S. D. Ch. S., Shiroyama T., Schrefl T., and Hono K. (2015). Effect of MgO underlayer misorientation on the texture and magnetic property of FePt–C granular film, *Acta Mater.*, **91**, p. 41.
74. Wicht S., Neu V., Schultz L., Mehta V., Jain S., Reiner J., Mosendz O., Hellwig O., Weller D., and Rellinghaus B. (2015). Modification of the structural and magnetic properties of granular FePt films by seed layer conditioning, *J. Appl. Phys.*, **117**, p. 013907.

coupling [4–9]. Together with the papers of Suess et al. and Victora et al., other recent theoretical papers predicted more advantages than the writability improvement, for example, high tolerance to easy-axis distribution, thus giving rise to a higher signal-to-noise ratio (SNR) compared to single-phase media, and faster and easier switching compared to single-phase media in microwave-assisted recording [1, 2, 10–14].

The ECC concept shows similarities with the coupled granular/continuous (CGC) principle originally proposed in 2001 [15]. Also the CGC medium is in fact a two-phase system but with a continuous capping layer with strong lateral exchange coupling deposited on top of a granular layer that acts as a pinning layer. The resulting system shows an increase in both SNR and thermal stability compared to the single granular layer [15, 16]. After the ECC proposal, a few authors compared the performances of ECC and CGC media and evaluated various combined ECC/CGC configurations [17–19].

Parallel to the extensive theoretical work on composite media, several experimental studies were realized about the optimization of perpendicular recording on conventional media by exploiting ECC, CGC, and combined ECC/CGC schemes (e.g., Refs. [16, 17, 19–22]). CoCrPt-based ECC media have been introduced in hard disk drives since 2007, as explained in Chapters 2 and 3. Anyhow, most of the experimental papers on hard–soft composite materials employ $L1_0$-FePt as the hard magnetic phase. Thanks to the promising results on FePt for next-generation recording media, several academic research groups focused on FePt-based ECC systems; in this case most of the published works show a materials science approach: model systems materials were studied with different soft phases and hard–soft interface properties analyzed with detailed structural and magnetic investigations (e.g., Refs. [23–28]).

Patterned composite dot arrays, where each dot is composed of a hard and a soft phase, were also realized and studied [29–32]. In this case both conventional and next-generation recording materials were explored, since advanced ECC-like bit-patterned media (BPM) could improve the design margin of BPM recording [33, 34].

Following the proposal to gradually decrease magnetic anisotropy from the hard to the soft layer [4], to have the largest-possible reduction in the switching field without decreasing thermal stability, several experimental works were realized on anisotropy-graded composite materials (e.g., Refs. [35–37]).

This chapter will recall the magnetic behavior and main advantages of hard–soft composite systems (Section 6.2) and review the most important results so far obtained on ECC conventional media (Section 6.3), ECC media based on FePt (Section 6.4), and anisotropy-graded media (Section 6.5).

6.2 Soft–Hard Composite Systems

6.2.1 *The Exchange-Spring Magnet*

The nanoscale interaction between a soft and a hard magnetic material can be exploited to tune the magnetic properties of a soft–hard composite system, a concept that was understood in 1965 but became popular within the magnetism community only in 1991, when Kneller and Hawig proposed to enhance the maximum energy product of permanent magnets by adding a nanostructured soft magnetic phase with high saturation magnetization to a hard magnet [3, 38].

The exchange length is the critical length-scale parameter to trace the magnetic phase diagram of a soft–hard composite system [38–40]. This is given by

$$l_\mathrm{w} - \sqrt{\frac{A}{K_\mathrm{h}}} \tag{6.1}$$

where A is the exchange stiffness constant, quantifying the coupling between the hard and soft phases and K_h is the hard-phase uniaxial anisotropy constant; l_w is typically a few nanometers for hard magnetic materials (e.g., $L1_0$–FePt, RE-TM hard magnetic alloys).

With the aim of maximizing the maximum energy product of a permanent magnet, the ideal magnetic regime for the composite system is the rigid magnet (RM) regime, occurring when the soft-phase size is less than l_w; in this regime the hard and soft phases simultaneously and irreversibly invert their magnetization at one critical field [40].

As shown in the magnetic phase diagram reported in Fig. 6.1a and in the scheme in Fig. 6.1c, when the soft-phase size is in

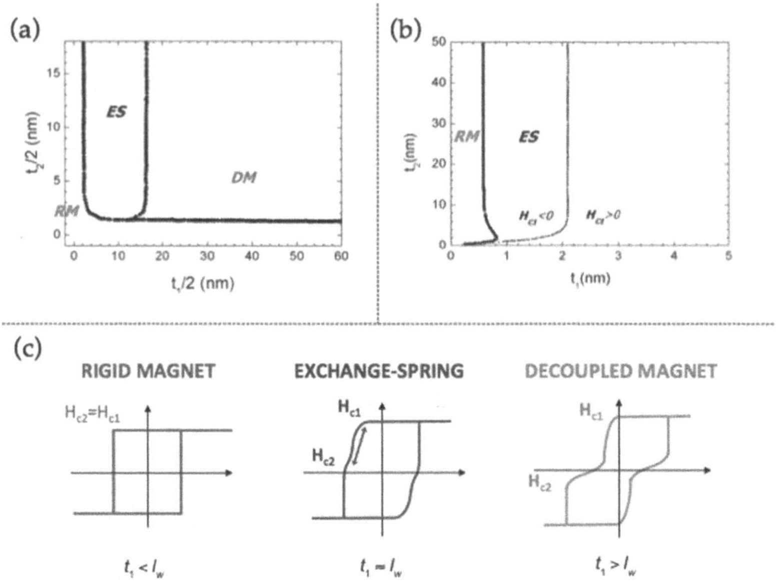

Figure 6.1 Magnetic phase diagram of the soft/hard composite bilayer as calculated for an easy-plane Fe/NdFeB bilayer in Ref. [40] (a). Reprinted with permission from Ref. [40]. Copyright (2004) by the American Physical Society. A perpendicular Fe/L1_0–FePt bilayer in Ref. [41] (b). t_1 and t_2 represent the thicknesses of the soft and hard layers, respectively; H_{C1} is the critical field for the soft-phase reversal (nucleation field). Reprinted with permission from Ref. [41]. Copyright (2006) by the American Physical Society. (c) Schematic representation of the hysteresis loop of a composite system in the different magnetic regimes; H_{C2} is the critical field for the hard-phase reversal (irreversibility field).

the order of few l_w, the composite system shows the exchange-spring (ES) regime; in this regime the system behaves as a single-phase material but with two critical fields: the soft phase firstly reverts its magnetization at the nucleation field, and drives the hard-phase reversal, which is irreversible and occurs at a different critical field, that is, the irreversibility field. The ES regime has been widely investigated theoretically and experimentally in bilayers and multilayers model systems, either for permanent magnet applications or simply for the inherent attractiveness of a simple but versatile approach, which allows tailoring the hysteresis loop

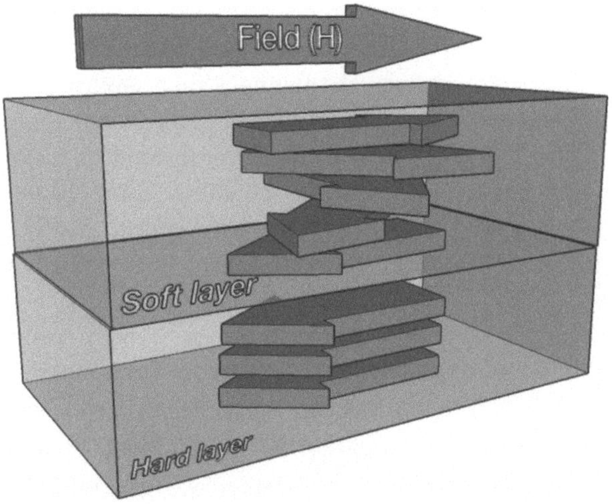

Figure 6.2 Schematic representation of the reversible rotation of soft-phase magnetic moments in the exchange-spring regime under an applied reversed field.

shape and varying coercivity in a wide range by suitably choosing the hard and soft magnetic materials and controlling the relative thickness of the two phases. The term "exchange spring" is due to Kneller and Hawig, who firstly recognized an analogy between the reversible rotation of the soft-phase magnetic moments in between the two critical fields and the mechanical motion of an ideally elastic spring (Fig. 6.2).

With increasing thickness of the soft phase well above l_w, the hysteresis loop of the composite system becomes typical of a two-phase material (decoupled magnet, DM), as the two phases switch independently and irreversibly at different critical fields (Fig. 6.1a,c).

When considering a bilayer system with perpendicular magnetic anisotropy (PMA), a magnetostatic energy term favoring the in-plane magnetization orientation in the soft layer has to be added to the magnetic free energy of the system. In this case, the magnetic phase diagram is modified, as shown in Fig. 6.1b; only two regimes are found with increasing the soft-phase thickness, RM and ES regimes [41]. As reported in Ref. [41], the magnetic phase diagram is strongly affected by the intensity of the exchange coupling at the interface between the two phases: weaker is the coupling, lower is the critical soft-layer thickness that separates RM and ES regimes. This result can be understood by considering that the strong anisotropy energy of the hard layer influences the orientation of soft-layer moments within a thickness comparable with the exchange length (Eq. 6.1). If the soft–hard coupling at the interface is reduced, this guidance role of the hard-layer anisotropy is less effective.

After soft–hard composite systems have been proposed and widely studied as magnetic recording media from 2005 to date (see next sections), further interest in these systems has been recently driven by the proposal of exploiting the ES effect in spin-transfer-torque (STT) oscillators [42]. In this case, obtaining a tunable magnetization tilt angle in the soft layer could allow the tailoring and fine-tuning of STT-driven magnetodynamic properties.

6.2.2 *Composite Media Advantages for Magnetic Recording*

The same physical models used in the first theoretical papers about ES magnets were employed for evaluating the performances of ECC systems for magnetic recording. The ECC system proposed in 2005 to solve the writability issue of perpendicular media with huge magnetic anisotropy is in fact the perfect analogue of the ES magnet [1–3, 38].

As clearly illustrated in the previous chapter, the huge magnetocrystalline anisotropy of $L1_0$–FePt ($K_1 = 7 \times 10^6$ J/m^3) could in principle allow the realization of thermally stable media with grain size down to 3–4 nm, corresponding to information densities up to 100 Tbit/in^2 but requiring writing fields above 3 T [43–45]. Heat-assisted magnetic recording (HAMR) technology is today considered a leading technology to solve the writability issue of next-generation recording media, thus increasing data storage density of hard disk drives (see Chapters 4 and 5). The ECC system was initially proposed as an alternative route to HAMR, since by coupling a soft phase to a hard phase, the switching field can be substantially reduced, while the thermal stability of the hard phase is maintained [1, 2].

The large amount of work developed on ECC media since 2005 has led to an improved knowledge of the physical behavior of the system. Furthermore, several new soft–hard systems have been studied in detail, which could be evaluated for BPM recording, and the effect of the nanoscale structure of the composite system has been investigated in more than one paper. In this and the following sections we will try to detail the above conclusions so that this review chapter could be useful for a wide audience active in the magnetic recording area.

Media writability can be improved in soft–hard composite media because of domain wall–assisted switching [5, 46]. In a composite grain where the soft layer is thick enough to fit the interfacial domain wall, magnetization reversal occurs through domain wall nucleation, compression in the applied field, depinning, and propagation through the soft–hard interface (Fig. 6.3). This incoherent rotational process is the same described by older models for the ES regime, as evidenced by Suess et al., who firstly proposed that the process could be exploited for a switching-field reduction [1]. The rigid-spin approximation used in the work by Victora and Shen, who coined the term "exchange-coupled composite" and proposed the ECC media in the same year as Suess et al., was successful in predicting a substantial switching-field reduction for an optimized exchange coupling at the interface [2]. However, a micromagnetic model can describe correctly the soft–hard composite system only allowing incoherent magnetization rotation in the soft phase [47].

One of the merits of the paper by Victora and Shen was also of introducing the gain factor ξ, to evaluate the advantage of ECC media compared to single-phase media, as the ratio of the stabilizing energy barrier to the switching field [2]. The gain factor was defined (in the CGS system) as

$$\xi = \frac{2\Delta E}{M_S H_{SW} V} \tag{6.2}$$

where ΔE is the energy barrier, M_S is the saturation magnetization, H_{SW} is the switching field, and V is the grain volume. We will recall this parameter in the next sections, since several experimental papers evaluated this gain factor.

The powerful tools today available for micromagnetic numerical calculations have allowed the accurate investigation of composite

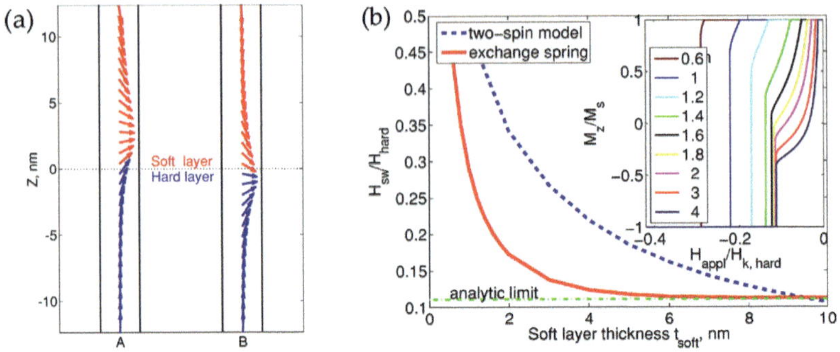

Figure 6.3 (a) Representation of the moment rotation during reversal in a composite grain. (b) Switching field as a function of soft-layer thickness for the exchange-spring and two-spin models (i.e., the rigid spin model of Ref. [2]); the inset reports $M(H)$ loops for different soft-layer thicknesses. Results in (a) and (b) were obtained with the following model parameters: saturation magnetization $M_{\text{soft}} = 4M_{\text{hard}} = 1600$ emu/cm^3; anisotropy field $H_{\text{hard}} = 200$ kOe, $H_{\text{soft}} = 6$ kOe in (a) and $H_{\text{soft}} = 0$ in (b); exchange stiffness $A_{\text{soft}} = A_{\text{hard}} = 1 \times 10^{-6}$ erg/cm; exchange energy density at the interface $J_{\text{exc}} = 100$ erg/cm^2; thickness of soft and hard layers = 20 nm in (a) and 10 nm in (b). Reprinted with permission from Ref. [46]. Copyright 2006, AIP Publishing LLC.

media phenomenology and the optimization of a wide spectrum of system parameters for improving the SNR and keeping high thermal stability [4–9]. The first parameter to be considered in the composite media design is the soft-layer thickness, which has to be large enough to allow domain wall reversal [5]. The domain wall reversal, or ES reversal, can be obtained when the soft-layer thickness is above l_w (Eq. 6.1), which for systems with an $L1_0$-FePt hard layer is a few nanometers. The hard-layer thickness does not influence the switching-field value when it is larger than l_w, so it is typically fixed at 10–15 nm. As shown by the switching-field dependence on soft-layer thickness reported in Fig. 6.3b, the largest reduction in the switching field can be obtained with large-aspect-ratio grains, which unfortunately are not so easy to realize experimentally. The typical hard–soft grain studied in Refs. [5, 6] has a 10–15 nm hard layer, a 10–15 nm soft layer, and a diameter of 3–6 nm. The graph in Fig. 6.3b also evidences that the switching-field reduction predicted by the rigid spin model (Ref. [2]) overestimates the effective reduction obtained through domain wall reversal.

The magnetic properties of the hard and soft phases, that is, anisotropy constant and saturation magnetization, are other important parameters to improve the performances of ECC media [5, 6, 8, 9]. Results for specific magnetic materials parameters can be found in later references, but we would like here to recall that increasing the soft-phase thickness and realizing a system with gradually decreasing anisotropy constant from the hard to the soft layer gives the largest possible reduction in the switching field [4]. The best performances can be obtained for a system with the anisotropy constant decreasing as a function of thickness z with a $1/z^2$ law [4, 6]. Following this result, several authors developed their own method for experimentally realizing these *multilayer ES media* or *anisotropy-graded media* (see Section 6.5). It has to be underlined that the gradually decreasing anisotropy recipe is correct for very-high-anisotropy hard layers, that is, FePt-based media, which are completely thermally stable. For conventional hard layers based on Co-alloys, the ECC principle can be useful to improve media performances but by adding semihard instead of soft layers to the hard phase to maintain sufficient thermal stability of the full stack structure [17].

The exchange coupling at the interface between the two phases is another key parameter for tailoring the media properties: increasing the soft-layer thickness reduces the hard-layer switching field and similarly does a soft–hard coupling reduction. However, a careful balance between the two parameters (i.e., soft-layer thickness and exchange coupling) allows maximizing the switchable hard-layer anisotropy for a fixed applied field [5].

As to the thermal stability, Fig. 6.4 shows that when the soft-layer anisotropy constant has the same sign as the anisotropy of the hard layer, the energy barrier for thermal reversal of a composite grain is the same as the barrier of the hard layer. On the contrary, a soft layer with a negative anisotropy constant determines a reduction of the medium thermal stability [6]. In Ref. [6], Suess et al. also discuss the correctness of the parameter defined by Victora et al. to evaluate the thermal stability gain (Eq. 6.2). The limit of a gain factor defined as in Eq. 6.2 is that ξ is thickness independent for single-phase

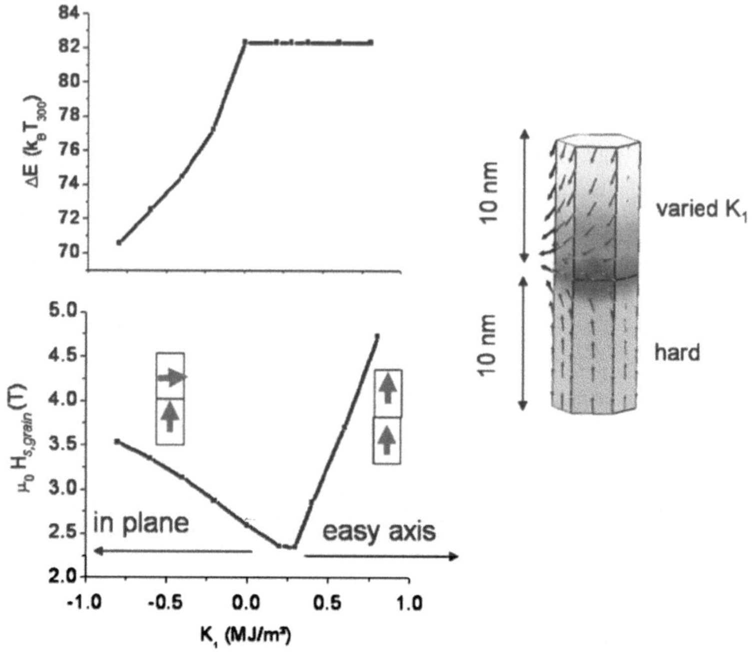

Figure 6.4 Switching field and energy barrier of a bilayer structure as a function of the soft-layer anisotropy (K_1). The magnetic parameters used in the simulations are $A = 10^{-11}$ J/m, $J_s = 0.5$ T, hard-layer anisotropy = 2×10^6 J/m^3. The soft-layer anisotropy is varied from a positive value to a large negative value, which leads to in-plane anisotropy. Reprinted with permission from Ref. [6], Copyright (2009), with permission from Elsevier.

systems that reverse homogenously, while a single-phase medium with higher thickness is definitely more stable than a thinner one. The calculations reported in Ref. [6] show that for thin single-phase media the energy barrier increases with thickness, since in this region the energy barrier follows from the Stoner–Wohlfarth theory: $\Delta E = KV$. Above a thickness of the order of l_w, the thermal stability cannot be improved by increasing the media thickness, because the system reverses via the formation of a domain wall: the energy barrier is proportional to the domain wall energy and thickness independent. The thermal stability instead increases with thickness in graded composite media and independently of the thickness range. Therefore, in graded media the optimum total thickness for domain wall reversal could also give a large advantage in thermal stability.

More advantages than the improvement in writability were predicted by several authors compared to single-phase media, that is, a reduction in switching-field distribution (SFD); high tolerance to easy-axis distribution, thus giving rise to a higher SNR; and faster and easier switching in microwave-assisted magnetic recording (MAMR) [6, 10–14]. In particular, the advantage for MAMR derives from the precessional reversal of the ECC grain, which determines a substantial reduction in the switching field and a largely faster switching [11–14].

It has to be noticed that most of the theoretical studies consider a single grain with a large aspect ratio, while experimentally realizing granular composite media with columnar grains and no lateral interactions is very challenging. Typically, in real systems the demagnetizing factor shows a value that increases with increasing the extent of lateral interactions between grains and reaches 4π for a perfectly interacting system. Hence, the soft phase in real systems usually experiences a certain magnetostatic energy favoring magnetization orientation in the film plane and the models for a single grain with large aspect ratios do not perfectly describe the real-system behavior. If similar considerations can be useful to give a phenomenological description of the magnetic behavior of a real system, Ref. [48] demonstrates that a multiscale micromagnetic approach is necessary to correctly take into account the magnetic interactions at grain boundaries. The work reported in Ref. [48] in fact evaluated the effect of lateral interactions between soft–hard composite grains and demonstrated that intergranular interactions affect not only the SNR but also thermal stability.

Clearly, ECC/BPM systems are simpler to be modeled than granular composite media [34]. In this case, simulations evidence that the ECC scheme could improve design margins of BPM recording, thanks to a tighter SFD and a lower required field gradient of the writing head [33, 34]. ECC/BPM schemes have been experimented upon in a few works [29–32], as discussed in the following sections.

More complex architectures have been also considered. Lomakin et al., for instance, proposed ledge-type nanocomposites, in which the soft phase has wider lateral extension compared to the hard phase, to further reduce the switching field and fully exploit media precessional reversal [49]. Goll et al. got back to this idea and evaluated the thermal stability of differently shaped ECC elements, concluding that mushroom-type ECC elements have the highest thermal stability ratio [7]. Mixed ECC/BPM/CGC schemes have been evaluated in Ref. [18], too. In this case the exchange coupling with a continuous soft capping layer was found to help in controlling the effect of dipolar interactions between ECC/BPM elements and gave improved writability compared to homogeneous BPM.

6.3 ECC Media Based on Conventional Media Materials

In this section we will summarize the results to date obtained by applying the ECC scheme to CoCrPt alloys, Co–Pt alloys, and Co-based multilayers. The reader can refer to Chapter 3 to have a comprehensive view on conventional perpendicular magnetic recording media.

Most of the studies realized on Co-based materials focused on the optimization of the media architecture to improve the recording performances. An exchange control layer (ECL) was often employed to control the hard–soft coupling, as originally proposed in the paper by Victora and Shen [2]. The effect of vertical and lateral exchange coupling was also evaluated through the comparison of combined capped and ECC configurations. All these results demonstrate that the introduction of one or more magnetically soft (or semihard) layers in the media is advantageous for improving the conventional media performances.

The first studies demonstrated improved media writability (Fig. 6.5), thanks to a reduction in the switching field via the domain wall nucleation and propagation process and better tolerance to angular dispersion of the write field; authors consequently obtained lower noise and increased the SNR compared to single-phase media and did not evidence thermal stability problems [20–22]. Improved areal recording density was also reported [21].

An accurate study on the composite media design demonstrated that the optimum anisotropy profile of a Co-alloy based ECC system

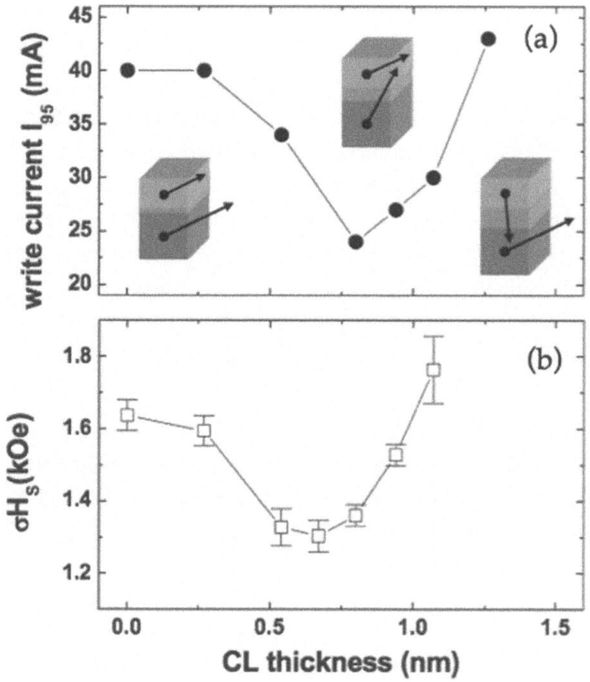

Figure 6.5 Composite media performances as a function of the ECL thickness: (a) write head current needed to achieve 95% of the media signal saturation; (b) switching-field distribution width. The schematic drawings in (a) represent the different reversal mechanisms as a function of ECL thickness. Reprinted with permission from Ref. [21]. Copyright 2008, AIP Publishing LLC.

is not the same as that previously reported for highly thermally stable hard layers by Suess et al. [4, 17]. A two-breaklayer ECC design was in fact reported in Ref. [17] to benefit from a convex anisotropy profile, which gives to the media significant switching-field reduction but sufficient thermal stability.

As predicted by Victora and Shen, the hard–soft coupling can be reduced by inserting an ECL between the hard and soft layers; the ECL is a layer of nonmagnetic material, interposed between the hard and soft layers to reduce the exchange interaction. The effect of a coupling reduction on the magnetic phase diagram of a composite system is the same as that evidenced for ES magnets [40, 41]: adding an ECL layer is equivalent to increasing the soft-layer thickness [50]. Tang et al. showed in Ref. [50] that an ECL can give similar overwrite advantages than thickening the soft layer; at similar overwrite, media with an ECL showed an increased SNR compared to media with a thickened soft layer, which instead showed higher thermal stability. The use of an exchange-coupling layer has, however, to be carefully evaluated, as reported in Ref. [17], since in a weakly coupled system the benefit of hard-phase switching assistance by soft–hard coupling could be lost because the two phases could switch independently. This is particularly critical when the ECC layer is also a CGC layer: the grains can be more strongly coupled to their neighbors laterally than they are, through a breaklayer, to the hard-layer grains below.

As conventional recording media have a CGC architecture [15, 16], researchers [17–19, 51] carefully evaluated the effect of vertical versus lateral coupling with a soft layer and compared different types of ECC, CGC, or combined ECC/CGC architectures. The correct amount of coupling between the soft and hard layers, of course, depends on the anisotropy difference between the two materials or anisotropy grading through the stack. Anyway, an increase in writability by controlling the vertical exchange coupling allows higher track densities to be achieved than controlling the lateral exchange [19].

The effect of an ECL layer on the angle-dependent reversal of ECC media was also numerically evaluated in [52]. With a semihard top layer and thin ECL layer the coupling between the layers was strong and no difference was found in the angular dependence of the switching field. A shift in the minimum value of the switching field as a function of the angle was, instead, found in the weaker coupling regime, obtained with a soft layer with lower anisotropy and larger soft-layer thickness compared to the strong coupling regime.

A more fundamental study was realized on a new kind of composite system, in which the soft layer was laminated and had a structure of the type $[CoPtCr–SiO_2/Pt]_N$ [53]. The idea was in this work to control the switching-field reduction by controlling the exchange coupling in the soft layer. Interestingly, X-ray magnetic circular dichroism experiments were performed on this system, which demonstrated the occurrence of the domain wall reversal process [54].

Other interesting results were reported on arrays of composite Co/CoPt dots, proving a reduction in switching field, SFD, and angular variation of switching field compared to arrays of single-phase dots [29, 30]. These studies demonstrate that the ECC scheme could prove advantageous for BPM recording with conventional media materials.

6.4 ECC Media Based on $L1_0$–FEPT

As previously explained, the ECC principle was originally proposed in 2005 to improve the writability of next-generation media, which, due their huge magnetocrystalline anisotropy, could allow a great increase in thermal stability but require, as a side effect, a very large writing field, larger than 3 T [1, 2]. A lot of experimental work has been thereupon realized on composite systems exploiting $L1_0$-FePt as the hard phase; most of the published works on this topic have been carried out by research groups from academic or public research institutions and thus present a fundamental materials science approach. Typically, room-temperature (RT) coercivity was taken as an indication of the switching field of the composite system.

Published results on ECC materials based on $L1_0$-FePt as the hard phase will be hereafter reviewed; the results on Fe/$L1_0$-FePt media will be presented in a specific section (Section 6.4.1) because of the number of papers published on this system. Section 6.4.2 will review the work published on $L1_0$-FePt-based composite systems, where other phases than Fe were chosen as the soft phase.

6.4.1 *Fe/$L1_0$–FePt Composite Media*

The magnetic phase diagram of the Fe/$L1_0$-FePt composite system was firstly traced after the analytical micromagnetic study by Asti et al. [41].

In this case, a bilayer system was considered where the hard $L1_0$-FePt layer has perpendicular easy magnetization direction because of its large magnetocrystalline anisotropy and crystallographic

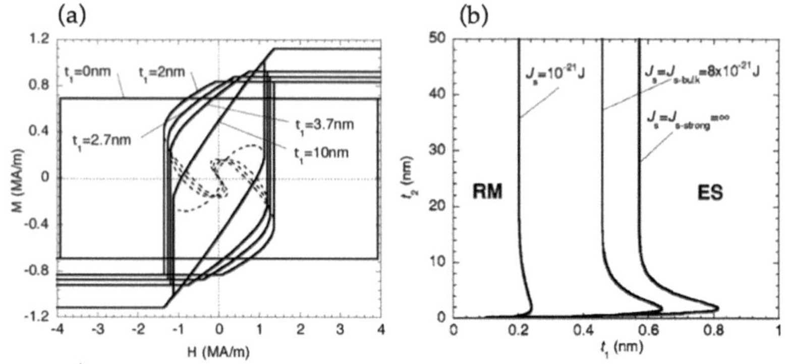

Figure 6.6 Simulated hysteresis loops for the Fe/L1_0–FePt bilayer (a) and magnetic phase diagram in the presence of reduced exchange coupling (J_S) at the soft/hard interface (b). t_1 is the soft-layer thickness and t_2 is the hard-layer thickness. Reprinted with permission from Ref. [41]. Copyright (2006) by the American Physical Society.

orientation, and the soft Fe layer has negligible magnetocrystalline anisotropy and experiences the exchange-coupling interaction with the hard layer. The obtained magnetic phase diagram is the one reported in Fig. 6.1b. The continuous line at soft-layer thickness (t_{soft}) $\approx$ 1 nm separates the RM from the ES regime. Due to the large demagnetizing energy experienced by the soft Fe layer, in the ES regime the nucleation field for the soft-phase reversal (H_{C1}) is in the second quadrant for t_{soft} below $\approx$ 2 nm (dotted critical line in Fig. 6.1b) and in the first quadrant above this critical thickness.

This is also well evidenced by the hysteresis loops calculated for the composite bilayer and reported in Fig. 6.6a. In the same paper, the effect of a reduction in the interface coupling was evaluated: as evidenced by the phase diagram here reproduced in Fig. 6.6b, reduced interface coupling shifts the critical line separating RM and ES regimes toward lower thickness values of the soft phase.

The theoretical results in Ref. [41] showed good qualitative agreement with the first experimental study on Fe/L1_0–FePt bilayers epitaxially grown on MgO(100); the coherence between the ES model and the magnetization reversal process of a real bilayer system was, in fact, demonstrated by measuring direct current demagnetization (DCD) remanence curves [41, 55]. The real weakness of analytic micromagnetic calculations in a comparison with experimental hysteresis loops is in the irreversibility (or depinning) field, that is, the critical field for hard-phase switching. Since analytical models do not take into account the presence of defects nor magnetic domains, the irreversibility field largely overestimates measured values of the hard-phase switching field. This discrepancy is generally reduced when numerical methods are used, since these methods allow the introduction in the modeled system of imperfections, such as soft-phase impurities inside the hard layer, and the formation of magnetic domains [56].

As already evidenced in Section 6.2.1, the above-described results were obtained by adding to the magnetic free energy of the system a term that takes into account the strong demagnetizing energy experienced by the soft layer. This is correct for a bilayer system, while for large-aspect-ratio grains, as those studied, for instance, in Refs. [5, 6], the shape anisotropy of the soft layer favors the magnetization alignment along the grain axis and not along the substrate plane. Therefore, the hysteresis loop of the composite system does not show positive nucleation fields for the soft-layer reversal, that is, nucleation fields in the first quadrant (see Refs. [4–6] and Fig. 6.3).

The single-grain geometry was considered by Goll et al., who used a similar modeling approach to Ref. [41] to study the Fe/$L1_0$–FePt system but considered a single composite nanoparticle, not a bilayer [8, 9]. The same authors also found good agreement between analytical and numerical models and, interestingly, evaluated different types of soft phases (e.g., Fe, Permalloy).

Figure 6.7 reports a graph taken from Ref. [9], which shows how the magnetic properties of the soft phase influence the critical fields of the composite system (i.e., H_N = nucleation field of the soft phase and H_C = depinning field of the hard phase). The graph also shows that the choice of the soft magnetic material affects the critical thickness separating RM and ES regimes, which corresponds to the passage from single-step to two-step hysteresis loops.

In the same year when the models by Goll et al. were published, accurate experimental studies on epitaxial Fe/$L1_0$–FePt systems appeared [23, 24]; further details on the correlation between nanoscale structure and magnetic properties of the same systems

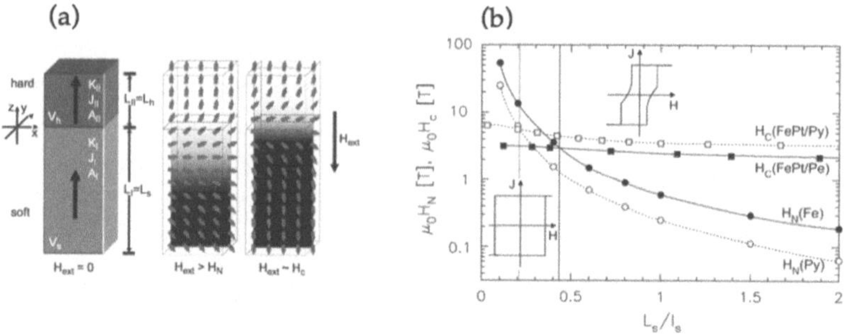

Figure 6.7 (a) Model of a composite particle with magnetization configurations after nucleation in the soft magnetic layer at two different values of the external field (H_{ext}); H_N is the nucleation field at which the soft-layer reversal starts, and H_C is the depinning field of the hard layer. (b) Thickness dependence of H_N and H_C for FePt/Fe and FePt/Py ECC particles. The crossing point of H_N and H_C curves separates one-step and two-step hysteresis regimes. Reprinted with permission from Refs. [8, 9], Copyright (2008), with permission from Elsevier.

were given in subsequent publications by the same authors [57, 58]. The studies by Casoli et al. evidenced that the magnetic phase diagram of the composite system is affected by soft–hard interface morphology, a feature that is difficultly taken into account in a micromagnetic model [23, 58]. These papers report a drastic coercivity reduction in Fe/$L1_0$–FePt(10nm) samples with increasing thickness of the Fe layer (t_{Fe}): the H_C/H_C^{hard} ratio reaches values between 0.37 and 0.4 for an Fe thickness of 3.5 nm. The coercivity reduction obtained at fixed t_{Fe} is only slightly influenced by the interface morphology, which, however, strongly affects the magnetic phase diagram. Authors, in fact, grew two $L1_0$–FePt epitaxial samples of 10 nm on MgO (100), exploiting slight changes in growth temperature and annealing conditions. The samples displayed different morphologies and were chosen as hard layers in two distinct Fe/FePt series with t_{Fe} = 2 and 3.5 nm. The magnetic properties of the two series are here summarized in Figs. 6.8 and 6.9.

It was found that the soft–hard coupling is optimized when FePt grows is quasi-continuous and Fe grows epitaxially on top of FePt, with orientation relationship $[110]Fe_{bcc}||[001]FePt_{fct}||[001]MgO$. In

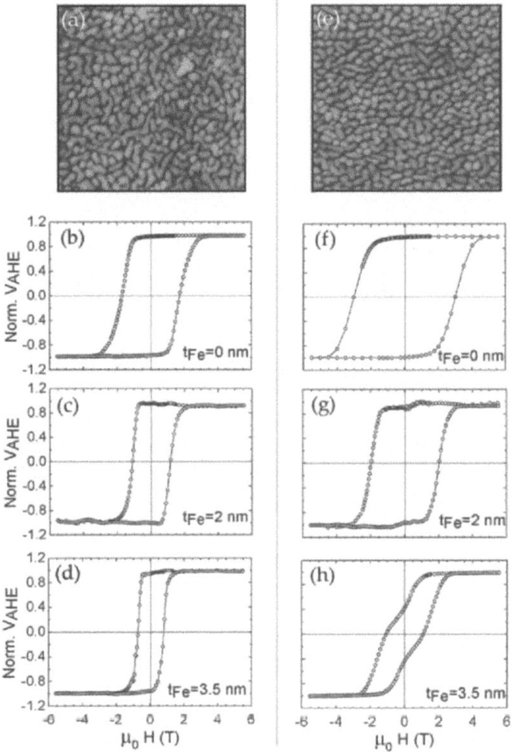

Figure 6.8 Perpendicular hysteresis loops measured by the anomalous Hall effect on two different Fe/$L1_0$–FePt bilayer series: with semicontinuous growth of FePt (b–d) and island-like growth of FePt (f–h). 1×1 μm^2 AFM images of the two types of FePt morphology are reported in (a) (semicontinuous growth) and (e) (island-like growth). Reprinted with permission from Ref. [23]. Copyright 2008, AIP Publishing LLC.

this case, for $t_{Fe} = 2$ nm the hysteresis loop shape is square and typical of the RM regime with the two phases behaving as a whole and showing an irreversible demagnetizing process (Fig. 6.8c). With increasing t_{Fe} to 3.5 nm no definite nucleation field was apparent, but a small reversible contribution was found in the recoil curves (not shown; see Ref. [23]). Very high values of squareness and remanence ratio were also obtained (Fig. 6.9), that is, 0.80 and 0.96, respectively.

Differently, when FePt has an island-like morphology Fe growth is more disordered and a considerable fraction of Fe grows directly

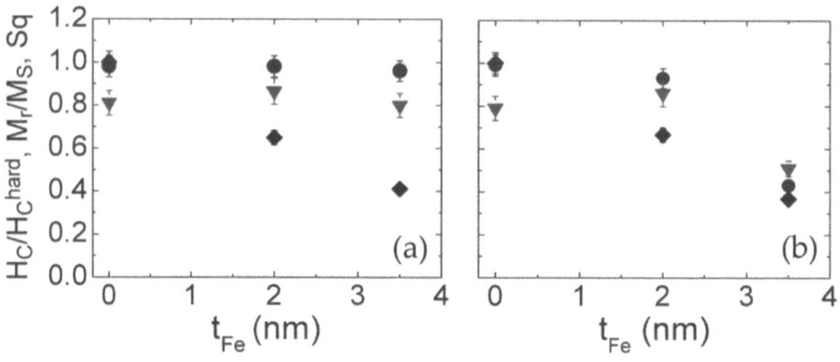

Figure 6.9 Normalized coercivity (H_C/H_C^{hard}, diamonds), remanence ratio (M_r/M_S, circles), and squareness (Sq, triangles) as a function of Fe thickness for semicontinuous growth of FePt (a) and island-like growth of FePt (b). Reprinted with permission from Ref. [23]. Copyright 2008, AIP Publishing LLC.

on MgO. As a consequence, hard–soft coupling is reduced, the hysteresis loop shape is modified and the magnetic regime at t_{Fe} = 2 nm and 3.5 nm is typical of the ES regime (Fig. 6.8g,h and Fig. 6.9). Considering Fig. 6.6b, this behavior can be understood as due to an exchange-coupling reduction in the case of island-like growth of FePt. In the theoretical phase diagram, an interface coupling reduction, in fact, shifts the boundary between different regimes toward lower Fe thickness.

In Ref. [24] Goll et al. studied the magnetic properties of Fe/$L1_0$-FePt(10 nm) composite bilayers, obtaining similar results to those presented in Ref. [23], but explored a larger portion of the magnetic phase diagram of the composite system and reached a thickness of the Fe layer of 10 nm. At this thickness, the coercivity of the composite system is approximately 20% of the hard-layer coercivity. The hysteresis loops measured on composite bilayers with increasing Fe thickness in Ref. [24] show a very good qualitative agreement with the loops calculated by the analytical micromagnetic model of Asti et al. (see Fig. 6.6a), evidencing that the Fe layer experiences a demagnetizing energy favoring magnetization orientation in the substrate plane.

A comparison between the experimental results on systems with the same nominal thickness but different growth characteristics indicates that the nanoscale structure importantly influences the magnetic phase diagram [23, 55, 57, 59, 60]. We have, for instance, evidenced above that the nanoscale structure of the composite system can strongly affect the soft–hard coupling and the demagnetizing energy. For this reason, only a model able to reproduce the complexity of a real system could ideally evaluate in the correct way the magnetic free energy. Therefore, a perfect quantitative agreement between models and experiments on composite systems is never obtained.

Anyhow, experimental model systems are often useful to understand the critical points; the powerful technologies today available for structural investigation at the nanoscale are very useful in this respect. Advanced synchrotron radiation techniques can additionally give to researchers the very special capability to "visualize" the magnetic moment's orientation on the microscopic scale. In a very interesting experimental paper on Fe/$L1_0$–FePt/MgO(100) bilayers the moment's orientation in the Fe layer during the reversal process was experimentally obtained, as a function of Fe thickness, via nuclear resonant scattering of synchrotron radiation [61]. An ES behavior on the microscopic scale was identified, and through a comparison with one-dimensional micromagnetic simulations, authors could find that the spring stiffness and critical thickness at which the transition from RM to ES occurs is determined by the strength of the exchange-coupling constant J within the Fe layer. The best agreement between model and experiment was found for $J = 2.0 \times 10^{-11}$ J/m. The evidenced moment's rotation in the Fe layer because of the ES effect, is in fairly good agreement with that predicted by several published papers showing the results of analytical and numerical micromagnetic simulations (see, for example, Refs. [9, 41, 62, 63]).

Epitaxial nanocomposite islands were also obtained on MgO(100) substrates by reducing the nominal thickness of the FePt layer to 3.2 nm [26, 64]. With this nominal thickness of the hard layer authors obtained epitaxial core–shell nanocomposites in which an Fe shell of increasing thickness covers the underlying $L1_0$–FePt core. The above-cited papers demonstrate that the ES reversal process can be obtained also in core–shell islands when the thickness of the soft layer is large enough compared to the exchange length. Fe/FePt

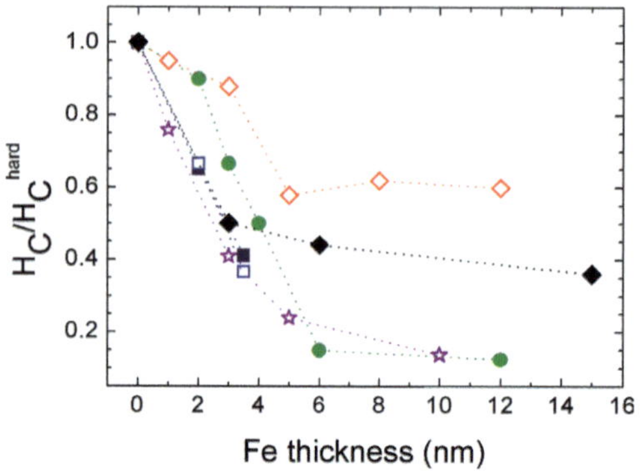

Figure 6.10 Coercivity reduction normalized to the hard-layer value as obtained by different authors on the Fe/FePt system: two Fe(t_{Fe})/FePt(10 nm)/MgO series grown with a very similar hard-layer morphology, open and solid-blue squares; Fe(t_{Fe})/FePt(10 nm)/MgO (as obtained by other authors; see text below), purple stars; Fe(t_{Fe})/FePt(3.2 nm)/MgO, green dots; Fe(t_{Fe})/FePt(10 nm)/glass, black diamonds; Fe(t_{Fe})/FePt(20 nm)/glass, orange diamonds.

nanocomposites consequently showed reduced coercivity compared to single-phase islands but also a lower energy barrier for thermal reversal and an increased gain factor ξ (see Eq. 6.2). In Refs. [26, 64] Ma et al. proposed a modified gain factor for the ECC media and demonstrated with this parameter an advantage in switching the composite media compared to single-phase media at comparable thermal stability.

After the Fe/FePt system was investigated in epitaxial composite bilayers grown on MgO single crystals, the same system was grown with good results onto glass substrates [25, 65, 66], which is an important step forward for applications. A similar magnetic phase diagram as a function of Fe thickness was traced, although the orientation and chemical order degrees of the $L1_0$–FePt phase were not as high as in the epitaxial systems. Coercivity reductions as a function of t_{Fe} obtained by different authors are compared in Fig. 6.10 [23–25, 64, 65], which shows that the largest reduction was obtained on epitaxial systems grown on MgO. The limit reduction is apparently slightly below 20% of the hard-layer value and is obtained for Fe thickness above 5 nm.

Contrary to what emerged from Section 6.3, the morphology of Fe/FePt systems published so far is different from the granular morphology of the ideal recording medium. Only few attempts have been made to realize granular composite media of Fe/FePt and the structural and morphological properties are still far to be optimized (see, for example, Ref. [59]). An ECL was employed only in a few papers [27, 67], differently from the case of ECC systems based on conventional media materials, in which the idea of the ECL to control the hard–soft coupling was largely explored on the basis of the proposal by Victora and Shen [2].

To conclude this section, we would like to draw the reader's attention to Refs. [68–71], in which more than one research group demonstrated the capability to realize regular arrays of nanocomposite Fe/FePt dots by block copolymer lithography [68, 69], nanoimprint lithography [70], and electron beam lithography (EBL) [71]. The first two techniques are suitable to realize regular patterns on large areas, while EBL is useful for model studies on the magnetization reversal process of composite dots with the aim of realizing ECC/BPM systems. We have already evidenced in Section 6.3 that the ECC scheme could prove advantageous for BPM recording with conventional media materials, reducing the distribution and angular variation of the switching field. In case the huge anisotropy constant of the $L1_0$-ordered FePt alloy was needed to improve the thermal stability of BPM, ECC dots could allow writability without thermal or microwave assistance. The minimum dot size was 50 nm in Ref. [71], 40 nm in Ref. [70], and 30 nm in Refs. [68, 69]. The magnetic properties of Fe/FePt dot arrays are consistent with that found on bilayers, showing strong coercivity reduction with increasing Fe thickness. Wang et al. also found an advantage in the gain factor ξ compared to single-phase dots [69].

6.4.2 *Composite Media with a Soft Phase Other Than Fe*

Other soft magnetic materials than Fe were chosen to realize soft/$L1_0$–FePt composite systems by researchers active in the field of magnetic recording. The disordered A1 alloy of FePt is the most simple and natural choice for growing a soft magnetic phase on top of $L1_0$–FePt (Fig. 6.11). This approach was used in Refs. [72–

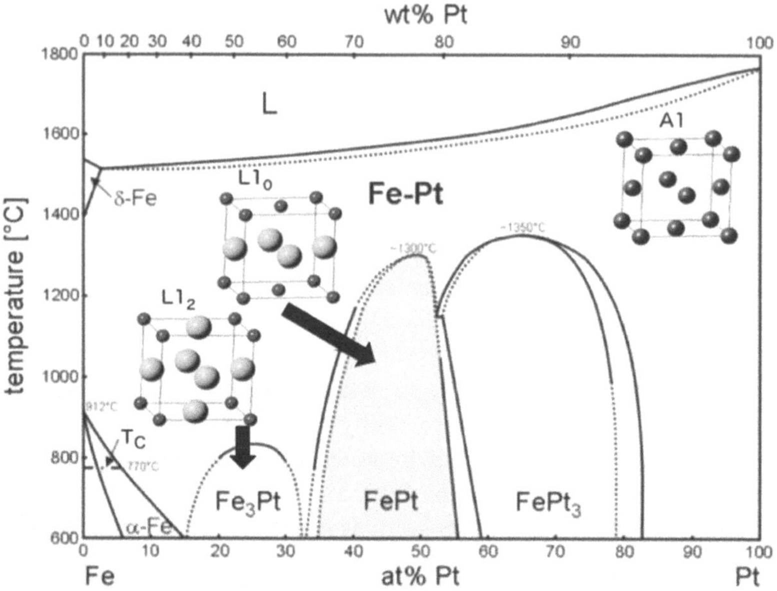

Figure 6.11 Equilibrium phase diagram of the binary Fe–Pt system. Unit cells of chemically disordered fcc FePt (A1–FePt), chemically ordered fct FePt ($L1_0$–FePt), and chemically ordered fcc Fe_3Pt ($L1_2$–Fe_3Pt) are also sketched. Reprinted from Ref. [74b], Copyright 1969, with kind permission from Springer Science and Business Media.

74a], in which the soft–hard bilayer media were grown on MgO(100) single crystals, and in Ref. [75], where Cr and MgO underlayers were used to induce the desired orientation onto a glass substrate. A similar, but in a certain sense even easier, approach as to the technological exploitation consists of epitaxially growing on top of $L1_0$–FePt and at the same growth temperature as for the $L1_0$ phase an Fe-richer soft phase of the Fe–Pt binary phase diagram, that is, Fe_3Pt (see Fig. 6.11) [76]. Different Co-based alloys have been also considered, such as CoCrPt [77], CoPt [78], and CoFe [79]. An interesting proposal, especially because this material is easy to be integrated in the production lines, is CoFeTaB; for the moment $L1_0$–FePt/CoFeTaB composite systems have been grown on SiO_2/Si and with thickness higher than the one needed for magnetic recording applications [80]. Another choice, which led to promising results, is to grow on top of $L1_0$–FePt Co-based multilayers with perpendicular magnetic anisotropy (PMA), that is, Co/Pt and Co/Ni multilayers. Several papers were realized on this type of composite systems by two different research groups [28, 81–83].

After this fast overview on the different systems proposed and studied in the last years, we will hereafter emphasize the most interesting results. In general, common conclusions of all the above-cited studies are the coercive field reduction and the occurrence of different magnetic regimes with increasing soft-phase thickness, similar to those obtained when choosing Fe as the soft magnetic phase of the composite system.

The results reported in Refs. [72–75] show that a FePt alloy with lower-order degree than the $L1_0$ phase is a good choice for realizing soft–hard composite systems. This choice allows to epitaxially grow the soft phase on top of the hard phase and is also technologically very appealing, as in this case the same sputtering targets are needed for both hard and soft phases and only the growth temperature has to be changed when passing from the growth of the hard layer (at $T \approx 400°C–700°C$) to the growth of the soft layer (at $T < 400°C$). This is the option presented in several papers on graded systems, too (see the next section). The properties of FePt(soft)/FePt(hard) composite systems can be controlled by varying the soft-layer anisotropy through variations in growth temperature in the range from RT to approximately 400°C; at fixed anisotropy of the soft layer, the soft-layer thickness is the key variable for optimizing the composite system writability. This approach can lead to an advantage in the gain factor ξ (Eq. 6.2) compared to single-phase media [74a, 75]. Increasing the soft-phase thickness above a certain value can be detrimental for the soft–hard epitaxy and thus determines a loss of coupling between the two phases [73]. Published research on this kind of systems also focused on granular media in which SiO_2 or TiO_2 were used for grain separation [72, 74a].

Another important result for the practical exploitation of these composite systems for magnetic media can be figured out: the same growth morphology of the hard layer is maintained during the soft-phase growth because of good lattice matching between the two phases and a lack of interruptions during the growth. This possibility is important for readjusting the best growth technology for FePt granular media (see Chapter 5) to the growth of composite media based on FePt.

A similar advantage is shown by the selection of Fe_3Pt as the soft phase for ECC materials. Fe_3Pt is an equilibrium phase of the binary Fe–Pt phase diagram reported in Fig. 6.11, showing in the bulk a chemically ordered cubic structure and soft magnetic properties [84].

The results reported by Casoli et al. in Ref. [76] show that the reduced misfit between Fe_3Pt and FePt favors the growth of fully epitaxial nanocomposites, with Fe_3Pt growing pseudomorphic and ordered on top of $L1_0$-FePt. In this work, Fe_3Pt/FePt composite nanoislands were prepared on two different single-crystalline substrates, that is, MgO(100) and $SrTiO_3$(100). FePt with 3.55 nm thickness was grown at 410°C–440°C; then the nominal composition of the alloy was changed abruptly to add a variable amount of Fe_3Pt corresponding to nominal thicknesses of 2, 3.5, and 5 nm. Fe_3Pt was grown at the same growth temperature as FePt.

The growth is epitaxial in the whole explored thickness range and the nanocomposites show a ledge [7, 49, 57] shape with a soft Fe_3Pt shell covering the hard FePt core. The nanocomposite morphology is similar to Fe/FePt core/shell nanoislands of Refs. [26, 64]. Figure 6.12 reports a couple of transmission electron microscopy

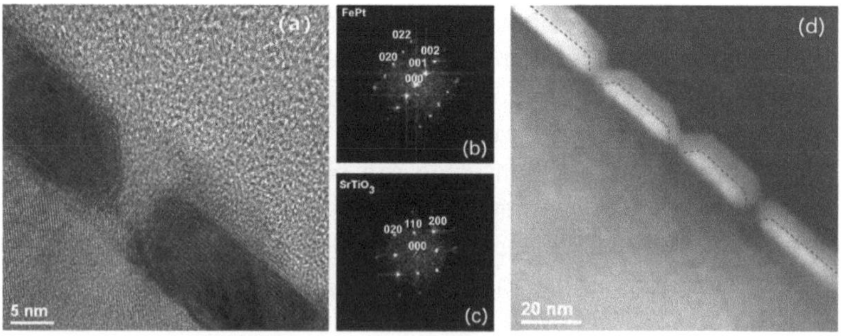

Figure 6.12 Cross-sectional high-resolution electron microscopy (HREM) image of a Fe_3Pt(5 nm)/FePt(3.55 nm)/$SrTiO_3$(100) sample (a) and corresponding FFT of the film (b) and substrate (c). (d) Cross-sectional high-angle annular dark-field (HAADF) images of the same sample. Reprinted with permission from Ref. [76]. Copyright 2009, AIP Publishing LLC.

(TEM) images of Fe_3Pt(5 nm)/FePt(3.55 nm) nanocomposites, showing the ledge shape evidenced by high-angle annular dark-field (HAADF) imaging; the epitaxial and pseudomorphic growth of the nanocomposites is evident in the high-resolution electron microscopy (HREM) image.

The growth on $SrTiO_3$(100) led to higher structural quality compared to MgO(100) because of the lower misfit between FePt and this substrate. Ledge nanocomposites on $SrTiO_3$(100) consequently display sharp faceting and an atomically flat surface (Fig. 6.12).

In the same work, the magnetic phase diagram was investigated as a function of the soft-phase thickness (t_{soft}) and for two different nominal compositions of the Fe_3Pt layer, that is, $Fe_{0.75}Pt_{0.25}$ and $Fe_{0.8}Pt_{0.2}$, which have slightly different magnetic anisotropies. For both compositions, after growing Fe_3Pt on top of FePt the coercivity is substantially reduced. With increasing t_{soft} the magnetic regime evolves from RM, showing the simultaneous reversal of both phases at one critical field, to ES magnet, characterized by a reversible part of the demagnetizing process, which involves the reversal of the soft phase only.

The soft-phase composition affects both the magnetic phase diagram and the coercivity reduction occurring with increasing soft-phase thickness. With higher content of Fe, and consequently lower anisotropy, in the soft phase ($Fe_{0.8}Pt_{0.2}$) the passage from RM to ES occurs at lower t_{soft} and the coercivity decreases faster with increasing soft-phase thickness. The effect is enhanced in the samples grown on $SrTiO_3$, for which coercivity values decrease down to 21% of the hard-layer value for the highest grown thickness of the soft phase, that is, 5 nm. On the same substrate, with increasing the atomic content of Fe in the soft phase of only 5% authors obtained a 30% reduction in coercivity at fixed t_{soft}. This reduction is evident in the graph reported in Fig. 6.13 [23, 24, 76].

Whereas a few models have calculated the analytical limit for the reduction in the critical field for hard-phase reversal in different FePt/soft systems, to the best of our knowledge Fe_3Pt has not been considered as the soft layer in theoretical works up to now. In Refs. [8, 9], the analytical limit for the irreversibility field reduction, which is reached when the soft-layer thickness is approximately 2–3 × l_w,

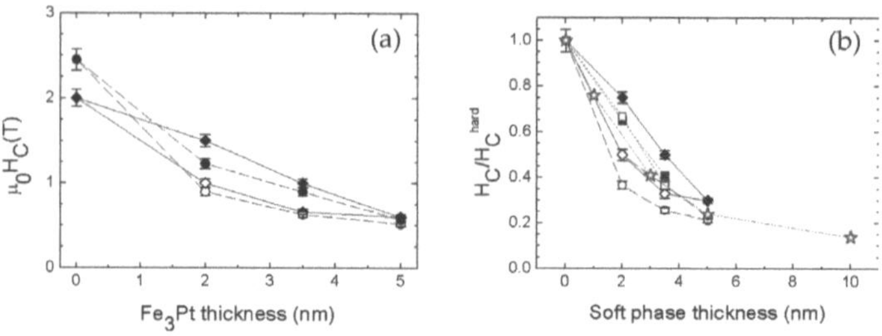

Figure 6.13 (a) Coercivity values vs. soft-phase nominal thickness in $Fe_{0.75}Pt_{0.25}$/FePt(3.55 nm)/MgO (solid diamonds), $Fe_{0.8}Pt_{0.2}$/FePt(3.55 nm)/MgO (open diamonds), $Fe_{0.75}Pt_{0.25}$/FePt(3.55 nm)/$SrTiO_3$ (solid circles), and $Fe_{0.8}Pt_{0.2}$/FePt(3.55 nm)/$SrTiO_3$ (open circles). (b) Normalized coercivity vs. soft-phase thickness for the same samples series reported in (a) and for different Fe/FePt/MgO bilayer series. Reprinted with permission from Ref. [76]. Copyright 2009, AIP Publishing LLC.

was calculated by taking into account the magnetic parameters of Fe, NiFe, and Co for the soft layer, giving values in the range 16%–33% (see Fig. 6.7b). We have already shown in Section 6.4.1 that literature data on Fe/FePt bilayers are consistent with these calculations, and we can here add that also the choice of Fe_3Pt as the soft phase leads to a limit reduction of the irreversibility field in this range [6, 8, 9].

Data reported in Ref. [76], however, also evidence that coercive field reduction proceeds faster with increasing soft-phase thickness in one of the ECC nanocomposite series grown on $SrTiO_3$ compared to Fe/FePt/MgO bilayers studied in previous works (data adapted from Refs. [23, 24] are also reported in Fig. 6.13). Both the soft-phase anisotropy and the substrate therefore importantly influence the magnetic properties. Coercivity decreases faster with increasing t_{soft} in samples with very high structural quality, where the soft–hard coupling is fully exploited, and in samples where the soft-phase anisotropy is lower.

As already outlined in Section 6.2.2, the effect of soft-phase anisotropy on the magnetic properties of a soft–hard-coupled system has been considered in a few theoretical works [5, 6, 8, 9, 40]. As to theoretical models, and coherently with the above-summarized experimental results, a decrease in the PMA of the soft phase affects the magnetic phase diagram of the system, causing the passage from RM to ES to occur at lower soft-phase thickness. In the experimental systems of Ref. [76] the soft Fe_3Pt phase grows pseudomorphically on top of the hard $L1_0$–FePt phase; soft-phase anisotropy is thus given by the sum of a perpendicular magnetocrystalline anisotropy contribution and an in-plane magnetostatic contribution due to the shape of the islands and lateral dipolar interactions. With increasing content of Fe in the soft phase the perpendicular contribution is reduced, thus favoring the in-plane orientation of the magnetic moments in the atomic layers that lie farther from the hard layer. Consequently, the ES regime occurs at lower t_{Fe3Pt}. For the same reason, the coercivity reduction occurring in ES systems and exploited for reducing the writing field in magnetic recording media is effective at lower soft-phase thickness. This is coherent with the results of the simulation by Suess et al. summarized in Fig. 6.4. A similar conclusion on the effect of soft-phase anisotropy on the properties of soft–hard composite systems was drawn in an experimental paper on CoPt-based materials [85].

In a couple of papers published in 2010 and 2014 Makarov et al. presented a detailed study on the [Co/Pt]/FePtCu system grown on SiO_2, with thickness of the hard layer fixed to 5 nm and variable thickness of the soft layer [28, 81]. The interlayer coupling was controlled through rapid thermal annealing (RTA); the comparison between magnetic characterization and micromagnetic simulations was usefully exploited to enhance interlayer coupling and adjust the layers' anisotropy. In this case, above 5 nm thickness of the soft layer the ECC media show two-step loops typical of the ES regime. In this regime the hard-phase reversal occurs at a critical field, which is proportional to the difference in anisotropy constant between the two phases (Fig. 6.14). This is coherent with that found by analytical and numerical micromagnetic calculations.

The apparent contradiction between the conclusions of Refs. [76, 85] on the one side and Refs. [28, 81] on the other side can be understood by carefully considering the analytical calculations summarized in Fig. 6.7b and numerical simulations giving the switching field in Fig. 6.4. The crossing point between the H_C and H_N curves in Fig. 6.7b separates one-step and two-step switching processes. This crossing point, coherently with that reported in Refs. [76, 85], is shifted toward lower thickness of the soft layer

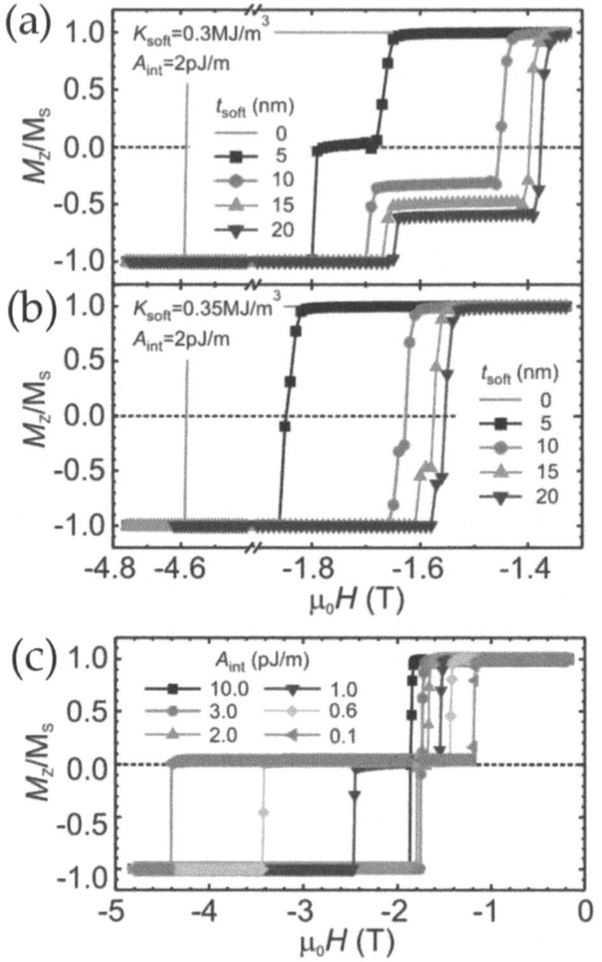

Figure 6.14 Calculated $M(H)$ curves for [Co/Pt]/FePtCu media using different soft-layer thicknesses (t_{soft}) for (a) $K_{soft} = 0.3 \times 10^6 J/m^3$ and (b) $K_{soft} = 0.35 \times 10^6 J/m^3$. (c) Calculated $M(H)$ curves for [Co/Pt]/FePtCu media with varying interlayer exchange constant (A_{int}) fixing $K_{soft} = 0.3 \times 10^6 J/m^3$ and $t_{soft} = 5$ nm. Reprinted with permission from Ref. [28]. Copyright 2010, AIP Publishing LLC.

for decreasing anisotropy of the soft phase. The switching field of one-step loops corresponds to H_N and decreases with decreasing magnetic anisotropy of the soft phase. Figures 6.4 and 6.7b are thus consistent with the experimental results of Refs. [76, 85]. The hard-phase switching field, corresponding to H_C in Fig. 6.7b for two-step processes, exhibits an opposite dependence on soft-phase anisotropy and increases with decreasing soft-phase anisotropy [28, 81]. It has to be additionally noticed that the coercive field of a specific composite system corresponds to the point where the system magnetization equals zero; this quantity therefore depends on H_N and H_C values but also on the relative magnetic moment of the hard and soft phases. Another important conclusion of Refs. [28, 81] for the possible exploitation of ECC media is that the switching field of the hard phase can be reduced by enhancing the strength of the interlayer coupling (see Fig. 6.14c).

In a series of papers published by Guo and coworkers between 2012 and 2014, ECC materials have been investigated with Co/Pt or Co/Ni multilayers with PMA as soft layers [82, 83, 86–88]. Besides finding a reduction of coercivity with increasing soft-layer thickness, narrow SFD, and large angular tolerance, these papers also present the results of detailed structural investigations giving important information for the potential application of PMA multilayers as soft layers in ECC recording media. Ref. [83] demonstrates the epitaxial growth on (001) oriented $L1_0$–FePt island-like grains of $[Co/Pt]_N$ with (200) orientation. Using Ag, the same authors could limit grain coarsening during the growth of composite media with $[Co/Ni]_N$ as the soft magnetic phase; the thermal stability was in this case improved compared to the same composite system grown without Ag [86]. The thermal stability of composite multilayers with different soft phases was furthermore compared in Ref. [86]. The conclusion is that [Co/Ni] and [Co/Pt] multilayers give better thermal stability compared to Fe, even though the switching-field reduction with increasing soft-phase thickness shows a similar trend in the three cases. This is coherent with that predicted by micromagnetic simulations because of the difference in anisotropy sign between PMA multilayers and Fe (see Fig. 6.4).

To conclude the present section, we would like to mention the original study of Zhang et al., who realized a femtosecond MOKE characterization of an ECC CoFe/FePt system in order to study the magnetization switching process and recording speed [79]. The precessional frequency was found to be independent of CoFe thickness, indicating that the dynamic magnetization process is dominated by a uniform Kittel precession mode. In contrast, the oscillation amplitude strongly depends on the soft-layer thickness and reaches a maximum at $t_{\mathrm{CoFe}} = 7$ nm. A rather large damping factor $\alpha \cong 0.13$ was obtained, governed by the hard magnetic FePt layer, which is an important parameter for the development of high speed magnetic data storage.

6.5 Anisotropy-Graded Media

As already highlighted in Section 6.2, Suess and coauthors estimated that the largest-possible reduction in the switching field could be realized by gradually decreasing the magnetic anisotropy constant of the composite system from the hard to the soft layer [4, 6]. More in detail, a system with an anisotropy constant decreasing as a function of thickness z with a $1/z^2$ law gives the largest reduction of the switching field with respect to layered media for similar values of thermal stability.

Following this result, several research groups developed their own method for experimentally realizing these *multilayer ES media* or *anisotropy-graded media*, that is, media in which the magnetic anisotropy can vary spatially with thickness. Successful approaches include exploiting growth temperature [35, 72, 89–93], composition gradients [94–96], postgrowth annealing and diffusion [32, 97–99], modulated layer thickness [36, 100], ion irradiation effects [37, 101], and ion implantation of specific ions [102]. Some of these fabrication procedures are rather complex, and for optimal performance, they have to be carefully tuned to the specific production system.

In the following section we will focus on the most significant experimental results to date published on graded media.

6.5.1 *Experimental Realization of Graded Media*

One of the first experimental works on anisotropy-graded media proposed to exploit the growth temperature of the soft phase to produce a mixed interface between hard and soft layers, thus giving rise to a variable anisotropy interface [35]. The specific system was in this case Fe/$L1_0$–FePt(5 nm) and the Fe layer was deposited in two steps: the first nanometers at temperatures between 200°C and 400°C and the remaining part of the soft layer at RT. Due to Fe interdiffusion into the FePt alloy, an interface layer with a gradually increasing amount of Fe toward the top surface forms. The coercivity of the system with a graded interface is reduced compared to a bilayer system with the same thicknesses but a sharp interface, that is, a bilayer in which the complete Fe layer was grown at RT. Moreover, coercivity further decreases with increasing thickness of the graded interface, a property that could be also used for fine-tuning of the switching field of ECC media.

A similar approach to that proposed by Goll et al. in Ref. [35] is the one chosen in Refs. [32, 97, 99]. In this case Fe/FePt bilayers are annealed after growth to create a graded interface by the interdiffusion of Fe at the interface with the FePt layer. A common conclusion of these papers on graded Fe/FePt systems is that in the presence of an intermixing layer, the switching-field reduction with respect to the hard layer is stronger compared to similar systems with a sharp interface. Other results as to the correlation between nanoscale structure and magnetic properties can, however, be deduced. In the paper by Huang et al., the soft phase is grown on top of an FePt hard layer with granular morphology. In this specific case, the comparison between the accurate characterization of the materials and the results of micromagnetic simulations evidences that postgrowth annealing creates Fe bridges between the hard FePt grains, which strongly contribute to coercivity reduction, and more importantly than the graded interface contribution [99].

Wang and coworkers evidenced in Ref. [32] that annealing the Fe/FePt samples after growth produces lower surface roughness than growing Fe at elevated temperatures, making their approach most appropriate for the realization of anisotropy-graded BPM. The same paper was also the first to demonstrate the feasibility of graded ECC dots with size down to 25 nm and on a large area, thanks to diblock copolymer lithography. The damage to $L1_0$ ordering produced by ion milling was solved with reannealing, which led to ordering improvement and narrowing of SFD.

Also Zhang et al. obtained anisotropy-graded dots with a reduced switching field, good thermal stability, and narrower SFD compared to a uniform anisotropy system [91]. In this case the employed lithography technique was EBL and the smallest obtained dot size was 87 nm. The gradual anisotropy variation was realized through a continuous variation of the MgO(100) substrate temperature from 650°C to 290°C during the growth. In this way, the chemical order parameter of the FePt alloy is gradually reduced from the substrate to the film surface, as demonstrated by X-ray diffraction experiments on 10 nm FePt films grown at different growth temperatures in the exploited range (Fig. 6.15).

The bottom panel of Fig. 6.15 evidences that the uniaxial anisotropy constant (K_u) of the FePt film increases proportionally to the ordering parameter (S), and both parameters almost linearly increase with the growth temperature. Media with linearly anisotropy grading can thus be obtained by varying the growth temperature of the FePt alloy, with the advantage of using the same sputtering targets and changing only a single parameter during growth: the substrate temperature.

Other authors exploited the same idea, but to realize epitaxial graded films onto MgO(100) [89], (001)-oriented graded films onto a glass substrate [92], (001)-oriented graded granular media onto glass [90, 93], and epitaxial graded granular media onto MgO(100) [72]. Despite the small differences in the details of the growth recipes used by different research groups, the results presented in these works demonstrate the possibility to realize graded materials with a granular morphology and on a suitable substrate for applications. Separation between graded grains was successfully reported with using TiO_2 [93] onto glass and SiO_2 [89] onto MgO single crystals, while the deposition of trilayer FePt–C films onto glass at variable temperature led to the formation of a graded media combined with a CGC media [90].

Postgrowth annealing was tried with soft phases other than Fe and FePt, that is, CoPt [97] and FeAu [103]. In both cases, atom diffusion at the interface with the hard $L1_0$–FePt layer produces a graded interface but with more complex compositional profile and microstructure compared to the choice of Fe or FePt as magnetically soft phases.

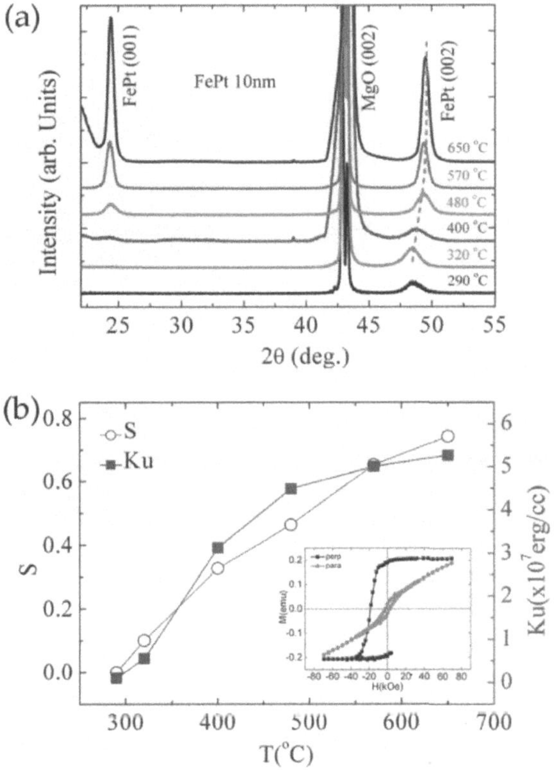

Figure 6.15 (a) X-ray diffraction patterns of 10 nm FePt thin films deposited at different temperatures. (b) Dependence of ordering parameter *S* and anisotropy constant on growth temperature. (Inset) Hysteresis loops (parallel and perpendicular to the film plane) of an FePt film grown at 570°C. Reprinted with permission from Ref. [91]. Copyright 2013, AIP Publishing LLC.

A valid alternative to temperature changes to realize an in-depth variation of the ordering degree of the FePt alloy is ion irradiation at variable dose and incident angle [37]. The work by di Bona and coauthors [37] shows that Ar^+ ion irradiation of well-ordered $L1_0$–FePt films, after the appropriate choice of ion beam parameters (e.g., ion dose, ion energy, and incidence angle), can result in anisotropy-graded media whose anisotropy profile can be tailored to suit the required purpose. Ion irradiation, in fact, induces atomic displacements and can consequently modify the chemical

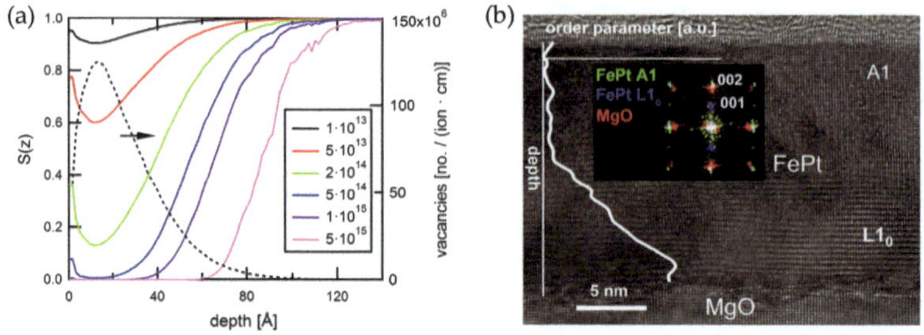

Figure 6.16 (a) Continuous lines: calculated spatial distribution of FePt chemical ordering parameter for different irradiation doses ($S_0 = 1$ for the unirradiated material); dashed line: vacancy production rate on FePt for Ar^+, 5 keV, 40° incidence angle. (b) Cross-sectional HRTEM image of a 20 nm thick FePt film irradiated from the top with a dose of 5×10^{14} ions/cm^2. (Inset) Superposition of the fast Fourier transform magnitude of MgO (red), bottom $L1_0$–FePt (blue), and top A1–FePt (green). Line graph: square root of the intensity of the Fourier-filtered images obtained by selecting the (001) superstructure spot, which is proportional to the chemical ordering parameter. Reprinted with permission from Ref. [37], Copyright (2013), with permission from Elsevier.

order of a material. Therefore, it can be effective in turning the $L1_0$ magnetically hard phase into the A1 chemically disordered phase. Moreover, by suitably choosing the irradiation parameters, an in-depth variable profile of the ordering degree can be obtained. In this way anisotropy-graded media are realized, which are made of a highly ordered layer, a layer with gradually decreasing degree of order, and, finally, a completely disordered layer at the top surface (Fig. 6.16).

Ion energy is the key parameter that determines the length scale of soft and graded regions, while for ion irradiation at 5 keV the thickness of the soft and graded regions can be varied by changing the ion incidence angle and dose. The width of the anisotropy transition region can be tuned to a length scale that is effective in reducing the switching field of the graded system, while the underlying magnetically hard material, which is not affected by ion irradiation, gives thermal stability to the media. The depth profile of the chemical order parameter that develops after ion irradiation was determined by high-resolution transmission electron microscopy (HRTEM) analysis (Fig. 6.16b). A simple mixing model mechanism

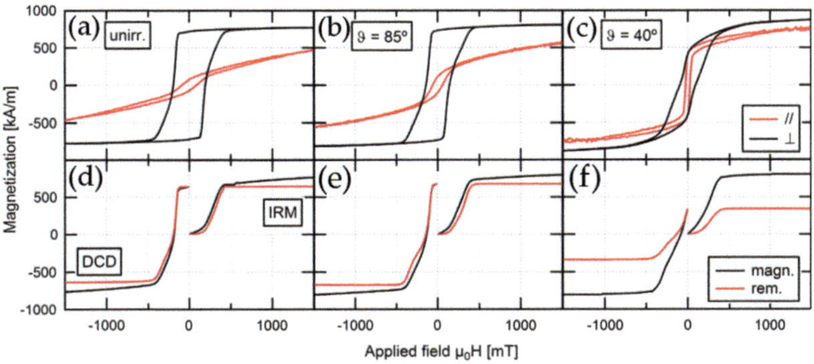

Figure 6.17 (a–c) Hysteresis loops of 15 nm FePt films measured with a field parallel (red/gray line) and perpendicular (black line) to the film plane, before (a) and after ion irradiation at 85° (b) and 40° (c) incidence angle. (d–f) Perpendicular magnetization curves (black line) and perpendicular isothermal remanent magnetization (red/gray line, 1st quadrant) and DC demagnetization remanence (red/gray line, 2nd and 3rd quadrants), before (d) and after irradiation at 85° (e) and 40° (f) incidence angle. Reprinted with permission from Ref. [37], Copyright (2013), with permission from Elsevier.

based on Monte Carlo simulations on ion interaction with matter was used to design the beam parameters and showed consistent results with the HRTEM results.

Using low-energy (5 keV) and low-dose (5×10^{15} ions cm^{-2}) Ar^+ beams, the authors could realize different in-depth profiles of chemical disorder and, consequently, of magnetocrystalline anisotropy by changing the beam incident angle. This can be understood from Fig. 6.17: after irradiation at 40°, a significant reduction (–50%) of the coercive field H_C and a positive nucleation field are observed (Fig. 6.17c). Changing the ion incidence angle to 85°, the ion penetration depth is reduced and the thicknesses of the soft and anisotropy-graded layers can be decreased (Fig. 6.17b). The magnetic behavior can thus be tuned: the irradiation at 85° still reduces H_C (–39%) but to a lower extent; the easy-plane component is also decreased compared to the sample irradiated at 40°. Moreover, irradiation at different incidence angles changes substantially the coupling regime and this affects the magnetization reversal process (see DC demagnetizing remanence and isothermal remanent magnetization curves in the bottom panel of Fig. 6.17). Authors also verified by atomic force microscopy that the optimized ion irradiation process does not modify the surface topography of the films, which is an important conclusion for magnetic recording.

The advantage of this method compared to temperature variation during growth, or during postgrowth annealing, is that the significant parameters of the ion irradiation process (i.e., ion energy, angle, and dose) can be measured with a great degree of accuracy and tightly controlled and reproduced in different fabrication systems. Moreover, ion irradiation is simple and fast and can be scaled up to large-area samples. In magnetic recording applications, this method can be extended to granular media and BPM based on ordered structures to further improve their performance (writability and thermal stability) without affecting the media morphology.

Ion irradiation was also exploited on Co/Pd multilayers [101]. In this case, He^+ ions' irradiation modified magnetic anisotropy by creating an intermixing at the Co/Pt interface. A similar approach to realize an anisotropy-graded system was employed in Ref. [102], which reports the results of ion implantation experiments on CoCrPt-SiO_2 media.

A gradual anisotropy variation was obtained also by progressively varying the composition during the growth, either in conventional media materials [94] or in FePt-based materials [95, 96]. In Ref. [94] single or multilayered granular oxides were capped with a CoPtCrB continuous layer, obtaining well-separated Co-rich grains and a laterally continuous cap layer. The anisotropy of the granular layer was controlled by adjusting the oxide thickness ratio of multilayered oxides with different anisotropy values. Authors concluded that the graded oxide media exhibit improved writability and resolution, sharper transition, and higher signal-to-noise ratio (SNR). Modeling the media behavior also allowed establishing that optimal tuning of the media anisotropy and exchange to the head design is critical in attaining good recording performances. The work by Dumas et al. is more a fundamental study on the structural and magnetic properties of graded FePtCu films [95]. A compositional gradient is first obtained by varying the Cu concentration from Cu-rich $(Fe_{53}Pt_{47})_{70}Cu_{30}$ to Cu-free $Fe_{53}Pt_{47}$. Then, an anisotropy gradient is realized with thermal annealing, thanks to the strong dependence of the $L1_0$ ordering temperature on composition. Interestingly, this work shows how a complete analysis of the magnetic behavior of a graded system can be obtained by combining alternating gradient magnetometry–first-order reversal curve (AGM-FORC) with the surface sensitive MOKE-FORC. At low annealing temperatures, FORC measurements clearly indicate the highly coupled reversal of soft and hard phases, while at elevated annealing temperatures significant interdiffusion results in virtually uniform films. A different method was employed in Ref. [96] to obtain a compositional variation as a function of thickness in $L1_0$–FePt(001) films on a glass substrate. The FePt film comprises in this case three layers with a different Fe composition: FePt (5 nm), Fe_2Pt (2 nm), and Fe_3Pt (3 nm). Interdiffusion due to high-temperature growth further helps in producing a graded material. All the graded samples show single-phase magnetization reversal behavior with significantly reduced coercivity compared to the hard layer and have a similar anisotropy field as the hard-layer film.

The last methodology that we will mention in this section to produce an anisotropy-graded material is based on the modulation of layer thickness in a multilayer structure. This procedure was usefully employed by Wang et al. with FePt-based media and by Kirby et al. with Co/Pd multilayers [36, 100]. In Ref. [36] Wang et al. intercalated the FePt layer with C layers of different thickness, thus obtaining well-isolated magnetic grains. In this case, authors concluded that adding C layers with different thickness not only tailors the anisotropy gradient but also refines the grain size and weakens the intergranular exchange interactions. A similar method was used by Kirby et al. but with a different material; they in fact realized Co/Pd multilayers with depth-dependent Co layer thickness [100]. The polarized neutron reflectometry experiments performed on these samples showed that Co thickness grading produces thickness-dependent anisotropy. The magnetization reversal process typical of composite media with graded anisotropy was also identified by magnetometry, confirming the validity of the method to achieve anisotropy grading.

References

1. Suess D., Schrefl T., Fähler S., Kirschner M., Hrkac G., Dorfbauer F., and Fidler J. (2005). Exchange spring media for perpendicular recording, *Appl. Phys. Lett.*, **87**, Article No. 012504.
2. Victora R. H., and Shen X. (2005). Composite media for perpendicular magnetic recording, *IEEE Trans. Magn.*, **41**, pp. 537–542.
3. Kneller E. F., and Hawig R. H. (1991). The exchange-spring magnet: a new material principle for permanent magnets, *IEEE Trans. Magn.*, **27**, pp. 3588–3600.
4. Suess D. (2006). Multilayer exchange spring media for magnetic recording, *Appl. Phys. Lett.*, **89**, Article No. 113105.
5. Dobin A. Y., and Richter H. J. (2007). Domain wall assisted magnetic recording, *J. Appl. Phys.*, **101**, Article No. 09K108.
6. Suess D., Lee J., Fidler J., and Schrefl T. (2009). Exchange coupled perpendicular media, *J. Magn. Magn. Mater.*, **321**, pp. 545–554.
7. Goll D., and Macke S. (2008). Thermal stability of ledge-type $L1_0$-FePt/Fe exchange-spring nanocomposites for ultrahigh recording densities, *Appl. Phys. Lett.*, **93**, Article No. 152512.
8. Goll D., and Kronmueller H. (2008). Critical fields of an exchange coupled two-layer composite particle, *Physica B: Condens. Matter*, **403**, pp. 1854–1859.
9. Goll D., Macke S., and Kronmueller H. (2008). Exchange coupled composite layers for magnetic recording, *Physica B: Condens. Matter*, **403**, pp. 338–341.
10. Lee J., Suess D., Schrefl T., Dean J., and Fidler J. (2009). Increases in effective head field gradients in exchange spring media, *Appl. Phys. Lett.*, **95**, Article No. 172509.
11. Tanaka T., Kashiwagi S., Furomoto Y., Otsuka Y., and Matsuyama K. (2013). Stable microwave-assisted magnetization switching for nanoscale exchange-coupled composite grain, *Nanoscale Res. Lett.*, **8**, Article No. 461.
12. Livshitz B., Inomata A., Neal Bertram H., and Lomakin V. (2007). Precessional reversal in exchange-coupled composite magnetic elements, *Appl. Phys. Lett.*, **91**, Article No. 182502.
13. Bashir M. A., Schrefl T., Suess D., Dean J., Goncharov A., Hrkac G., et al. (2009). Exchange coupled bit patterned media under the influence of RF-field pulses, *IEEE Trans. Magn.*, **45**, pp. 3851–3854.

14. Li S., Livshitz B., Bertram H. N., Fullerton E. E., and Lomakin V. (2009). Microwave-assisted magnetization reversal and multilevel recording in composite media, *J. Appl. Phys.*, **105**, 07B909.
15. Sonobe Y., Weller D., Ikeda Y., Schabes M., Takano K., Zeltzer G., Yen B. K., Best M. E., Greaves S. J., Muraoka H., and Nakamura Y. (2001). Thermal stability and SNR of coupled granular/continuous media, *IEEE Trans. Magn.*, **37**, pp. 1667–1670.
16. Yasumori J., Sonobe Y., Greaves S. J., and Tham K. K. (2009). Approach to high-density recording using CGC structure, *IEEE Trans. Magn.*, **45**, pp. 850–855.
17. Nolan T. P., Valcu B. F., and Richter H. J. (2011). Effect of composite designs on writability and thermal stability of perpendicular recording media, *IEEE Trans. Magn.*, **47**, pp. 63–68.
18. Lubarda M. V., Li S., Livshitz B., Fullerton E. E., and Lomakin V. (2011). Reversal in bit patterned media with vertical and lateral exchange, *IEEE Trans. Magn.*, **47**, pp. 18–25.
19. Richter H. J., Choe G., and Terris B. D. (2011). Recording behavior of exchange coupled composite media, *IEEE Trans. Magn.*, **47**, pp. 4769–4774.
20. Wang J.-P., Shen W., and Hong S.-Y. (2007). Fabrication and characterization of exchange coupled composite media, *IEEE Trans. Magn.*, **43**, pp. 682–686.
21. Berger A., Supper N., Ikeda Y., Lengsfield B., Moser A., and Fullerton E. E. (2008). Improved media performance in optimally coupled exchange spring layer media, *Appl. Phys. Lett.*, **93**, Article No. 122502.
22. Qiu L. J., Shi J. Z., Piramanayagam S. N., Chen J. S., and Ding J. (2008). Nanocomposite magnetic films for high-density perpendicular magnetic recording media, *Thin Solid Films*, **516**, pp. 5381–5385.
23. Casoli F., Albertini F., Nasi L., Fabbrici S., Cabassi R., Bolzoni F., and Bocchi C. (2008). Strong coercivity reduction in perpendicular FePt/Fe bilayers due to hard/soft coupling, *Appl. Phys. Lett.*, **92**, Article No. 142506.
24. Goll D., Breitling A., and Macke S. (2008). Magnetic properties of exchange-coupled $L1_0$-FePt/Fe composite elements, *IEEE Trans. Magn.*, **44**, pp. 3472–3475.
25. Tsai J.-L., Tzeng H.-T., and Lin G.-B. (2010). Magnetization reversal process in Fe/FePt films, *Appl. Phys. Lett.*, **96**, Article No. 032505.
26. Ma B., Wang H., Zhao H., Sun C., Acharya R., and Wang J.-P. (2010). $L1_0$-FePt/Fe Exchange coupled composite structure on MgO substrates, *IEEE Trans. Magn.*, **46**, pp. 2345–2348.

27. Feng C., Gong K., Xu C.-C., Yang M., Li X., Li B., and Yu G. (2014). Synthesis and property tunability of interparticle exchange-decoupled $L1_0$-FePt:Au/Fe perpendicular ECC films, *J. Alloys Compd.*, **590**, pp. 289–293.
28. Makarov D., Lee J., Brombacher C., Schubert C., Fuger M., Suess D., et al. (2010). Perpendicular FePt-based exchange-coupled composite media, *Appl. Phys. Lett.*, **96**, Article No. 062501.
29. Mitsuzuka K., Shimatsu T., Muraoka H., Aoi H., Kikuchi N., and Kitakami O. (2008). Magnetic properties of Co-Pt/Co hard/soft stacked dot arrays, *J. Appl. Phys.*, **103**, art.no. 07C504.
30. Sbiaa R., Aung K. O., Piramanayagam S. N., Tan E.-L., and Law R. (2009). Patterned media with composite structure for writability at high areal recording density, *J. Appl. Phys*, **105**, Article No. 073904.
31. McCallum A. T., Krone P., Springer F., Brombacher C., Albrecht M., Dobisz E., et al. (2011). $L1_0$ FePt based exchange coupled composite bit patterned films, *Appl. Phys. Lett.*, **98**, Article No. 242503.
32. Wang H., Zhao H., Rahman T., Isowaki Y., Kamata Y., Maeda T., et al. (2013). Fabrication and characterization of FePt exchange coupled composite and graded bit patterned media, *IEEE Trans. Magn.*, **49**, pp. 707–712.
33. Shiroishi Y., Fukuda K., Tagawa I., Iwasaki H., Takenoiri S., Tanaka H., et al. (2009). Future Options for HDD Storage, *IEEE Trans. Magn.*, **45**, pp. 3816–3822.
34. Krone P., Makarov D., Schrefl T., and Albrecht M. (2010). Exchange coupled composite bit patterned media, *Appl. Phys. Lett.*, **97**, Article No. 082501.
35. Goll D., Breitling A., Gu L., Van Aken P. A., and Sigle W. (2008). Experimental realization of graded $L1_0$-FePt/Fe composite media with perpendicular magnetization, *J. Appl. Phys.*, **104**, Article No. 083903.
36. Wang F., Zhang J., Zhang J., Wang C., Wang Z., Zeng H., et al. (2013). Graded/soft/graded exchange-coupled thin films fabricated by $[FePt/C]_5$/Fe/$[C/FePt]_5$ multilayer deposition and post-annealing, *Appl. Surf. Sci.*, **271**, pp. 390–393.
37. di Bona A., Luches P., Albertini F., Casoli F., Lupo P., Nasi L., et al. (2013). Anisotropy-graded magnetic media obtained by ion irradiation of $L1_0$ FePt, *Acta Mater.*, **61**, pp. 4840–4847.
38. Goto E., Hayashi N., Miyashita T., and Nakagawa K. (1965). Magnetization and Switching Characteristics of Composite Thin Magnetic Films, *J. Appl. Phys.*, **36**, pp. 2951–2958.

39. Leineweber T., and Kronmuller H. (1997). Micromagnetic examination of exchange coupled ferromagnetic nanolayers, *J. Magn. Magn. Mater.*, **176**, pp. 145–154.
40. Asti G., Solzi M., Ghidini M., and Neri F. M. (2004). Micromagnetic analysis of exchange-coupled hard-soft planar nanocomposites, *Phys. Rev. B*, **69**, Article No. 174401.
41. Asti G., Ghidini M., Pellicelli R., Pernechele C., Solzi M., Albertini F., et al. (2006). Magnetic phase diagram and demagnetization processes in perpendicular exchange-spring multilayers, *Phys. Rev. B*, **73**, Article No. 094406.
42. Anh Nguyen T. N., Fang Y., Fallahi V., Benatmane N., Mohseni S. M., Dumas R. K., and Akerman J. (2011). [Co/Pd]–NiFe exchange springs with tunable magnetization tilt angle, *Appl. Phys. Lett.* **98**, Article No. 172502.
43. Ivanov O. A., Solina L.V., Demshina V.A., and Magat L.M. (1973). Determination of the anisotropy constant and saturation magnetization, and magnetic properties of powders of an iron-platinum alloy, *Fiz. Metal. Metalloved.*, **35**, pp. 81–85.
44. Varaprasad B. S. D. C. S., Takahashi Y. K., and Hono K. (2013). Microstructure control of $L1_0$-ordered FePt granular film for heat-assisted magnetic recording (HAMR) media, *JOM*, **65**, pp. 853–861.
45. Kryder M. H., Gage E. C., Mc Daniel T. W., Challener W. A., Rottmayer R. E., Ju G., et al. (2008). Heat assisted magnetic recording, *Proc. IEEE*, **96**, pp. 1810–1835.
46. Dobin A. Y., and Richter H. J. (2006). Domain wall assisted magnetic recording, *Appl. Phys. Lett.*, **89**, art.no. 062512.
47. Aharoni A. (2000) *Introduction to the Theory of Ferromagnetism*, 2nd Ed. (Oxford University Press, UK).
48. Suess D., Lee J., Fidler J., Jung H. S., Velu E. M. T., Jiang W., et al. (2009). Effect of intergranular exchange on the thermal stability and coercive field of perpendicular, single phase, exchange spring, and coupled granular continuous (CGC) perpendicular recording media, *IEEE Trans. Magn.*, **45**, pp. 88–99.
49. Lomakin V., Choi R., Livshitz B., Li S., Inomata A., and Bertram H. N. (2008). Dual-layer patterned media "ledge" design for ultrahigh density magnetic recording, *Appl. Phys. Lett.*, **92**, Article No. 022502.
50. Tang K., Takano K., Choe G., Wang G., Zhang J., Bian X., and Mirzamaani M. (2008). A study of perpendicular magnetic recording media with an exchange control layer, *IEEE Trans. Magn.*, **44**, pp. 3507–3510.

51. Choe G., Ikeda Y., Ghaderi A., Zhang K., and Park J. (2010). Switching behavior and writeability in perpendicular magnetic dual oxide media with capping layer, *IEEE Trans. Magn.*, **46**, pp. 1802–1805.
52. Saharan L., Morrison C., Miles J. J., Thomson T., Schrefl T., and Hrkac G. (2013). Modelling interfacial coupling in thin film magnetic exchange springs at finite temperature, *J. Appl. Phys*, **114**, Article No. 153908.
53. Hou H.-C., Lin M.-S., Liao J.-W., Wu T.-L., Lai C.-H., Chen R.-Z., et al. (2009). Effects of laminated soft layer on magnetization reversal of exchange coupled composite media, *J. Appl. Phys*, **105**, Article No. 07B729.
54. Hou H.-C., Suess D., Liao J.-W., Lin M.-S., Lin H.-J., Chang F.-H., and Lai C.-H. (2011). Direct probing magnetization reversal of exchange-coupled-composite media by x-ray magnetic circular dichroism, *Appl. Phys. Lett.*, **98**, art.no. 262507.
55. Casoli F., Albertini F., Fabbrici S., Bocchi C., Nasi L., Ciprian R., and Pareti L. (2005). Exchange-coupled FePt/Fe bilayers with perpendicular magnetization, *IEEE Trans. Magn.*, **41**, pp. 3877–3879.
56. Varvaro G., Albertini F., Agostinelli E., Casoli F., Fiorani D., Laureti S., et al. (2012). Magnetization reversal mechanism in perpendicular exchange-coupled Fe/$L1_0$–FePt bilayers, *New J. Phys.*, **14**, Article No. 073008.
57. Goll D., and Breitling A. (2009). Coercivity of ledge-type $L1_0$-FePt/Fe nanocomposites with perpendicular magnetization, *Appl. Phys. Lett.*, **94**, Article No. 052502.
58. Casoli F., Albertini F., Nasi L., Fabbrici S., Cabassi R., Bolzoni F., et al. (2010). Role of interface morphology in the exchange-spring behavior of FePt/Fe perpendicular bilayers, *Acta Mater.*, **58**, pp. 3594–3601.
59. Tsai J. L., Tai H. W., Huang J. C., Lin C. S., and Chen L. H. (2011). Magnetic properties and microstructure of exchange coupled Fe/(FePt-Ag_2Se) particulate films, *IEEE Trans. Magn.*, **47**, pp. 3625–3628.
60. Liu L., Sheng W., Bai J., Cao J., Lou Y., Wang Y., et al. (2012). Magnetic properties and magnetization reversal process of $L1_0$ FePt/Fe bilayers magnetic thin films, *Appl. Surf. Sci.*, **258**, pp. 8124–8127.
61. Laenens B., Planckaert N., Demeter J., Trekels M., L'abbe C., Strohm C., et al. (2010). Spin structure in perpendicularly magnetized Fe-FePt bilayers, *Phys. Rev. B*, **82**, Article No. 104421.
62. de Sousa N., Apolinario A., Vernay F., Monteiro P. M. S., Albertini F., Casoli F., et al. (2010). Spin configurations in hard/soft coupled bilayer

systems: transitions from rigid magnet to exchange-spring, *Phys. Rev. B*, **82**, Article No. 104433.

63. Song Y., Zhang Z., Duan N., Wang J., Chen Y., Tong B., et al. (2014). Composition and size dependence of magnetic properties of FePt/Fe exchange-spring films, *J. Magn. Magn. Mater.*, **37**, pp. 100–105.
64. Ma B., Wang H., Zhao H., Sun C., Acharya R., and Wang J.-P. (2011). Structural and magnetic properties of a core-shell type $L1_0$ FePt/Fe exchange coupled nanocomposite with tilted easy axis, *J. Appl. Phys.*, **109**, Article No. 083907.
65. Sun A.-C., Yuan F.-T., Hsu J.-H., Lin Y. H., and Kuo P. C. (2009). Magnetic reversal behaviors of perpendicular exchange-coupled Fe/FePt bilayer films, *IEEE Trans. Magn.*, **45**, pp. 2709–2715.
66. Tsai J.-L., Tzeng H.-T., and Liu B.-F. (2010). Coercivity variation in exchange-coupled Fe/FePt bilayer with perpendicular magnetization, *Thin Solid Films*, **518**, pp. 7271–7274.
67. Gayen A., Biswas B., Singh A. K., Saravanan P., and Perumal A. (2013). High temperature magnetic properties of indirect exchange spring FePt/M(Cu,C)/Fe trilayer thin films, *J. Nanomater.,* **2013,** Article No. 718365.
68. Wang H., Rahman M. T., Zhao H., Isowaki Y., Kamata Y., Kikitsu A., and Wang J.-P. (2011). Fabrication of FePt type exchange coupled composite bit patterned media by block copolymer lithography, *J. Appl. Phys.*, **109**, Article No. 07B754.
69. Wang H., Li W., Rahman M. T., Zhao H., Ding J., Chen Y., and Wang J.-P. (2012). Characterization of $L1_0$-FePt/Fe based exchange coupled composite bit pattern media, *J Appl. Phys.*, **111**, art.no. 07B914.
70. Goll D., and Bublat T. (2013). Large-area hard magnetic $L1_0$-FePt and composite $L1_0$-FePt based nanopatterns, *Phys. Status Solidi A,* **210**, pp. 1261–1271.
71. Huang L. S., Hu J. F., Zong B. Y., Zeng S. W., Ariando and Chen J. S. (2014). Magnetic properties of $L1_0$-FePt/Fe exchange-coupled composite nanodots, *J. Phys. D: Appl. Phys.*, **47**, Article No. 245001.
72. George T. A., Yu Y., Yue L., Skomski R., and Sellmyer D. J. (2010). Control of coercivity in exchange-coupled graded (001) FePt : SiO_2 nanocomposite films, *IEEE Trans. Magn.*, **46**, pp. 2435–2437.
73. Wan J., Bonder M. J., Huang Y., Hadjipanayis G. C., and Ni C. (2010). (001) Exchange-coupled FCC/$L1_0$ FePt bilayers, *J. Magn. Magn. Mater.*, **322**, pp. 1811–1815.
74. (a) Zhou T.-J., Cher K. M., Lwin P. W., and Hu J. F. (2013). Energy barrier measurement and optimization in exchange coupled FePt/TiO_2 nano-

composite thin films, *J. Magn. Magn. Mater.*, **331**, pp. 187–192; (b) Predel, B. (1969). *Fe-Pt (Iron-Platinum)* (Springer).

75. Sun C.-J., Stafford D., and Acharya R. (2010). FePt-based exchange-coupled composite perpendicular recording media, *IEEE Trans. Magn.*, **46**, pp. 1795–1797.

76. Casoli F., Lupo P., Nasi L., Cabassi R., Fabbrici S., Bolzoni F., Ranzieri P., and Albertini F. (2015). Exploiting full epitaxy in the magnetic nanocomposite system Fe_3Pt/FePt, *J. Appl. Phys.*, **117**, Article No. 083906.

77. Xu Z., Zhou S. M., Ge J. J., Du J., and Sun L. (2009). Magnetization reversal mechanism of perpendicularly exchange-coupled composite $L1_0$-FePt/CoCrPt bilayers, *J. Appl. Phys.*, **105**, Article No. 123903.

78. Pandey K. K. M., Chen J. S., and Chow G. M. (2010). $L1_0$ Phase CoPt-TiO_2/FePt-TiO_2 exchange coupled media with small switching field, *IEEE Trans. Magn.*, **46**, pp. 1955–1958.

79. Zhang Z., Cui B., Wang G., Ma B., Jin Q. Y., and Liu Y. (2010). Ultrafast laser-induced magnetization precession dynamics in FePt/CoFe exchange-coupled films, *Appl. Phys. Lett.*, **97**, Article No. 172508.

80. Kaushik N., Sharma P., Yubuta K., Makino A., and Inoue A. (2010). Domain wall assisted magnetization switching in (111) oriented $L1_0$ FePt grown on a soft magnetic metallic glass, *Appl. Phys. Lett.*, **97**, Article No. 072510.

81. Lee J., Makarov D., Brombacher C., Dymerska B., Suess D., Albrecht M., and Fidler J. (2014). Scaling dependence and tailoring of the pinning field in FePt-based exchange coupled composite media, *Nanotechnology*, **25**, Article No. 045604.

82. Guo H. H., Liao J. L., Ma B., Zhang Z. Z., Jin Q. Y., Rui W. B., et al. (2012). $L1_0$-FePt based exchange coupled composite films with soft $[Co/Ni]_N$ multilayers, *J. Appl. Phys.*, **111**, Article No. 103916.

83. Guo H. H., Liao J. L., Ma B., Zhang Z. Z., Jin Q. Y., Wang H., and Wang J. P. (2012). Microstructure and magnetization reversal of $L1_0$-FePt/$[Co/Pt]_N$ exchange coupled composite films, *Appl. Phys. Lett.*, **100**, Article No. 142406.

84. Menshikov A., Tarnoczi T., and Kren E., (1975). Magnetic structure of ordered FePt and Fe_3Pt alloys, *Phys. Status Solidi A*, **28**, pp. K85–K87.

85. Tang R., Chua S., Zhang W., and Li Y. (2011). Effects of soft layer softness on the magnetic properties of perpendicular exchange-coupled nanocomposite films, *J. Magn. Magn. Mater.*, **323**, pp. 2569–2574.

86. Guo H., Chu H., Liao J., Ma B., Zhang Z., and Jin Q. (2013). Thermal stability of FePt-based exchange coupled composite films, *IEEE Trans. Magn.*, **49**, pp. 3683–3686.

87. Guo H. H., Chu H. G., Ma B., Zhang Z. Z., Jin Q. Y., Wang H., and Wang J. P. (2013). Influence of Ag on the microstructure and magnetic properties of perpendicular exchange coupled composite $L1_0$-FePt-Ag/(Co/Ni)$_N$ films, *Appl. Phys. Lett.*, **102**, Article No. 252410.

88. Chu H. G., Guo H. H., Xie H. L., Ma B., Zhang Z. Z., Wang Y., and Jin Q. Y. (2014). Magnetic properties of perpendicular exchange coupled composite with a synthesis antiferromagnetic sandwich structure, *Appl. Phys. Lett.*, **104**, Article No. 102402.

89. Alexandrakis V., Speliotis T., Manios E., Niarchos D., Fidler J., Lee J., and Varvaro G. (2011). Hard/graded exchange spring composite media based on FePt, *J. Appl. Phys.*, **109**, Article No. 07B729.

90. Chen J. S., Huang L. S., Hu J. F., Ju G., and Chow G. M. (2010). FePt-C graded media for ultra-high density magnetic recording, *J. Phys. D: Appl. Phys.*, **43**, Article No. 185001.

91. Zhang J., Sun Z., Sun J., Kang S., Yu S., Han G., et al. (2013). Structural and magnetic properties of patterned perpendicular media with linearly graded anisotropy, *Appl. Phys. Lett.*, **102**, p. 152407.

92. Yuan F.-T., Lin Y.-H., Mei J. K., Hsu J.-H., and Kuo P. C. (2012). Structure and magnetic properties of FePt(001) graded films deposited on glass substrates, *J. Appl. Phys.*, **111**, Article No. 07B715.

93. Jiang C. J., Chen J. S., Hu J. F., and Chow G. M. (2010). FePt-TiO_2 exchange coupled composite media with well-isolated columnar microstructure for high density magnetic recording, *J. Appl. Phys.*, **107**, Article No. 123915.

94. Choe G., Park J., Ikeda Y., Lengsfield B., Olson T., Zhang K., et al. (2011). Writeability enhancement in perpendicular magnetic multilayered oxide media for high areal density recording, *IEEE Trans. Magn.*, **47**, pp. 55–62.

95. Dumas R. K., Zha C., Fang Y., Bonanni V., Lau J. W., Nogues J., and Akerman J. (2011). Graded anisotropy FePtCu films, *IEEE Trans. Magn.*, **47**, pp. 1580–1586.

96. Lin Y.-H., Hsu J.-H., Yuan F.-T., Kuo P. C., and Mei J. K. (2013). Microstructure and magnetic performance of perpendicularly magnetic anisotropic $Fe_3Pt/Fe_2Pt/L1_0$-FePt(001)/MgO(002) graded films, *IEEE Trans. Magn.*, **49**, pp. 3679–3682.

97. Alexandrakis V., Niarchos D., Mergia K., Lee J., Fidler J., and Panagiotopoulos I. (2010). Magnetic properties of graded A1/$L1_0$ films obtained by heat treatment of FePt/CoPt multilayers, *J. Appl. Phys.*, **107**, Article No. 013903.
98. Tsai J.-L., Tzeng H.-T., and Liu B.-F. (2010). Magnetic properties and microstructure of graded Fe/FePt films, *J. Appl. Phys.*, **107**, Article No. 113923.
99. Huang L. S., Hu J. F., Chow G. M., and Chen J. S. (2013). Annealing effect on the FePt/Fe exchange-coupled granular bilayer, *J. Appl. Phys.*, **114**, Article No. 173903.
100. Kirby B. J., Davies J. E., Kai Liu, Watson S. M., Zimanyi G. T., Shull R. D. Kienzle P. A., and Borchers J. A. (2010). Vertically graded anisotropy in Co/Pd multilayers, *Phys. Rev. B*, **81**, Article No. 100405(R).
101. Chang L. V., Nasruallah A., Ruchhoeft P., Khizroev S., and Litvinov D. (2012). Graded bit patterned magnetic arrays fabricated via angled low-energy He ion irradiation, *Nanotechnology*, **23**, Article No. 275705.
102. Gaur N., Pandey K. K. M., Maurer S. L., Piramanayagam S. N., Nunes R. W., Yang H., and Bhatia C. S. (2011). Magnetic and structural properties of CoCrPt-SiO_2-based graded media prepared by ion implantation, *J. Appl. Phys.*, **110**, Article No. 083917.
103. Wang F., Xu X., Liang Y., Zhang J., and Wu H. (2009). FeAu/FePt exchange-spring media fabricated by magnetron sputtering and postannealing, *Appl. Phys. Lett.*, **95**, Article No. 022516.

Chapter 7

Bit-Patterned Magnetic Recording

Denys Makarov,[a,b] Philipp Krone,[c] and Manfred Albrecht[d]

[a]*Institute for Integrative Nanosciences, IFW Dresden, Helmholtzstraße 20, Dresden, 01069, Germany*
[b]*Institute of Ion Beam Physics and Materials Research, Helmholtz-Zentrum Dresden-Rossendorf e. V., 01328 Dresden, Germany*
[c]*Institut für Physik, Technische Universität Chemnitz, Reichenhainer Strasse 70, Chemnitz, 09126, Germany*
[d]*Institut für Physik, Universität Augsburg, Universitätsstraße 1 Nord, Augsburg, 86159, Germany*
d.makarov@hzdr.de, manfred.albrecht@physik.uni-augsburg.de

7.1 Introduction

The capacity of hard disk drives (HDDs) is determined by the physical size of an information bit. In conventional HDDs an individual bit consists of about 10 to 15 weakly magnetically coupled CoCrPt grains [1]. To further increase the areal storage density (number of bits per unit area), a reduction of the dimensions of the grains is required. However, this reduction has a lower boundary, as with reduction of the grain size, the magnetic anisotropy energy $K_{\mathrm{eff}} \cdot V$ (K_{eff} being the effective magnetic anisotropy constant and V being the volume of a grain) will come closer to the thermal activation

Ultrahigh-Density Magnetic Recording: Storage Materials and Media Designs
Edited by Gaspare Varvaro and Francesca Casoli

ISBN 978-981-4669-58-0 (Hardcover), 978-981-4669-59-7 (eBook)
www.panstanford.com

energy. Thus, the recording medium reaches the superparamagnetic limit and becomes thermally instable [1–4].

In modern magnetic recording schemes, the bit-patterned media (BPM) concept implies that one single bit of information is stored in an individual nanostructure with perpendicular magnetic anisotropy (PMA), giving the possibility to increase the areal density significantly. Theoretical predictions show a possible areal density of up to 50 Tbit/in^2 [2]. However, even though there is enormous progress in lithography techniques [5–11], it is rather challenging to create BPM with areal densities of about 1 Tbit/in^2, as for this, a bit pitch of 25 nm has to be achieved. At such small distances, the magnetostatic interactions between neighboring nanostructures increase drastically, influencing substantially the magnetic properties of the BPM [12, 13]. In this respect, the understanding of the interplay between exchange and magnetostatic interactions in such arrays of nanostructures is far from being complete. For example, it is well established that with decreasing size of a bit, the transition from the incoherent to the coherent magnetization reversal mechanism will occur at a critical size [14], accompanied by a change of the energy barrier for the magnetization reversal process, which is responsible for long-term thermal stability of the media. These fundamental aspects of the magnetization reversal in a single-domain magnetic nanostructure as well as in ordered arrays are in the scope of this chapter.

In addition to the larger magnetic volume of an individual magnetic entity in BPM compared to granular media, the increase of the uniaxial magnetic anisotropy constant K_u is an efficient way to enhance the thermal stability of the media. In this respect, hard magnetic alloys (i.e., $L1_0$-ordered FePt) have been suggested as candidates for a recording layer [15, 16]. According to the Stoner–Wohlfarth model for magnetization reversal in a single-domain magnetic nanostructure [17], it is expected that the nanostructure reverse coherently resulting in a high switching field close to the anisotropy field. For an $L1_0$ chemically ordered $Fe_{50}Pt_{50}$ alloy, the anisotropy field is about 110 kOe, which is substantially higher compared to available write fields of 10 kOe provided by conventional magnetic recording heads [18]. This difference in write and switching field is known as the *writability issue* in magnetic data storage [1]. To solve this issue, the concept of exchange-coupled composite (ECC) media was suggested [19–22]. A bit of such a medium consists of a soft part in which the reversal nucleates and a hard part that provides thermal stability. The lowering of the switching field of FePt-based BPM employing the ECC media approach will be addressed in this chapter as well.

7.2 Magnetization Reversal of a Single-Domain Particle with Uniaxial Anisotropy

The long-term stability of storage media is given by the product $K_{\text{eff}} \cdot V$; thereby, K_{eff} is the effective magnetic anisotropy constant and V the switching volume of an individual magnetic entity. In the case of perpendicular granular media, V is related to a single grain, while in the BPM concept, V corresponds to the volume of a bit assuming coherent rotation as a magnetization reversal process. However, this expression is written for a single-domain magnetic particle that is homogenously magnetized and reverses its magnetization quasi-coherently [23]. The latter is a rather strong constraint that is not necessarily valid for magnetic bits in BPM.

Therefore, first of all, we address the field-driven reversal mechanism of the bit, either quasi-coherent or incoherent rotation of magnetization [24]. This is of high importance for the read/write process of the bits as the reversal mechanism influences the required write field and thermal stability. Whereas for an ellipsoid of revolution, a self-consistent reversal theory exists [25], for more complex geometries, no complete theories are available due to the inhomogeneous distribution of magnetization. Hence, we describe the dependence of the energy barrier for magnetization reversal on the geometry of individual bits. Therefore, the *shape*, *size*, and *thickness* of one bit are varied. Additionally, the uniaxial magnetic anisotropy value, K_u, is systematically changed. The energy barrier is calculated using the nudged elastic band method [26, 27], implemented in the micromagnetic solver Finite Element MicroMagnEtics (FEMME) [28].

The model consists of one bit in the form of either a prism with a square base plane with edge length s and thickness t (from now

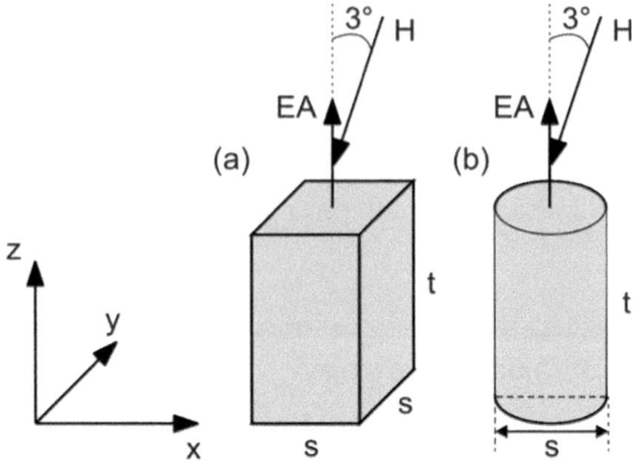

Figure 7.1 Structures under study: (a) prism and (b) cylinder. The orientation of the easy axis, EA, and the external magnetic field, H, is shown. Reprinted from Ref. [24], Copyright (2010), with permission from Elsevier.

on referred to as *prism*) or with a circular base plane with diameter s and thickness t (from now on referred to as *cylinder*). Figure 7.1 shows the structures under study with all relevant geometrical properties.

The base plane of the structures shall be the x-y plane. As required for BPM, the bit possesses a uniaxial anisotropy with the easy axis of magnetization pointing along the z axis. The magnetic anisotropy value was varied in the applications relevant range between 0.5×10^7 ergs/cm^3 and 1.0×10^8 ergs/cm^3; the thickness, t, between 3 nm and 15 nm; and the size, s, between 5 nm and 20 nm. The saturation magnetization, M_S, was fixed at 14 kG and the exchange constant, A, at 1×10^{-6} ergs/cm, corresponding to the value of a chemically ordered $L1_0$ FePt alloy [15, 29, 30]. For each set of the parameters s, t, and K_u, the magnetization reversal mode was simulated. The cell size of the finite element mesh for the simulations was set to 1 nm for $K_u < 6.0 \times 10^7$ ergs/cm^3 and 0.5 nm for $K_u > 6.0 \times 10^7$ ergs/cm^3. The external magnetic field was applied under an angle of 3° relative to the easy axis to avoid numerical instabilities related to the vanishing torque from the applied magnet field acting on the magnetization. Additionally, the energy barrier ΔE between the two states with $M_z = +1$ and $M_z = -1$ (M_z being the normalized z component of the magnetization) was calculated at $T = 0$ K using the nudged elastic band method [27]. Thereby, a so-called *minimum energy path* between two magnetic states is calculated. At first, an initial path corresponding to certain arbitrary chosen magnetic states is assumed and represented by a number of magnetic states that "connect" the initial magnetization state with the final magnetization state. These magnetic states are then altered to find the optimal path, that is, the minimum energy path, if the gradient of the energy only points along the path.

To consider the energy barrier for magnetization reversal processes, ΔE, first the lifetime, τ, of a magnetic state is introduced, given by the Arrhenius law

$$\tau = \tau_0 \exp\left(\Delta E / k_B T\right) \tag{7.1}$$

Thereby, τ_0 corresponds to the resonance relaxation time of the spin system, T to temperature, and k_B to the Boltzmann constant. The energy barrier of a magnetic bit with uniaxial magnetic anisotropy and in a single-domain magnetic state without an applied external magnetic field is written as [25]

$$\Delta E = K_{\text{eff}} V \tag{7.2}$$

with

$$K_{\text{eff}} = K_u + 1/2 N_{\text{eff}} \mu_0 M_S^2 \tag{7.3}$$

V being the volume of the bit and N_{eff} being the effective demagnetization factor. Using the relation $N_{\text{eff}} = 1/2 \left(1 - 3\tilde{N}\right)$, it finally follows

$$\Delta E = \left(K_u + 1/4 \mu_0 M_S^2 \left(1 - 3\tilde{N}\right)\right) \times V \tag{7.4}$$

The demagnetization factor $\tilde{N}$ can be calculated using analytical formulas derived by Aharoni [23] for both shapes. However, approximate expressions are also available [31]:

$$\tilde{N}_{\text{prism}} \approx \frac{1}{1 + 2n} \tag{7.5}$$

$$\tilde{N}_{\text{cylinder}} \approx \frac{1}{1 + 4n/\sqrt{\pi}} \tag{7.6}$$

with $n = t/s$ being the aspect ratio of the bit. Thus, in the case of the geometry defined in Fig. 7.1, the demagnetization factor of the prism is larger than the one of the cylinder for all aspect ratios.

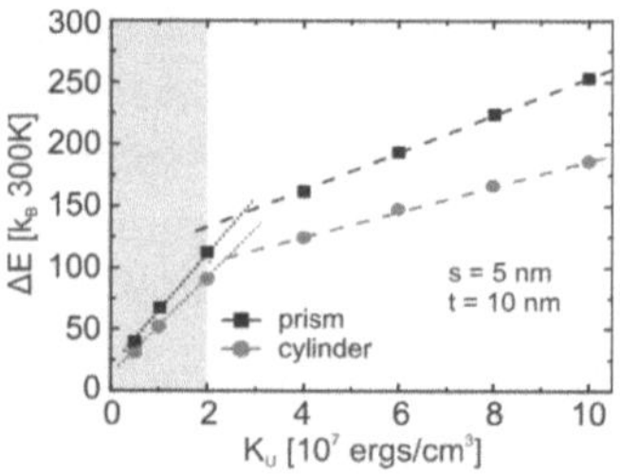

Figure 7.2 Energy barrier ΔE as a function of the magnetic anisotropy constant, K_u, for $s = 5$ nm and $t = 10$ nm. "Prism" stands for the model with a square base plane and "cylinder" for the one with a circular base plane. In the gray-marked area, the field-driven magnetization reversal process happens through quasi-coherent rotation (see discussion in the text). Reprinted from Ref. [24], Copyright (2010), with permission from Elsevier.

In the following discussion, a reversal process in which the magnetic moments of all cells are aligned with a deviation smaller than 2° is defined as quasi-coherent rotation. Respectively, the magnetization rotation is incoherent if this is not the case. Figure 7.2 shows the energy barrier as a function of the magnetic anisotropy for $s = 5$ nm and $t = 10$ nm for both shapes.

With increasing magnetic anisotropy, the energy barrier increases, as the first summand of Eq. 7.3 increases (Fig. 7.2). In the gray-marked area, the reversal of the bit happens through quasi-coherent rotation of the magnetization, as due to the low anisotropy, the energy loss because of all moments being perpendicular to the easy axis during the quasi-coherent rotation is not as high compared to the gain in exchange energy. At the same time, the data can be fitted using Eq. 7.4 with the volume as free parameter. Here, the parameter $\eta = V_{act}/V_{bit}$ is introduced, which is the ratio between fitted activation volume, V_{act}, and real physical volume of the bit, V_{bit}. Please note that the activation volume is not defined as the first derivative of the energy barrier with respect to the magnetic field but as the volume satisfying Eq. 7.4. The fitting of the data in the gray-marked region (Fig. 7.2, dotted line) results in $\eta = 1$. However, the increase of K_u results in a transition from quasi-coherent to incoherent reversal behavior and Eq. 7.4 cannot be used as is. The increase of ΔE is still found to be linear with K_u; however, the slope of the fitted line (Fig. 7.2, dashed line), and thus the ratio η, is reduced to about 0.25, meaning that the activation volume for magnetization reversal is 25% of the real volume of the bit for the specific case of $s = 5$ nm and $t = 10$ nm. Additionally, the intersection of both fits gives an approximation for the highest value of the magnetic anisotropy, K_u^{max}, for which the reversal is still considered to be quasi-coherent. In the discussed case of the model with $s = 5$ nm and $t = 10$ nm, K_u^{max} is 2.2×10^7 ergs/cm^3 for a cylinder and 2.4×10^7 ergs/cm^3 for a prism. Moreover, the energy barrier of the model with a square base plane is higher than the one of the model with a circular base plane. The difference between the ΔE values for the two models is attributed to a different volume for the same s and t values ($V_{cylinder}/V_{prism} = \pi/4 < 1$), and a different demagnetization factor, influencing the effective anisotropy value. Moreover, the slope in the inhomogeneous region is smaller for the cylinder than for the prism, leading to increased stability of the prism. Therefore, the following discussion will be limited to the model with a square base plane, as no qualitative difference in reversal behavior was found. Figure 7.3 shows a pseudo color image of $\eta = V_{act}/V_{bit}$ and K_u^{max} as a function of the size and thickness of the bit [24].

Both, K_u^{max} and $\eta = V_{act}/V_{bit}$ have a similar dependency from the size and thickness. For small sizes and thicknesses ($s < 6$ nm and $t < 6$ nm), the parameters reach their maximum values of 6.0×10^7 ergs/cm^3 and 0.5, respectively. Please note that this magnetic anisotropy value is about the value of the $L1_0$-chemically-ordered FePt alloy. For increasing dimensions of the bit, K_u^{max} and $\eta = V_{act}/V_{bit}$ gradually decrease down to about 1.0×10^7 ergs/cm^3 and 0.1, respectively, for sizes larger than about 17 nm and all investigated thicknesses. It is important to mention that the ratio between activation volume and bit volume does not exceeded 50% in the case of inhomogeneous magnetization reversal.

Figure 7.4 shows the energy barrier dependency on the size and thickness of the bit for a fixed $K_u = 2.0 \times 10^7$ ergs/cm^3.

In both cases, the energy barrier increases with increasing geometrical parameters as the volume increases. Additionally, the increase of the dimensions of the bit changes the demagnetization factor, influencing the effective anisotropy value. In this respect, it

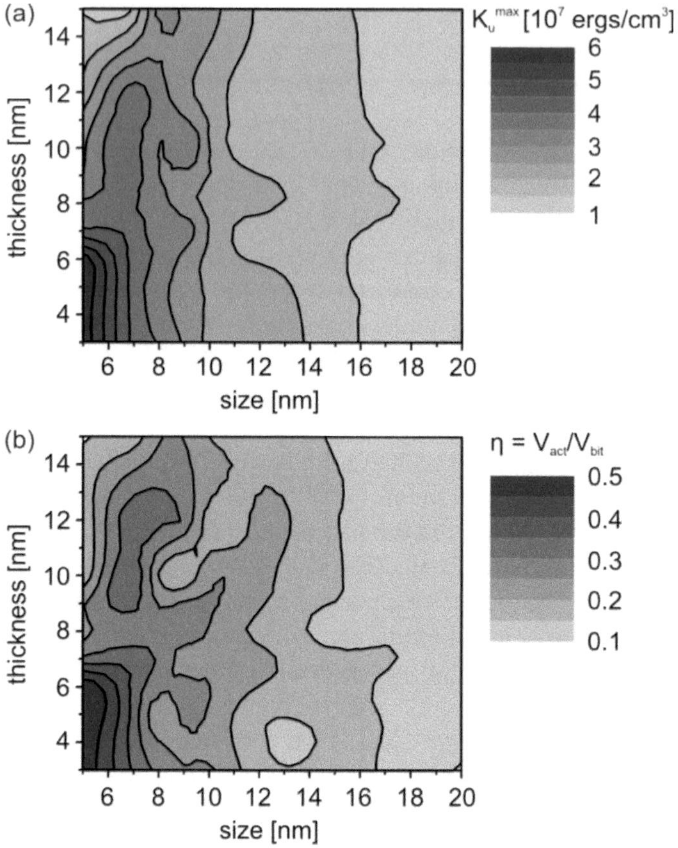

Figure 7.3 (a) Highest value of the magnetic anisotropy, K_u^{max}, for which the field-driven reversal happens through quasi-coherent rotation and (b) ratio $\eta = V_{act}/V_{bit}$ as a function of the size and thickness of the bit for the prism. Reprinted from Ref. [24], Copyright (2010), with permission from Elsevier.

was found that in the case of a fixed t of 10 nm (Fig. 7.4a), the quasi-coherent magnetization reversal process is preferable for $s < 10$ nm, while it is incoherent for larger sizes [24]. The dependences in both regions show a linear increase of the energy barrier with increasing size with smaller slope in the inhomogeneous region. Furthermore, a saturation effect in the dependence of the energy barrier from the thickness was observed for bit sizes of 5 nm and 8 nm. This saturation effect can be explained in terms of a change of the reversal

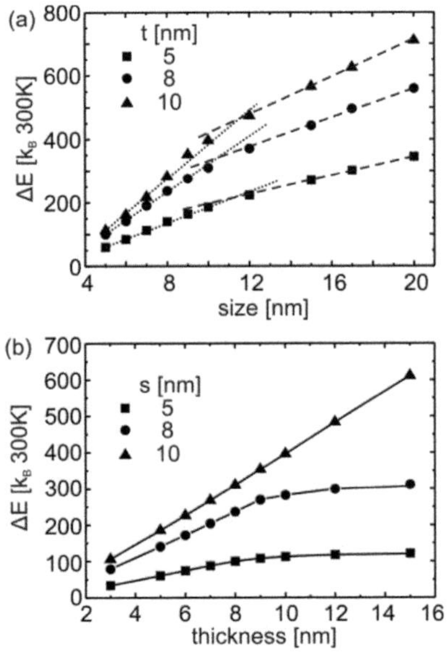

Figure 7.4 Energy barrier ΔE as a function of the (a) size and (b) thickness of a prism with a square base plane for $K_u = 2.0 \times 10^7$ ergs/cm^3. Reprinted from Ref. [24], Copyright (2010), with permission from Elsevier.

mechanism from a quasi-coherent rotation process to an incoherent one by accumulation of a multidomain vortex state [32]. However, please note that the multidomain vortex states are no equilibrium states of the system. Indeed, the Bloch wall width $\delta_{\text{Bloch}} = \pi\sqrt{A/K_u}$ is approximately 9 nm for the given parameters. For a bit size of 10 nm (and larger), a saturation effect was not observed in the range of the investigated bit thickness. Also the fact that such a saturation effect is not observed for $\Delta E(s)$ (Fig. 7.4a) is related to the relatively small range of investigated bit sizes. With further increasing the size of the bit, a saturation effect is expected to occur as well due to a change of the magnetization reversal mechanism to domain wall nucleation and propagation.

From both an application and a fundamental point of view, the transition from a quasi-coherent reversal regime to a regime of incoherent magnetization reversal with increasing bit size and thickness is of interest. For an ellipsoid, the transition happens at a certain critical diameter, d_{crit}, proportional to the exchange length, where both, the nucleation field for quasi-coherent rotation and the one for incoherent rotation (i.e., curling) are equal [25]. Figure 7.5 shows images depicting the reversal mode of the models under study. The upper row shows the results for the model with a square base plane (prism), the lower row for the model with a circular base plane (cylinder). The white color represents the regime of incoherent magnetization reversal, whereas the gray color corresponds to the quasi-coherent rotation of magnetization.

By comparing Fig. 7.5a and Fig. 7.5c, it turns out that the cylinder has a higher critical coherence size than the prism. This is related to the different demagnetization factor and volume of the bit with the same geometrical properties s and t. The analysis suggests that for small lateral sizes and high thicknesses, the reversal becomes incoherent (see incoherent region in the upper-left corner in Fig. 7.5a,c. With increasing magnetic anisotropy constant, the

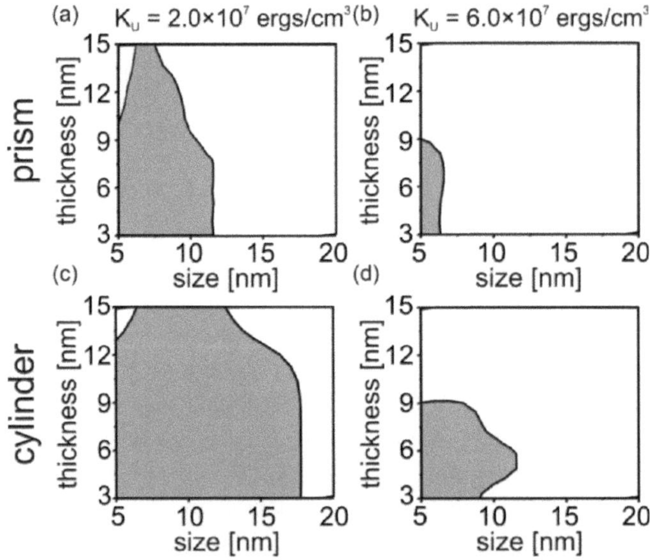

Figure 7.5 Pseudo color image of the reversal mechanism of a single bit with a square (upper row) or a circular (lower row) base plane as a function of size and thickness of the bit. Additionally, the magnetic anisotropy value, K_u, was varied. The gray areas represent quasi-coherent magnetization reversal and the white areas incoherent magnetization reversal. Reprinted from Ref. [24], Copyright (2010), with permission from Elsevier.

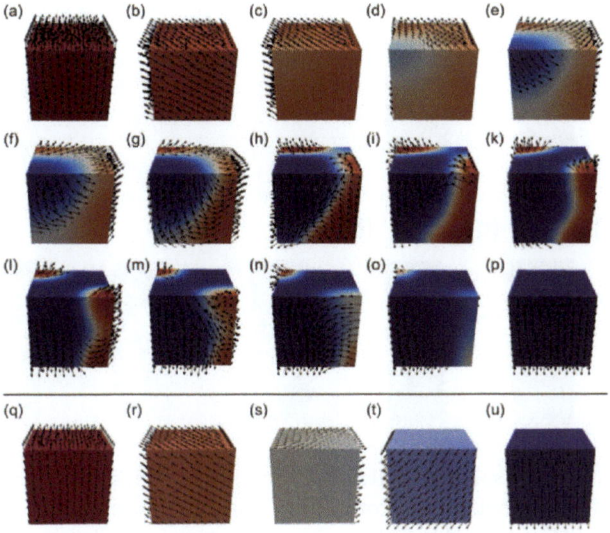

Figure 7.6 Temporal evolution of the magnetization reversal of an individual bit with a square base plane with thickness $t = 10$ nm and size $s = 10$ nm. (a–p) Magnetic anisotropy constant $K_u = 6 \times 10^7$ ergs/cm^3 (incoherent reversal process) and (q–u) magnetic anisotropy constant $K_u = 2 \times 10^7$ ergs/cm^3 (quasi-coherent reversal process). Reprinted from Ref. [24], Copyright (2010), with permission from Elsevier.

reversal process becomes preferably incoherent (compare Fig. 7.5a and Fig. 7.5b). To illustrate the temporal evolution of the reversal process of the bit, a series of magnetic states for the prism model with $t = s = 10$ nm corresponding to incoherent magnetization reversal (Fig. 7.6a–p) for $K_u = 6.0 \times 10^7$ ergs/cm^3 and quasi-coherent magnetization rotation (Fig. 7.6q–u) for $K_u = 2.0 \times 10^7$ ergs/cm^3 is presented [24].

Figure 7.6a–p shows that the reversal of a hard bit with $K_u = 6 \times 10^7$ ergs/cm^3 happens by domain wall nucleation at a corner (Fig. 7.6d) and consecutive domain wall propagation through the entire bit. This incoherent reversal process is preferable for high K_u values as the total energy gets minimized by avoiding the collective tilt of the magnetization to high angles compared to the easy axis. However, please note that the domain walls built during the reversal process are not stable. For a softer bit (Fig. 7.6q–u), the reversal

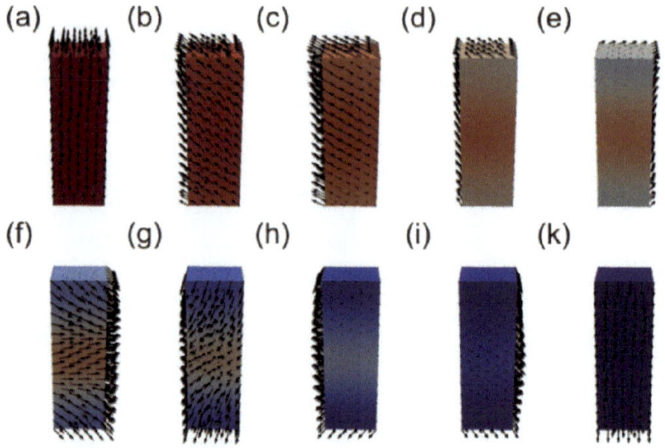

Figure 7.7 Reversal of an individual bit with a thickness of 15 nm, a size of 5 nm, and a magnetic anisotropy constant of 2×10^7 ergs/cm^3. Reprinted from Ref. [24], Copyright (2010), with permission from Elsevier.

happens through quasi-coherent rotation, and all spins are collinear within 2° for the complete reversal process.

In addition, it is important to emphasize the area of incoherent rotation for a thickness of 15 nm and a size of 5 nm (very-left-upper corner of Fig. 7.5a,c). Figure 7.7 summarizes the temporal evolution of the magnetic states of a prism with $t = 15$ nm, $s = 5$ nm, and $K_u = 2 \times 10^7$ ergs/cm^3. For those very thin and long bits, the magnetization reversal process happens through curling with stabilization of a $\pi/2$ magnetic domain state [32].

7.3 BPM: Array of Magnetic Nanodots (Bits)

A crucial aspect of the usability of BPM is the distribution of media parameters, as, for example, given by the size and position of the magnetic units, as well as the variation of magnetic properties [33–37]. Thus, for instance Thomson et al. [33] and Pfau et al. [37] demonstrated that the distribution of magnetic anisotropy within an individual nanodot is vital to understand the magnetization reversal in nanopatterned magnetic arrays.

Here the magnetization reversal in square arrays of magnetic nanostructures, taking into account a distribution of K_u values within the array, was investigated [12]. The goal is to address the influence of K_u in value and distribution on the switching field as well as on the coercive field of BPM. In addition, a detailed study on the magnetostatic interaction in the array of magnetic nanostructures and its influence on the switching-field distribution (SFD) will be presented. The simulations were performed using MAGPAR, an open source, parallel, finite element method (FEM)-based micromagnetic simulation software developed by Scholz et al. [38].

The model under study is an array consisting of 8 × 8 bits, each bit having a volume of 10 × 10 × 5 nm^3 (Fig. 7.8). The center-to-center distance, *D*, is varied between 11 and 110 nm, which covers areal densities of up to 3 Tbit/in^2. To achieve such high areal densities and satisfy the criteria of thermal stability [1, 39, 40], hard magnetic materials, that is, $L1_0$-ordered FePt alloys, have to be used owing to their high magnetic anisotropy constant [15, 16, 29, 41]. Therefore, magnetic materials with parameters close to the one of FePt alloys are considered: saturation magnetization, M_S, of 14 kG, and an exchange parameter, *A*, of 10^{-6} ergs/cm [30, 42]. To account for the aforementioned distribution of media properties, the magnetic anisotropy constant was varied with a Gaussian distribution with a mean value K_u between 0.6 × 10^7 and 2.0 × 10^7 ergs/cm^3 and a standard deviation σ_{K_u} (normalized on K_u) between 0 and 0.2. However, note that for each bit, the magnetic

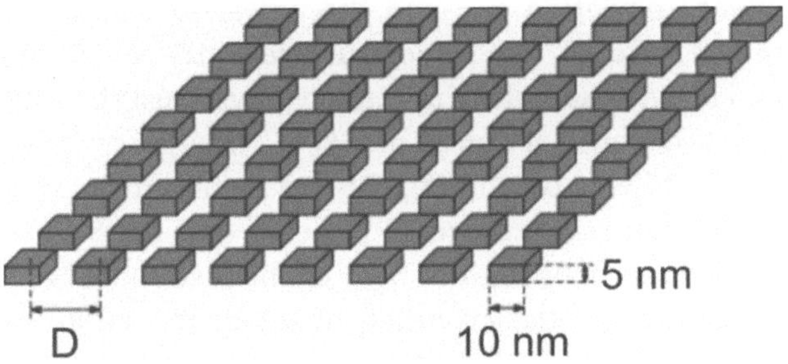

Figure 7.8 Sketch of the model under study. Reproduced with permission from Ref. [12]. Copyright 2009, AIP Publishing LLC.

anisotropy value is fixed and the distribution is realized by assigning a different K_u value to every bit randomly. Assuming these magnetic parameters, the exchange length and the Bloch domain wall width are both larger than 2 nm. Thus, a mesh size of 2 nm was chosen for all simulations. The external magnetic field was applied under an angle of 2° to avoid numerical instabilities. However, note that there is no dispersion in the easy axis considered. To exclude the influence of the fact that the array has a limited size of 8×8 bits only, the simulations were cross-checked with simulations for an array of 20×20 bits, but no substantial differences were found ($< 7\%$).

Each simulation with a given set of parameters (D, M_S, K_u, σ_{Ku}) was performed 30 times, whereas for each run, a new set of 8×8 magnetic anisotropy values was chosen randomly and assigned to the individual bits. The magnetic parameters of interest (coercive field H_C and corresponding SFD) were determined using statistical averaging over all 30 runs of one set, resulting in a distribution of the magnetic parameters (shown as error bars in Figs. 7.10, 7.13, and 7.14). From an application point of view, the highest value of SFD/H_A (with $H_A = 2K_u/M_S$ being the anisotropy field) is of interest as this shows the worst case that has to be handled, for example, the hardest bit at the corner of the array and the softest at the center. However, real BPM consist of an enormous amount of bits, lowering the influence of those worst cases.

To investigate the magnetization reversal in the array of magnetic nanostructures, the series of simulations was performed by varying systematically the value of the mean magnetic anisotropy constant, the standard deviation σ_{Ku}, and the center-to-center distance D, keeping the size of the nanostructures ($10 \times 10 \times 5$ nm^3) constant [12]. An example of a simulated hysteresis half-loop for an interparticle distance of $D = 20$ nm, $K_u = 10^7$ ergs/cm^3, $M_S = 14$ kG, and σ_{Ku} of 0.05 is shown in Fig. 7.9a to point out the way H_C and SFD are determined. It is worth noticing that the SFD is defined as the difference between the field of the last and first switching events. Additionally, the evolution of the switching events in the array corresponding to discrete steps in the hysteresis loop with increasing reverse field is presented in Fig. 7.9b–k.

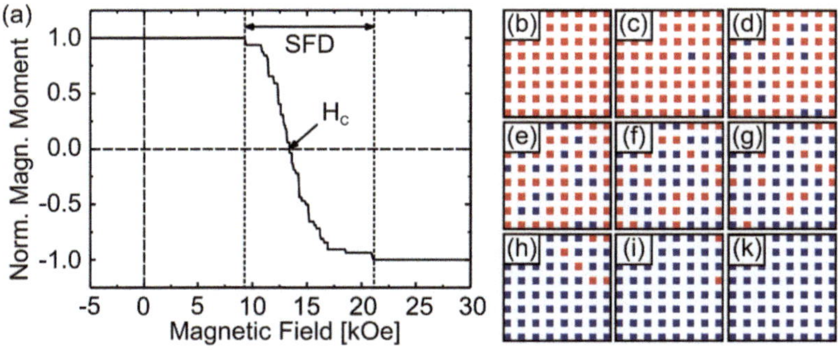

Figure 7.9 (a) Hysteresis half-loop with marked switching-field distribution (SFD) and coercive field, H_C. Magnetic states of the system at (b) 0 Oe, (c) 9.4 kOe, (d) 11 kOe, (e) 12.6 kOe, (f) 14 kOe, (g) 14.9 kOe, (h) 16.2 kOe, (i) 18.6 kOe, and (k) 21.2 kOe are shown. Parameters of this simulation: $D = 20$ nm, $K_u = 10^7$ ergs/cm^3, and $\sigma_{Ku} = 0.05$. Reproduced with permission from Ref. [12]. Copyright 2009, AIP Publishing LLC.

7.3.1 *SFD: Dependence on the Magnetic Anisotropy Value*

The dependence of the coercive field on the mean magnetic anisotropy value for a given distribution σ_{Ku} (0.05 and 0.1) and a fixed center-to-center distance of 20 nm is presented in Fig. 7.10a, revealing a linear increase of H_C with K_u following the Brown equation [43–45]:

$$H_C = \alpha H_A + H_D \tag{7.7}$$

with $H_A = 2K_u/M_S$ and $H_D = -N_{eff} \cdot M_S$ being the demagnetization field; N_{eff} is the effective demagnetization factor. The α parameter is related to the type of magnetization reversal process (nucleation or pinning dominated) [45, 46]. Thereby, α describes the presence of defects of an inhomogeneous magnetic region on the surface and the influence of misalignment of the easy axis. In our case, α was determined from simulations by analyzing the H_C over H_A dependence to be approximately 0.9, which is typical for magnets with a nucleation-dominated reversal mechanism [45, 46]. The difference compared to the classical model considered by Brown, where α is assumed to be 1 [43], results from the effective angle between the total magnetic field (applied field plus magnetostatic interaction field) and the magnetocrystalline anisotropy axis.

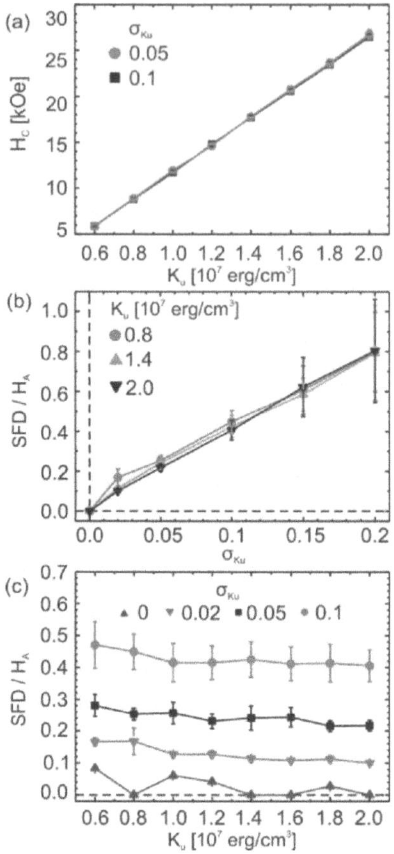

Figure 7.10 (a) Dependency of the coercive field on the magnetic anisotropy constant for different σ_{Ku} values. (b) Variation of SFD/H_{A} as a function of the standard deviation σ_{Ku} for different values of K_{u}. (c) Variation of SFD/H_{A} as a function of the magnetic anisotropy constant using various standard deviations, σ_{Ku}. Note that the lines are guides to the eye. The distance between two bits was set to 20 nm (1.5 TBit/in^2). Reproduced with permission from Ref. [12]. Copyright 2009, AIP Publishing LLC.

According to Stoner–Wohlfarth theory, the switching field reduces by a factor of 0.9 to 0.86 if the field angle changes from 1° to 2°. In addition, it is interesting to note that the magnetization shows a tilt from the easy-axis direction in remanence after an initially saturated state. This tilt arises from the local demagnetizing fields

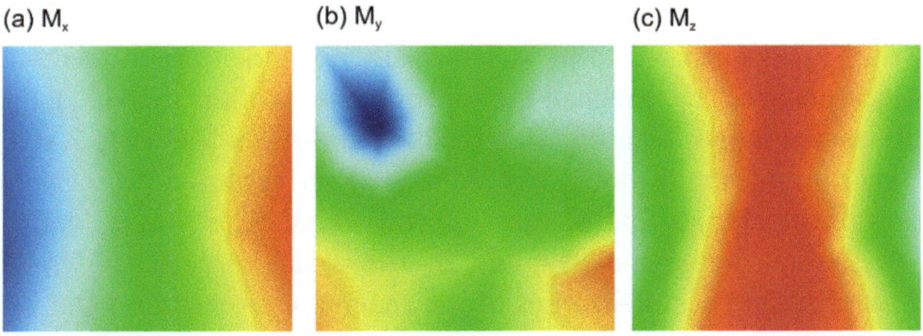

Figure 7.11 (a) M_x, (b) M_y, and (c) M_z component of the magnetization direction of the center bit of a 3 × 3 bit array after relaxation from an initially saturated state. The magnetic anisotropy constant, K_u, was set to 10^7 ergs/cm^3 and the magnetization to 14 kG. In the color scheme the red color corresponds to a maximum, the blue to a minimum, and the green to an intermediate value of the component along the *x*, *y*, or *z* direction. Note that the absolute difference between minimum and maximum values is about 3%, which defines the tilt of the magnetization from the easy axis of a bit. Reproduced with permission from Ref. [12]. Copyright 2009, AIP Publishing LLC.

and contributes to N_{eff} [47]. Note that although we have an initially uniaxial system, the relaxation of magnetic moments results in a tilt of the magnetic moments at the edges of the single bits due to the strong influence of the stray fields.

This effect is shown in Fig. 7.11 presenting the top view of the three components of the magnetization for the center bit of a 3 × 3 array ($K_u = 10^7$ ergs/cm^3) in a remanent state after initial saturation. In the color scheme the red color corresponds to a maximum, the blue to a minimum, and the green to an intermediate value of the component along the *x*, *y*, or *z* direction. However, note that the absolute difference between the minimum and the maximum value is about 3%, which defines the tilt of the magnetization from the easy axis. The presence of this tilt was further evidenced by an increase of the uniaxial magnetic anisotropy constant, resulting in a more parallel alignment of the moments at the edges of the nanostructures, and thus a reduction of the tilt of the magnetization. Also, the tilt increases with decreasing distance between two individual bits, resulting from the increased stray field.

Moreover, a variation of σ_{Ku} does not influence the coercive field, as the coercivity is defined by the mean value of the magnetic anisotropy constant and not by its distribution. Contrary to that, the SFD is expected to be substantially dependent on σ_{Ku} as with increasing σ_{Ku} the range of the switching fields of the individual bits becomes larger, resulting in a broadening of the SFD (Fig. 7.10b). It is important to note, that even with an assumed σ_{Ku} of zero, a nonvanishing value of the SFD was found. The reason for this is the magnetostatic coupling in the array of bits, leading to stabilization/destabilization of the magnetic moments of the bits through magnetic dipole–dipole interaction. With increasing σ_{Ku}, the normalized SFD/H_A increases linearly, as the probability to find bits with very high and very low magnetic anisotropy constants becomes higher. Note that a normalization of the SFD on H_A instead of H_C was chosen to eliminate the influence of the H_C variation and thus be able to concentrate on the changes of the SFD. The result in Fig. 7.10b allows the estimation of the upper limit for σ_{Ku} for a recording system with a given field gradient of the recording head, assuming all other system properties are not influencing the recording process. For instance, the typical field gradient of a recording head is 300 Oe/nm, resulting in an allowed normalized standard deviation of only 3% for an average switching field of 11 kOe [48]. However, this value is challenging to achieve experimentally, as the pattern preparation is usually influencing the magnetic properties of the bits. A reported experimental value of σ_{Ku} achieved by electron beam lithography (EBL) for Co/Pt multilayers, for example, is about 5% [33].

Additionally, the dependency of the SFD on the magnetic anisotropy constant was extracted (Fig. 7.10c). Analysis has shown that the SFD gradually increases with K_u. However, the SFD normalized on H_A is independent of the magnetic anisotropy constant for all simulated standard deviations σ_{Ku} larger than 0.02 (Fig. 7.10c). In this respect, for $\sigma_{Ku} = 0$, a nonlinear dependence was observed. For high magnetic anisotropy constants ($K_u > 1.6 \times 10^7$ ergs/cm^3), the normalized SFD is found to be zero, whereas for smaller magnetic anisotropy constants, a deviation from zero was encountered. The latter is attributed to the softening of the system with the decrease of K_u, resulting in an enhancement of the influence of external magnetic fields, including magnetostatic interactions between the individual bits in the array. Thus, for

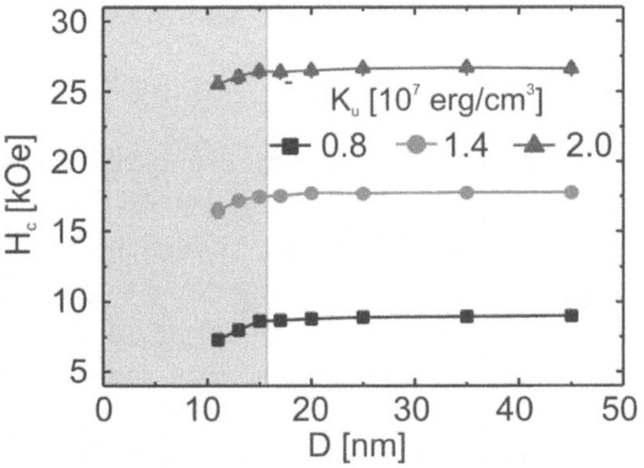

Figure 7.12 Dependency of H_C on the center-to-center distance, D, for varying magnetic anisotropy values. The standard deviation, σ_{Ku}, was set to 0.05. The gray area defines the region with areal densities beyond 1 Tbit/in^2. Reproduced with permission from Ref. [12]. Copyright 2009, AIP Publishing LLC.

achieving an SFD/H_A value smaller than, for example, 5% for an areal density of 1 Tbit/in^2, the BPM must consist of bits harder than 1.8×10^7 ergs/cm^3 with a σ_{Ku} smaller than 0.02.

7.3.2 *SFD: Dependence on Dipole–Dipole Interaction between Bits*

Dipole–dipole interaction in an array of bits was studied by varying the distance between bits [12]. In Fig. 7.12, the changes of H_C with the distance between two squares shall be discussed for a σ_{Ku} of 0.05. With decreasing distance between bits, H_C at first decreases in the limit of ultrahigh recording densities (gray-marked area in Fig. 7.12), independent of the magnetic anisotropy constant, and remains nearly constant for larger distances.

This can be explained in terms of the dipolar field influence from one bit to the others as investigated experimentally by Pfau et al. [13]. With increasing center-to-center distance, D, the magnetostatic interaction becomes weaker and thus a higher field is required for switching half of the squares in the array. The $H_C(D)$ curve can therefore be fitted with a function of the form

$$H_C = (H_A - H_D) - m/(4\pi D^3) \tag{7.8}$$

where $m = M_S \cdot V_{square}$ is the magnetic moment of an individual bit, which is in agreement with recent data presented by Basu et al. [49].

The normalized SFD/H_A shows a similar behavior on the separation distance, at first decreasing down to 0.2 for $D = 20$ nm and then being constant within the margin of error (Fig. 7.13a). A strong dependence on the anisotropy constant is found in the ultrahigh density limit ($D < 25$ nm), where SFD/H_A drops substantially with the increase of D. For larger distances the ratio SFD/H_A stays nearly independent of the magnetic anisotropy constant. This finding can be explained in terms of dipolar fields: As the energy of a system is minimized for antiparallel aligned dipoles, such a dipole–dipole interaction results in an early switching of some bits and a late switching of some others, as those are then stabilized by the surrounding already switched bits. Additionally, the dependence of the SFD/H_A on the center-to-center distance for different values of the saturation magnetization, M_S, was analyzed (Fig. 7.13b). It was found that with decreasing M_S, SFD/H_A drops due to less magnetostatic interaction between the bits, opening a way to decrease the SFD below the necessary 5% for recording media,

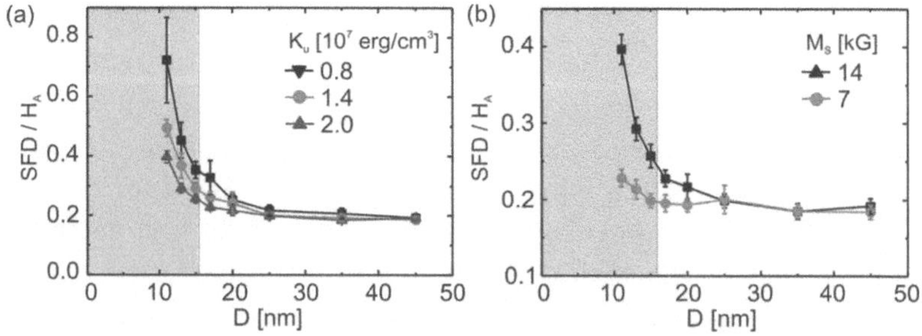

Figure 7.13 Dependence of SFD/H_A on the center-to-center distance, D, between two bits for (a) different values of the magnetic anisotropy constant ($M_S = 14$ kG) and (b) different values of M_S ($K_u = 2 \times 10^7$ ergs/cm^3). The standard deviation, σ_{Ku}, was set to 0.05. The gray areas define the region with areal densities beyond 1 Tbit/in^2. Reproduced with permission from Ref. [12]. Copyright 2009, AIP Publishing LLC.

which can be realized by using *antiferromagnetically coupled media* [50].

7.3.3 *SFD: Role of the Angular Orientation of the Magnetic Anisotropy Axes in BPM*

To obtain more information on the magnetization reversal in the system, an analysis of the angular dependence of the switching field, H_S, should be performed [51–56]. Here, this is done by computing the hysteresis loops of the system exposed to an external magnetic field applied under an angle relative to the easy axis for a fixed K_u value of 10^7 ergs/cm^3. The angular dependence of the switching field is presented in Fig. 7.14a, revealing a clear minimum at 45° and then an increase toward the initial value [12]. Both features

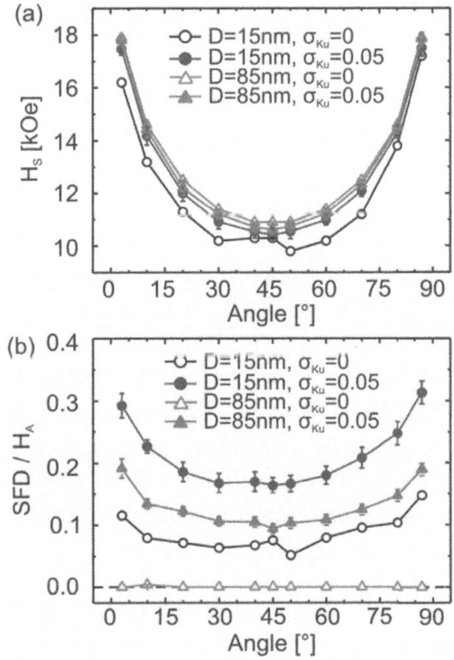

Figure 7.14 (a) Angular dependence of the (a) switching field and (b) normalized switching-field distribution SFD/H_A for different distances between two bits and different σ_{Ku} ($K_u = 10^7$ ergs/cm^3). Reproduced with permission from Ref. [12]. Copyright 2009, AIP Publishing LLC.

are characteristic for a reversal process in a magnetic material with the nucleation-dominated reversal mechanism, as described in the Stoner–Wohlfarth model for the reversal of single-domain particles with uniaxial magnetic anisotropy [17].

The analysis shows that the magnetization reversal is independent within the margin of error of the center-to-center distance between two bits and on the intrinsic distribution of the magnetic anisotropy constant, following the discussion above (Fig. 7.14a).

Moreover, the angular dependence of the normalized SFD was investigated for different distances and different σ_{Ku} values (Fig. 7.14b). For large center-to-center distances and vanishing σ_{Ku}, there is no SFD, as there is neither magnetostatic interaction between the individual bits nor an intrinsic distribution of the magnetic anisotropy constant. However, with both, increasing σ_{Ku} and decreasing center-to-center distance, the normalized SFD increases and shows a minimum at 45°. The reduction of SFD/H_{A} toward 45° is related to the decreased switching fields that favors earlier switching. According to Stoner–Wohlfarth theory both the switching field of the first bit and the switching field of the last bit are reduced by the same factor, which also reduces the difference between these fields. In addition, magnetostatic interactions not only change the absolute value of the total magnetic field acting on a bit but also vary the angle of the total magnetic field. Near 45°, any variation of the angle only leads to a small change of the switching field, according to Stoner–Wohlfarth theory. Thus, if the external field is applied at 45°, the magnetostatic interaction is contributing less to the SFD. This finding of a narrowing of SFD/H_{A} of the array might be of interest for the tilted-media concept [57–60], where reading/writing is performed in the inclined geometry of the applied magnetic field with respect to the magnetization orientation.

7.4 Experimental Realization of BPM via Lithographic Patterning

Ordered arrays of magnetic nanostructures can be achieved using various lithography techniques [61–65] combined with nanoimprint or self-assembly approaches [9–11, 66–68]. Here, we present a study on the magnetic properties of a dot array consisting of Co/Pt multilayers patterned by EBL with dot sizes between 200 and 40 nm, keeping the filling factor (number of magnetic dots per area) constant at about 0.16. The evolution of the nucleation, switching, and saturation fields will be discussed as a function of the dot size [69].

7.4.1 *Sample Preparation and Experimental Techniques*

Co/Pt multilayers were deposited on thermally oxidized Si(100) wafers with a 100 nm thick SiO_2 layer in a molecular beam epitaxy (MBE) chamber equipped with an electron gun loaded with Co and Pt (purity better than 99.99%). The base pressure was 4×10^{-11} mbar and the pressure during deposition did not exceed 5×10^{-9} mbar. The multilayer stack consisting of $[Co(0.3\ nm)/Pt(0.8\ nm)]_8$ was deposited on a 5 nm thick Pt buffer layer to provide the required texture for the multilayer growth. An additional 3 nm thick Pt cover layer was deposited to protect the sample from oxidation. The magnetic characterization was performed using a superconductive quantum interference device (SQUID) magnetometer, revealing a saturation magnetization of about $M_S = 480$ emu/cm^3 and a uniaxial magnetic anisotropy constant K_u of 3×10^6 ergs/cm^3. Subsequently, the multilayer film was patterned using EBL. EBL was carried out using a LEICA 5000+ electron beam nanowriter working at 100 kV. A conventional process using the resist poly(methyl methacrylate) (PMMA) and lift-off technique was applied. The pattern transfer in the magnetic layer after the lithographic process was performed by an ion beam etching process. Scanning electron microscopy (SEM) images of the samples after patterning are presented in Fig. 7.15. Hexagonal arrays with different dot sizes and different pitches were created, while holding the filling factor constant at 0.16. A hexagonal arrangement was chosen as this arrangement is the preferred one from an application point of view as it provides the highest areal density.

The integral magnetic properties of the patterned samples were characterized by polar magneto-optical Kerr effect (p-MOKE) magnetometry using a focused beam (spot size of about 100 μm) operating at a wavelength of 670 nm. The external magnetic field

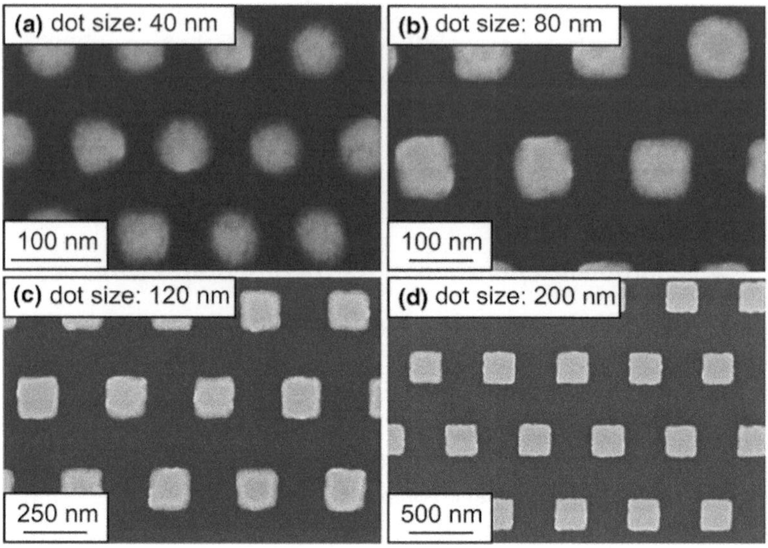

Figure 7.15 SEM images taken after the patterning process. Reprinted from Ref. [69], Copyright 2011, with kind permission from Springer Science and Business Media.

was applied perpendicular to the substrate plane. In addition to hysteresis loops, revealing the presence of an out-of-plane easy axis of magnetization, remanence curves were taken. The shape of the hysteresis loops and remanence curves was found to be identical, suggesting irreversible magnetization reversal processes. Therefore, in the following section, we concentrate on the discussion of remanence curves. In addition, magnetic force microscopy (MFM) imaging was performed to investigate the magnetic properties of individual dots both at remanence and in an applied magnetic field.

7.4.2 *Magnetization Reversal Process and SFD in BPM*

MFM images of the patterned samples were taken in the alternating current (ac)-demagnetized state and after setting a reverse field H_R of −2.75 kOe to the initially saturated samples (Fig. 7.16). The study performed on the dot array with the smallest dot size of 40 nm reveals that individual dots are in a magnetic single-domain state

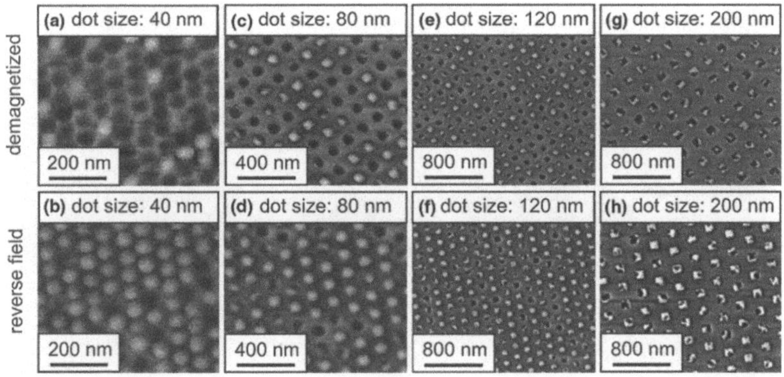

Figure 7.16 MFM images taken in the ac-demagnetized state (a, c, e, g) and after setting a reverse field $H_R = -2.75$ kOe to an initially saturated dot array (b, d, f, h). Reprinted from Ref. [69], Copyright 2011, with kind permission from Springer Science and Business Media.

independent of the magnetic history (Fig. 7.16a,b). Please note that for this dot array, the applied field of −2.75 kOe is already larger than the coercive field, and thus, more than 50% of all dots are already switched. In contrast, MFM data taken on an array with 200 nm dot sizes reveal that the nanostructures are in a clear multidomain state (Fig. 7.16g,h). The patterns with dot sizes of 80 and 120 nm are at the border between single- and multidomain states, as indicated by the presence of dots that are showing either a single-domain or a multidomain state (Fig. 7.16c–f).

To study the magnetization reversal process of individual dots in the multidomain state, in-field MFM measurements were carried out on a dot array with a dot size of 200 nm. Figure 7.17 shows the evolution of the magnetization reversal in different applied magnetic fields perpendicular to the sample plane [69].

It can be seen that the reversal of the dots starts with the nucleation of a magnetic domain at the edges of the dots, most likely due to defects resulting from the patterning process (see white circle in Fig. 7.17a). However, there are also dots where the magnetization reversal occurs around the center of the dots due to the larger demagnetizing field in the dot center (see white circle in Fig. 7.17b). With an increasing magnetic field, the nucleated domains spread over a certain distance and eventually are pinned on structural

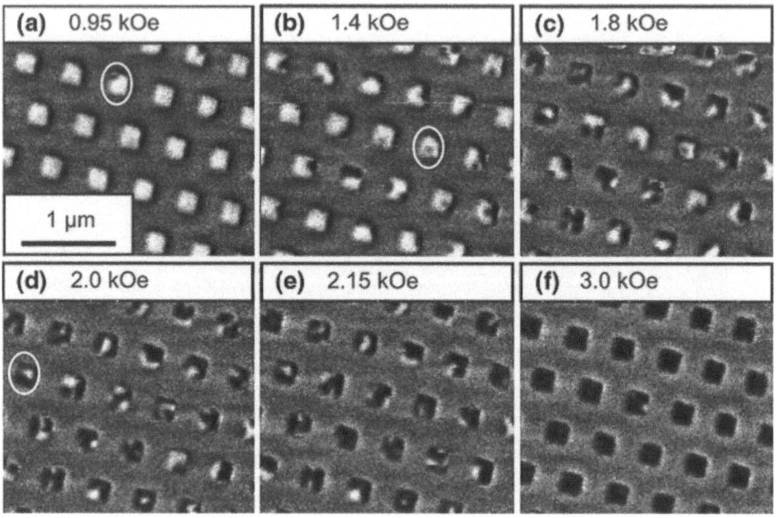

Figure 7.17 In-field MFM images of an array with a dot size of 200 nm taken at (a) 0.95 kOe, (b) 1.4 kOe, (c) 1.8 kOe, (d) 2.0 kOe, (e) 2.15 kOe, and (f) 3.0 kOe. Note that the images were taken at the same location to follow the evolution of the magnetization reversal. Reprinted from Ref. [69], Copyright 2011, with kind permission from Springer Science and Business Media.

inhomogeneities, that is, at the boundaries between neighboring Co/Pt grains. This process is accompanied either by depinning of the magnetic domains or by further nucleation of domains in the same dot and its spread, leading to complete magnetization reversal. Similar domain patterns were already observed in patterned CoCrPt perpendicular media [70]. This study shows that the magnetization reversal process in the array of dots with a size of 200 nm is nucleation dominated, whereas the coercive field is determined by pinning of magnetic domain walls on structural inhomogeneities. Moreover, it is worth mentioning that the nucleation process is deterministic, always starting at the same location of the dots when the same reverse field is applied.

A series of remanence curves as measured by p-MOKE is shown in Fig. 7.18a. From these curves, the first derivative of the normalized magnetization with respect to the magnetic field was extracted (Fig. 7.18b). In this case, the field corresponding to the maximum of the first derivative represents to the averaged switching field,

H_S, while the full width at half maximum corresponds to the SFD. Moreover, the nucleation field H_{nuc} was defined as the field where the magnetization of the array reaches 90% of the saturation magnetization (Fig. 7.18a) [69]. Accordingly, the saturation field, H_{sat}, is defined as the field where the magnetization of the array

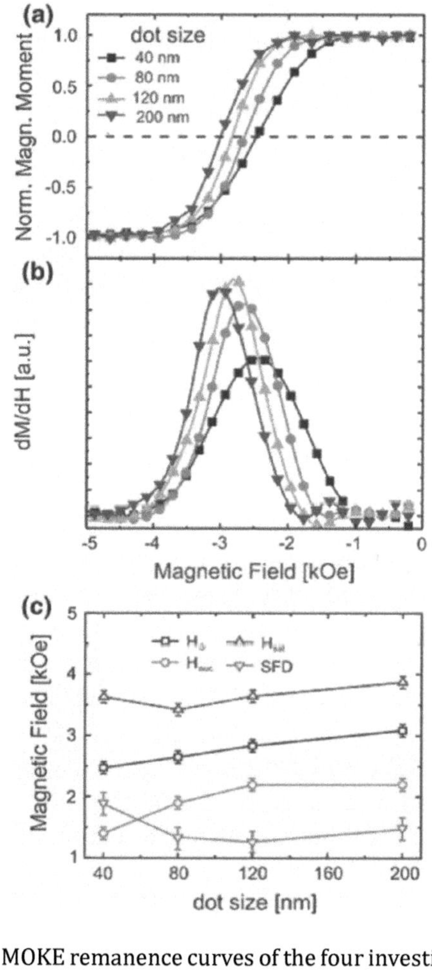

Figure 7.18 (a) MOKE remanence curves of the four investigated dot arrays, (b) derivative of the remanence curve with respect to the magnetic field, and (c) extracted nucleation field H_{nuc}, switching field H_S, saturation field H_{sat}, and switching-field distribution (SFD). Reprinted from Ref. [69], Copyright 2011, with kind permission from Springer Science and Business Media.

reaches ~90% of the saturation magnetization. Figure 7.18c shows the dependence of the switching field, nucleation field, saturation field, and SFD on the dot size. The analysis shows that H_S, as well as H_{nuc}, increases with increasing dot size, whereas H_{sat} does not vary much with the dot size (Fig. 7.18c). The saturation field is determined by the magnetization reversal of the magnetic hardest parts of the dots.

The patterning divides the initially exchange coupled film into exchange-decoupled regions. The magnetic hard parts are then distributed over the individual dots. Assuming that the magnetic hard parts of the Co/Pt multilayers are smaller than the dot sizes, this will lead to the observed independency of the saturation field on the dot size.

The reduction of the dot size below 200 nm results in a modification of the magnetization reversal mechanism from domain wall nucleation and propagation to coherent/incoherent rotation of the magnetization (see micromagnetic simulations [69]). It is important to note that a coherent reversal process is accompanied by a substantial increase in H_S, compared to the incoherent reversal mechanism [71–73]. Therefore, further lowering of the switching and nucleation fields with a reduction of the dot size can be explained by an incoherent magnetization reversal process. In addition, degradation of the magnetic properties induced by the patterning process is expected. With decreasing dot size, the relative area that is affected by the patterning process increases, resulting in a lowering of the magnetic anisotropy value of the magnetic material, finally leading to a reduction of the switching field [74]. Indeed, the SEM images show an increased roughness at the edges of the individual bits at smaller dot sizes, revealing even a change of the shape of the dots from a rectangular to a more round shape. Moreover, the reduction of the switching field is accompanied by a substantial broadening of the SFD, in particular for the array with 40 nm dots.

7.5 Exchange-Coupled Composite BPM Concept

BPM at densities larger than 1 Tbit/in^2 require hard magnetic materials to guarantee thermal stability of the individual bits

[1, 2, 16]. However, the magnetic field required to switch the magnetization of one individual bit typically exceeds the maximum available write field provided by a magnetic recording head. To solve this so-called writability issue in magnetic data storage, Suess et al. and Victora et al. proposed a recording concept known as *exchange-coupled composite* (ECC) media, *exchange-spring media*, or *graded media*, aiming to lower the switching field of individual bits [19–22, 75]. In this approach, one bit consists of two or more layers with decreasing magnetic anisotropy value from the bottom to the top whereby the layers are fully exchange coupled. Micromagnetic simulations have shown that the magnetic energy barrier is determined by the magnetic and geometrical properties of the hardest layer in the stack [21]. First experimental realizations of ECC BPM have been presented [76–80], showing the potential of the ECC concept for future generations of magnetic storage devices. From an application point of view, a narrow SFD of BPM is of high importance to guarantee precise writing of discrete bits [81, 82].

Here, we present results of micromagnetic simulations of arrays of magnetostatically interacting bits consisting of various ECC stacks, including graded media where layers with a quadratically decreasing magnetic anisotropy constant were employed. It was shown that the combination of ECC with BPM reveals superior properties compared to single-layer BPM, as it leads to a significant reduction of the switching field, H_S, and substantial narrowing of the SFD [83].

Micromagnetic simulations were performed at zero temperature using FEMME [28]. The size of the finite element mesh was set to 1 nm. We consider an array of 8 × 8 bits with a center-to-center distance of 20 nm between two bits, leading to an areal density of about 1.5 Tbit/in^2. Each bit has a lateral size of 10 × 10 nm^2 and possesses uniaxial magnetic anisotropy with a magnetic easy axis parallel to the surface normal. To account for the unavoidable distribution of media properties, the magnetic anisotropy constant was varied with a Gaussian distribution (standard deviation of 0.05 K_u). Here, we concentrate on bits composed of two and three layers with different magnetic anisotropy values assigned to each layer and compare the results with bits consisting of a hard magnetic single layer. Furthermore, a graded layer stack

Table 7.1 Magnetic and geometrical properties of bit arrays with various layer stacks. The thickness of the individual layers in the stack, t, and the corresponding magnetic properties are listed. K_u is the uniaxial magnetic anisotropy constant and M_S the saturation magnetization of the layer

#	Layer stack		K_u (10^7 ergs/cm^3)	M_S(kG)	t (nm)
1a	Single layer	Hard	1.0–4.0	14	5
1b	Single layer	Hard	1.0–4.0	14	15
2	Bilayer	Hard	1.0–4.0	14	5
		Soft	1.0	6	10
3	Trilayer	Hard	1.0–4.0	14	5
		Medium	1.0	6	5
		Soft	0.6	6	5
4	10 layers	Graded	1.0–4.0	14	15

consisting of 10 layers with a quadratically decreasing magnetic anisotropy value from the bottom to the top was investigated. Table 7.1 summarizes the geometrical and magnetic properties of the models under study. Both, interlayer and intralayer exchange-coupling constants were set to 1×10^{-6} ergs/cm to ensure strong exchange coupling between the layers in the ECC stack. It was shown that, in the assumption of the uniform magnetization in the layer, weakening of the interlayer exchange coupling is beneficial with respect to lowering the switching field of the ECC bit [19]. However, detailed investigations taking into account a spatially inhomogeneous magnetization in the bit [75, 84] revealed that the energy barrier for magnetization reversal is maximal for a fully exchange coupled layer stack. Therefore, here, we focus on the fully exchange coupled case only. Figure 7.19 shows a sketch of the bit array with stack #3 (Table 7.1). In this case, a dark color represents a high magnetic anisotropy value, while a brighter color corresponds to lower magnetic anisotropy values.

Simulations for each model were carried out 30 times, whereas for each run, a new set of magnetic anisotropy values was chosen randomly from the Gaussian distribution and assigned to the individual layers of the bits. Furthermore, only the descending branch of the hysteresis loop (from +10 kOe to −100 kOe) was simulated with a magnetic field sweep rate of 1 kOe/ns. The

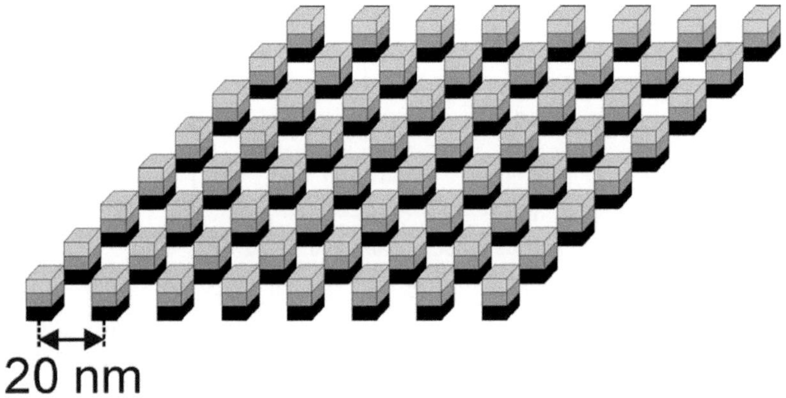

Figure 7.19 Model of 8 × 8 bits with each bit consisting of three layers with different magnetic properties (corresponds to stack #3, Table 7.1). Reproduced with permission from Ref. [83]. Copyright 2010, AIP Publishing LLC.

magnetic field was applied under an angle of 3° to avoid numerical instabilities. The magnetic parameters of interest (H_S and SFD) were determined using statistical averaging over all 30 runs for each model, resulting in a distribution shown as error bars. Thereby, the SFD was determined as the field difference between 90% and −90% of the saturation magnetization.

Figure 7.20 shows descending branches of exemplary hysteresis loops for the different layer stacks of Table 7.1. The magnetic anisotropy value of the hard layer was set to 4×10^7 ergs/cm^3. With an increasing number of layers in the stack, the switching field of the bit array decreases and the SFD becomes narrower. To further analyze the influence of the different stacks of a bit on H_S and SFD, these parameters were plotted as a function of the magnetic anisotropy value of the hard layer K_u^{hard} (Fig. 7.21). With increasing K_u^{hard}, the switching field, H_S, increases (Fig. 7.21a). For single-layer bits (stacks #1a and #1b), this increase is linear and a larger switching field was obtained for the thicker single layer (stack #1b). The latter is due to a modification of the shape anisotropy contribution, leading to a more incoherent magnetization reversal process. For the other investigated layer stacks (stacks #2–#4), there is a clear trend to saturation in the dependence $H_S(K_u^{hard})$. Please

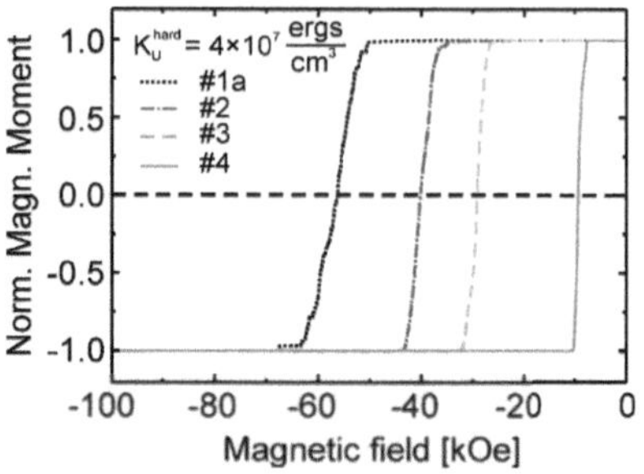

Figure 7.20 Descending branches of hysteresis loops for various layer stacks given in Table 7.1. The uniaxial magnetic anisotropy constant of the hard layer was set to 4×10^7 ergs/cm^3. Reproduced with permission from Ref. [83]. Copyright 2010, AIP Publishing LLC.

note that for $K_u^{hard} > 5 \cdot K_u^{medium}$, a further increase of H_S is expected as the combined effect of the applied magnetic field and torque from the soft layer is insufficient to switch the magnetization of the hard layer [19]. However, due to computational limitations, simulations of magnetically harder layers were not performed. It is shown that with an increasing number of layers in the ECC stack, the switching field successively decreases, reaching a value of about 9 kOe for graded media (stack #4; $K_u^{hard} = 4 \times 10^7$ ergs/cm^3), considering a quadratic decrease of K_u over 10 layers, which is in agreement with a previous study [22].

Contrary to the behavior of $H_S(K_u^{hard})$, the normalized SFD, that is, SFD/H_S, decreases with increasing K_u^{hard}, which is attributed to both a change of the magnetization reversal mechanism of the media and a decrease of the influence of the dipolar fields from neighboring bits with increasing magnetic anisotropy value (Fig. 7.21b) [12]. With increasing numbers of layers in the stack, SFD/H_S decreases from about 0.23 for a bit composed of a single hard layer down to about 0.14 for graded media ($K_u^{hard} = 4 \times 10^7$ ergs/cm^3). Moreover, the total thickness of the single layer for stack #1 has almost no influence on SFD/H_S.

Another important aspect was recently highlighted by Hellwig et al. [78] and Suess et al. [21], namely a correlation between the distributions of the magnetic anisotropy values of different layers has a severe influence on SFD. Simulations with uncorrelated

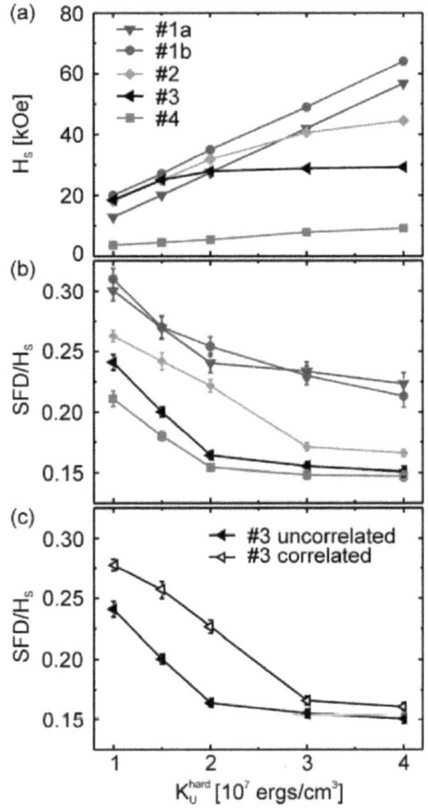

Figure 7.21 (a) Switching field and (b) normalized SFD as a function of the magnetic anisotropy value of the hard layer for various layer stacks given in Table 7.1 for the case when the distributions of magnetic anisotropy values of the different layers are uncorrelated. (c) SFD/H_S of the trilayer stack (stack #3, Table 7.1) with correlated and uncorrelated magnetic anisotropy values of the different layers in one bit. Please note that the error bars in (a) are within the size of the symbols. Moreover, the legend of (a) applies to (b) as well. Reproduced with permission from Ref. [83]. Copyright 2010, AIP Publishing LLC.

and correlated distributions of magnetic anisotropy values were exemplary performed for layer stack #3. Thereby, the correlation was introduced by using the same change of the magnetic anisotropy value of the individual layers of one bit relative to their respective mean values. For example, if the anisotropy of the hard layer of one bit is 5% larger than the average K_u^{hard}, the magnetic anisotropy value of the soft layer of this bit is also 5% larger than the average K_u^{soft}. By introducing this correlation to layer stack #3, the SFD/H_S increases substantially as shown in Fig. 7.21c, whereas H_S remains unaffected (not shown). The correlation between the distributions of the magnetic anisotropy values of the different layers results in an increased SFD. This can be explained in terms of the appearance of bits with a relatively high K_u^{soft} and high K_u^{hard} that have an increased switching field and the appearance of bits with a small K_u^{soft} and relatively small K_u^{hard} that have a decreased switching field [83].

Moreover, the energy barrier for the magnetization reversal, ΔE, at 300 K was simulated for the reversal process of one bit located in the central area of the array starting from an initially saturated state [83]. Please note that this magnetic configuration is the most energetically unfavorable with respect to the magnetostatic interaction between the bits in the array. The magnetic anisotropy value of the hardest layer was set to 4×10^7 ergs/cm^3. The results for the different layer stacks are summarized in Table 7.2. As it follows from the data presented in Table 7.2, the stability of the ECC bits is only slightly reduced compared to the single-layer bit (stack #1b). Moreover, the stability of the ECC bits is higher than the one of the single-layer bits with a thickness of 5 nm (stack #1a), which can be explained in terms of an increased total thickness of the bits in stacks #2 and #3, where the added material is still magnetically relatively hard. By comparing the magnetic energy barriers and switching fields (Table 7.2), it can be seen that by tailoring the

Table 7.2 Energy barriers, ΔE, for magnetization reversal of the central bit in the array for the case when the distributions of the magnetic anisotropy values of the layers are not correlated. H_S and SFD/H_S are also given. The magnetic anisotropy value of the hard layer was set to 4×10^7 ergs/cm^3

	#1a	#1b	#2	#3	#4
ΔE (k_B300 K)	276	731	706	698	673
H_S(kOe)	56	65	44	29	9
SFD/H_S	0.22	0.21	0.17	0.15	0.14

magnetic properties of the layer stacks of the bits, the switching field can be substantially reduced while keeping the magnetic energy barrier nearly constant.

7.5.1 *Correlation of Magnetic Anisotropy Distributions in Layered ECC BPM*

In this section, we present a more detailed study on the effect of correlations as mentioned in the last section on the SFD (Fig. 7.21c). We show that by tailoring the correlation between the values of the distributions of the magnetic anisotropy values of the hard and soft layers, the SFD of the bit array can be reduced significantly, while the switching field, H_S, remains unaffected [85].

Micromagnetic simulations were performed at zero temperature using FEMME [28]. The size of the finite element mesh was set to 1 nm. We consider an array of 8 $\times$ 8 bits with a center-to-center distance of 20 nm between two bits, leading to an areal density of about 1.5 Tbit/in^2. Each bit has a lateral size of 10 $\times$ 10 nm^2 and possesses uniaxial magnetic anisotropy with a magnetic easy axis parallel to the surface normal. The saturation magnetization of all layers was set to 14 kG.

The intralayer and interlayer exchange-coupling constant was set to 1 $\times$ 10^{-6} ergs/cm. Please note that previous studies [75, 84] revealed that the energy barrier for magnetization reversal is maximal for a fully exchange coupled layer stack. Therefore, here, we focus on the fully-exchange-coupled case only. The left illustration of Fig. 7.22 shows a bit array, whereas the right image summarizes the geometrical properties of a single bit.

To account for the unavoidable distribution of media properties, the magnetic anisotropy values of the bits in the array were varied with a Gaussian distribution with mean values $\overline{K_u^{soft}}$ and $\overline{K_u^{hard}}$ and corresponding standard deviations $\sigma_{K_u}^{soft}$ and $\sigma_{K_u}^{hard}$ for the soft and hard layers, respectively. Thereby, a correlation between the values of the two distributions expressed by a correlation factor C was introduced: At first for each bit, a magnetic anisotropy value of the hard layer, K_u^{hard}, was randomly chosen from the Gaussian distribution (mean value $\overline{K_u^{hard}}$, standard deviation $\sigma_{K_u}^{hard}$) and assigned to the respective hard layer of the bit. From K_u^{hard},

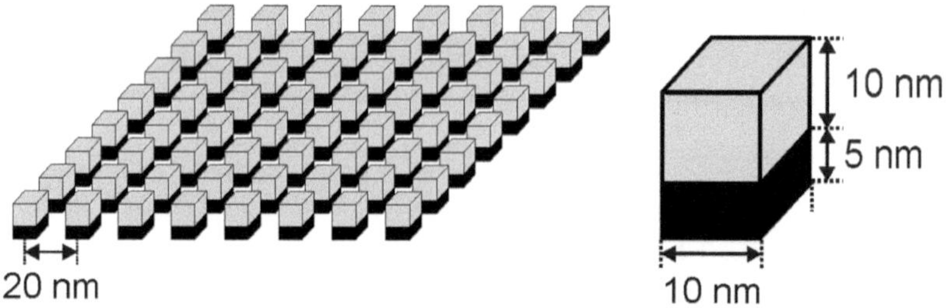

Figure 7.22 Model of 8x8 bits with each bit consisting of a hard (dark color) and a soft (bright color) magnetic layer. Reproduced with permission from Ref. [85]. Copyright 2011, AIP Publishing LLC.

the fully correlated value ($C = 1$) for the soft layer, $K_{\text{u}}^{\text{soft, cor}}$, was calculated as follows:

$$K_{\text{u}}^{\text{soft, cor}} = \frac{\overline{K_{u}^{\text{soft}}}}{\overline{K_{u}^{\text{hard}}}} K_{\text{u}}^{\text{hard}} \tag{7.9}$$

Then, for each bit, an uncorrelated ($C = 0$) magnetic anisotropy value of the soft layer, $K_{\text{u}}^{\text{soft,uncor}}$, was randomly chosen from the Gaussian distribution (mean value $\overline{K_{\text{u}}^{\text{soft}}}$, standard deviation $\sigma_{K_{\text{u}}}^{\text{soft}}$). Finally, the magnetic anisotropy value of the soft layer of each bit, $K_{\text{u}}^{\text{soft}}$ was determined:

$$K_{\text{u}}^{\text{soft}} = (1 - C) \cdot K_{\text{u}}^{\text{soft, uncor}} + C \cdot K_{\text{u}}^{\text{soft, cor}} \tag{7.10}$$

Simulations for each model were carried out 30 times, whereas for each run, a new set of magnetic anisotropy values was chosen, as described above, and assigned to the bits. Furthermore, only the descending branch of the hysteresis loop (from +10 kOe to -100 kOe) was simulated with a magnetic field sweep rate of 1 kOe/ns. The magnetic field was applied under an angle of 3° to avoid numerical instabilities. The magnetic parameters of interest (H_{S} and SFD) were determined using statistical averaging over all 30 runs for each model, resulting in a distribution shown as error bars. The SFD was determined as the field difference between 90% and −90% of the saturation magnetization.

Figure 7.23a shows the normalized SFD of the bit array as a function of the correlation factor C for different values of $\overline{K_{\text{u}}^{\text{hard}}}$. The magnetic anisotropy value of the soft layer was set to 1×10^{7} ergs/cm^{3}. Both $\sigma_{K_{\text{u}}}^{\text{soft}}$ and $\sigma_{K_{\text{u}}}^{\text{hard}}$ were chosen to be 0.05 of their

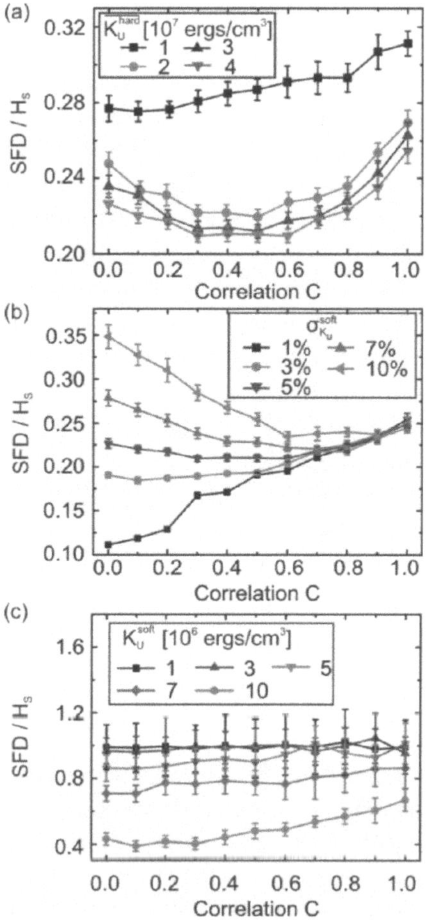

Figure 7.23 Normalized switching-field distribution as a function of the correlation factor C between the magnetic anisotropy values of the individual layers for (a) different K_u^{hard}, (b) different σ_{Ku}^{soft} and (c) different K_u^{soft}. Reproduced with permission from Ref. [85]. Copyright 2011, AIP Publishing LLC.

respective mean value. Please note that the switching field, H_S, was not influenced by a correlation of the values of the distributions of the magnetic anisotropy values (not shown). For values of the magnetic anisotropy of the hard layer close to the one of the soft layer ($\overline{K_u^{soft}} = 1 \times 10^7$ ergs/cm^3; $\overline{K_u^{hard}} = 2 \times 10^7$ ergs/cm^3), the smallest SFD/H_S is achieved at $C = 0$, while an increasing correlation leads to a gradual increased SFD/H_S. In this case, the bit can nearly be treated as a single-phase bit and an increased correlation leads to an increased appearance of bits with (high K_u^{hard})/(high K_u^{soft}) combination or a (low K_u^{hard})/(low K_u^{soft}), which results in a broadening of the SFD [83]. However, in the case of high magnetic anisotropy values of the hard layer (3–5 $\times$ 10^7 ergs/cm^3), the shown function SFD/H_S exhibits a pronounced minimum at about $C = 0.5$. The reason for this minimum is an interplay of two effects: On the one hand, an increased correlation leads to increased appearance of hard/hard and soft/soft bits, as described above, whereas on the other hand, an intermediate correlation leads to a reduction of the possible K_u values of the soft magnetic layer (Eq. 7.10) and thus indirectly narrows $\sigma_{K_u}^{soft}$.

Figure 7.23b shows the normalized SFD as a function of the correlation factor C for different $\sigma_{K_u}^{soft}$. The distribution of the hard magnetic layer, $\sigma_{K_u}^{hard}$, was set to 5%, $\overline{K_u^{hard}}$ to 4 $\times$ 10^7 ergs/cm^3, and $\overline{K_u^{soft}}$ to 1 $\times$ 10^7 ergs/cm^3. The switching field, $H_S \approx 40$ kOe, is not influenced by the correlation.

7.5.2 *Magnetic Recording on ECC BPM*

ECC films that consist of a high-anisotropy layer (HAL) coupled to a low-anisotropy layer (LAL) have been shown to combine both thermal stability and low coercivity for good writability. This concept has been applied for BPM [9, 10, 76, 80, 82, 86]. In cooperation with the group of O. Hellwig (HGST, San Jose, USA), ECC films were fabricated having HAL and LAL layers separated by an interlayer (IL), which allows varying of the coupling strength (Fig. 7.24a) [80].

An L1_0-chemically ordered FePt (10 nm) is sputter-deposited at 580°C on single-crystal MgO(001) substrates using CrRu(3 nm)/Pt(6 nm) seed layers and a Pt(1 nm) cap layer. Owing to its strong perpendicular magnetic anisotropy (PMA), this FePt alloy film is acting as the HAL. A [Co/Pd]–[Co/Ni] multilayer system is used as the LAL. Two sample series are prepared with thin [Pd(0.5 nm)] and thick [Pt(1 nm)/Ta(0.75 nm)/Pd(0.9 nm)] ILs enabling strong and weak interlayer coupling in the ECC stack, respectively.

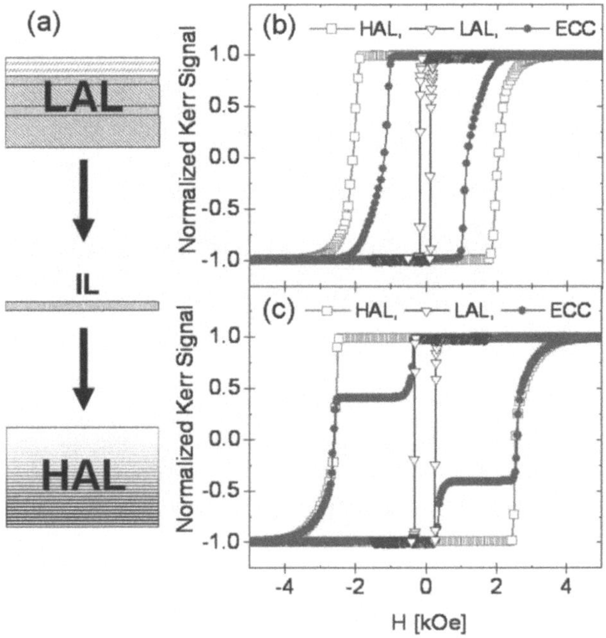

Figure 7.24 (a) A schematic illustration of the ECC layer structure. (b and c) Polar MOKE hysteresis loops for the thin (b) and thick (c) IL ECC full-film samples as well as for individual HAL and LAL full-film samples. Reproduced with permission from Ref. [80]. Copyright 2011, AIP Publishing LLC.

For the sample with thick IL, LAL, and HAL in the composite reverse almost independently and possess the same reversal fields as the separate LAL and HAL systems (Fig. 7.24c). This magnetization reversal behavior is in line with the small coupling between the components of the ECC stack. In contrast, the sample with a thin IL reveals a very different reversal behavior. Here, the reversal occurs in a single step, characteristic of strong coupling between the LAL and the HAL (Fig. 7.24b).

To realize ECC BPM, the extended films with strong and weak coupling in the ECC stack were patterned by EBL into $200 \times 200\ \mu m^2$ square arrays of magnetic pillars with a size of 180 nm and a pitch of 300 nm. A 2.5-fold reduction of the coercive field of the HAL from 9.3 kOe in the weakly coupled sample to 3.7 kOe in

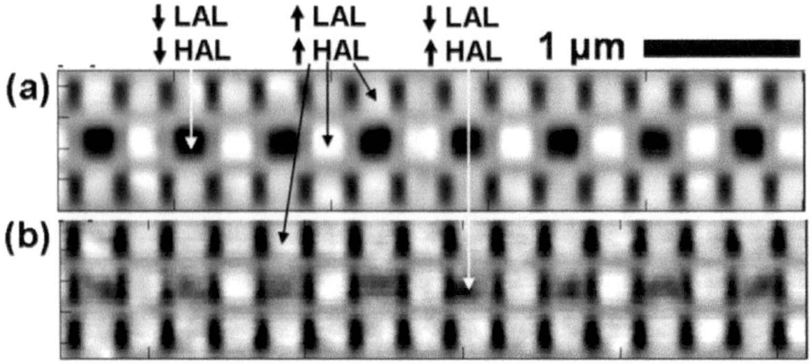

Figure 7.25 SMRM images of 1 T-tones written in the thin (a) and the thick (b) IL ECC BPM structures. Original sample state HAL/LAL was up/up. Equalization was performed line by line, thus resulting in the strongest contrast for alternating up/up and down/down HAL/LAL states. Reproduced with permission from Ref. [80]. Copyright 2011, AIP Publishing LLC.

the strongly coupled sample is observed, demonstrating the ECC concept.

The significant reversal field reduction in this ECC BPM structure is independently confirmed by static read/write testing using a scanning magnetoresistive microscope (SMRM) at HGST (San Jose, USA). It was demonstrated that a conventional write head is able to switch the entire magnetic stack, including the FePt $L1_0$ HAL for the strongly coupled IL ECC structure (Fig. 7.25a), while for the weakly coupled IL structure only the soft layer can be reversed (Fig. 7.25b).

7.6 Experimental Realization of BPM via Self-Assembly

Most of the lithography-based methods to fabricate BPM (Fig. 7.26a) may not be suitable for large-scale device fabrication because they are multistep and expensive and involve time-consuming procedures. Therefore, there is an intensive search for other ways to produce patterned templates are self-assembly techniques such as anodized alumina [87, 88], phase separation in block copolymers (Fig. 7.26b) [79, 89–91], surface instability induced by

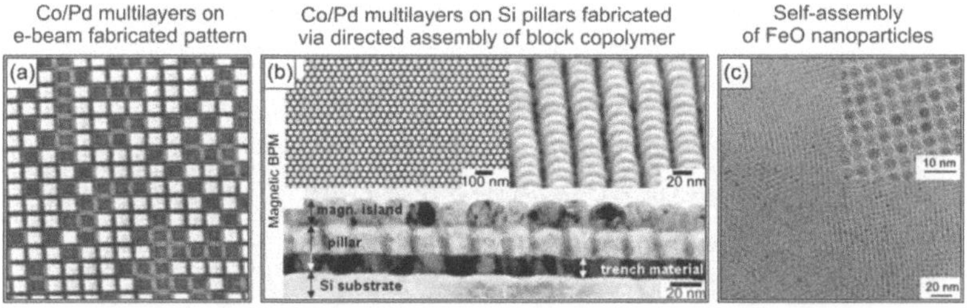

Figure 7.26 (a) Magnetic force microscopy (MFM) image revealing the magnetic state in the array of Co/Pd islands with a size of 50 nm prepared by e-beam patterning. Reproduced with permission from Ref. [62]. Copyright 2004, AIP Publishing LLC. (b) BPM array consisting of Co/Pd multilayers deposited onto Si pillar substrates fabricated via e-beam-directed assembly of block copolymer films. Reproduced with permission from Ref. [79]. Copyright 2010, AIP Publishing LLC. (Left) Top-view scanning electron microscopy (SEM) micrograph. (Right) Section view at 85° angle. (Bottom) Bright-field transmission electron microscopy (TEM) cross-sectional image. (c) Self-assembly of 4 nm Fe_3O_4 or γ-Fe_2O_3 nanocrystals forming superlattices. (Inset) Higher magnification of the image. Reprinted with permission from Ref. [95]. Copyright 2004 American Chemical Society.

ion bombardment [92, 93], and self-assembled magnetic particles (Fig. 7.26c) [94–98].

An alternative method was recently introduced, which combines self-assembly of nonmagnetic spherical particles with defined magnetic properties provided by a magnetic film deposited onto particle arrays (Fig. 7.27a) [60, 99]. This enables an elegant way to create magnetic nanostructure arrays (cap arrays) with strong PMA as required for high thermal stability [36, 55, 100–104]. In this regard, Kappenberger et al. [103] employed a template-directed self-assembly approach to arrange spherical particles into a polymer resist template with regular hole arrays, produced using extreme ultraviolet interference lithography (EUV-IL) (Fig. 7.27b,c).

This approach allows fabrication of ordered arrays of magnetic nanostructures with PMA over large areas, as required for the BPM concept. In addition to the issues with producing arrays of magnetic entities, methods need to be found for successful writing and reading of information in BPM.

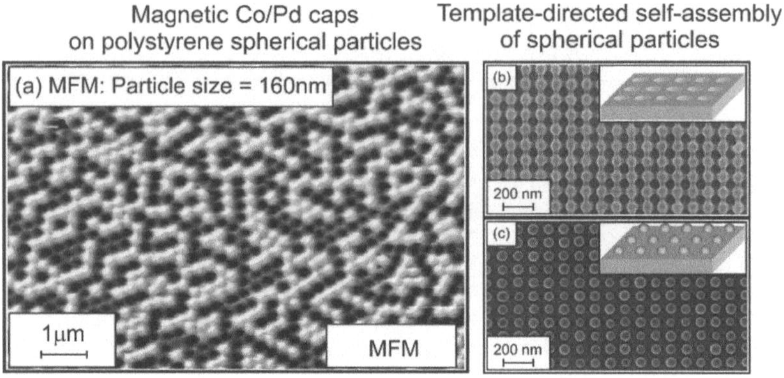

Figure 7.27 (a) MFM image of Co/Pd multilayers deposited on an array of spherical particles with a size of 160 nm. SEM images of (b) a patterned hole array in a PMMA resist created with EUV-IL with a hole size of 60 nm and a period of 100 nm on a Si substrate. (c) Nanospheres with a diameter of 60 nm captured in the regular square hole array shown in (b). The ordering is schematically shown in the insets. Reproduced with permission from Ref. [103]. Copyright 2009, AIP Publishing LLC.

7.6.1 *Magnetic Probe Recording on Nanocaps*

Writing imposes a careful lateral adjustment of the recording head position on the track and requires the synchronization of write pulse to the media pattern [81, 105–108]. For the probe recording scheme, an atomic force microscope cantilever tip coated with a magnetic material used to read/write information to each magnetic nanostructure is considered to be promising [109]. Probe recording has been achieved in nonmagnetic materials by oxidizing the surface of the storage material, by storing charge in semiconductors, or by forming indentations in polymers [110, 111]. However, these techniques suffer from relatively slow reading and writing speed. Therefore, large-scale parallelization of the tips is required to boost performance, as demonstrated, for example, in the Millipede system [109, 110].

To perform probe recording on arrays of magnetic cap structures, MFM tips have to fulfill two critical requirements: The stray field of the MFM tip has to be sufficiently large to reverse the magnetization direction of the individual magnetic cap, and the spatial distribution of the stray field has to be rather narrow (i.e., the stray field gradient has to be sufficiently steep) so that the magnetic state of the neighboring magnetic nanostructures is not affected. A stray field of less than 10 mT of commercially available MFM tips is generally too weak for probe recording applications. Therefore, to increase the tip's stray field, a series of MFM tips was specially manufactured by the group of H. J. Hug at EMPA (Dübendorf, CH) [103]. The largest stray field of 67.5 mT was achieved by using a 400 nm long Co rod with a diameter of 100 nm grown by electrochemical deposition inside an aluminum membrane and attached near the apex of the cantilever tip. To focus its stray field, the rod's apex was further shaped into a cone by a focused ion beam (inset in Fig. 7.28d), resulting in an apex diameter of less than 25 nm.

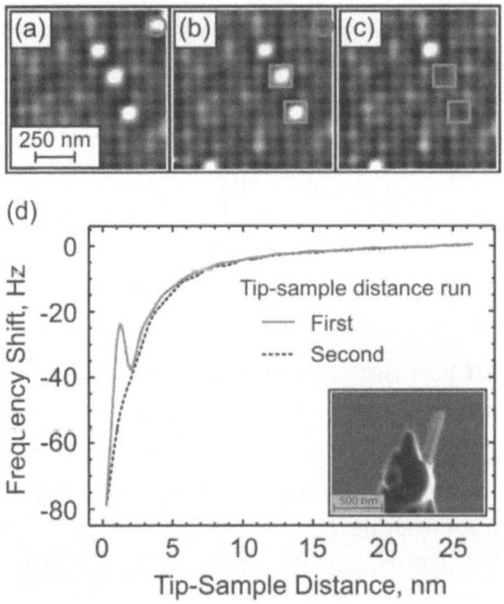

Figure 7.28 Probe recording on an array of magnetic nanocaps on 60 nm nanoparticles with a period of 100 nm. (a–c) The series of MFM images reveals that the magnetization of the dots (marked with a frame) switch by the tip field from an antiparallel (bright) to a parallel configuration (dark). The graph in (d) shows force–distance curves displaying a kink when the magnetization of the nanocap changes its direction. A custom made rod-like tip used for probe recording experiments is shown as an inset in (d). Reproduced with permission from Ref. [103]. Copyright 2009, AIP Publishing LLC.

In this study, a template-directed self-assembly process was employed to arrange spherical particles into a polymer resist template with regular hole arrays, produced at the Paul Scherrer Institute (PSI Villigen, CH) using EUV-IL [103], as already presented in Fig. 7.27. An array of magnetic caps was created by depositing $[Pt(0.8\ nm)/Co(0.3\ nm)]_8/Pt(5\ nm)$ multilayers onto these patterns, forming single-domain cap structures with out-of-plane magnetization.

The proof-of-concept probe recording measurements were performed using a variable temperature ultrahigh vacuum (UHV)-MFM at a base temperature of 8 K operated by the group of H. J. Hug (EMPA, Dübendorf, CH). The magnetic state of the sample in an applied bias magnetic field is presented in Fig. 7.28a revealing that the sample is almost magnetically saturated with only a few magnetic caps left with a reverse magnetization (in white), which we use for the probe recording trials. To initiate a magnetic writing event, the MFM tip was positioned over the nanocap of choice. Then a force–distance curve was acquired by approaching the cantilever slowly in the vertical direction (Fig. 7.28d, solid curve). The spectra show a well-pronounced peak at a tip–sample separation distance of about 1.2 nm, indicating a magnetic switching event (island in circular frame in Fig. 7.28a), which is confirmed by the next MFM scan performed at larger tip–sample separation (Fig. 7.28b). Moreover, in an additional measurement taken after the switching event at the same location (Fig. 7.28d, dashed curve) no peak is present. Indeed, the presence of the peak in a force–distance curve is useful for tracking successful switching events in a magnetic probe recording concept. In addition, two further switching events are indicated by the square frames in Fig. 7.28b,c, clearly demonstrating that using the field-assisted probe recording approach, isolated single nanocaps can be switched in a controlled fashion [103].

7.6.2 *Magnetic Probe Recording on Cap Arrays: Micromagnetic Simulations*

Micromagnetic simulations presented in this section are performed in cooperation with A. Goncharov (HGST, San Jose, USA). The MFM

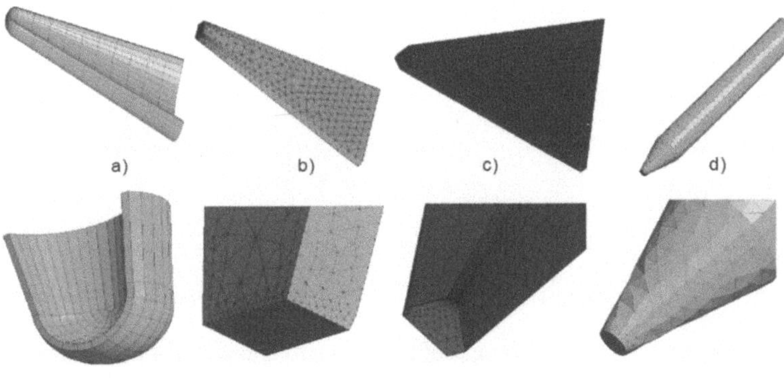

Figure 7.29 Four shapes of MFM tips used for the micromagnetic simulations. The shape in (a) does not show the element mesh used in the simulation but solely the element cells.

tips of various shapes coated with magnetic materials possessing different properties are calculated to assess the maximum achievable magnetic stray field, which could be used to switch magnetization of magnetic cap structures.

Figure 7.29 shows various tip geometries used in the simulation. The perpendicular component of the stray field of the tip in Fig. 7.29d is about twice as large as the field of tips in Fig. 7.29a,b. The tip in Fig. 7.29c shows the smallest perpendicular component of the stray field.

First, we assume that there is no external field acting on the tip and there is no interaction between the tip and magnetic caps. In this case, decreasing the anisotropy of the magnetic material of the tip does not influence the stray field of the tip. For tips with a high aspect ratio, the magnetic state of the MFM tip is mainly governed by shape anisotropy. In addition, the strong exchange interaction of the tip material ensures that the magnetization at the apex of the tip is perpendicular to the surface. Therefore, an increase of the saturation magnetization of the tip material results in almost linear increase of the perpendicular component of the magnetic stray field (Fig. 7.30a). It should be noted that this is true only for small tip spherical bases (10 nm in these calculations), where the parallel alignment of the ferromagnetic spins has smaller energy. For larger tip bases ($>$20 nm), the magnetic state at the apex is driven by the shape

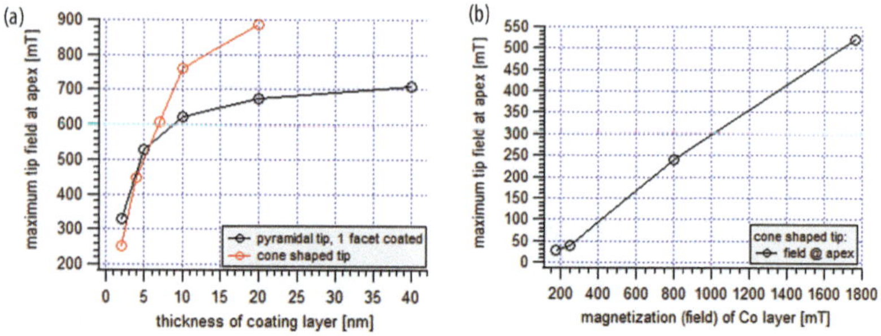

Figure 7.30 Maximum strength of the perpendicular stray field component at a distance of 1.5 nm below the apex of the tip versus (a) saturation magnetization of the film and (b) thickness of the magnetic coating layer. The data shown with black and red symbols correspond to the calculations performed for the MFM tips shaped, as shown in Fig. 7.29a and Fig. 7.29b, respectively.

anisotropy and the magnetization starts to curl around the spherical surface of the tip. The base sphere diameter of the magnetic tip does have a significant impact on the generated stray field. Due to the curling of the magnetization for large sphere diameters, the net vertical component of the stray field is reduced. This can only be compensated by increasing the thickness of the magnetic material used to produce the MFM tip. Figure 7.30b shows how the tip's stray field changes with the layer thickness.

During the recording process, the magnetostatic interaction between a tip and a cap can disturb the magnetization configuration of the tip. This impact is strongly dependent on the magnetization reversal mechanism of the cap, that is, coherent rotation or domain wall propagation. If the cap switches coherently, which holds for a 20 nm cap, for example, the magnetization of the tip is hardly disturbed (Fig. 7.31). If a domain wall is formed in the cap (50 nm cap, Fig. 7.32) during the magnetization reversal process, then the stray field of the tip can be lowered. The latter can result in the failure in the writing process. The impact of the cap magnetization on the magnetic state of the tip can be minimized by increasing the magnetocrystalline anisotropy of the material of the tip. For instance, reliable switching of a cap of a diameter of 50 nm can be achieved by using tips with the magnetic anisotropy constant of 4.5

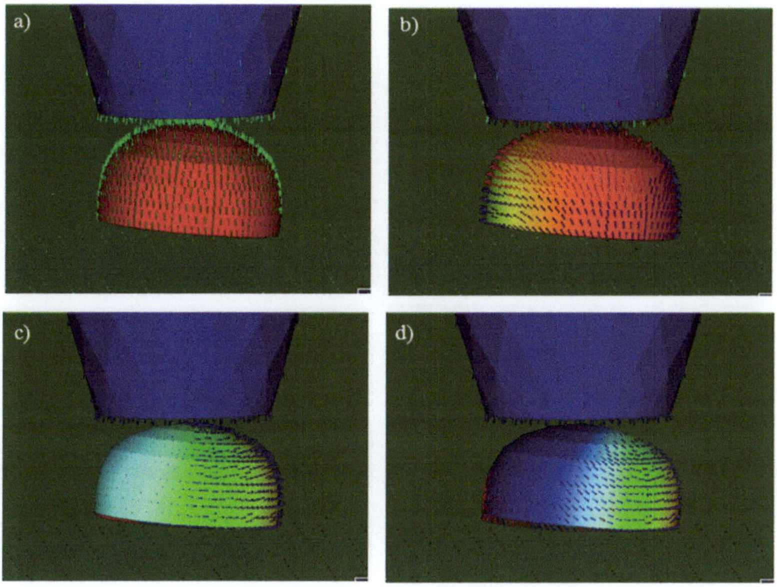

Figure 7.31 Snapshots of the magnetization distribution revealing the switching event of a 20 nm cap. The images are taken at (a) 0 ns, (b) 0.5 ns, (c) 1 ns, and (d) 1.5 ns of the magnetization reversal process.

$\times$ 10^6 ergs/cm^3. In contrast, magnetization of the cap will not be reversed if the anisotropy of the tip is 0.45×10^6 ergs/cm^3. Thus, the magnetocrystalline anisotropy of the material of the MFM tip must be kept high.

The simulations reveal that magnetization reversal of individual cap structures is possible using tips with geometry shown in Fig. 7.29d. Selective switching of 20 and 50 nm sized individual nanocaps using tips of the shape shown in Fig. 7.29a is possible but an additional magnetic bias field is needed to lower the energy barrier of the media. The sum of the bias field and the stray field from the MFM tip can exceed the coercive field of the cap and enable the switching event. However, as the entire cap array with the unavoidable distribution of the magnetic and structural properties will be exposed to this bias field, its strength must be chosen carefully to avoid unintentional loss of information.

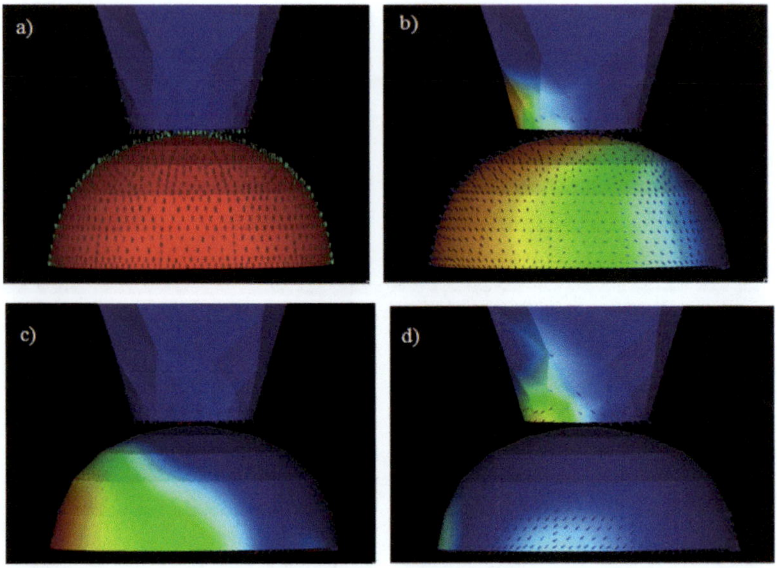

Figure 7.32 Snapshots of the magnetization distribution revealing the switching event of a 50 nm cap. The images are taken at (a) 0 ns, (b) 0.05 ns, (c) 0.1 ns, and (d) 0.2 ns of the magnetization reversal process.

Probe recording using MFM tips to address individual magnetic cap structures has been experimentally demonstrated by Kappenberger et al. [103] (see discussion in the previous section). The technique with one MFM tip suffers from the relatively slow reading and writing speed. Still, its performance can be boosted by relying on Millipede systems with multiple tips [109, 110]. To demonstrate the working principle of the Millipede approach applied to the array of magnetic nanocaps, here, we consider an array of MFM tips (shape is shown in Fig. 7.29d) magnetized in opposite directions along z axis pointing along the tip axis (Fig. 7.33a). These MFM tips are resting at 10 nm distance above the array of magnetic nanocaps with a diameter of 20 nm and can be lowered down to 1 nm to initialize writing event. When the tip is lowered, the stray field of the tip switches the magnetization of the target nanocap. Figure 7.33b shows a magnified picture of the relative position of the target cap and the tip. The target cap has magnetization pointing opposite to the magnetization of the tip (red-colored cap and blue-colored tip).

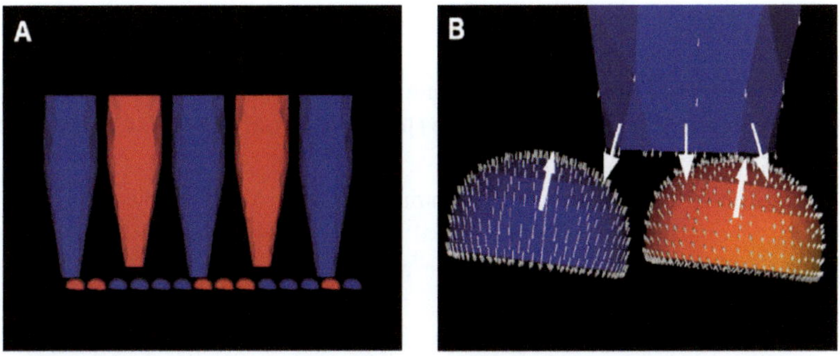

Figure 7.33 (a) Picture showing arrays of 5 MFM tips over the array of magnetic nanocaps. Blue-colored tips are magnetized in the negative z direction, and red ones are magnetized in the positive z direction. Position of the blue-colored tips is lowered down to 1 nm spacing. Red-colored tips are at 10 nm above the caps. (b) A magnified area of the tip apex. Introducing the tilt of the caps of 5° with respect to the tip axis favors switching of the right cap and prevents switching of the left one.

If the magnetization of the cap is aligned along the stray field of the tip, this cap will not switch its magnetization. To ensure that the neighboring cap colored in blue (Fig. 7.33b) will not switch its magnetization, the caps are slightly tilted with respect to the tip axis. The tilt should be adjusted to the local stray field configuration of the tip. In the example shown in Fig. 7.33, tilting the caps 5° to the tip axis allows achieving of reliable switching of the red-colored cap, while maintaining the magnetization orientation of the blue-colored one.

Acknowledgments

We would like to thank our long-standing collaborators T. Schrefl, D. Suess (TU Vienna, Austria), and A. Goncharov (HGST, San Jose, USA) for supporting us with micromagnetic simulations. Furthermore, we are thankful to our research partners O. Hellwig and M. Grobis (HGST, San Jose, USA), L. J. Heyderman (PSI Villigen, CH), and M. Marioni and H. J. Hug (EMPA, Dübendorf, CH) for fruitful discussions and cooperation in joint projects.

References

1. Piramanayagam S. N., and Chong T. C., eds. (2012). *Development in Data Storage, Materials Perspective* (John Wiley and Sons, New York), ISBN 978-0-470- 50100-9.
2. Weller D., Brändle H., Gorman G., Lin C., and Notarys H. (1992). Magnetic and magneto optical properties of cobalt platinum alloys with perpendicular magnetic anisotropy, *Appl. Phys. Lett.*, **61**, pp. 2726–2729.
3. Weller D., and McDaniel T. (2006). Media for extremely high density recording, in *Advanced Magnetic Nanosctructures*, eds., Sellmyer D., and Skomski R. (Springer, USA).
4. Richter H. J. (2009). Density limits imposed by the microstructure of magnetic recording media, *J. Magn. Magn. Mater.*, **321**, p. 467.
5. Ross C. A. (2001). Patterned magnetic recording media, *Annu. Rev. Mater. Sci.*, **31**, p. 203.
6. Adeyeye A. O., and Singh N. (2008). Large area patterned magnetic nanostructures, *J. Phys. D: Appl. Phys.*, **41**, p. 153001.
7. Terris B. D. (2009). Fabrication challenges for patterned recording media, *J. Magn. Magn. Mater.*, **321**, pp. 512–517.
8. Albrecht T. R., Hellwig O., Ruiz R., Schabes M. E., Terris B. D., and Wu X. Z. (2009). Bit-patterned magnetic recording: nanoscale magnetic islands for data storage, in *Nanoscale Magnetic Materials and Apllications*, pp. 237–274, eds., Liu J. P., Fullerton E. E., Gutfleisch O., and Sellmyer D. J. (Springer Science+Business Media), DOI:10.1007/978-0-387-85600-1_9.
9. Grobis M. K., Hellwig O., Hauet T., Dobisz E., and Albrecht T. R. (2011). High-density bit patterned media: magnetic design and recording performance, *IEEE Trans. Magnetics*, **47**, p. 6.
10. Albrecht, T. R., Dobisz E., Gao H., Grobis M., Hellwig O., Kercher D., Lille J., Marinero E., Patel K., Ruiz R., Schabes, M. E., Wan L., Weller D., and Wu, T. W. (2013). Bit patterned media at 1 Tdot/in^2 and beyond, *IEEE Trans. Magn.*, **49**, p. 773.
11. Albrecht T. R., Arora H., Ayanoor-Vitikkate V., Beaujour, J.-M., Bedau D., Berman D., Bogdanov A. L., Chapuis Y.-A., Cushen J., Dobisz E. E., Doerk G., Gao H., Grobis M., Gurney B., Hanson W., Hellwig O., Hirano T., Lille J., Jubert P.-O., Kercher D., Liu Z., Mate C. M., Oboukhov I., Patel K. C., Rubin K., Ruiz R., Schabes M., Wan L., Weller D., Wu T.-W., and Yang E.,

(2015). Bit patterned magnetic recording: theory, media fabrication, and recording performance, *IEEE Trans. Magn.*, **51**(5), pp. 1–42.

12. Krone P., Makarov D., Schrefl T., and Albrecht M. (2009). Effect of the anisotropy distribution on the coercive field and switching field distribution of bit patterned media, *J. Appl. Phys.*, **106**, p. 103913.
13. Pfau B., Günther C. M., Guehrs E., Hauet T., Hennen T., Eisebitt S., Hellwig O. (2014). Influence of stray fields on the switching-field distribution for bit-patterned media based on pre-patterned substrates, *Appl. Phys. Lett.*, **105**, p.132407.
14. Skomski R. (2003). Nanomagnetics, *J. Phys.: Condens. Matter*, **15**, pp. R841–R896.
15. Lyubina J., Rellinghaus B., Gutfleisch O., and Albrecht M. (2011). *Handbook of Magnetic Materials*, Vol. 19, pp. 291–395, ed., Buschow K. H. J. (Elsevier, Amsterdam), ISBN: 978-0-444- 53780-5.
16. Weller D., Mosendz O., Parker G., Pisana S., and Santos T. S. (2013). $L1_0$ FePtX–Y media for heat-assisted magnetic recording,*Phys. Status Solidi A*, **210**, p. 1245.
17. Stoner E. C., and Wohlfarth E. P. (1948). A mechanism of magnetic hysteresis in heterogeneous alloys, *Trans. R. Soc. A*, **240**, pp. 599–642.
18. Batra S., Hannay J. D., Hong Z., and Goldberg J. S. (2004). Investigations of perpendicular write head design for 1 Tb/in_2, *IEEE Trans. Magn.*, **40**, p. 319.
19. Suess D., Schrefl T., Fähler S., Kirschner M., Hrkac G., Dorfbauer F., and Fidler J. (2005). Exchange spring media for perpendicular recording, *Appl. Phys. Lett.*, **87**, p. 012504.
20. Victora R. H., and Xiao S. (2005). Composite media for perpendicular magnetic recording, *IEEE Trans. Magn.*, **41**, pp. 537–542.
21. Suess D. (2006). Multilayer exchange spring media for magnetic recording, *Appl. Phys. Lett.*, **89**, p. 113105.
22. Suess D., Lee J., Fidler J., and Schrefl T. (2009). Exchange-coupled perpendicular media, *J. Magn. Magn. Mater.*, **321**, pp. 545–554.
23. Aharoni A. (1996). *Introduction to the Theory of Ferromagnetism* (Oxford Science).
24. Krone P., Makarov D., Albrecht M., Schrefl T., and Suess D. (2010). Magnetization reversal processes of single nanomagnets and their energy barrier, *J. Magn. Magn. Mater.*, **322**, pp. 3771–3776.
25. Kronmüller H., and Fähnle M. (2003). *Micromagnetism and the Microstructure of Ferromagnetic Solids* (Cambridge Press).

26. Henkelmann G., Uberuaga B. P., and Jonsson H. (2000). A climbing image nudged elastic band method for finding saddle points and minimum energy paths, *J. Chem. Phys.*, **113**, p. 9901.
27. Dittrich R., Schrefl T., Sucss D., Scholz W., Forster H., and Fidler J. (2002). A path method for finding energy barriers and minimum energy paths in complex micromagnetic systems, *J. Magn. Magn. Mater.*, **250**, p. 12.
28. Schrefl T., Schabes, M. E., Suess D., Ertl O., Kirschner M., Dorfbauer F., Hrkac G., and Fidler J. (2005). Partitioning of the perpendicular write field into head and SUL contributions, *IEEE Trans. Magn.*, **41**, p. 3064.
29. Farrow R. F. C., Weller D., Marks R. F., Toney M. F., Cebollada A., and Harp G. R. (1996). *J. Appl. Phys.*, **79**, p. 5967.
30. Weller D., Moser A., Folks L., Best M. E., Lee W., Toney M. F., Schwichert M., Thiele J., and Doerner M. F. (2000). High Ku materials approach to 100 Gbits/in^2, *IEEE Trans. Magn.*, **36**, pp. 10–15.
31. Sato M., and Ishii Y. (1989). Simple and approximate expressions of demagnetizing factors of uniformly magnetized rectangular rod and cylinder, *J. Appl. Phys.*, **66**, p. 983.
32. Rave W., Fabian K., and Hubert A. (1998). Magnetic states of small cubic particles with uniaxial anisotropy, *J. Magn. Magn. Mat.*, **190**, pp. 332–348.
33. Thomson T., Hu G., and Terris B. D. (2006). Intrinsic distribution of magnetic anisotropy in thin films probed by patterned nanostructures, *Phys. Rev. Lett.*, **96**, p. 257204.
34. Hellwig O., Berger A., Thomson T., Dobisz E., Bandic Z. Z., Yang H., Kercher D. S., and Fullerton E. E. (2007). Separating dipolar broadening from the intrinsic switching field distribution in perpendicular patterned media, *Appl. Phys. Lett.*, **90**, p. 162516.
35. Kikitsu A. (2009). Prospects for bit patterned media for high-density magnetic recording, *J. Magn. Magn. Mater.*, **321**, pp. 526–530.
36. Ulbrich T. C., Bran C., Makarov D., Hellwig O., Risner-Jamtgaard, J. D., Yaney D., Rohrmann H., Neu V., and Albrecht M. (2010). The effect of magnetic coupling on the magnetization reversal in arrays of magnetic nanocaps, *Phys. Rev. B*, **81**, p. 054421.
37. Pfau B., Günther C. M., Guehrs E., Hauet T., Dobisz E., Kercher D., Yang H., Vinh L., Xu X., Yaney D., Rick R. L., Eisebitt S., Hellwig O. (2011). Origin of magnetic switching field distribution in bit patterned media based on pre-patterned substrates, *Appl. Phys. Lett.*, **99**, p. 062502.

38. Scholz W., Fidler J., Schrefl T., Suess D., Dittrich R., Forster H., and Tsiantos V. (2003). Scalable parallel micromagnetic solvers for magnetic nanostructures, *Comput. Mater. Sci.*, **28**, pp. 366–383.
39. Piramanayagam S. N. (2007). Perpendicular recording media for hard disk drives, *J. Appl. Phys.*, **102**, p. 011301.
40. Piramanayagam S. N., and Srinivasan K. (2009). Recording media research for future hard disk drives, *J. Magn. Magn. Mater.*, **321**, pp. 485–494.
41. Sakuma A. (1994). First principle calculation of the magnetocrystalline anisotropy energy of FePt and CoPt ordered alloys, *J. Phys. Soc. Jpn.*, **63**, p. 3053.
42. Scholz W., Suess D., Schrefl T., and Fidler J. (2004). Micromagnetic simulation of magnetization reversal in small particles with surface anisotropy, *J. Appl. Phys.*, **95**, pp. 6807–6809.
43. Brown W. F., Jr. (1945). Virtues and weaknesses of the domain concept, *Rev. Mod. Phys.*, **17**, p. 15.
44. Aharoni A. (1962). Theoretical search for domain nucleation, *Rev. Mod. Phys.*, **34**, p. 227.
45. Kronmüller H., Durst K. D., and Martinek G. (1987). Angular dependence of the coercive field in sintered Fe77Nd15B8 magnets, *J. Magn. Magn. Mater.*, **69**, p. 149.
46. Kronmuller H. (1987). Theory of nucleation fields in inhomogeneous ferromagnets, *Phys. Status Solidi B*, **144**, p. 385.
47. Schabes M. E. (1991). Micromagnetic theory of non-uniform magnetization processes in magnetic recording particles, *J. Magn. Magn. Mater.*, **95**, p. 249.
48. Kato T., Miura K., Aoi. H., Muraoka H., Nakamura Y. (2008). Analysis of magnetization transition in perpendicular magnetic recording for narrow transition width, *J. Magn. Magn. Mater.*, **320**, pp. 2900–2903.
49. Basu S., Fry P. W., Gibbs M. R. J., Schrefl T., and Allwood D. A. (2009). Control of the switching behavior of ferromagnetic nanowires using magnetostatic interactions, *J. Appl. Phys.*, **205**, p. 083901.
50. Hellwig O., BergerA., Kortright J. B., and Fullerton E. E. (2007). Domain structure and magnetization reversal of antiferromagnetically coupled perpendicular anisotropy films, *J. Magn. Magn. Mater.*, **319**, p. 13.
51. Kondorsky E. (1940). On hysteresis in ferromagnetics, *J. Phys. (Moscow)*, **II**, p. 161.

52. Coffey K. R., Thomson T., and Thiele J. U. (2002). Angular dependence of the switching field of thin-film longitudinal and perpendicular magnetic recording media, *J. Appl. Phys.*, **92**, p. 4553.
53. Coffey K. R., Thomson T., and Thiele J. U. (2003). Angle dependent magnetization reversal of thin film magnetic recording media, *J. Appl. Phys.*, **93**, p. 8471.
54. Dittrich R., Hu G., Schrefl T., Thomson T., Suess D., Terris B. D., and Fidler J. (2005). Angular dependence of the switching field in patterned magnetic elements, *J. Appl. Phys.*, **97**, p. 10J705.
55. Ulbrich T. C., Makarov D., Hu G., Guhr I. L., Suess D., Schrefl T., and Albrecht M. (2006). Magnetization reversal in a novel gradient nanomaterials, *Phys. Rev. Lett.*, **96**, p. 077202.
56. Lee J., Brombacher C., Fidler J., Dymerska B., Suess D., and Albrecht M. (2011). Contribution of the easy axis orientation, anisotropy distribution and dot size on the switching field distribution of bit patterned media, *Appl. Phys. Lett.*, **99**, p. 062505.
57. Gao K. Z., and Bertram H. N.(2002). Magnetic recording configuration for densities beyond 1 Tb/in^2 and data rates beyond 1 Gb/s, *IEEE Trans. Magn.*, **38**, p. 3675.
58. Wang J. P., Zou Y. Y., Hee C. H., Chong T. C., and Zheng Y. F. (2003). Approaches to tilted magnetic recording for extremely high areal density, *IEEE Trans. Magn.*, **39**, p. 1930.
59. Wang J. P. (2005). Magnetic data storage: tilting for top, *Nat. Mater.*, **4**, p. 191.
60. Albrecht M., Hu G., Guhr I. L., Ulbrich T. C., Boneberg J., Leiderer P., and Schatz G. (2005). Magnetic multilayers on nanospheres, *Nat. Mater.*, **4**, p. 203.
61. Hyndman R., Warin P., Gierak J., Ferré J., Chapman, J. N., Jamet, J. P., Mathet V., and Chappert C. (2001). Modification of Co/Pt multilayers by gallium irradiation: part 1; the effect on structural and magnetic properties, *J. Appl. Phys.*, **90**, p. 3843.
62. Hu G., Thomson T., Albrecht M., Best M. E., Terris B. D., Rettner C. T., Raoux S., McClelland G. M., and Hart M. W. (2004). Magnetic and recording properties of Co/Pd islands on prepatterned substrates, *J. Appl. Phys.*, **95**, p. 7013.
63. Lodder J. C. (2004). Methods for preparing patterned media for high-density recording, *J. Magn. Magn. Mater.*, **272–276**, p. 1692.
64. Luo F., Heyderman, L. J., Solak, H. H., Thomson T., and Best, M. E. (2008). Nanoscale perpendicular magnetic island arrays fabricated by extreme ultraviolet interference lithography, *Appl. Phys. Lett.*, **92**, p. 102505.

65. Brombacher C., Grobis M., Lee J., Fidler J., Eriksson T., Werner T., Hellwig O., and Albrecht M. (2012). $L1_0$ FePt:Cu bit patterned media, *Nanotechnology*, **23**, p. 025301.

66. McClelland G. M., Hart, M. W., Rettner C. T., Best M. E., Carter K. R., and Terris B. D. (2002). Nanoscale patterning of magnetic islands by imprint lithography using a flexible mold, *Appl. Phys. Lett.*, **81**, p. 1483.

67. Dong Q., Li G., Ho C. L., Faisal M., Leung C. W., Pong, P. W. T., Liu K., Tang B. Z., Manners I., and Wong W. Y. (2012). Nanoimprint lithography: a polyferroplatinyne precursor for the rapid fabrication of $L1_0$-FePt-type bit patterned media by nanoimprint lithography, *Adv. Mat.*, **24**, p. 1033.

68. Hellwig O., Marinero, E. E., Kercher D., Hennen T., McCallum A., Dobisz E., Wu, T. W., Lille J., Hirano T., Ruiz R., Grobis, M. K., Weller D., and Albrecht, T. R. (2014). Bit patterned media optimization at 1 Tdot/in^2 by post-annealing, *Appl. Phys. Lett.*, **116**, p. 123913.

69. Krone P., Makarov D., Cattoni A., Faini G., Haghiri-Gosnet, A.-M., Knittel I., Hartmann U., Schrefl T., and Albrecht M. (2011). Investigation of the magnetization reversal of a magnetic dot array of Co/Pt multilayers, *J. Nanopart. Res.*, **13**, pp. 5587–5593.

70. Rettner C. T., Anders S., Thomson T., Albrecht M., Ikeda Y., Best M. E., and Terris B. D. (2002). Magnetic characterization and recording properties of patterned $Co_{70}Cr_{18}Pt_{12}$ perpendicular media, *IEEE Trans. Magn.*, **38**, pp. 1725–1730.

71. Frei E. H., Shtrikman S., and Treves D. (1957). Critical size and nucleation field of ideal ferromagnetic particles, *Phys. Rev.*, **106**, pp. 446–455.

72. Ishii Y. (1991). Magnetization curling in an infinite cylinder with a uniaxial magnetocrystalline anisotropy, *J. Appl. Phys.*, **70**, p. 3765–3770.

73. Aharoni A. (1997). Angular dependence of nucleation by curling in a prolate spheroid, *J. Appl. Phys.*, **82**, pp. 1281–1288.

74. Makarov D., Tibus S., Rettner C. T., Thomson T., Terris B. D., Schrefl T., and Albrecht M. (2008). Magnetic strip patterns induced by focused ion beam irradiation, *J. Appl. Phys.*, **103**, p. 063915.

75. Suess D. (2007). Micromagnetics of exchange spring media: optimization and limits, *J. Magn. Magn. Mater.*, **308**, p. 183.

76. Hauet T., Dobisz E., Florez S., Park J., Lengsfield B., Terris B. D., and Hellwig O. (2009). Role of reversal incoherency in reducing switching field and switching field distribution of exchange coupled composite bit patterned media, *Appl. Phys. Lett.*, **95**, p. 262504.

77. Breitling A., Bublat T., and Goll D.(2009). Exchange-coupled $L1_0$-FePt/Fe composite patterns with perpendicular magnetization, *Phys. Status Solidi-R*, **3**, pp. 130–132.
78. Hellwig O., Hauct T., Thomson T., Dobisz E., Risner-Jamtgaard J. D., Yaney D., Terris B. D., and Fullerton E. E.(2009). Coercivity tuning in Co/Pd multilayer based bit patterned media, *Appl. Phys. Lett.*, **95**, p. 232505.
79. Hellwig O., Bosworth J. K., Dobisz E., Kercher D., Hauer T., Zeltzer G., Risner-Jamtgaard J. D., Yaney D., and Ruiz R.(2010). Bit patterned media based on block copolymer directed assembly with narrow magnetic switching field distribution, *Appl. Phys. Lett.*, **96**, p. 052511.
80. McCallum A. T., Krone P., Springer F., Brombacher C., Albrecht M., Dobisz E., Grobis M., Weller D., and Hellwig O. (2011). $L1_0$ FePt based exchange coupled composite bit patterned films, *Appl. Phys. Lett.*, **98**, p. 242503.
81. Albrecht M., Rettner C. T., Moser A., Best M. E., and Terris B. D. (2002). *Appl. Phys. Lett.*, **81**, p. 2875.
82. Grobis M., Dobisz E., Hellwig O., Schabes M. E., Zeltzer G., Hauet T., and Albrecht T. R. (2010). Measurements of the write error rate in bit patterned magnetic recording at 100–320 Gb/in_2, *Appl. Phys. Lett.*, **96**, p. 052509.
83. Krone P., Makarov D., Schrefl T., and Albrecht M. (2010). Exchange coupled composite bit patterned media, *Appl. Phys. Lett.*, **97**, p. 082501.
84. Makarov D., Lee J., Brombacher C., Schubert C., Fuger M., Suess D., Fidler J., and Albrecht M. (2010). Perpendicular FePt-based exchange-coupled composite media, *Appl. Phys. Lett.*, **96**, p. 062501.
85. Krone P., Makarov D., Schrefl T., and Albrecht M. (2011). Correlation of magnetic anisotropy distributions in layered ECC bit patterned media, *J. Appl. Phys.*, **109**, p. 103901.
86. Wang H., Zhao H., Rahman T., Isowaki Y., Kamata Y., Maeda T., Hieda H., Kikitsu A., and Wang J. P. (2013). Fabrication and characterization of FePt exchange coupled compositeand graded bit patterned media, *IEEE Trans. Magn.*, **49**, p. 707.
87. Wang T., Wang Y., Fu Y., Hasegawa T., Oshima H., Itoh K., Nishio K., Masuda H., Li F. S., Saito H., and Ishio S. (2008). Magnetic behavior in an ordered Co nanorod array, *Nanotechnology*, **19**, p. 455703.
88. Piraux L., Antohe, V. A., Abreu Araujo F., Srivastava S. K., Hehn M., Lacour D., Mangin S., and Hauet T. (2012). Periodic arrays of magnetic

nanostructures by depositing Co/Pt multilayers on the barrier layer of ordered anodic alumina templates, *Appl. Phys. Lett.*, **101**, p. 013110.

89. Bita I., Yang J. K. W., Jung Y. S., Ross C. A., Thomas E. L., and Berggren K. K. (2008). Graphoepitaxy of self-assembled block copolymers on two-dimensional periodic patterned templates, *Science*, **321**, p. 939.

90. Ruiz R., Kang, H. Detcheverry F. A., Dobisz E., Kercher D. S., Albrecht T. R., de Pablo J. J., and Nealey P. F. (2008). Density multiplication and improved lithography by directed block copolymer assembly, *Science*, **321**, pp. 936–939.

91. Ruiz R., Dobisz E., and Albrecht T. R. (2011). Rectangular patterns using block copolymer directed assembly for high bit aspect ratio patterned media, *ACS Nano*, **5**(1), pp. 79–84.

92. Facsko S., Dekorsy T., Koerdt C., Trappe C., Kurz H., Vogt A., and Hartnagel H. L. (1999). Formation of ordered nanoscale semiconductor dots by ion sputtering, *Science*, **285**, p. 1551.

93. Ball D. K., Lenz K., Fritzsche M., Varvaro G., Günther S., Krone P., Makarov D., Mücklich A., Facsko S., Fassbender J., and Albrecht M. (2014). Magnetic properties of granular CoCrPt:SiO2 thin films deposited on GaSb nanocones, *Nanotechnology*, **25**, p. 085703.

94. Sun S., Murray, C. B., Weller D., Folks L., Moser A. (2000). Monodisperse FePt nanoparticles and ferromagnetic FePt nanocrystal superlattices, *Science*, **287**, p. 1989.

95. Redl, F. X., Black, C. T., Papaefthymiou, G. C., Sandstrom, R. L., Yin M., Zeng H., Murray, C. B., and O'Brien, S. P. (2004). Magnetic, electronic, and structural characterization of nonstoichiometric iron oxides at the nanoscale. *J. Am. Chem. Soc.*, **126**, pp. 14583–14599.

96. Terris, B. D., and Thomson T. (2005). Nanofabricated and self-assembled magnetic structures as data storage media, *J. Phys. D: Appl. Phys.*, **38**, pp. R199–R222.

97. Zhu M. W., Li Y. M., Meng T., Zhan P., Sun J., Wu J., Wang Z. L., Zhu S. N., and Ming N. B. (2006). Controlled strain on a double-templated textured polymer film: a new approach to patterned surfaces with Bravais lattices and chains, *Langmuir*, **22**, p. 7248.

98. Sun S. (2006). Recent advances in chemical synthesis, self-assembly, and applications of FePt nanoparticles. *Adv. Mater.*, **18**, pp. 393–403.

99. Albrecht M., and Makarov D. (2012). Magnetic films on nanoparticle arrays, *Open Surf. Sci. J.*, **4**, p. 42.

100. Makarov D., Brombacher C., Liscio F., Maret M., Parlinska M., Meier S., Kappenberger P., and Albrecht M. (2008). FePt films on self-assembled SiO_2 particle arrays, *J. Appl. Phys.*, **103**, p. 053903.

101. Makarov D., Bermúdez-Ureña E., Schmidt, O. G., Liscio F., Maret M., Brombacher C., Schulze S., Hietschold M., and Albrecht M. (2008). Nanopatterned CoPt alloys with perpendicular magnetic anisotropy, *Appl. Phys. Lett.*, **93**, p. 153112.

102. Makarov D., Klimenta F., Fischer S., Liscio F., Schulze S., Hietschold M., Maret M., and Albrecht M. (2009). Nonepitaxially grown nanopatterned Co-Pt alloys with out-of-plane magnetic anisotropy, *J. Appl. Phys.*, **106**, p. 114322.

103. Kappenberger P., Luo F., Heyderman, L. J., Solak H. H., Padeste C., Brombacher C., Makarov D., Ashworth T. V., Philippe L., Hug H. J., and Albrecht M. (2009). Template-directed self-assembled magnetic nanostructures for probe recording, *Appl. Phys. Lett.*, **95**, p. 023116.

104. Günther C. M., Hellwig O., Menzel A., Pfau B., Radu F., Goncharov A., Makarov D., Albrecht M., Schlotter W. F., Rick R., Lüning J., and Eisebitt S. (2010). Microscopic reversal behavior of magnetically capped nanospheres, *Phys. Rev. B*, **81**, p. 064411.

105. Lohau J., Moser A., Rettner C. T., Best M. E., and Terris B. D. (2001). Writing and reading perpendicular magnetic recording media patterned by a focused ion beam, *Appl. Phys. Lett.*, **78**, p. 990.

106. Barbic M., Schultz S., Wong J., and Scherer A. (2001). Recording processes in perpendicular patterned media using longitudinal magnetic recording heads, *IEEE Trans. Magn.*, **37**, p. 1657.

107. Albrecht M., Rettner C. T., Best M. E., and Terris B. D. (2003). Magnetic coercivity patterns for magnetic recording on patterned media, *Appl. Phys. Lett.*, **83**, p. 4363.

108. Albrecht M., Rettner C. T., Moser A., and Terris B. D. (2007). Magnetic reversal of sub-100 nm nanostructures studied by a FIB-trimmed recording head, *Microsys. Technol.*, **13**, p. 129.

109. Mamin H. J., Terris B. D., Fan, L.-S., Hoen S., Barrett R. C., and Rugar D. (1995). High-density data storage using proximal probe techniques, *IBM J. Res. Dev.*, **39**, pp. 681–700.

110. Vettiger P., Cross G., Despont M., Drechsler U., Dürig U., Gotsmann B., Häberle W., Lantz M. A., Rothuizen H. E., Stutz S., and Binning G. K. (2002). The “millipede”: nanotechnology entering data storage, *IEEE Trans. Nanotechnol.*, **1**, p. 39.

111. Gidon S., Lemonnier O., Rolland B., Bichet O., Dressler C., and Samson Y. (2004). Electrical probe storage using Joule heating in phase change media, *Appl. Phys. Lett.*, **85**, p. 6392.

Chapter 8

Magnetic Characterization of Perpendicular Recording Media

Gaspare Varvaro, Alberto Maria Testa, Elisabetta Agostinelli, Davide Peddis, and Sara Laureti
Istituto di Struttura della Materia—CNR, Area della Ricerca Roma 1, Monterotondo Scalo (Roma), 00015, Italy
gaspare.varvaro@ism.cnr.it

8.1 Introduction

Previous chapters reported on the status of current perpendicular magnetic recording (PMR) media and the recent advancements, as well as the possible research directions for next-generation PMR systems. The criteria for the choice of the best material have been thoroughly discussed within the framework of the magnetic recording "trilemma" that imposes the search for the best compromise among three conflicting requirements, that is, thermal stability, writability, and signal-to-noise ratio [1].

Developing a nanostructured material to fit the design criteria for PMR media requires a multiple approach, which involves the investigation of structural, microstructural, and magnetic

Ultrahigh-Density Magnetic Recording: Storage Materials and Media Designs
Edited by Gaspare Varvaro and Francesca Casoli

ISBN 978-981-4669-58-0 (Hardcover), 978-981-4669-59-7 (eBook)
www.panstanford.com

properties, analytical and numerical simulations, and recording tests to evaluate the material's read/write performance. Studying the magnetic properties represents a challenging task as the magnetic behavior of materials for PMR media is governed by a complex interplay among intrinsic material properties, magnetic interactions, and various sources of microscopic inhomogeneity, the overall properties being further complicated by thermal effects that lead to time-dependent magnetic phenomena. Moreover, the magnetic behavior of systems with perpendicular anisotropy is significantly affected by the presence of strong demagnetizing fields that, in principle, should be corrected in all magnetic measurements for a proper investigation.

Several techniques covering a wide range of length (from macroscopic to local) and time (from 10^{-9} to 1 s) scales are available for measuring the magnetic properties of PMR media, including both laboratory and large-scale facilities. Here the focus will be on magnetometric methods [2, 3], which are widely used for laboratory characterization, and only a few specific examples on methodologies based on nonmagnetometric techniques will be reported. The main advantages and drawbacks of the most common magnetometers are collected in Table 8.1. Typically, they operate on the long time-scale regime and allow macroscopic averaged magnetic properties to be accessed, since the magnetic field is uniformly applied to the whole sample. Thus, advanced protocols are needed to extract the magnetic properties of interest for recording, which is a local process (i.e., the write field is applied in a tiny region) occurring at a subnanosecond time scale.

This chapter will provide an overview of both basic and advanced magnetic characterization methods commonly used to investigate the magnetic behavior of materials for PMR media in relation to the requirements for high-density recording. In particular, Section 8.2 will provide a selection of methodologies for the investigation of granular thin films with perpendicular anisotropy that are used in currently available PMR media, being also the basis of future recording systems. This section will introduce a number of basic concepts that are generally valid for all PMR technologies. All the methods discussed in Section 8.2 are also suitable for the characterization of exchange-coupled composite (ECC) systems and next-generation bit-patterned magnetic recording (BPMR) media, the analysis of

Table 8.1 Main advantages and drawbacks of the most common magnetometric techniques. Standard equipment allows scalar measurements to be performed, that is, the moment is measured only along the field direction, with a rotating stage possibly present to vary the angle between the applied field and the sample. Equipment for vector magnetic measurements is also available (mainly VSM and MOKE magnetometers); in this case, the moment is simultaneously measured along different directions, proving particularly useful for anisotropic systems suitable for PMR media

Technique	Advantages	Drawbacks	Sensitivity (Am^2)
VSM	Easy to use Fast data collection	Limited sensitivity	10^{-9}
SQUID magnetometer	High sensitivity	Costly operation Slow data collection	10^{-12}
MOKE magnetometer	Ultrafast data collection Local measurements Layer-selective measurements	Lack of quantitative magnetic information	-
AGFM	High sensitivity Fast data collection	Not suitable for remanence and coercivity measurements of soft materials	10^{-11}
Torque magnetometer	High sensitivity High accuracy	Only suitable for material anisotropy measurements	10^{-11}
AHE magnetometer	High sensitivity Local measurements Layer-selective measurements	Lack of quantitative magnetic information	-

VSM: vibrating sample magnetometer; SQUID: superconductive quantum interference device; MOKE: magneto-optic Kerr effect; AGFM: alternate gradient force magnetometer; AHE: anomalous Hall effect

their most relevant magnetic properties being discussed in Sections 8.3 and 8.4, respectively. It is worth mentioning that in this chapter "exchange-coupled composite" will refer to the whole class of soft-hard exchange-coupled systems; different classifications can be found in the literature and also in the present book. Section 8.5 will focus on the energy-assisted magnetic recording (EAMR) technology by reporting on some specific measurements needed to verify if a material is able to match the peculiar requirements for the application.

8.2 Perpendicular Granular Media

The present section deals with the issue of characterizing magnetically hard granular thin films made of single-domain ferromagnetic grains with perpendicular uniaxial magnetic anisotropy separated by a nonmagnetic grain boundary material [1, 4, 5]. Such granular films were used in the first-generation PMR media, consisting of only one CoCrPt:SiO_2 single recording layer [1, 4, 5], and in the following media where complex stacked structures made of exchange-coupled layers with different anisotropy and intergranular interaction are used [6–12]. Granular thin films are furthermore suitable for next-generation PMR media, that is, ECC systems based on high-anisotropy materials (e.g., $L1_0$–FePtX alloys, X = Cu, Au, Ag, Cu) [13–16] and EAMR media [16–18].

The magnetic properties of granular thin films for PMR media can be described, to a first approximation, by the *Stoner–Wohlfarth* (*S–W*) model that refers to an assembly of noninteracting, uniaxial single-domain particles with perfectly aligned anisotropy easy axes [19]. However, the magnetic behavior of real systems generally deviates from that of an S-W system because of the presence of intergranular magnetostatic and exchange interactions as well as of a distribution of grain properties (i.e., alloy composition, size, shape, crystalline orientation, etc.) [5]. In the following section, we first report on the magnetic analysis for the evaluation of primary magnetic parameters (e.g., switching field, demagnetizing field, anisotropy, etc.); hence we present advanced methods for a thorough characterization of granular materials for PMR media, which include the determination of the switching-field distribution (SFD), the study of the reversal mechanism, and the investigation of thermal effects and time-dependent phenomena. Note that magnetic interactions, which are a key issue for the performance of recording media [1], will not be treated in a special section since they are commented on across the entire chapter.

In PMR media, the hard magnetic layer is grown on top of a soft underlayer (SUL) [1], which swamps its signal when a classical magnetometric technique (e.g., vibrating sample magnetometer [VSM], superconductive quantum interference device [SQUID], alternate gradient force magnetometer [AGFM]) is used. The contribution of the SUL can be subtracted from the measurements by applying an external in-plane field to pin the SUL in a constant magnetic state [20] or by using either a magneto-optic Kerr effect (MOKE) or anomalous Hall effect (AHE) magnetometers that allow the signal to be collected only from the hard layer [21, 22]. Note that when the properties of the hard magnetic layer are investigated, films are generally deposited without the SUL.

It is worth mentioning that the methodologies that will be illustrated in this section are also suitable for the characterization of magnetic properties and physical processes underlying the magnetism of ECC systems and next-generation BPMR media. This section will also introduce a number of basic concepts that are generally valid for all PMR technologies.

8.2.1 *Investigation of Primary Magnetic Properties*

8.2.1.1 Magnetization and remanence curves

The preliminary characterization of any magnetic material, including granular films for PMR media, starts with the measure of the magnetization curve, which is collected by recording at constant temperature the magnetization M while sweeping the magnetic field H from positive to negative values. Figure 8.1 (left) shows a typical major hysteresis loop of a granular film for PMR media measured along the normal to the film plane, that is, the easy-axis direction along which the magnetization spontaneously aligns. The term "major" indicates that the maximum applied field is large enough to bring the system in a fully saturated state where M is a single-valued function of H.

The hysteresis loop provides an overall picture of the macroscopic magnetic behavior, including both reversible and irreversible contributions, and can be used as a rough guide to the recording performance of a given material. The main parameters extracted from the hysteresis loop are:

- *Saturation magnetization* (M_S). It is the magnetization value corresponding to the state where all the magnetic moments are aligned along the applied field direction. Large M_s values

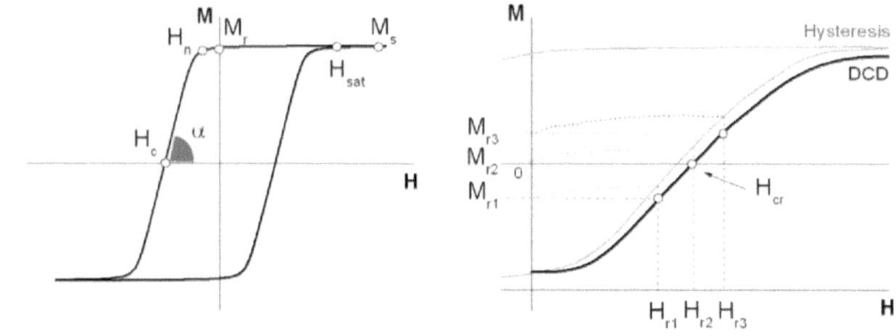

Figure 8.1 (Left) Schematic representation of a typical major hysteresis loop of a granular film for PMR media measured along the easy-axis direction (i.e., normal to the film plane). (Right) Schematic representation of the protocol used to measure a DCD remanence curve. Critical parameters are indicated in the figures.

are beneficial to reduce the writing field needed to switch the magnetization of a grain but can also result in undesired high magnetostatic effects that cause a tendency of the film to demagnetize, thus negatively affecting thermal stability [23, 24]. Typical values of M_s are currently in the 400–800 kA/m range.

- *Remanence* (M_r). It is the value of magnetization at $H = 0$ after saturation. High values of remanent magnetization and a squareness ratio $M_r/M_s = 1$ are required to obtain an adequate read-back signal over the medium noise. The quantity $M_r t$ (i.e., areal moment density), where t is the film thickness, is a useful design parameter to define the amplitude of the read-back signal.
- *Saturation field* (H_{sat}). It refers to the field that brings the system to a fully saturated state. It is related to the maximum writing field that must be applied to reverse all the grains in a bit. Lower H_{sat} values indicate that the recording bit can be written more easily.
- *Nucleation field* (H_n). It describes the onset of reversal in the presence of an external field and it is commonly defined as the reverse field that causes the magnetization to drop to 95% of the saturation magnetization [25]. Optimized materials show a nucleation occurring in the

second quadrant; that is, the more negative H_n is, the more stable the remanent state is as a higher reverse field is required to affect the magnetization.

- *Coercive field* (H_c), *switching field* (H_{sw}). The coercive field measures the resistance of a ferromagnetic material to demagnetize and is defined as the applied reverse field necessary to bring the macroscopic net magnetization from saturation to zero. If no reversible contributions are present, H_c corresponds to the mean switching field $\langle H_{sw} \rangle$ that is only related to the irreversible changes of the magnetization. The maximum value of the switching field for a given material is defined by the anisotropy field H_K equal to $2K_u/\mu_0 M_s$ for an S-W system, where K_u is the uniaxial magnetic anisotropy constant and μ_0 the permeability of the free space ($= 4\pi \times 10^{-7}$ Tm/A). The switching field at subnanosecond writing time must not exceed the magnetic (writing) field that can be generated by the current write head ($H_w^{max} = 1.5$ T) to avoid inefficient writing in the medium; typical values are around 1 T. Note that very hard materials (e.g., $L1_0$–FePtX alloys) proposed for future recording technologies can possess larger switching fields that can be reduced without affecting thermal stability, by coupling them with a softer phase (ECC media) or using an external source of energy (EAMR media) [13, 16].
- *Coercive squareness* (S^*). It is a design parameter defined as $S^* = 1 - M_r/\alpha H_c$, where $\alpha = dM/dH|_{H_c}$ is the hysteresis slope at coercivity that is, ignoring thermal effects, a function of the intrinsic SFD (related to the dispersion of intrinsic grain material properties) and of both exchange and magnetostatic intergranular interactions [26–29]. $S^* = 1$ indicates a square M–H loop, which is desirable to reduce the medium noise and the maximum writing field required to reverse all the grains. However, since stronger intergranular exchange interactions and narrower intrinsic SFD may both cause square loops, care should be taken in using the S^* parameter to define the performance of a recording medium. For the same intrinsic SFD, either S^*

or H_c can provide a rough indication of the intergranular exchange interactions; that is, increasing the exchange interactions leads to an increase of S^* and a reduction of H_c [30].

In addition to the hysteresis measurements, which encompass both reversible and irreversible processes, important information can be gained from remanence measurements that allow purely irreversible magnetization changes, which are closely related to the recording process, to be investigated. The remanence curve of primary interest for the characterization of granular thin films for PMR media is the direct current demagnetization (DCD) remanence curve, which is directly connected to the writing process and the retention of stored bits. A DCD curve is measured by applying a progressively higher DC reverse field H_r to a sample previously saturated in the opposite direction and by recording, for each step, the value of remanence M_r, which is then plotted as a function of H_r (Fig. 8.1, right). Comparing the hysteresis and remanence curves measured along the easy-axis direction is a measure of how much the magnetic behavior of a system deviates from that of an ideal granular film for PMR media where only irreversible switching occurs.[a] When reversible processes are present, remanence curves are preferred to investigate the switching behavior and the main parameter that can be evaluated is the remanent coercivity (H_cr), defined as the reverse field giving zero irreversible magnetization, which well represents the mean switching field of the medium.

It is worth mentioning that for an accurate evaluation of the parameters previously mentioned both the thermal effects (see Section 8.2.4) and the demagnetizing fields must be considered (see Section 8.2.1.2).

8.2.1.2 Demagnetizing fields

The evaluation of the magnetic properties in thin films with perpendicular anisotropy is complicated by the presence of strong

---

[a]It is worth mentioning that for ECC media, which are characterized by an inhomogeneous reversal mechanism [31], the two curves deviate in the region where reversible processes occurs [32–35].

demagnetizing fields H_d, created by self-magnetostatic interactions, which modify the actual internal field H_{int} experienced by the material according to $H_{int} = H + H_d$ [36]. If the demagnetizing fields are not properly accounted for or, even worse, are neglected, the results have to be dealt with care, especially when H_d is of the same order of magnitude of the anisotropy field.

For a uniformly magnetized ellipsoid, the strength of the demagnetizing fields along a given axis is equal to $-NM$, where $0 \leq N \leq 1$ is the demagnetizing factor along this axis. In thin films, as those used in PMR media, the thickness is many orders of magnitude smaller than the other dimensions; hence, assuming an infinite uniformly magnetized layer (*continuous medium approximation*), the demagnetizing factor is 0 in the film plane and maximum along the orthogonal direction (i.e., $N_\perp = 1$). Since out-of-plane measurements are required in thin films for PMR media, the strong demagnetizing fields induce a large variation of H_{int}, which must be taken into account to properly assess the intrinsic magnetic properties. A tangible evidence of the demagnetizing fields is a shearing of the $M(M_r)$–H curves, which assume a slope equal to $1/N$ in the magnetization reversal part [1, 37, 38] (Fig. 8.2, left). The effect is more important in the case of DCD curves since H_d affects not only the reverse field but also the actual zero field at which the remanence is measured. In principle, for measurements along the easy axis (i.e., perpendicular $\perp$ to the film plane) and under the *mean-field approximation* (i.e., N is independent of M), H_{int} can be

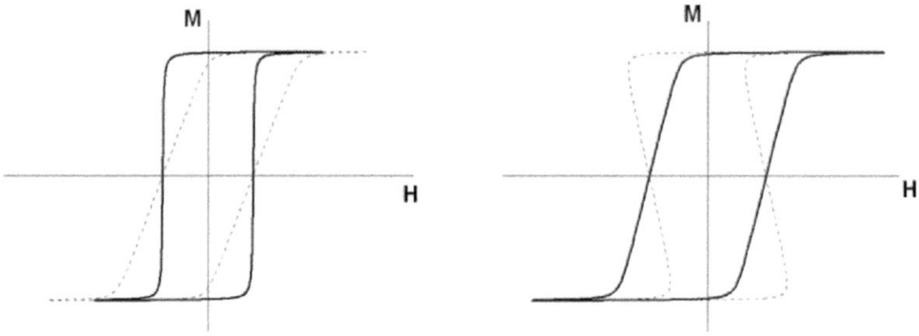

Figure 8.2 (Left) Effect of demagnetizing fields on the hysteresis loop shape; (continuous line): $H_d = 0$, (dashed line): $H_d \neq 0$. (Right) Comparison between uncorrected (continuous line) and overcorrected S-shaped (dashed line) hysteresis loops.

obtained by assuming a mean demagnetizing field $H_d = -N_\perp M_\perp = -M_\perp$ (i.e., $N_\perp = 1$)[b,c] [1, 37, 38]. However, correcting the $M(M_r)$-H curves (*deshearing procedure*) by a demagnetization factor of 1 often causes the curve to shear and acquire a negative slope (S-shaped curves), thus indicating that the actual demagnetizing factor is smaller than 1 (Fig. 8.2, right). The deviation of $N_\perp$ from unity is first ascribed to the granular nature of thin films used in PMR media, which don't consist of a single layer but rather of an assembly of single-domain grains. Basically, in such granular films, the continuous medium approximation cannot be applied due to the discrete nature of magnetization reversal of each grain. To take into account the granularity of the material, the total demagnetizing field on a given grain is separated into two contributions [1, 37, 38]: a first term corresponding to the grain's self-demagnetizing field, which is proportional to M_s and generally included in the magnetic anisotropy constant as a shape contribution, that is, $K_s^{grain} = \mu_0 M_s^2(n_{//} - n_\perp)$ where $n_{//}$ and $n_\perp$ are the in-plane and perpendicular demagnetizing factors of a cylindrical grain [40], and a second term that depends on the magnetization M, which accounts for the effect of all the other grains. For ideally fully exchange-decoupled granular thin films with a magnetic volume fraction x close to 100%[d], and assuming that magnetostatic collective effects in magnetization reversal are negligible, simple considerations show that along the normal to the film plane, the M-dependent contribution to the demagnetizing fields can be approximated to $H_d(M) = -(xN_\perp - n_\perp)M$. The existence of intergranular exchange interactions opposing to the demagnetizing fields makes the switching more complex, as they lead to collective reversal processes where the switch of a grain triggers the reversal of one or more neighboring grains. If the interactions are weak enough not to significantly affect the reversal

---

[b] The correction has no effect on H_c because the magnetization and hence the mean demagnetizing fields are zero at the coercivity.

[c] When the external field is applied at an angle ϕ with respect to the easy axis, the vector sum of $H(\phi)$ and $H_d = -N_\perp M_\perp$ causes a change not only in the magnitude but also in the direction of the internal field, especially at low H values [39].

[d] The parameter x accounts for the presence of a nonmagnetic matrix separating the magnetic grains, such as SiO_2 in the CoCrPt-based PMR media.

mechanism and assuming a larger $n_\perp$ factor, corresponding to a cluster of grains switching collectively, all the above considerations can be assumed to be still valid and a further reduction of the demagnetizing factor can be observed [1]. If the exchange interactions can be treated as mean fields, their effect is generally assessed by introducing the concept of an effective demagnetizing factor N_e (also referred to as *mean interaction factor*), which encompasses both magnetostatic and exchange interactions. For exchange-decoupled or weakly coupled granular systems, N_e can be experimentally evaluated by using the methods described later in Section 8.2.2. However, one has to keep in mind that large cooperative effects caused by either strong exchange or dipolar interactions significantly modify the reversal mechanism; in this case, the mean-field approximation fails and the demagnetizing factor is no more a constant, but rather it becomes a function of the specific magnetic domain configuration. In this case, numerical simulations are necessary for a rigorous analysis [37].

8.2.1.3 Magnetic anisotropy

Studying magnetic anisotropy is a key issue for the characterization of materials for PMR media. It must be uniaxial and directed normal to the film plane (i.e., perpendicular magnetic anisotropy [PMA] [41])[e] besides being large enough to support a perpendicular component of magnetization against the large sample demagnetizing fields that give rise to a shape anisotropy[f] whose constant is given by $K_s = \mu_0 N_e M_s^2/2$. The uniaxial anisotropy constant (K_u, of the order of 10^5 J/m^3 in CoCrPt-based PMR media and as large as 10^6 J/m^3 for next-generation $L1_0$–FePtX storage media [16]) is a critical parameter defining the recording performance since it determines (i) the maximum switching field, that is, $H_{sw} \propto H_K = 2K_u/\mu_0 M_s$, and (ii) thermal stability, which requires $\Delta E_0 \geq 60\ k_B T$ for 10 years'

[e] The main source of PMA in granular films for PMR media is the magneto-crystalline anisotropy (MCA) that depends on the spin–orbit coupling [41]. However, as highly packed grains are achieved in future ultrahigh-density recording media, a surface anisotropy term, which opposes MCA, should be taken into account, to determine the effective uniaxial anisotropy [42].

[f] It is possible to define a quality factor $Q = K_u/K_s$ that must be bigger than 1 for a medium capable of supporting a perpendicular component of demagnetization [43].

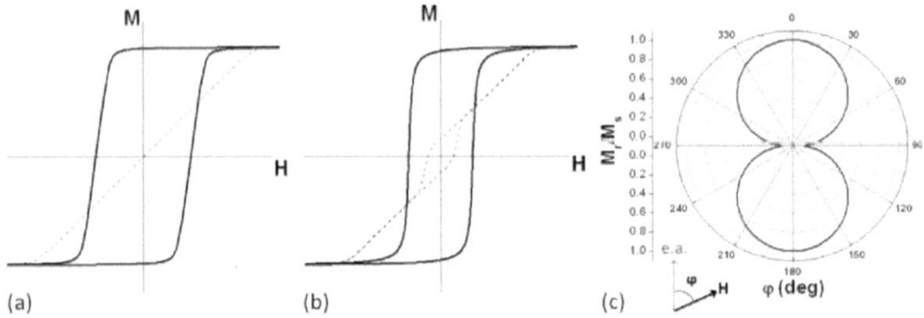

Figure 8.3 (a, b) Typical magnetization curves of a granular film with uniaxial PMA measured along the film plane, that is, the hard axis (dashed line), and the normal to the surface, that is, the easy axis (continuous line): (a) perfect easy-axis orientation and (b) easy axis with a finite angular distribution and/or not perfectly aligned along the film normal. (c) Typical angular remanence curve (i.e., M_r/M_s vs. φ) in a polar plot showing the twofold symmetry of a system with uniaxial PMA.

data retention [44], where $\Delta E_0 = K_u V$ is the energy barrier at a zero external field (see also Section 8.2.4), V the volume of one grain in the absence of interactions, and k_B the Boltzmann constant.

As a first step, the uniaxial anisotropy symmetry must be verified. This can be done in different ways, including M–H curves, torque measurements and angular remanence curves, to name just a few. The most common approach consists of comparing the magnetization curves measured along the directions parallel and perpendicular to the easy axis (Fig. 8.3a). The M–H curve along the easy-axis direction (normal to the surface) is expected to exhibit an almost square loop (at least after correcting for H_d), while the curve measured along the hard-axis direction (i.e., on the film plane) should increase almost linearly before reaching the saturation [45]. If the easy axis is not perfectly aligned along the perpendicular direction and/or a certain angular dispersion is present, the hard-axis magnetization curve can show a finite hysteresis [30] as depicted in Fig. 8.3b. Information about the anisotropy symmetry can be also derived by torque measurements [46] (see below) or by collecting the angular dependence of the reduced remanence M_r/M_s (Fig. 8.3c). For uniaxial systems, the latter can be described, to a first approximation, by a $|\cos\varphi|$ function (where φ is the angle between H and the easy axis) showing a maximum (ideally $M_r/M_s = 1$) and a minimum (ideally $M_r/M_s = 0$) in correspondence to the easy- and hard-axis directions, respectively [43 47].

Owing to the relevant role of the magnetic anisotropy, several experimental methods have been proposed to determine the value of the anisotropy constant, which exploit either the quasi-static (e.g., magnetometric measurements) [15, 41, 45, 46, 48–54] or the dynamic response (e.g., ferromagnetic resonance measurements [55, 56], Brillouin light scattering [57]) of a magnetic system. In the following section, we will briefly introduce some of the most common magnetometric approaches exploiting the quasi-static response, that is, analysis of easy- and hard-axis magnetization curves [41, 45, 46, 48–50], torque curve measurements [46, 48, 51], and the *Miyajima* (also referred to as *45-degree*) method [15, 52] (Fig. 8.4).

The uniaxial magnetic anisotropy constant K_u can be directly obtained by the area enclosed between the anhysteretic magnetization curves (simply obtained by averaging the upper and lower branches of the magnetization curve) measured along the easy (e) and hard (h) axes [41, 45, 49] (Fig. 8.4a), that is, $K_u' = \int_0 M_s \mu_0 \left(H^h - H^e\right) dM$, where K_u' is an effective anisotropy constant including also the shape anisotropy constant, which must be properly removed to obtain K_u. Besides determining the total uniaxial anisotropy, the second- and fourth-order K_1 and K_2 anisotropy constants (of which K_u is composed) can be individually evaluated by replotting the hard-axis curve as [45, 48, 50]

$$\frac{H}{m} = \frac{2K_1'}{\mu_0 M_s} + \frac{4K_2}{\mu_0 M_s} m^2 \tag{8.1}$$

where $m = M/M_s$ is the reduced magnetization and K_1' is the second-order anisotropy constant including the shape anisotropy (*Sucksmith–Thompson* method). The H/m versus m^2 plot is expected to be linear and the constants K_1 and K_2 can be determined by the intercept and the slope, respectively. The knowledge and control of higher-order uniaxial anisotropy terms is an important issue in modern magnetic recording technology as it was found that increasing K_2 allows the resistance to thermal agitation to be improved without significantly increasing the switching field [58].

The main disadvantage of these methods is related to the need of a strong external magnetic field to saturate the film along the hard-

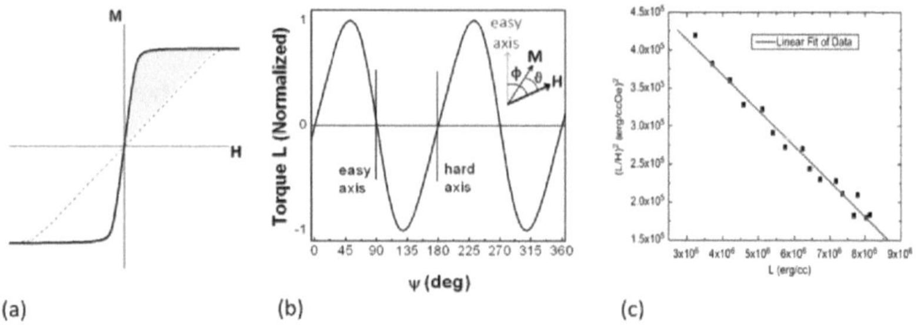

Figure 8.4 Comparison of three different methods for anisotropy measurements. (a) Area method; continuous and dashed lines represent the anhysteretic easy- and hard-axis magnetization curves, respectively. (b) Torque curves. (c) Miyajima method; plot of $(L/\mu_0 H)^2 (= M_\perp^2)$ versus L for a granular $L1_0$–FePt:TiO_2 film from which K_u is evaluated by extrapolation to high fields. Reprinted from Ref. [15], Copyright (2013), with permission from Elsevier.

axis direction (especially for high-anisotropy materials such as the $L1_0$–FePtX alloys).

One of the most popular methods to measure magnetic anisotropy involves the use of torque curves, which consist of a plot of the torque required to move the magnetization away from the easy direction as a function of the applied field angle (Fig. 8.4b) [46, 48, 51]. Under the framework of the single-domain model for uniaxial systems, the total energy density can be written as a sum of the Zeeman and anisotropy energy terms:

$$E/V = -\mu_0 H M_s \cos\vartheta + K_1' \sin^2(\varphi - \vartheta) + K_2 \sin^4(\varphi - \vartheta) \quad (8.2)$$

where ϑ is the angle between M and H and φ is the angle between H and the easy axis (see inset in Fig. 8.4b). At equilibrium, the total energy density is minimized, that is, $\frac{1}{V}\partial E/\partial\vartheta = 0$; hence

$$\mu_0 H M_s \sin\vartheta = K_1' \sin 2(\varphi - \vartheta) + 2K_2 \sin 2(\varphi - \vartheta) \sin^2(\varphi - \vartheta) \quad (8.3)$$

where the first term is the torque (per unit of volume) exerted on **M** by **H**, while the second term is the anisotropy torque acting to rotate **M** toward the easy axis. A torque magnetometer is generally used to measure the torque L that balances the anisotropy torque, from which one derives $\vartheta = \sin^{-1}(L/\mu_0 H M_s)$ and the angle $\psi = \varphi - \vartheta$ between M and the easy axis, φ being known from measurements. Plotting L as a function of ψ and fitting the curve to Eq. 8.3 allows the anisotropy constants to be determined. If the external field is sufficiently high to bring the sample to saturation for each angle (i.e., $\vartheta = 0 \rightarrow \psi = \varphi$), Eq. 8.3 reduces to

$$L = -K_1' \sin 2\psi - 2K_2 \sin 2\psi \sin^2 \psi \tag{8.4}$$

thus simplifying the analysis of the torque curves and making the single-domain assumption more reasonable.

Note also that a torque curve provides information about the position of the easy- (hard-) axis, which corresponds to the point at $L = 0$ where the slope of the curve is negative (positive).

It is worth mentioning that the term $M_s \sin \vartheta$ in Eq. 8.3 corresponds to the component of magnetization perpendicular to the applied field ($M_\perp$). This means that measuring $M_\perp$ by using a vector magnetometer allows evaluating $\mu_0 H M_\perp$, which is numerically equal to the torque. The main weakness of this approach is that in the high-field region, where the best results are expected, ϑ and then $M_\perp$ are small and therefore more difficult to be precisely measured.

The requirement of strong external fields, which is a characteristic of the above-mentioned methodologies, can be in principle bypassed by using the Miyajima [15, 52] and rotational methods [53, 54]. In the former case (Fig. 8.4c), the system is first brought into its remanent state by applying a high saturating field along the easy-axis direction and subsequently reducing the applied field to zero; hence, an external field is applied at $\varphi = 45°$. The applied field is supposed to exert a torque, which forces the rotation of the magnetization toward the field direction, while the perpendicular component of the field (also referred to as stabilizing component) is assumed to keep the system in a single-domain state. Starting from Eq. 8.3 and neglecting the K_2 anisotropy constant, a simple analysis for $\varphi = 45°$ shows that

$$L = \mu_0 H M = K_1' \left(1 - 2\frac{M_\perp^2}{M_s^2}\right) \tag{8.5}$$

In the ideal case, the torque L has a linear dependence on M, and K_1' can be evaluated by extrapolating to $M^2 = (L/\mu_0 H)^2 = 0$; that is, high saturation fields are not necessary, as the response of the magnetization at low fields can be used to obtain K_1' by extrapolation. Micromagnetic analyses have shown that such a method, which is based on the single-domain model, is valid only in systems with sufficiently strong anisotropy or intergranular exchange interactions [52]. Indeed, they require lower stabilizing fields to suppress the formation of domains and maintain a uniform magnetization, as assumed in Eq. 8.5. In all the other cases, the linear relation between L and $M_\perp$ is not fully respected anymore and the Miyajima method should be used prudently.

8.2.2 *Determination of Switching-Field Distribution*

One of the key factors determining the overall performance of granular films for PMR media is the SFD, that is, the dispersion of the fields necessary to switch the magnetization of individual grains. The presence of grains with a high switching field reduces the resolution of the written transitions, thus limiting the achievable storage density, while grains with a low switching field that can reverse easily will reduce the overall thermal stability [4, 59]. It follows that the SFD must be as narrow as possible to ensure good transition resolution and high thermal stability.

Recording being a local process, since a highly localized field is used to switch a recording bit, the relevant quantity is the local SFD $D_l(H_{sw})$ observed in a narrow region defined by the write head [60]. A local distribution of switching fields can arise from either a dispersion of the intrinsic material properties, such as the volume of the grains, the magnetic anisotropy constant, and the angle of the easy axis[g], or the local interactions that also depend on the nearby magnetic structure, that is, the specific bit pattern. $D_l(H_{sw})$ cannot be directly determined by macroscopic magnetometric measurements, since in this case a uniform field is applied to the whole system; however, under certain assumptions, macroscopic magnetic measurements (including ferromagnetic resonance analysis [56], recoil curve measurements [60, 64–87], and complex transverse alternating current (AC) susceptibility technique [88]) can be used

[g]The distribution of the easy-axis orientation (around 3° in the most recent devices [16]) is a central factor in determining the SFD since even a small deviation from the perpendicular direction causes a large variation of H_{sw}. To a first approximation, it can be evaluated by studying the angular dependence of remanence [61], or even better by the variation of coercivity with the angle [62, 63].

to extract a parameterized intrinsic SFD which is independent on the mean interaction fields and the specific bit pattern itself. Once the $D_i(H_{sw})$ parameter is determined, it can be inserted in the micromagnetic simulations to fully investigate the recording process.

Among the different methodologies proposed to determine the intrinsic SFD, here we will review some of the most popular magnetometric methods that exploit recoil loop measurements [60, 65–87]. Starting from a positive saturated state, a recoil loop is collected by applying a negative reverse field H_r and measuring the magnetization by cycling the field from H_r to saturation and back to H_r; the procedure is then repeated with increasing H_r to collect a series of recoil loops[h] (Fig. 8.5 left). In the literature, these measurements are also referred to as minor loops measures, while the return curve from the reverse point is called recoil curve or first-order reversal curve (FORC).

Note that for proper determination of the SFD, thermal effects, which cause a broadening of the SFD [27, 72], must be considered, as discussed later in Section 8.2.4.

Assuming that the interactions can be treated as mean fields, the intrinsic SFD can be determined, to a first approximation, from the $M(M_r)$–H curves after correction for a mean interaction field $H_i\,(M) = -N_e M$. Differentiating the desheared curves allows $D_i(H_{sw})$ (Fig. 8.5, right), which can be quantified by the standard deviation σ, to be determined from the full width at half maximum (FWHM) of the curve or, even better, by the ratio σ to the mean switching field, that is, $\sigma_{H_{sw}}/H_{sw}$. Typical values of $\sigma_{H_{sw}}/H_{sw}$ are around 0.2 [75].

One of the approaches proposed to determine N_e [79, 80] is based on the assumption of a log-normal distribution of $D_i(H_{sw})$ and exploits a convergent iterative procedure with feedback (explained in detail in Ref. [79]), which repetitively interpolates the recoil

[h] Although the recoil loop repeatability is a generally assumed property of a magnetic material, one must be aware that not all the systems fulfill this condition. For example, it does not apply to very thin perpendicular oriented films and multilayers with strong lateral exchange coupling, for which the protocol must be applied with caution [5, 89].

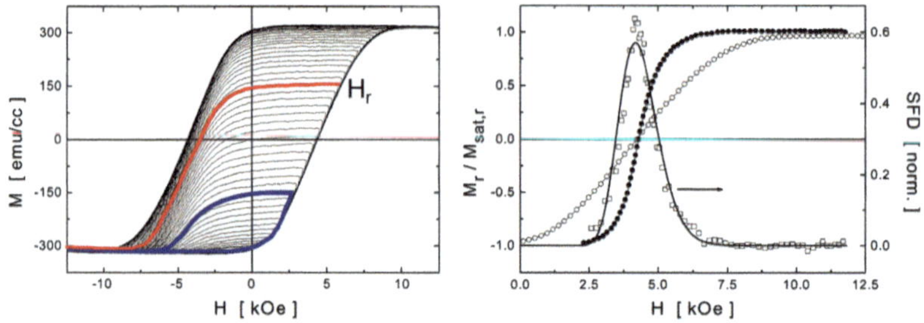

Figure 8.5 (Left) Set of recoil curves of a CoCrPt film. For sake of clarity, a recoil (or minor) loop (blue line) and a recoil (or FORC) curve (red line) are highlighted. (Right) Normalized DCD curves extracted from the original (○) and corrected (•) recoil curves together with the SFD curve (□) obtained from the remanence curve corrected for the demagnetizing fields according to the approach proposed in Ref. [79]. © 2003 IEEE. Reprinted, with permission, from Ref. [79].

curves to found the value N_e that provides the best fitting of the intrinsic SFD (Fig. 8.5).

To relax the requirement of a predetermined functional form of $D_i(H_{sw})$, a different approach can be used [81, 82] (Fig. 8.6). The method is based on the assumption of a granular system consisting of individual grains, each one reversing at a given field h_c. Within this framework, each recoil loop starting at a given reverse field H_r is associated with the reversal of P_1 grains with coercivity $h_c \leq H_r$, while remaining unchanged the magnetization of the rest of the grains (P_2) with coercivity $h_c > H_r$. For a given recoil loop, at the points A^+ and A^-, which divide the minor loop into two equal parts, half of the P_1 grains has a moment $+m$, while the other half has a moment equal to $-m$; as a consequence, the mean interaction field, $H_i(M)$, at these points due to the P_1 grains is zero and is only determined by the P_2 grains, whose magnetization is unchanged during the reversal, thus leading to the same value of $H_i(M^A)$ in A^+ and A^-, where M^A is the magnetization at the A points. Since H_{int} is the same (with opposite in sign) in A^+ and A^-, it follows that $H\left(A^+\right) + H\left(A^-\right) = -2H_i\left(M\right) = 2N_eM$; hence, measuring a set of recoil loops and plotting $H_i(M^A)$ (in Fig. 8.6c reported as H_D^{eff}) as a function of M^A allows N_e to be determined from the slope. This

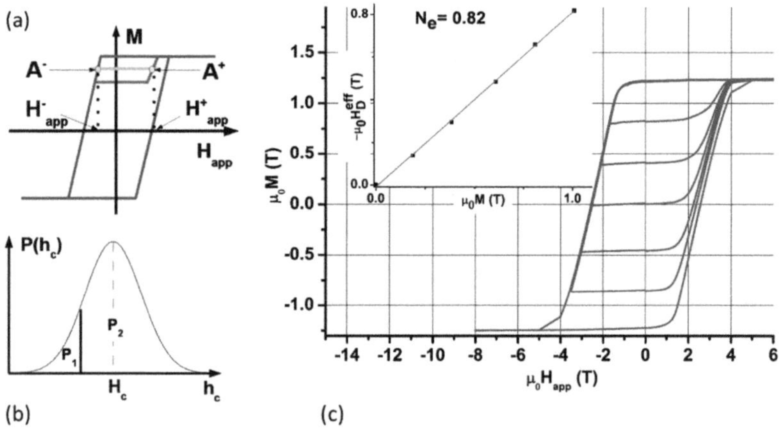

Figure 8.6 Overview of the experimental method proposed in Ref. [81] for the determination of the SFD. (a) Schematic representation of major and minor loops reporting the magnetic states A^+ and A^- and the corresponding fields. (b) Schematic drawing of the coercivity distribution $P(h_c)$, the left part representing the P_1 grains that reverse during a recoil measurement. (c) Experimental major and recoil loops for a hard film with PMA; inset: extracted dependence of the mean interaction field H_i (in the figure reported as H_D^{eff}) on the magnetization. Reprinted with permission from Ref. [81]. Copyright 2010, AIP Publishing LLC.

method has also the intrinsic advantage that the assumption of a constant value of N_e can be verified.

The above methodologies are based on the mean-field approximation and assume a constant N_e factor, thus restricting its applicability to exchange-decoupled or weakly coupled systems.

A more refined measure of $D_i(H_{sw})$ can be achieved by using the $\Delta H(M, \Delta M)$ methodology [60, 66, 68–76], which is a generalization of the ΔH_c method proposed by Tagawa and Nakamura [77]. The method is based on the interacting hysteron model[i] and assumes a mean interaction field $H_i(M)$, which is a function of the magnetization M. It relays on the measure of the field difference ΔH at constant magnetization M between the major hysteresis loop

[i]The hysteron is a mathematical representation of a bistable switching unit corresponding to an individual grain in granular films for PMR media.

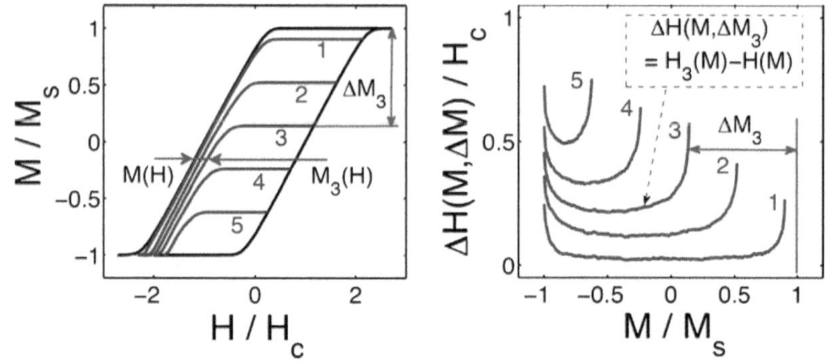

Figure 8.7 Overview of the $\Delta H(M, \Delta M)$ method for the determination of the SFD. (Left) Schematic representation of the major hysteresis loop and a series of recoil curves obtained after reversing the applied field at distance ΔM_i ($i = 1, \ldots, 5$) from saturation. (Right) Corresponding $\Delta H(M, \Delta M)$ curves. Reprinted with permission from Ref. [76]. Copyright 2009, AIP Publishing LLC.

branch and a set of recoil curves, each of them starting at a given distance ΔM from the saturated state (Fig. 8.7).

Assuming that M is normalized to M_s, the upper branch of the major hysteresis loop recorded under an external field H can be represented as

$$M(H) = 1 - 2\int_{-\infty}^{-(H+H_i)} D_i(H_{sw})dH_{sw} \tag{8.6}$$

while the recoil curve starting at $M(H_r, H \geq H_r) = 1 - \Delta M$ is given by

$$M(H_r, H) = 1 - \Delta M - 2\int_{-\infty}^{-(H_r+H_i)} D_i(H_{sw})dH_{sw} \tag{8.7}$$

Defining a distribution integral function $I(H) = \int_{-\infty}^{H} D_i(H_{sw})dH_{sw}$, one derives

$$H(M) = -I^{-1}\left(\frac{1-M}{2}\right) - H_i(M) \tag{8.8}$$

$$H_r(M) = -I^{-1}\left(\frac{1-M-\Delta M}{2}\right) - H_i(M) \tag{8.9}$$

where I^{-1} is the inverse of the integral function.

Within the frame of the mean-field approximation, the $H_i(M)$ terms in Eqs. 8.8 and 8.9 are the same for identical M values, and a field separation ΔH can be derived as

$$\begin{aligned}\Delta H(M, \Delta M) &= H_\mathrm{r}(M) - H(M)\\ &= I^{-1}\left(\frac{1-M}{2}\right) - I^{-1}\left(\frac{1-M-\Delta M}{2}\right)\end{aligned} \quad (8.10)$$

which is independent of the main intergranular interactions, thus allowing a direct experimental access to $D_i(H_{sw})$. Although the $\Delta H(M, \Delta M)$ method relaxes the linear assumption between the mean interaction field and the magnetization (assumed in the previously reported approaches), it is still based on the mean-field approximation, as it assumes that each M value corresponds to the same mean internal field independently of the specific microscopic magnetization configuration.

From the $\Delta H(M, \Delta M)$ curves, several key features of the intrinsic SFD can be determined by a least-squares fitting against the inverse of a parameterized Lorentzian or Gaussian distribution function of the form σF, where F is a function that includes detailed information about the shape of D_i (H_{sw}), as faithfully described in Ref. [74].

The reliability range of the mean-field approximation can be checked by analyzing the quality of the fitting through the multiple correlation coefficient R^2, which is equal to 1 for perfect data fitting. Besides using the conventional quality measures for numerical fit, a self-consistent procedure was also proposed, which is based on data redundancy among multiple recoil curves, as well explained in [69, 70, 74, 75]. Numerical simulations have proved that the $\Delta H(M, \Delta M)$ approach is capable of accurately identifying the intrinsic SFD as long as the exchange and magnetostatic interactions are of comparable strength. However, the approach fails if the two opposite interactions do not compensate, the effect being worse when the exchange coupling dominates [68, 69, 71, 75] (Fig. 8.8). On the other hand, an extension of the approach, referred to as the *reference function* $\Delta H(M, \Delta M)$ method, was demonstrated to allow an accurate identification of $D_i(H_{sw})$ also in the case of large exchange interactions up to the point they become strong enough to lead to an almost fully correlated magnetization reversal [74, 75].

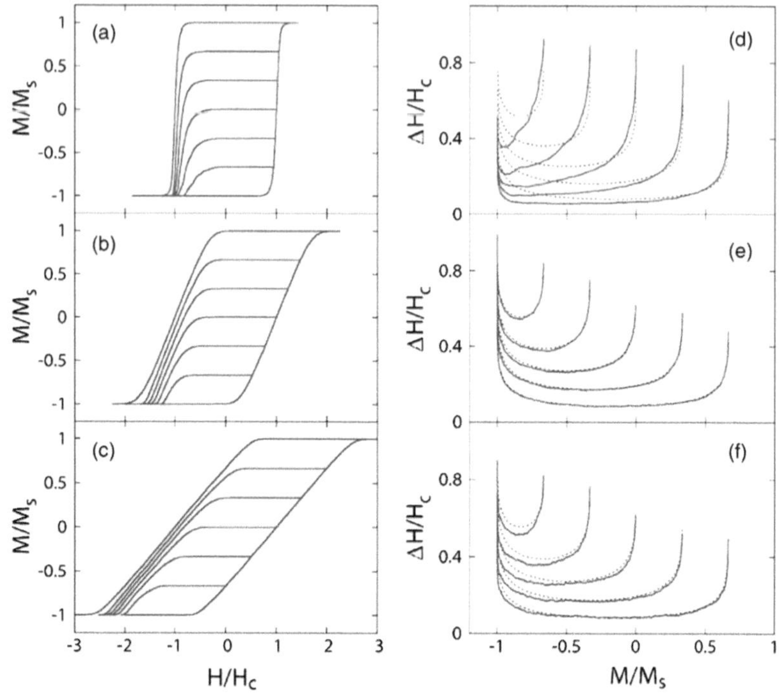

Figure 8.8 Role of exchange (J_{ex}) and magnetostatic (J_{ms}) interactions on the reliability of the $\Delta H(M, \Delta M)$ method: (a, d) $J_{ms} = 0$, $J_{ex} \neq 0$; (b, e) $J_{ex} = J_{ms}$; (c, f) $J_{ms} > J_{ex}$ (Left) Numerical simulated major hysteresis loop and recoil curves. (Right) Corresponding $\Delta H(M, \Delta M)$ curves: numerical results (solid line), mean-field approximation (dotted line). Reprinted with permission from Ref. [71]. Copyright 2008, AIP Publishing LLC.

An alternative approach to extract information on the intrinsic SFD is the FORC method [83, 86, 87], which exploits a mathematical procedure to generate from a large set of FORCs a 2D (FORC) diagram, which provides a mapping of reversal events in hysteretic systems. Due to the complexity of the FORC data analysis, we illustrate here the fundamental aspects of the technique only, showing a few examples; the reader is referred to Refs. [90–95] for a detailed discussion.

A FORC diagram is obtained from a family of FORCs (Fig. 8.9a), of the order of 100, collected starting from different reverse fields H_r, with equal spacing, then sweeping the field back to saturation in a

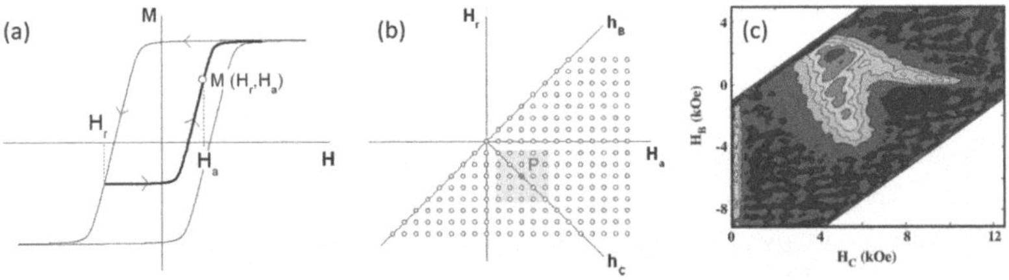

Figure 8.9 Overview of the FORC technique. (a) Definition of a FORC and related parameters. (b) Matrix of $M(H_r, H_a)$ values in the (H_r, H_a) and 45° -rotated (h_C, h_B) field space, where $h_C = (H_a - H_r)/\sqrt{2} = H_C\sqrt{2}$ and $h_B = (H_a + H_r)/\sqrt{2} = H_B\sqrt{2}$. (c) Typical FORC diagram of a granular film for PMR media in the (H_C, H_B) reference system. © 2011 IEEE. Reprinted and adapted, with permission, from Ref. [96].

series of regular H_a steps. The field steps ΔH_r and ΔH_a are chosen in order to find the best compromise between the FORC diagram precision and the measuring time[j] [94]. The magnetization value at the applied field H_a on the FORC with a reverse field H_r is indicated by $M(H_r, H_a)$, where $H_a \geq H_r$. The set of the $M(H_r, H_a)$ values forms a magnetization matrix in the (H_r, H_a) field space (Fig. 8.9b), from which a FORC distribution is determined as the mixed second-order derivative of $m(H_r, H_a) = M(H_r, H_a)/M_s$, that is,

$$\rho(H_r, H_a) = -\frac{1}{2}\frac{\partial^2 m(H_r, H_a)}{\partial H_r \partial H_a}, \quad \text{with} \quad H_a \geq H_r \tag{8.11}$$

Note that Eq. 8.11 does not capture the reversible contributions and provides a full mapping of only irreversible switching events. However, it has been shown that the reversible contributions can be accounted for by means of the *extended* FORC method described in Ref. [90].

To calculate $\rho(H_r, H_a)$ at any point P in the (H_r, H_a) plane, a least-squares fitting of $M(H_r, H_a)$ is accomplished on a square grid of points around P (Fig. 8.9b). Therefore, the magnetization is commonly fitted with a second-order polynomial function of the form

[j]To optimize the measure, the values of ΔH_r and ΔH_a can be changed to improve the accuracy in different regions, but the ratio $\Delta H_r/\Delta H_a$ must be maintained between 1 and 2.5 [94].

$$a_1 + a_2 H_a + a_3 H_a^2 + a_4 H_r + a_5 H_r^2 + a_6 H_a H_r \quad (8.12)$$

where $-a_6/2$ is the value of $\rho(H_r, H_a)$ at P. The size of the grid is given by $(2SF + 1)^2$, where SF is the smoothing factor, which is typically set in between 2 and 5, in order to reduce the experimental noise (the larger the noise, the higher the SF must be) without causing a smoothing of important features of the distribution. The final FORC diagram is a contour plot representation of $\rho(H_r, H_a)$, which is commonly presented in a more convenient set of coordinates, that is, $H_C = (H_a - H_r)/2$ and $H_B = (H_a + H_r)/2$ (Fig. 8.9c).

For a granular film for PMR media, consisting of a set of grains with a narrow perpendicular easy-axis distribution (within a few degrees, $\sim 3°$), the FORC diagram can be interpreted, to a first approximation, within the frame of the *classical Preisach* model [91], where the granular film is assumed to behave as a collection of interacting hysterons. According to the model, H_C and H_B are related to the intrinsic switching field and the local interaction field among individual hysterons, respectively. Any distribution of the switching and local interaction fields (the latter assumed constant at a given location but different from site to site) will cause a spread in the diagram contour along the H_C and H_B axes, respectively. Although the assumptions of the classical Preisach model are compatible with the properties of granular films for PMR media, a more extensive picture is obtained by using the *generalized moving Preisach* model, which includes also a mean interaction term, incorporating both magnetostatic and exchange interactions [87, 91, 97]; in this case H_B represents the quantity by which the interaction field locally deviates from the mean interaction field.

Before discussing how a FORC diagram can be used to extract information about $D_i(H_{sw})$, it is instructive to have a look first at typical contours for selected simple cases under the mean-field approximation (Fig. 8.10). In the absence of interactions but with a finite distribution of switching fields, the FORC diagram would be a narrow horizontal band on the $H_B = 0$ axis (Fig. 8.10a). When only magnetostatic interactions are present, for a very narrow SFD the FORC diagram would be a vertical band parallel to the H_B axis (Fig. 8.10b), while it would evolve toward a wishbone shape with the increase of the SFD (Fig. 8.10c), showing a strong peak at a positive

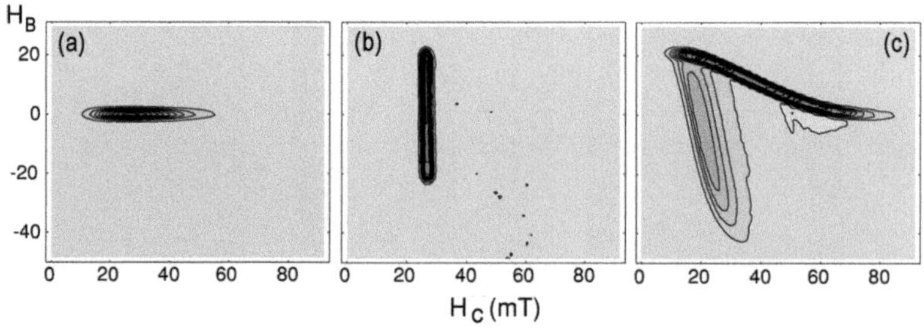

Figure 8.10 Simulated FORC contours for a collection of interacting hysterons under the mean-field approximation. (a) Magnetostatic interaction $J_{\mathrm{ms}} = 0$ and broader SFD. (b) $J_{\mathrm{ms}} \neq 0$ and narrow SFD. (c) $J_{\mathrm{ms}} \neq 0$ and broader SFD. Reprinted figure with permission from Ref. [90]. Copyright (2005) by the American Physical Society.

H_B value in correspondence of the intercept between the upper and vertical branches. Note that the presence of exchange interactions generally causes a shift of the contour toward negative H_{B} values [98].

For granular films with moderate intergranular interactions, if the factor N_{e} is known, the effect of the mean interaction field can be removed by the measured FORCs, applying the deshearing procedure previously described (Section 8.2.1.2) [87]. Numerical simulations and simple analyses show that compensating for the mean interaction effects, the FORC diagram would assume the shape expected in the frame of the classical Preisach model, that is, a contour plot symmetric with respect to the H_{C} reference axis (or equivalently the second bisector in the (H_{r}, H_{a}) plane, Fig. 8.11). This statement suggests that the effective N_{e} value can be determined by correcting the experimental FORCs for different N_{e} values, until the value that causes a symmetric behavior is found. A refined procedure to exactly evaluate N_{e} is described in Ref. [87]. When the correction is performed, the FORC diagram allows determining of D_i (H_{sw}), corresponding to the projection of the distribution along the H_{C} reference axis (or equivalently along the second bisector in the (H_{r}, H_{a}) plane), that is, $\int \rho(H_{\mathrm{C}}, H_{\mathrm{B}}) dH_{\mathrm{B}}$. Moreover, the FORC diagram can be also used to extract information about the distribution of local interaction fields corresponding to the projection along the H_{B} reference axis (or equivalently the

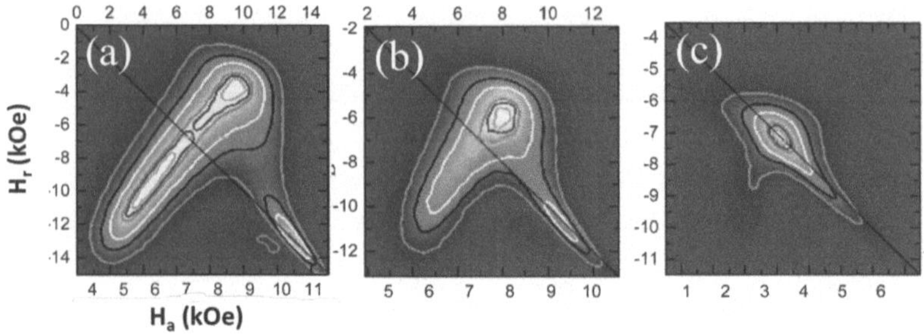

Figure 8.11 Evolution of the FORC diagram in the (H_r, H_a) plane of a CoCrPt:oxide granular film for PMR media after correction for different N_e values. The contour changes from a wishbone (a, b) to an elliptical shape (c) along the second bisector when N_e is varied from 0 to the right value. Reprinted with permission from Ref. [87]. Copyright 2011, AIP Publishing LLC.

first bisector in the (H_r, H_a) plane), that is, $\int \rho(H_C, H_B) dH_C$. In a granular film for PMR media, the distribution of local interaction fields originates from local fluctuations on the intergrain boundaries as well as from a variance of the geometrical arrangement of neighboring grains. Besides the intrinsic SFD, the local interaction field distribution is a useful parameter to characterize the magnetic behavior of a granular film, as, for example, it is expected to be larger for systems with smaller intergrain spacing where the effects of local inhomogeneity are enhanced [65, 87].

Although the FORC technique has recently become more and more popular, it must be stressed that it mainly represents a mathematical data transformation tool and it does not allow a self-consistent determination of microscopic information as it can be done by using the previously reported methods. In particular, a correct interpretation of the FORC diagram to get Preisach-type information (i.e., distribution of coercive and interaction fields) needs a careful investigation of the FORC structure (e.g., by using simulation tools) to correctly unfold the information it contains [95].

8.2.3 *Study of the Reversal Mechanism*

To improve the materials properties of a data storage medium a full comprehension of the magnetization reversal process is mandatory.

The reversal of granular films for PMR media basically occurs following a *coherent* switching mechanism [4], but the presence of intergrain exchange interactions [1] gives rise to *incoherent* reversal contributions whose extent depends on the strength of the coupling [5, 10, 20, 30, 99–106].

The most popular method to investigate the magnetization reversal mechanism is through the analysis of the angular dependence of the switching field (i.e., H_{sw} vs. φ, φ being the angle between **H** and the easy axis), which should exhibit, in a simplified fashion, a behavior between the two limits of the S-W (coherent reversal) [19] and *Kondorsky* (incoherent reversal) [107, 108] models (Fig. 8.12a). According to the S-W model, which applies to noninteracting single-domain and uniaxial particles having a common anisotropy axis, the angular dependence of the S-W switching field H_0 is given by $H_0(\varphi) = H_0(0°)/\left[\cos(\varphi)^{2/3} + \sin(\varphi)^{2/3}\right]^{3/2}$, that is, H_0 is maximum when the applied field is parallel or perpendicular to the easy axis ($H_0\,(0°) = H_0\,(90°) = H_K$) and minimum when it is directed at 45° with respect to the easy axis ($H_0(45°)/H_K = 0.5$). For continuous or strongly interacting granular systems, the switching occurs by nucleation and propagation of domain walls, and it can be described by the Kondorsky model [107, 108] if the reversal is dominated by domain wall propagation, that is, the field necessary to move a domain wall (i.e., the pinning field, H_p) is higher than the nucleation field H_n. The model assumes that the external field is not large enough to induce the rotation of magnetization away from the easy axis, but its component along the easy axis can allow the movement of the domain walls among pinning sites. The model is then valid only if $H_p \ll H_K$ and held to a monotonic increase of the switching field ($\equiv H_p$) with $\cos(\varphi)^{-1}$, that is, $H_p(\varphi) = H_p(0°)/\cos(\varphi)$ with $H_p(60°)/H_p(0°) = 2$) [100, 107, 108]. It is worth mentioning that an S-W angular dependence of the switching field by itself does not univocally prove that the whole system is made of uniaxial and single-domain isolated grains; indeed, if $H_n > H_p$ (i.e., nucleation-dominated reversal), the reversal has its origin in a nucleation event with S-W characteristics immediately followed by domain wall motion [108]. This is found for example in bit-patterned systems with a dot size larger than the maximum value supporting a single-domain state (see Section 8.4.2).

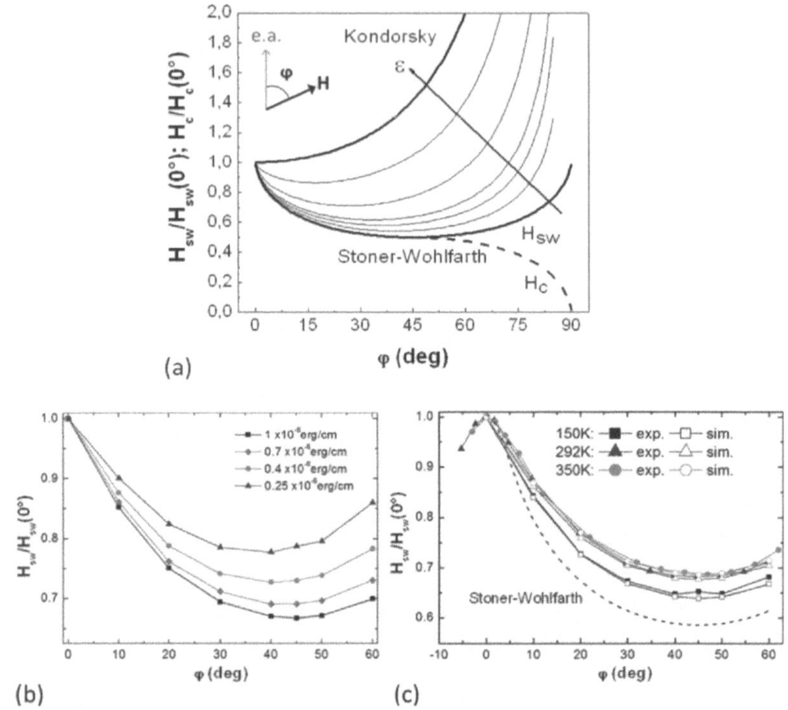

Figure 8.12 (a) Angular trend of normalized switching field ($H_{sw}(\varphi)/H_{sw}(0°)$) for S-W ($\equiv H_0(\varphi)/H_0(0°)$) and Kondorsky ($\equiv H_p(\varphi)/H_p(0°)$) switching models. Reversal curves obtained from a weighted average of the two-limit switching regimes are also shown ($\varepsilon \cos(\varphi)^{-1} + (1-\epsilon)\, H_{sw}^{S-W}(\varphi)$). For comparison, the angular trend of H_c for an S-W system is also reported. All the curves are calculated for $T = 0$ K and a 0° dispersion of easy axis. (b, c) Effect of the intergranular exchange coupling strength (in ergs/cm) (b) and temperature (c) on the angular dependence of switching field of CoCrPt:SiO_2 films. Reprinted with permission from Ref. [104]. Copyright 2011, AIP Publishing LLC.

Note that the different reverse modes do not result only in a change of the angular dependence of H_{sw} but also result in a reduction of the switching-field value along the easy-axis direction in the case of the Kondorsky reversal mechanism.

The angular dependence of the switching field can be determined by measuring a series of hysteresis loops at a different angle φ and estimating $\langle H_{sw} \rangle$ from the maximum of the susceptibility evaluated by the field derivative of the hysteresis loop [109]. It is worth noting that the switching field does not necessary correspond to H_c, especially in the high-φ region; for example, for an S-W system, the angular variations of H_c and H_0 are the same up to $\varphi = 45°$, while for $45° < \varphi < 90°$, $H_c(\varphi)$ and $H_0(\varphi)$ are mirrored with respect to the horizontal line at $H_K/2$ (Fig. 8.12a). An alternative and very common method to determine $H_{sw}(\varphi)$ consists of measuring a set of DCD curves at a different angle φ from which $\langle H_{sw} \rangle$ is evaluated from the point of zero remanent magnetization (i.e., H_{cr}) [10, 102–104, 106]. However, when using conventional DCD curves, where the remanence is measured along the applied field direction, $H_{cr}(\varphi)$ can lead to some nonintuitive behavior [100]. Indeed, the equivalence between H_{cr} and H_{sw} is in principle ensured at $\varphi = 0°$, while they are not necessarily the same at other angles, especially in the high-φ-angle region; for example, for an S-W system, H_{cr} goes to 0 as φ tends toward 90°, while H_0 increases up to its maximum value. A better correspondence between experimental data and simple models can be achieved by using a nonconventional procedure, where $H_{sw}(\varphi)$ is obtained by measuring the $H_{cr}(\varphi)$ trend from a set of so-called easy-axis DCD curves [20, 30, 99–101]. They are collected by using vector magnetometers that allow the component of the remanent magnetization along the easy-axis direction to be measured at each angle φ, starting from the same initial magnetization state that corresponds to the remanent state along the easy-axis direction. It is worth mentioning that for a better comparison between experimental and theoretical curves both thermal effects (see Section 8.2.4) and demagnetizing fields [43] (see Section 8.2.1.2) must be considered.

In granular films for PMR media, deviations from the S-W toward the Kondorsky switching behavior is generally assumed as an indication of the presence of exchange interactions [10, 20, 30, 99–106]. In particular, it was demonstrated that exchange interactions cause a symmetry breaking in magnetic reversal due to incoherency, together with a shift of the angle where the minimum switching field occurs (Fig. 8.12b). On the other hand, thermal effects and a finite SFD result in a decrease of the depth of the S-W minimum, without modification of the overall trend (Fig. 8.12c). Note that a symmetry breaking can be also observed in hard materials where the fourth-order anisotropy term cannot be neglected (e.g., $L1_0$–FePt alloy). In such a case a modified S-W functional form should be used to describe the reversal mechanism [32, 110].

Besides the investigation of thc angular dependence of the switching field, a wealth of useful information about the switching process can be achieved by analysis of the FORC diagram [86, 87, 98, 111, 112]. However, it is worth mentioning that a FORC contour cannot be correctly understood without the simulation of the diagram. An example of this will be given in Section 8.3.1 for a CoCrPt-based stacked system.

8.2.4 *Thermal Effects and Time-Dependent Measurements*

As learned from the previous paragraphs the magnetic properties of granular films for PMR media are governed by a complex interplay among intrinsic properties, magnetic interactions, and various sources of inhomogeneity present in the material. The magnetic behavior, and its comprehension, is further complicated by thermal effects, which may assist in overcoming the reversal energy barrier ΔE between two magnetization states [1, 4, 113]. According to the *Arrhenius–Néel formalism*, the probability per time unit of successfully crossing the energy barrier is a function either of the external field, which reduces ΔE, or of the temperature according to $\tau^{-1}(H, T) = f_0 e^{-\Delta E(H)/k_B T}$, where τ is the relaxation time and f_0 is the attempt frequency for barrier crossing evaluated between 10^{11} and 10^{12} Hz for granular films in PMR media [44]. It follows that at finite temperature the magnetization will experience, even in a zero external field, a time (t) variation, which is expressed, for a single energy barrier as $M(t) = A + Be^{(-t/\tau)}$, where A and B are time-independent constants. Thermally activated relaxation phenomena are particularly relevant in magnetic recording as they are responsible for the worsening of stability of stored information, thus limiting the maximum achievable recording density [1, 4]. The thermal energy ($k_B T$) and the magnetic energy stored in a grain (ΔE_0, that is, the energy barrier at $H = 0$) are encompassed in the so-called thermal stability factor $\Delta E_0 / k_B T$, which must be as large as 60 to ensure a 10-year data retention [44]. In addition, the time-dependent behavior of magnetization also leads to a variation of magnetic behavior with the field sweep-rate [14, 27, 30, 72–74, 82, 100, 114–120]. As a consequence, the relevant properties derived from measurements performed at typical laboratory time scales (of the order of seconds)[k] are different than those observed at the short time scales of the recording process (of subnanosecond order), and methods for extrapolating between these two very different time scales are necessary to properly evaluate the recording performance of a given material. Ultimately, thermal effects can also limit the accuracy of the methods proposed in the previous paragraphs, most of them being based on models that ignore thermal relaxation phenomena.

Dynamic magnetic measurements are exploited to gain insight into thermal effects as well as to determine parameters of interest for magnetic recording. A technique frequently used to investigate the dynamic behavior consists of measuring a family of hysteresis loops at different sweep rates or, equivalently, a set of recoil loops and remanence curves for various waiting times at the reversal points [14, 27, 30, 73, 82, 100, 114–117, 120, 121] (Fig. 8.13 left). The recorded data are then analyzed under the model developed by Sharrock [120], which follows the Arrhenius–Néel formalism and assumes an S-W model of reversal, for which $\Delta E = \Delta E_0[1 - \mathrm{H}/\mathrm{H}_0(\varphi)]^n = K_\mathrm{u} V[1 - \mathrm{H}/\mathrm{H}_0(\varphi)]^n$, where the exponent n is a parameter that depends on the orientation of the easy axis with respect to the external field. Within the framework of the Sharrock model, one can derive a working equation of the form $\mathrm{H}_\mathrm{sw}(t) = \mathrm{H}_0 \left\{ 1 - \left[\frac{k_\mathrm{B}T}{\Delta E_0} \ln\left(\frac{f_0 t}{\ln 2}\right)\right]^{1/n} \right\}$, which can be used to fit the trend of switching fields obtained from dynamic measurements to extract both the thermal stability factor $\Delta E_0/k_\mathrm{B}T$ and the H_0 values (Fig. 8.13, right). Note that in real systems, the parameter H_0, also referred to as thermal-independent (or short-time-scale) switching field $H_{\mathrm{sw}0}$, and the factor $\Delta E_0/k_\mathrm{B}T$ generally differ from that expected for an S-W system because of microscopic inhomogeneity and intergranular interactions. When the fitting is performed, care needs to be taken in the choice of the n value, since it affects the

[k]Magnetic properties at the recording time scale can be directly determined using a pulse field magnetometer [114, 115].

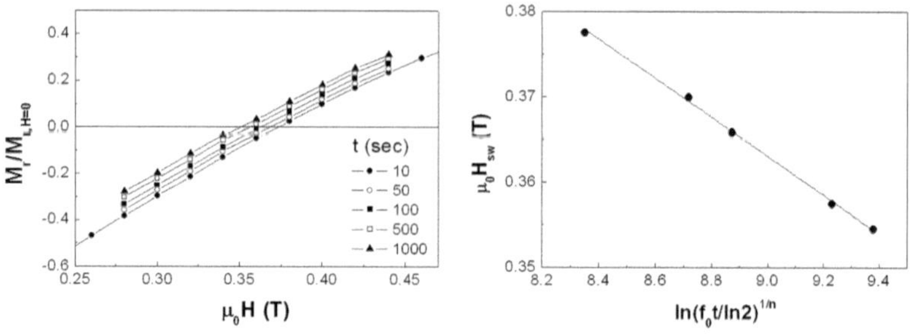

Figure 8.13 Dynamic measurements of a typical CoCrPt:SiO_2 film with PMA. (Left) Room-temperature DCD curves around the remanence coercivity H_{cr} (corresponding to the mean H_{sw}) for different waiting times (10 $<t<$1000 s); the field was applied at 45° with respect to the easy axis. (Right) Switching field as a function of waiting time (symbols) together with Sharrock's fitting curve (solid line) from which the thermal-independent (or short-time-scale) switching field H_{sw0} and the factor $\Delta E_0/k_B T$ can be determined.

proper evaluation of the stability factor [15, 121, 122]. In granular films for PMR media with the easy axes well oriented within a few degrees (~3°), the parameter n can be assumed, to a first approximation, equal to 2. However, it has been reported that the exponent n is not generally constant and is a function of the strength of the field when it is applied along the easy axis; on the contrary, when it is directed at 45° with respect to the easy axis, n is almost independent on the field strength and changes between 1.4 and 1.5. It follows that the so-called 45-degree method is more appropriate to extract information from dynamic magnetic measurements. Note that Sharrock's formalism is strictly valid for S-W systems, and it was found to produce good results in case of relatively low exchange coupling.

A different way to explore the dynamic properties is via viscosity measurements [10, 30, 109, 117, 89, 123–129]. Here, the magnetization is recorded as a function of the elapsed time t (commonly set between 10^3 and 10^4 s) while applying a constant reverse field H_r after saturation in the opposite direction (Fig. 8.14a,b). For a single-domain particle, the magnetization decays (or increases) exponentially, as previously reported, and for a set of particles (or grains in a granular film) with a distribution of

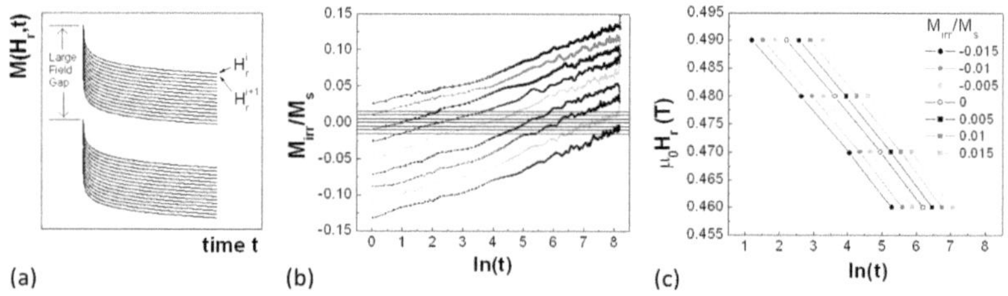

Figure 8.14 (a) Family of $M(H_r, t)$ curves measured over a period t for different reverse fields $H_r^{i+1} = H_r^i + \Delta H_r$; closely spaced measurements separated by larger field steps are reported. (b) Set of M_{irr}versus $\ln(t)$ curves of a CoCrPt:SiO_2 film measured at room temperature for different reverse fields H_r around $\langle H_{sw} \rangle$; the M_{irr}/M_s = const. lines around $\langle H_{sw} \rangle$ (i.e., $M_{irr}/M_s = 0$) are reported (const. = 0, 0.005, 0.01, 0.015). (c) Set of H versus $\ln(t)$ at different M_{irr}/M_s values. Reprinted from Ref. [30], Copyright (2013), with permission from Elsevier.

energy barriers, a logarithmic time dependence is observed, that is, $M(H_r, t) = \text{const.} \pm S \ln(t)$, where the constant is related to the magnetization at the beginning of the measurement, S is the magnetic viscosity, and t is the elapsed time since the measurement starts. In principle, S is a function of the reverse field H_r at which the magnetization decay is recorded. The parameter S is related, through the irreversible susceptibility χ_{irr} at $H = H_r$ (determined, for example, as the field derivative of a DCD curve), to the fluctuation field $H_f = S(H)/\chi_{irr}(H)$, that is, a fictitious field describing the effect of thermal activation on the magnetization reversal processes (i.e., a field description of $k_B T$). The fluctuation field is commonly used to evaluate the activation volume $v_{act}(H) = k_B T/\mu_0 M_s H_f$ that corresponds to a certain fraction of material involved in a thermally activated reversal process [109, 124, 127, 130]. The activation volume is used to estimate the size of the magnetic clusters (or the magnetic correlation length) that increases with the increase of v_{act} and intergranular exchange interactions [30, 102, 109, 123, 131, 132]. The magnetic cluster size is a critical parameter correlated to the media noise that can be directly determined by the analysis of magnetic force microscopy images [133, 134] and read-back media noise spectra [135], as well as by minor loop measurements, even if indirectly [65, 85]. In principle, v_{act} can be

8.3.1 Reversal Mechanism

The reversal mechanism of ECC media is schematically described by the domain wall–assisted switching model [31, 141, 143, 144]. In exchange-spring composites, the reversal begins at a field H_n with the nucleation in the soft phase of a domain wall that is pinned at the soft–hard interface, where it is compressed until the external field becomes large enough to overcome the pinning field H_p ($H_p > H_n$ for an optimized exchange-spring system). Then, the domain wall irreversibly moves into the hard region, thus leading to the reversal of the whole system. Note that both nucleation and compression of domain walls are reversible processes. The strong pinning force, which is caused by the abrupt change of the anisotropy at the soft–hard interface, is strongly reduced in graded media by gradually varying the anisotropy across the interface.

The first indication of the occurrence of an inhomogeneous reversal mechanism is given by the analysis of the major hysteresis loop that would be characterized by the presence of two critical fields (corresponding to H_n and H_p), with the switching of the whole system occurring at a value lower than that required for the reversal of the single hard phase (Fig. 8.15, left). However, the two critical fields are not always clearly distinguishable [153–155] and then an in-depth analysis is necessary to disclose the reversal

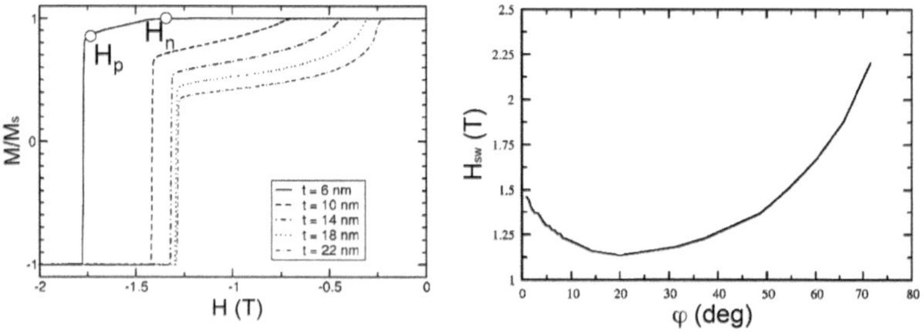

Figure 8.15 Simulation of the magnetic behavior of a soft–hard ECC system. (Left) Demagnetizing curves for different thicknesses t of the soft layer. Reprinted and adapted from Ref. [136], Copyright (2007), with permission from Elsevier. (Right) Switching field as a function of the angle φ between the external field and the easy axis. Reprinted and adapted from Ref. [160], Copyright (2005), with permission from Elsevier.

mechanism and find the optimum design for an ECC system. This can be done by using different complementary approaches, including the comparison among hysteresis loops and DCD curves [33, 34, 147], analysis of recoil curves [33, 153, 156], study of the transverse signal during hysteresis measurements along the easy axis [47, 154], angular dependence of switching curves [20, 32, 153, 154], and FORC diagrams [96, 155, 157–159]. In the following a couple of examples concerning the last two methods are reported.

The angular dependence of the switching field in composite systems is expected to experience a symmetry breaking (Fig. 8.15, right) [20, 32, 153, 154] reflecting the nucleation of a domain in the softer magnetic part and the domain wall pinning at the soft–hard interface, whose corresponding critical switching fields (i.e., H_n and H_p) are characterized by an angular dependence described by the S-W and Kondorsky relations (see Section 8.2.3), respectively. However, in granular ECC media, a symmetry breaking also arises from intergranular exchange interactions, which can cause a lateral domain wall displacement [47], thus indicating that the above method is especially valid for laterally exchange-decoupled ECC systems, where only a vertical domain wall displacement can take place.

Measuring the FORC diagrams and comparing the results with numerical micromagnetic simulations represent an alternative and effective method to investigate the reversal mechanism in ECC media [96, 155, 157–159].

Numerical micromagnetic simulations have shown that the wishbone shape, typical of a single granular hard film, significantly modifies depending on the strength of intergranular and interlayer coupling. In particular, as reported in Fig. 8.16 for a CoCrPt-based exchange-spring system [96], for weak intergranular coupling and by increasing the interlayer vertical coupling (Fig. 8.16 left), the FORC diagram moves from a two-island contour, which is typical of two phases reversing independently, to a narrow vertical band shape with significant switching events at a negative H_B field, which was indicated as characteristic of an exchange-spring system. Finally, further increasing the interface interactions, the system evolves toward a wishbone shape, which is typical of a system behaving as a single phase. Keeping constant the vertical interlayer

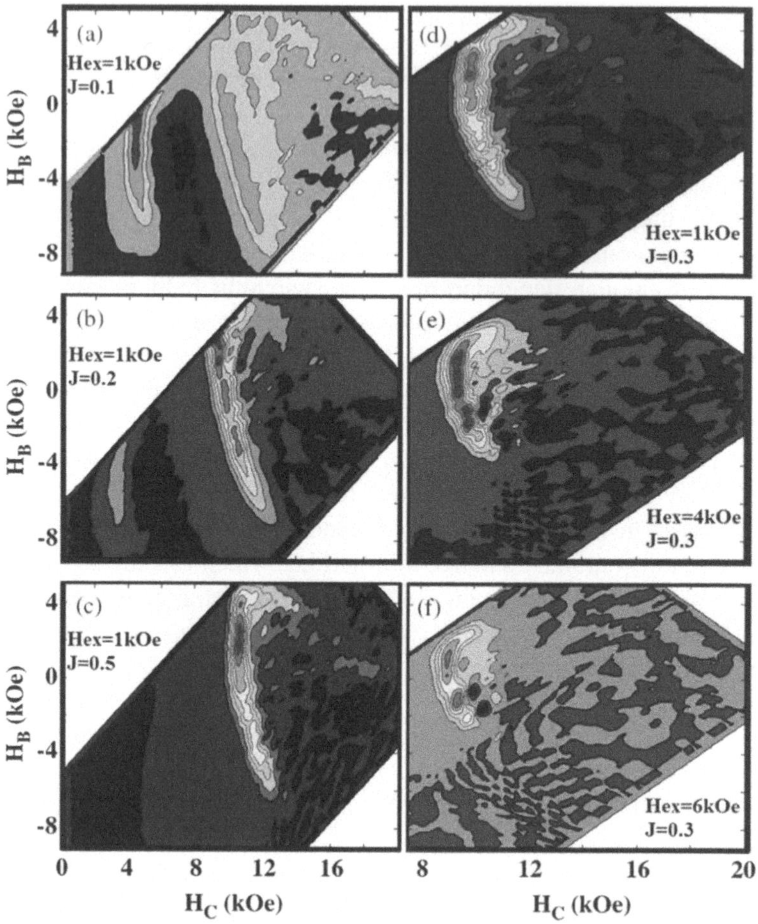

Figure 8.16 Simulated FORC diagrams for a two-layer CoCrPt-based system for different intergranular exchange field H_{ex} and interlayer vertical coupling J (here defined as a fraction of the bulk value). (Left) $H_{ex} = 0$, $0.1 < J < 0.5$. (Right) $J = 0.3$, $1 < H_{ex} < 6$ kOe. © 2011 IEEE. Reprinted, with permission, from Ref. [96].

coupling (Fig. 8.16, right), the simulation indicates that, varying the intergranular coupling, the peak moves from positive to negative H_B values, and the contour splits in two peaks for a very strong coupling that qualitatively indicates a tendency of the grains to clusterize [87].

8.3.2 *Thermal Stability and Writability*

The benefit of ECC media with respect to single-phase systems, in terms of writability and thermal stability, can be best quantified by measuring the ratio between the energy barrier of an ECC medium and that of a single hard phase with the same magnetization, switching field, and thickness [31]. Due to the practical difficulty in respecting the above-mentioned constraints, a different figure of merit is commonly used, also referred to as gain factor $\xi = 2\Delta E_0/H_{\text{sw0}}(\varphi)\mu_0 M_s V^*$, where V^* corresponds to the volume of the whole ECC grain in granular media or dot in BPMR media with no interactions [139]. The parameter ξ is equal to 1 for single-phase systems, while it assumes a maximum theoretical value of 2 for exchange-spring composites [138] and increases up to 4 for graded media [144]. This figure of merit has the disadvantage of being independent on the thickness for single phases, although a thicker phase is definitely more stable. However, by fixing the thickness of the hard layer, its effect can be eliminated, thus allowing a comparison among samples with different soft-layer thickness [15] (Fig. 8.17). The values of the energy barrier ΔE_0 and H_{sw0} can be obtained using Sharrock's approach described in Section 8.2.4 [15, 148, 159, 160]. In particular for a soft–hard system a value of $n = 1.5$ was proposed for the best fitting using the 45-degree method [15], while for graded media, because of the linear dependence of the energy barrier on the field, a value of 1 is considered appropriate [144, 161–163].

8.4 Bit-Patterned Magnetic Recording Media

In BPMR technology, data are stored in arrays of magnetic nano-elements, each one corresponding to a single bit whose magnetization can switch between two well-defined states (binary digit) under the application of a writing field. Recording densities up to 5 Tbit/in^2 can be attained simultaneously addressing the writability and thermal stability issues by ordered arrays of single-domain islands made of CoCrPt alloy or $[Co/Pt(Pd)]_n$ multilayer materials (where n is the number of layers), while storage densities between

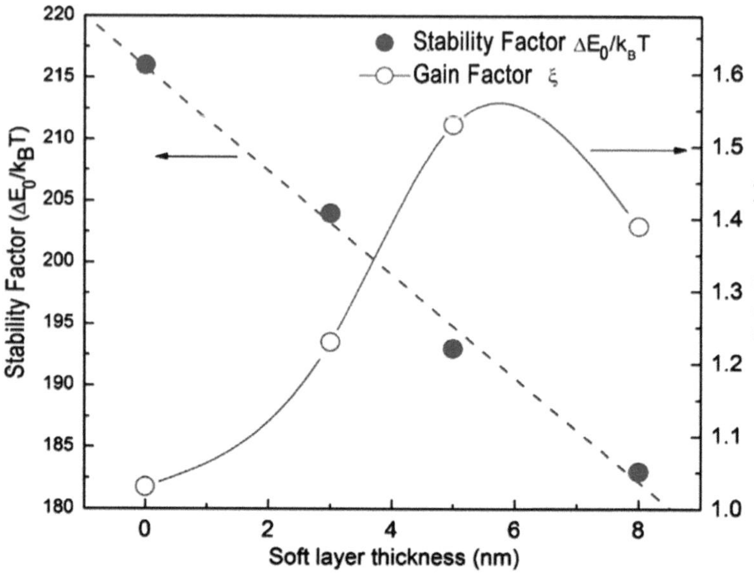

Figure 8.17 Stability factor ($\Delta E_0/k_B T$) and gain factor (ξ) of FePt:TiO_2 hard–soft films measured using the 45-degree method, as a function of soft-layer thickness. Reprinted from Ref. [15], Copyright (2013), with permission from Elsevier.

20 and 100 Tbit/in^2 can be achieved by using hard L1_0–FePtX alloys in combination with other recording technologies (e.g., ECC and EAMR) [164].

Optimizing a system for BPMR implies a strict control on fabrication process, as any structural and/or microstructural defects present in the magnetic dots will induce a distribution of magnetic properties and then a broadening of the SFD, which represents a key factor for the overall performance of the medium. A broad SFD would mean that the magnetization flipping induced by the writing field might not occur in all the bits but is limited to the low-anisotropy ones. However, while it may be assumed that the strong exchange coupling between the grains inside the bit would lead to a narrow SFD, this is not what has been experimentally observed. Indeed, in addition to the intrinsic part of the SFD, which is due to the variation of the intrinsic magnetic properties (e.g., anisotropy constant, angle of the easy axis, magnetization) and to the fabrication irregularities (e.g., dots shape and size), the magnetostatic interactions from adjacent bits give rise to an extrinsic factor that contributes to the SFD broadening [165–167]. For this reason, a complete magnetic characterization of the switching behavior of a patterned sample (i.e., SFD, reversal mechanism) is needed in order to optimize the fabrication process and to pose as the basis for a complete understanding of the recording potential of the system.

Among the different methodologies used to characterize BPMR media, magnetic force microscopy (MFM) resulted to be the most useful technique to investigate the magnetic behavior of single elements, thanks to its high lateral resolution that allows a dot-by-dot analysis of the magnetization under an external applied field or in the remanent state [168, 169]. As an example, the ascending branch of the irreversible hysteresis loop can be typically derived by counting the number of dark and bright islands in the remanent state after saturating the sample in a perpendicular negative field (all islands bright) and applying increasing switching fields (Fig. 8.18).

The experimental curve is obtained from the remanence ratio values (i.e., switched dots/total dot) and its derivative can be used to evaluate the experimental value of the total SFD [169]. The reliability of this method for studying the magnetization process of a patterned

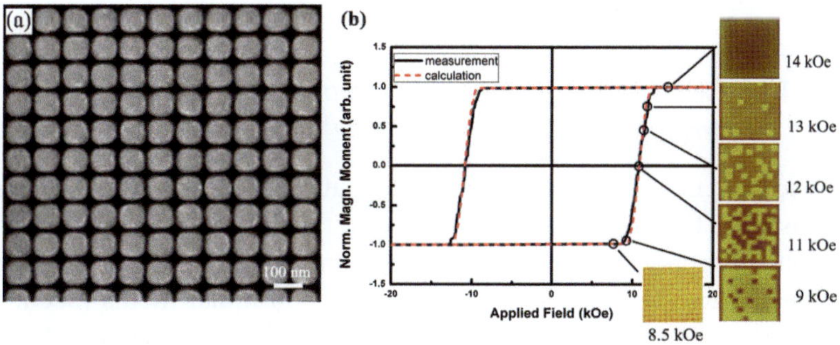

Figure 8.18 Hysteresis loop derived by the MFM patterns at different fields applied along the direction perpendicular to the media surface; (solid line): experimental data; (dashed line): calculation (see below). Reprinted with permission from Ref. [169]. Copyright 2011, AIP Publishing LLC.

system was demonstrated, for example, through the correspondence between a DCD curve obtained by this technique and the one measured by using a conventional magnetometer [168]; moreover, the possibility to directly observe the magnetic configuration of single islands under the applied field allows a detailed study of the magnetization process to be performed. However, although the MFM technique allows locally probing the magnetization of the dots, the method is not able to define the magnetic behavior during a recording process, since the application of a uniform magnetic field affects the pattern as a whole (i.e., the observed dot and the surrounding neighbors), unlike a bit writing event, where the switching process takes place thanks to a high-gradient magnetic field of a recording head applied to a limited region of the media (in principle, one dot) and on a much shorter time scale [170].

In the following sections, an overview of the most used methodologies (mainly based on MFM imaging) to determine the SFD and the magnetization reversal mechanisms will be presented.

Besides providing significant information on magnetic properties useful for technological applications, these methods are also a powerful tool to confirm theoretical models. Indeed, by exploiting the strict control over size, shape, and interdot distance offered by advanced techniques, dot arrays with finely tuned features are now available and can be used as model systems to test and verify the predictions.

8.4.1 *Switching-Field Distribution and Switching-Field Probability*

The most used methodology to investigate the SFD in BPMR media is based on the analysis of magnetization reversal curves acquired on a macroscopic scale, that is, averaged on a large bit array. They are the same as those used for characterizing granular media (Section 8.2.2), although some experiments may be also carried out by MFM imaging, by counting the number of switched islands at each step, as discussed above. However, the recording process takes place on a microscopic scale, induced by a nonuniform magnetic writing field that, in principle, should be able to switch the magnetization of a single element without affecting the magnetic stability of its adjacent neighbors. For this reason, to study the recording potential of a patterned media, separating the behavior of a single dot from long-range dipolar interactions is of fundamental importance. This can be achieved by applying methods that allow the measured SFD to be separated into intrinsic and extrinsic contributions, that is, by evaluating the contribution to the SFD given by the dipolar interactions between adjacent islands within the array.

The $\Delta H(M, \Delta M)$ method, discussed in Section 8.2.2, allows the dipolar contribution to be determined by measuring a series of recoil curves corresponding to the magnetization reversal of islands in different magnetostatic environments. Neglecting thermal effects, the critical field for the switching process of an individual magnetic island can be written as [169]

$$H_{\mathrm{sw}} = \beta H_{\mathrm{K}} - nM_{\mathrm{s}} - \sum M_{\mathrm{s}} V_{\mathrm{bit}}/r^3 \qquad (8.13)$$

where the first term represents the anisotropy field H_{K} reduced by a factor β, which accounts for material defects, n is the dot's demagnetizing factor, M_{s} is the saturation magnetization, and the last term is the sum of dipolar interactions from the neighboring dots (extrinsic part), V_{bit} being the bit volume and r the reciprocal distance. The intrinsic contribution of the island's switching field is given by the first two terms in Eq. 8.13 and its theoretical value ($\beta = 1$) can be compared to the experimental value to derive the reduction factor β for the given system. As an example, in the calculation reported in Ref. [169] the intrinsic SFD experimentally derived by the $\Delta H(M, \Delta M)$ method (Fig. 8.19, bottom) was used to generate 121 single values through a Gaussian distribution. To calculate the theoretical total SFD, the obtained values were randomly attributed to an array of 121 dots, and for each island, the dipole–dipole interactions from the first two neighboring shells (8 islands) were added. The calculated total SFD, defined in the paper by the parameter 2σ is equal to 0.17 T, resulted in being comparable to the distribution obtained by differentiating the demagnetizing curve measured by MFM ($2\sigma = 0.2$ T), as illustrated in Fig. 8.19. The example shows that the dipole–dipole interaction among dots is responsible for a significant broadening of the SFD with respect to

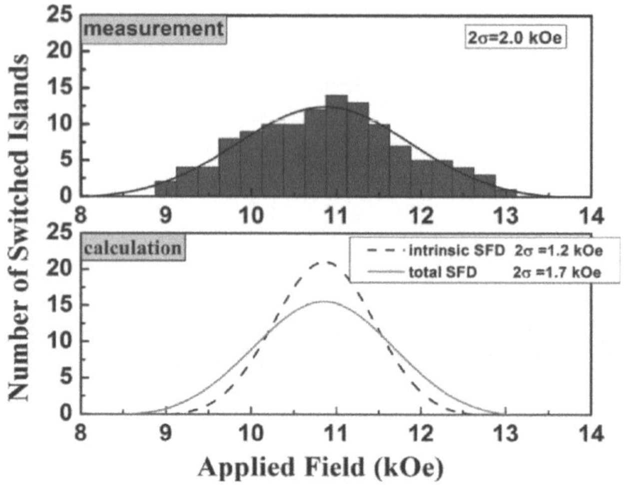

Figure 8.19 (Top) Distribution of reversal fields in columns and the corresponding SFD, as obtained by differentiating the experimental MFM magnetization curve in Fig. 8.18. (Bottom) Comparison between the intrinsic SFD, derived by the $\Delta H(M, \Delta M)$ method and the total SFD obtained by adding the dipole–dipole contribution to the intrinsic SFD. Reprinted with permission from Ref. [169]. Copyright 2011, AIP Publishing LLC.

its intrinsic value, although a further broadening has to be attributed to additional factors not considered in Eq. 8.13.

Among these factors, thermal fluctuation processes have been demonstrated to play an important role in the switching process of individual islands. Indeed, even if the magnetic anisotropy energy stored in one dot is much larger than the thermal energy, the strong magnetostatic contribution due to the surrounding elements affects the switching process by increasing or decreasing the energy barrier ΔE for magnetization reversal, depending on whether the net dipolar interactions from neighboring islands are in the same or in the opposite direction with respect to the dot magnetization. In this case, the thermal activation may affect the magnetization reversal, resulting in a finite probability p, for a given element, to switch under the external applied field. For this reason, besides studying the SFD, determining the switching probability distribution (SPD) in patterned systems is crucial for a thorough characterization

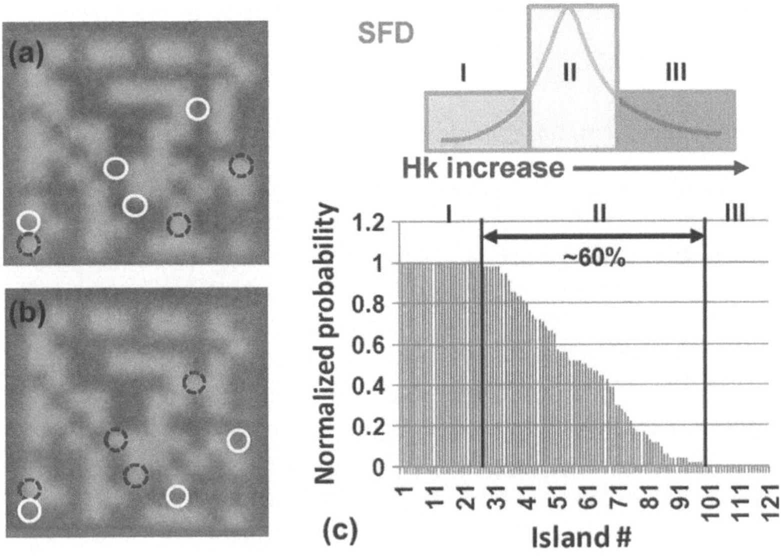

Figure 8.20 (a, b) Typical MFM images of the same pattern after repeated tests (i.e., the same H_r); circles indicate the differences between the dot magnetization direction in the two images. (c) Histogram of switching probability p based on data for 60 repetitions. © 2010 IEEE. Reprinted, with permission, from Ref. [171].

of the media, being a parameter directly related to the bit writing process. Experiments aimed at determining the SPD are based on repeated MFM images of the same pattern in the remanent state after the same sequence of applied fields, that is, saturation followed by the application of a reverse field H_r along the opposite direction [171]. Two typical MFM images showing a finite number of dots (circled) in different remanent states are reported in Fig. 8.20.

For these islands, there is a probability p ($0 < p < 1$) to reverse the magnetization under the application of H_r. For a given H_r value, for islands with a switching field H_{sw} well below the external field, p tends to 1 (always switching) while no switching is expected if H_r is lower than H_{sw} (p tends to 0). This statistical behavior is described in a histogram where p for a given number of repetitions is reported as a function of the number of islands. The value of SPD rises as H_r is increased, leading to a lower switching error probability; however, since the maximum writing field cannot be increased, because of practical limitations, a narrow SFD and a sharp writing field gradient are strongly required in order to optimize the media performance.

8.4.2 *Reversal Mechanism*

In BPMR media with dot size below a critical value (e.g., around 100 nm for $[Co/Pd]_n$ [166a]) the characteristic high anisotropy of the ferromagnetic materials currently used and the strong exchange coupling between their grains ensure a single-domain state unlike what happens in the unpatterned granular film of the same material, where a multidomain state is attained. However, the reversal mechanism of these systems cannot be unequivocally determined and MFM imaging, combined with the analysis of the angular dependence of the switching field, was proven to be a useful tool to understand the switching process in patterned systems. Indeed, in a single-domain island the magnetization is expected to be always collinear, resulting in a typical S-W reversal mechanism characterized by coherent rotation of the dot as a whole. In this case, the switching field is only determined by the dot anisotropy, while a strong reduction is observed in reversal processes accomplished by pinning and propagation of domain wall as it can occur in strongly exchange-coupled thin films (Kondorsky reversal mechanism). Actually, higher values of the switching field usually found in patterned systems may be considered as a fingerprint of the transition between the two mechanisms; however, a closer investigation shows an unexpected behavior, as illustrated below.

A typical method to determine the reversal mechanism is based on the measure of the angular dependence of the switching field H_{SW}, as described in Section 8.2.3. As an example, the investigation of the switching process in $[Co/Pd]_n$ patterned systems as a function of the dot size by means of MFM analysis and magneto-optic Kerr effect (MOKE) magnetometry [166b, 172] will be discussed. The MFM images (Fig. 8.21a) clearly reveal a multidomain configuration for a dot size above 200 nm; these islands may support domain walls and a behavior similar to the reversal in a continuous film is expected. However, a surprising result was obtained by the angular dependence of remanent coercivity, showing an S-W-like behavior even in 1 μm islands (Fig. 8.21b). The dot size being too large

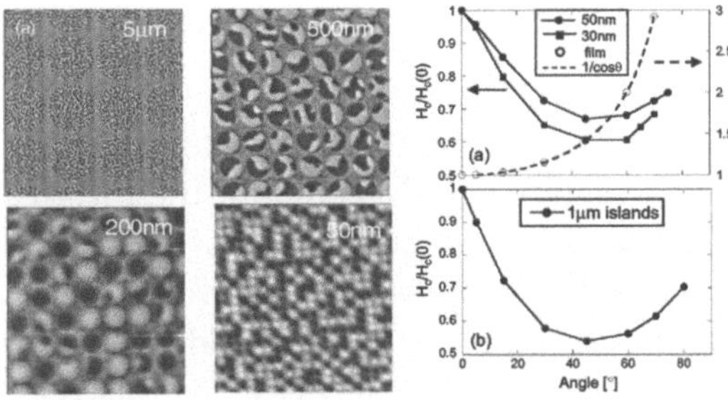

Figure 8.21 (a) MFM images of the AC-demagnetized state of $[Co/Pd]_n$ islands with different sizes; a multidomain state develops in dots of 200 nm and larger. Reprinted figure with permission from Ref. [166b]. Copyright (2006) by the American Physical Society. (b) Remanent switching field of islands of different size as a function of the angle between the applied field and film normal. Reprinted with permission from Ref. [172]. Copyright 2010, AIP Publishing LLC.

to switch via coherent rotation in a single reversal event, this behavior was explained as being due to a nucleation event with S-W characteristics occurring within the dots, followed by domain wall motion that causes the reversal of the whole dot.

8.5 Energy-Assisted Magnetic Recording Media

Heat- and microwave-assisted magnetic recording technologies (HAMR and MAMR, respectively) have been recently proposed as promising approaches to simultaneously address writability and thermal stability issues [173–176]. Although the EAMR technology is based on the same materials used in other magnetic recording systems (typically the $L1_0$–FePtX alloys), the magnetic characterization of the media needs specific measurements in order to verify if the material is able to match the peculiar requirements for this application. In this section, an overview of some characteristic measurements recently developed for studying the material behavior under specific conditions (heat pulse for HAMR and radio frequency [RF] field pulse for MAMR) is presented.

8.5.1 *Heat-Assisted Magnetic Recording Media*

The peculiarity of HAMR technology is that the data writing temperature is different from the storage temperature; indeed, the writing process requires local heating of the bit area in order to drastically decrease the switching field of the media. This effect allows the write head to selectively switch the magnetization direction of the bit with a magnetic field much lower than the one used in conventional PMR devices. For this reason, a deeper understanding of the magnetic behavior at high temperature, that is, close to the Curie temperature (T_C), becomes fundamental for characterizing these media.

In particular, in addition to the room-temperature stability factor $\Delta E_0/k_B T$, a strict control on the relative distributions of the Curie temperatures $D(T_C)$ and of the switching fields is necessary. As an example, compared to conventional magnetic recording technologies, additional contributions to the room-temperature SFD are present in HAMR media (i.e., $D(T_C)$–induced SFD and thermal fluctuation–induced SFD) [174, 177] and the control on the switching mechanism at high temperature represents one of the main challenges for the recording media to be optimized. In general, attaining a good control on T_C, $D(T_C)$ and SFD is of utmost importance since these parameters are critical for HAMR performance as they determine not only the writing process but also the medium noise [173]. In granular media the actual value of T_C is scaled toward lower values with respect to the bulk value $T_C(\infty)$ and experimental and atomistic approaches have demonstrated a finite-size scaling dependence [178]

$$\frac{T_C(\infty) - T_C(L)}{T_C(\infty)} = \left(\frac{L}{L_0}\right)^{-1/\nu} \qquad (8.14)$$

where L is the grain size, L_0 is a microscopic constant length, and ν is the universal critical exponent of the bulk correlation length [179]. Once $T_C(\infty)$, L_0, and ν values are calculated from the analysis of the differentiated $M(T)$ data, $D(T_C)$ can be directly derived from the grain size distribution, the log-normal form of the distribution function being preserved under the finite-size scaling low regime [180].

From a practical point of view, the major difficulty in measuring HAMR media around T_C is that the magnetic behavior is affected by strong magnetization thermal fluctuations due to the proximity to the phase transition [181]; moreover, standard laboratories facilities for magnetic characterization acquire data in a much longer time scale than the effective time scale for writing. Many attempts have been made to develop new setups [17, 182], most of them based on pump and probe systems, for studying the temperature dependence of magnetic properties, while taking advantage of the high versatility of modern laser sources for the local heating. The correlation between the data obtained by common magnetometric techniques (i.e. vibrating sample magnetometer [VSM], superconductive quantum interference device [SQUID], alternate gradient force magnetometer [AGFM]) and fast dynamic methods is not straightforward, but the use of theoretical models to support data interpretation proved to be a useful tool to study the material in recording conditions.

8.5.1.1 Determination of T_C and $D(T_C)$

Conventional VSM measurements at high temperature can be used to study the temperature dependence of magnetic properties, such as saturation magnetization and magnetocrystalline anisotropy, and, finally, to obtain the Curie temperature of the material. A systematic work was carried out by Thiele et al. on $Fe_{55-x}Ni_xPt_{45}$ thin films as a function of the Ni content [183]. The Curie temperature, calculated according to the Curie–Weiss formalism [46], as well as the second-order anisotropy constant, obtained by using the 45-degree method, are strongly dependent on the composition, both decreasing at higher Ni content. The study pointed out that a trade-off between the Curie temperature and the anisotropy field needs to be considered when designing a material for HAMR technology.

Once the mean value of the Curie temperature $\langle T_C \rangle$ is calculated for a given system, the determination of its distribution $D(T_C) = \sigma_{T_C}/\langle T_C \rangle$, where σ_{T_C} is the standard deviation of T_C, is also strongly required since its control represents one of the key challenges for the development of HAMR media [173]. Although analytical approaches succeeded in deriving the $D(T_C)$ from the grain size distributions [180] (also accounting for the chemical ordering in the case of $L1_0$

materials [178]), a method has been recently proposed to directly measure $D(T_C)$ by exploiting a pump and probe setup for the characterization of hard granular FePt thin films [182]. The approach is based on a laser source generating pulses at variable energy to temporary heat the grains up to their Curie temperature under a constant applied field H. The value of H is set to a value below the threshold for the grain switching at room temperature but still high enough to reverse the magnetization of the grains after the heating pulse, while cooling through T_C. Before each pulse, the sample is brought to an initially saturated state with grains uniformly magnetized along the direction opposite to H; then, the system is rapidly (<10 ns) heated by the laser pulse under H and the fraction of switched grains is measured by Kerr rotation as a function of the pulse energy (Fig. 8.22). The key point of the proposed experiment is that the dispersion of the laser energies σ_{laser} necessary to switch the media grains is a measure of the $D\,(T_C)$, according to $\sigma_{T_C}/\langle T_C\rangle =$

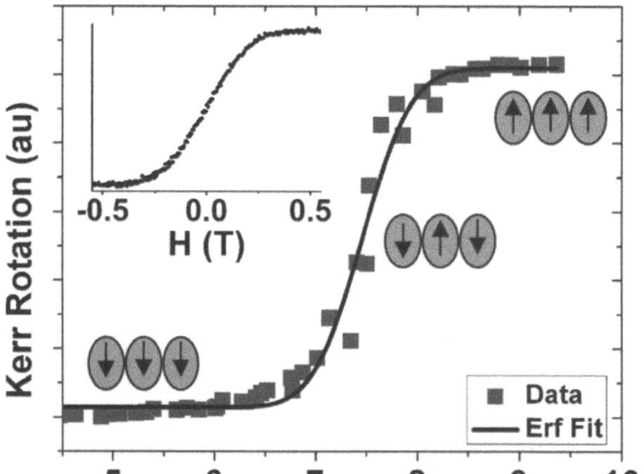

Figure 8.22 Kerr rotation measurement of switched grains as a function of the pump pulse energy. The schematics represent the grain magnetization direction as the temperature is increased. The inset shows an M–H magnetization curve measured by Kerr rotation where for each field value a pump pulse of saturating energy was applied. Reprinted with permission from Ref. [182]. Copyright 2014, AIP Publishing LLC.

$\sigma_{laser}[(\langle T_C \rangle - T_{amb})/\langle T_C \rangle]$, where $\langle T_C \rangle$ is measured independently, for example, by magnetometric measurements.

A similar approach was proposed in Ref. [17], where a specific pump–probe experimental setup was optimized to study the media magnetic properties in proximity of the Curie temperature in a short time scale, simulating the HRMR writing process. In such a *thermal erasure* experiment, the sample is initially saturated, and after removing the external magnetic field, the same laser diode is used first to heat the sample and, at reduced power, to measure the MOKE. Then, the laser power and the associated temperature are increased monotonically and the room-temperature magnetization remanence is measured (Fig. 8.23). The conversion between the laser power and the corresponding temperature was obtained by carrying out a series of remancence measurements as a function of temperature by means of a VSM magnetometer. Although the two experiments are characterized by a huge difference in the time scale (of the order of seconds or microseconds for the VSM or the pump and probe method, respectively), the authors came out with the comparison and data interpretation by developing a

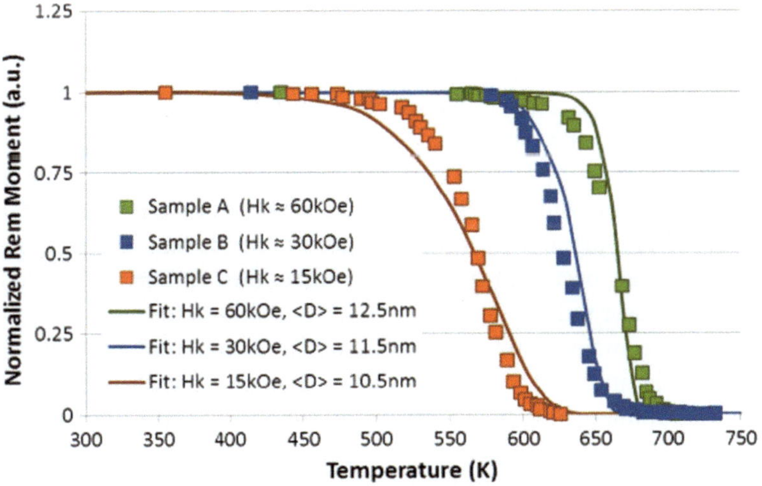

Figure 8.23 Thermal erasure experiments for FePt-based films with a different anisotropy field H_K. © 2013 IEEE. Reprinted, with permission, from Ref. [17].

semi-analytical model based on Arrhenius–Néel thermal relaxation in a system of noninteracting S-W particles. The same experimental input parameters (i.e., M_S, H_K, demagnetization factor, and grain size and H_K distributions) were used for the model to simultaneously fit both the VSM and the MOKE data and find out the conversion factor from the laser power to temperature.

8.5.1.2 Determination of SFD

During the writing process, as the temperature increases, the magnetization can be reversed by an external field that is lower than the room-temperature switching field of the medium. However, this does not necessary imply that the magnetization perfectly aligns along the field direction, since, close to T_C, the energy barrier is strongly reduced by the high thermal energy, with an increased probability for the magnetization to switch along all the possible directions, thus resulting in a broadening of the SFD. For this reason, conventional methods, as the one recently presented in Ref. [184] and based on the measure of the major and minor loops, for studying the room-temperature SFD in granular HAMR media are not exhaustive for a complete determination of the media viability; however, they are essential since they provide both qualitative and quantitative evaluation of the main magnetic characteristics.

For specific determination of the high-temperature switching behavior, the pump–probe experimental setup proposed by Chernyshov et al. [17] allows the thermoremanent major/minor loops to be measured, where the data reported on the M–H curves (Fig. 8.24) represent the room-temperature snapshot of a switching process that occurs at high temperature, that is, during the field-assisted laser pulse.

In the major loop measurements at room temperature, the sample is initially saturated to $-M$, a laser pulse with fixed power (corresponding to a temperature of $\sim$650 K) is used to heat the system while applying the field H, and then the room-temperature magnetization remanence is acquired for each value of H from -0.5 T to $+0.5$ T (ascending branch); the same protocol is applied for the descending branch. For the measurement of minor loops, the system is initially saturated to $+M$ and a reverse field H_r is

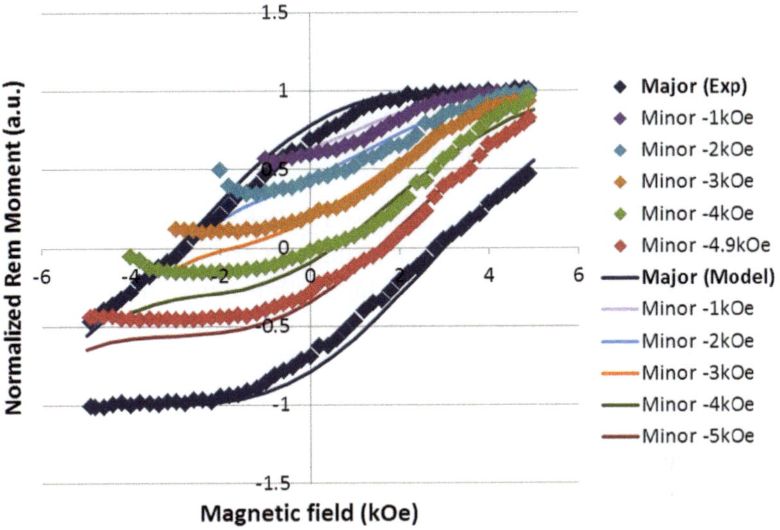

Figure 8.24 Experimental (dot) and fitted (line) major and minor hysteresis loops measured at 650 K for a FePtX medium. © 2013 IEEE. Reprinted, with permission, from Ref. [17].

then applied under the fixed laser pulse; the ascending branch is then measured with the same method used for the major loop (i.e., switching during the pulse and data acquired at room temperature) and the measurements are repeated for different values of H_r. The curves are then fitted according to the same model developed for the determination of T_C and $D(T_C)$ (see the thermal erasure experiment). A good data fitting was obtained by the authors for a switching process occurred at 650 K and using the same value of SFD as that obtained by conventional major/minor loops measurements performed at 473 K by the MOKE technique. At least up to 650 K (i.e., 50 K below T_C), this result shows that the value of the SFD does not change significantly, resulting in a large dH/dT gradient, which is one of the main requirements for HAMR media [185].

8.5.2 *Microwave-Assisted Magnetic Recording Media*

A basic characterization of MAMR media is the investigation of the magnetization reversal mechanism under microwave excitation. In the presence of a small uniform microwave RF field (applied along a direction perpendicular to the magnetization easy axis) whose frequency matches the ferromagnetic resonance frequency (FMR) of the media, the magnetization starts a precession mode that allows lowering the coercivity when a DC external field is applied along the easy direction.

The FMR basic theory is discussed in Ref. [186], where calculations of resonance conditions for common crystalline symmetries (tetragonal and/or cubic) occurring in thin films are reported. A description of experimental setups can be found in Ref. [187] for conventional FMR microwave cavity method, where the resonance is probed by scanning the external field around the resonant field at fixed frequency. The broadband (i.e., multiple frequencies) vector network analyzer-coplanar waveguide technique, where the external field value is fixed and the resonance is probed by scanning the frequency of the pumping RF field, is described in [188].

Although the RF-induced switching was first experimentally demonstrated in 2003 [189], the majority of the experiments were limited to soft magnetic materials [190]. Following the increasing interest in this effect for applications in the high-density HDD storage, many recent papers reported on theoretical calculations simulating the magnetic behavior of hard materials or exchange-coupled systems [175, 191], while few experiments have been carried out on perpendicular magnetic materials, such as $[Co/Pt]_n$ multilayers, CoPt alloys and CoCrPt-based granular films [192].

8.5.2.1 Frequency dependence of H_{sw}

In a typical experimental setup [193] the magnetic film is patterned on a cross shaped electrode and the magnetization along the perpendicular direction is detected by the Anomalous Hall Effect (AHE). A microwave pulse is applied along the y direction in the field plane (Fig. 8.25a) while the external direct current (DC) magnetic field is provided by an electromagnet, as illustrated in Fig. 8.25.

A common experiment is the measurement of the hysteresis loop as a function of the frequency f_{RF} (typically in the range of 0–30 GHz), alternating current (AC) field amplitude h_{RF} (around 50 mT), and temperature. In Fig. 8.26 (top), the comparison between the

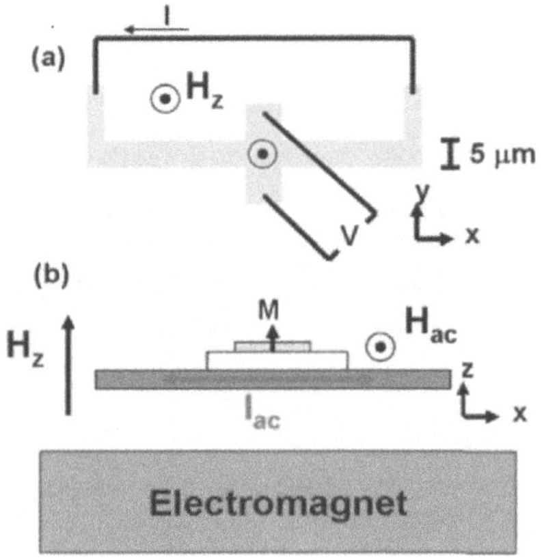

Figure 8.25 Scheme of a typical experimental setup for characterizing the microwave-assisted magnetization reversal effect: (a) top view and (b) side view. Reprinted with permission from Ref. [193]. Copyright 2012, AIP Publishing LLC.

hysteresis loops with and without the applied RF field measured on a perpendicular CoPtCr patterned system is reported [193]. The effect of the microwave field in assisting the magnetization reversal is clearly visible when the coercivity H_c (corresponding to H_{sw}) and the coercive squareness S^* are reported as a function of frequency (Fig. 8.26, bottom). The decrease of the switching field is observed over a broad range of f_{RF} values down to a minimum near 18 MHz. Above a critical frequency value f_c, the magnetization curve coincides with the one measured without the application of an RF field, since the system is not able to absorb energy out of the resonance regime. Note that the recovery of the same initial value of H_c is the proof that the thermal energy due to the Joule heating occurring during the measurements does not affect the magnetization reversal that is exclusively assisted by the microwave pulse.

The transition from the in- and off-resonance regimes is distinctly visible in the experiment presented by Okamoto et al. [192] carried out on a $[Co/Pt]_n$ dot arrays with different size. The hysteresis loops measured at different values of frequencies f_{RF} (h_{RF} = 50 mT) for a 50 nm dot show a relevant decrease of the switching field (Fig. 8.27) at increasing frequency up to a value

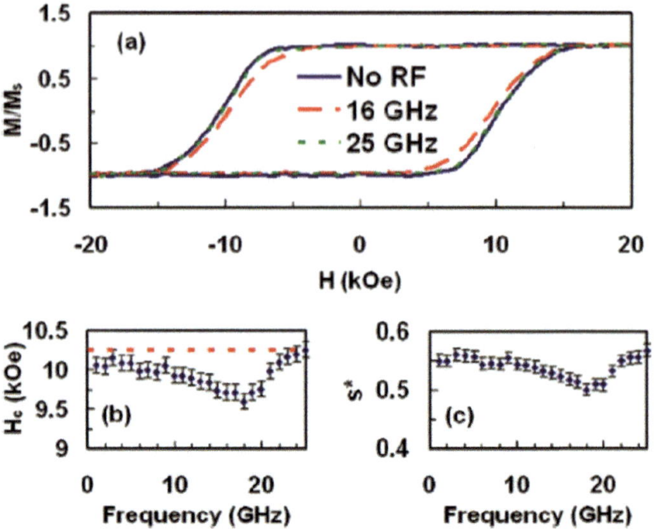

Figure 8.26 (a) AHE hysteresis loops of a CoCrPt patterned system measured at 4 K for different values of f_{RF} ($h_{\mathrm{RF}} \sim 50$ mT). (b, c) Coercive field $\mathrm{H_c}$ (b) and coercive squareness S^* (c) as a function of f_{RF}. Reprinted with permission from Ref. [193]. Copyright 2012, AIP Publishing LLC.

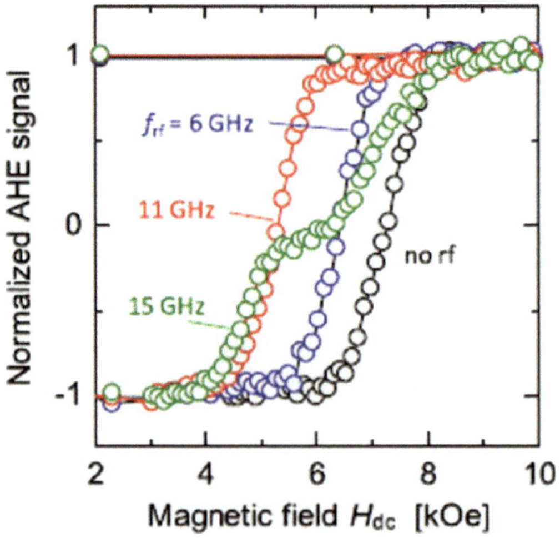

Figure 8.27 AHE curves (half loops) of 50 nm [Co/Pt]$_n$ dots at different f_{RF} (h_{RF} = 50 mT). © 2014 IEEE. Reprinted, with permission, from Ref. [192].

$f_{RF} = 11$ GHz. On the contrary, the loop measured at 15 GHz (Fig. 8.27, red curve), which is close to the critical frequency f_c, shows a clear double reversal; in such conditions because of the dispersion of critical frequencies in a real patterned system, only the dots with $f_c \leq 15$ GHz are effectively assisted in the magnetization reversal, while the behavior of the off-resonance dots resembles that of the same system without the application of the RF field.

References

1. Richter H. J. (2007). The transition from longitudinal to perpendicular recording, *J. Phys. D. Appl. Phys.*, **40**, pp. R149–R177.
2. Ripka P. (2001). Magnetic Sensors and Magnetometers (Artech House).
3. Fiorillo F. (2005). Characterization and Measurement of Magnetic Materials (Academic Press).
4. Piramanayagam S. N., and Srinivasan K. (2009). Recording media research for future hard disk drives, *J. Magn. Magn. Mater.*, **321**, pp. 485–494.
5. Berger A. (2012). Magnetization reversal in granular thin films, *Phys. B*, **407**, pp. 1322–1329.
6. Sonobe Y., Weller D., Ikeda Y., Schabes M., Takano K., Zeltzer G., Yen B. K., Best M. E., Greaves S. J., Muraoka H., and Nakamura Y. (2001). Thermal stability and SNR of coupled granular/continuous media, *IEEE Trans. Magn.*, **37**, pp. 1667–1670.
7. Sonobe Y., Weller D., Ikeda Y., Takano K., Schabes M. E., Zeltzer G., Do H., Yen B. K., and Best M. E. (2001). Coupled granular/continuous medium for thermally stable perpendicular magnetic recording, *J. Magn. Magn. Mater.*, **235**, pp. 424–428.
8. Greaves S. J., Muraoka H., Sonobe Y., Schabes M., and Nakamura Y. (2001). Pinning of written bits in perpendicular recording media, *J. Magn. Magn. Mater.*, **235**, pp. 418–423.
9. Tham K. K., Saito S., Hasegawa D., Itagaki N., Hinata S., Ishibashi S., and Takahashi M. (2012). Effect of inhomogeneous microstructure of granular layer on inter granular/inter layer exchange coupling in stacked perpendicular recording media, *J. Appl. Phys.*, **112**, p. 093917(8).
10. Jung H. S., Velu E. M. T., Malhotra S. S., Kwon U., Suess D., and Bertero G. (2007). CoCrPtO-based granular composite perpendicular recording media, *IEEE Trans. Magn.*, **43**, pp. 2088–2090.

11. Nolan T. P., Valcu B. F., and Richter H. J. (2011). Effect of composite designs on writability and thermal stability of perpendicular recording media, *IEEE Trans. Magn.*, **47**, pp. 63–68.
12. Choe G., Ikeda Y., Zhang K., Tang K., and Mirzamaani M. (2009). Control of exchange coupling between granular oxide and highly exchange coupled cap layers and the effect on perpendicular magnetic switching and recording characteristics, *IEEE Trans. Magn.*, **45**, pp. 2694–2700.
13. Varvaro G., Laureti S., and Fiorani D. (2014). $L1_0$ FePt-based thin films for future perpendicular magnetic recording media, *J. Magn. Magn. Mater.*, **368**, pp. 415–420.
14. Zhang L., Zhong W. W., Yu S. S., Xue S. X., Liu Y. P., Li Z. G., and Chen W. P. (2013). FePtCu–C granular film for perpendicular magnetic recording media, *J. Alloys Compd.*, **560**, pp. 177–180.
15. Zhou T.-J., Cher K. M., Lwin P. W., and Hu J. F. (2013). Energy barrier measurement and optimization in exchange coupled FePt/TiO_2 nano-composite thin films, *J. Magn. Magn. Mater.*, **331**, pp. 187–192.
16. Weller D., Mosendz O., Parker G., Pisana S., and Santos T. S. (2013). $L1_0$ FePtX-Y media for heat-assisted magnetic recording, *Phys. Status Solidi A*, **210**, pp. 1245–1260.
17. Chernyshov A., Treves D., Le T., Papusoi C., Yuan H., Ajan A., and Acharya R. (2013). Measurement of magnetic properties relevant to heat-assisted-magnetic-recording, *IEEE Trans. Magn.*, **49**, pp. 3572–3575.
18. Ishida N., Soeno Y., Sekiguchi K., and Nozaki Y. (2013). Frequency dependence of critical switching asteroid of CoCrPt–SiO_2 granular film under 50-ns microwave impulse, *J. Appl. Phys.*, **114**, p. 043915(5).
19. Stoner E. C., and Wohlfarth E. P. (1991). A mechanism of magnetic hysteresis in heterogeneous alloys, *IEEE Trans. Magn.*, **27**, pp. 3475–3518.
20. Thomson T., Lengsfield B., Do H., and Terris B. D. (2008). Magnetic anisotropy and reversal mechanisms in dual layer exchanged coupled perpendicular media, *J. Appl. Phys.*, **103**, p. 07F548(3).
21. Wu J., Holloway L., Laidler H., O'Grady K., Khizroev S., Howard J. K., Gustafson R. W., and Litvinov D. (2002). Magnetic characterization of perpendicular recording media, *IEEE Trans. Magn.*, **38**, pp. 1682–1686.
22. Lindemuth J., and Dodrill B. (2004). Measurement of the magnetic properties of double-layered PMRM using an AHE magnetometer, *IEEE Trans. Magn.*, **40**, pp. 2191–2193.
23. Richter H. J., and Harkness IV S. D. (2006). Media for magnetic recording beyond 100 Gbit/in^2, *MRS Bull.*, **31**, pp. 384–388.

24. Winkelmann A., Przybylski M., Luo F., Shi Y., and Barthel J. (2006). Perpendicular magnetic anisotropy induced by tetragonal distortion of FeCo alloy films grown on Pd(001), *Phys. Rev. Lett.*, **96**, p. 257205(4).

25. Shi J. Z., Piramanayagam S. N., Mah C. S., Zhao H. B., Zhao J. M., Kay Y. S., and Pock C. K. (2005). Influence of dual-Ru intermediate layers on magnetic properties and recording performance of CoCrPt–SiO_2 perpendicular recording media, *Appl. Phys. Lett.*, **87**, p. 222503(3).

26. Honda N., and Ouchi K. (2001). Low noise design of perpendicular magnetic recording media, *J. Magn. Magn. Mater.*, **235**, pp. 289–296.

27. Srinivasan K., Acharya B. R., and Bertero G. (2010). Thermal effects on the magnetization reversal process and its interpretation in perpendicular magnetic recording media, *J. Appl. Phys.*, **107**, p. 113912(7).

28. Saito S., Hoshi F., Itagaki N., and Takahashi M. (2003). Characterization to realize CoCr-based perpendicular magnetic recording media with high squareness and normalized coercivity, *J. Appl. Phys.*, **93**, pp. 6775–6777

29. Yan M. L., Skomski R., Kashyap A., Gao L., Liou S. H., and Sellmyer D. J. (2004). Hysteresis of granular FePt:Ag films with perpendicular anisotropy, *IEEE Trans. Magn.*, **40**, pp. 2495–2497.

30. Varvaro G., Testa A. M., Agostinelli E., Fiorani D., Laureti S., Springer F., Brombacher C., Albrecht M., Del Bianco L., Barucca G., Mengucci P., and Rinaldi D. (2013). Study of microstructure and magnetization reversal mechanism in granular CoCrPt:SiO_2 films of variable thickness, *Mater. Chem. Phys.*, **141**, pp. 790–796.

31. Suess D., Lee J., Fidler J., and Schrefl T. (2009). Exchange-coupled perpendicular media, *J. Magn. Magn. Mater.*, **321**, pp. 545–554.

32. Goll D., and Bublat T. (2013). Large-area hard magnetic $L1_0$-FePt and composite $L1_0$-FePt based nanopatterns, *Phys. Status Solidi A*, **210**, pp. 1261–1271.

33. Casoli F., Albertini F., Nasi L., Fabbrici S., Cabassi R., Bolzoni F., Bocchi C., and Luches P. (2010). Role of interface morphology in the exchange-spring behavior of FePt/Fe perpendicular bilayers, *Acta Mater.*, **58**, pp. 3594–3601.

34. Casoli F., Albertini F., Fabbrici S., Bocchi C., Nasi L., Ciprian R., and Pareti L. (2005). Exchange-coupled FePt/Fe bilayers with perpendicular magnetization, *IEEE Trans. Magn.*, **41**, pp. 3877–3879.

35. Sun A.-C., Yuan F.-T., Hsu J., Lin Y. H., and Kuo P. C. (2009). Magnetic reversal behaviors of perpendicular exchange-coupled Fe/FePt bilayer films, *IEEE Trans. Magn.*, **45**, pp. 2709–2715.

36. Brown W. F. (1962). Magnetostatic Principles in Ferromagnetism (North-Holland).
37. Miles J. J., Mc Kirdy D. M., Chantrell R. W., and Wood R. (2003). Parametric optimization for terabit perpendicular recording, *IEEE Trans. Magn.*, **39**, pp. 1876–1890.
38. Dobrynin A. N., Barthem V. M. T. S., and Givord D. (2009). Revisiting magnetization processes in granular hard magnetic materials, *Appl. Phys. Lett.*, **95**, p. 052511(3).
39. Wu X. W., Zhou H., Weller D., and Richter H. J. (2006). Measurement of angular-dependent hysteresis loops in perpendicular magnetic recording media, *J. Magn. Magn. Mater.*, **303**, pp. 5–8.
40. Chen D.-X., Brug J. A., Goldfarb R. B., and Member S. (1991). Demagnetizing factors for cylinders, *IEEE Trans. Magn.*, **27**, pp. 3601–3619.
41. Johnson M. T., Bloemen P. J. H., den Broeder F. J. A., and de Vries J. J. (1996). Magnetic anisotropy in metallic multilayers, *Rep. Prog. Phys.*, **59**, pp. 1409–1458.
42. Goh J. Q., Yuan, Z.-M., Shen L., Zhou T., and Feng Y. P. (2011). FePt grain size limit and required switching field in the presence of surface anisotropy, *IEEE Trans. Magn.*, **47**, pp. 3809–3812.
43. Ghidini M., Zangari G., Prejbeanu I. L., Pattanaik G., Buda-Prejbeanu L. D., Asti G., Pernechele C., and Solzi M. (2006). Magnetization processes in hard Co-rich Co–Pt films with perpendicular anisotropy, *J. Appl. Phys.*, **100**, p. 103911(10).
44. Gao K. Z., Heinonen O., and Chen Y. (2009). Read and write processes, and head technology for perpendicular recording, *J. Magn. Magn. Mater.*, **321**, pp. 495–507.
45. Inoue K., Shima H., Fujita A., Ishida K., Oikawa K., and Fukamichi K. (2006). Temperature dependence of magnetocrystalline anisotropy constants in the single variant state of $L1_0$-type FePt bulk single crystal, *Appl. Phys. Lett.*, **88**, p. 102503(3).
46. Cullity B. D., and Graham C. D. (2008). *Introduction to Magnetic Materials* (Wiley—IEEE Press).
47. Varvaro G., Albertini F., Agostinelli E., Casoli F., Fiorani D., Laureti S., Lupo P., Ranzieri P., Astinchap B., and Testa, A. M. (2012). Magnetization reversal mechanism in perpendicular exchange-coupled Fe/$L1_0$–FePt bilayers, *New J. Phys.*, **14**, p. 073008(14).
48. Richter H. J., Hellwig O., Florez S., Brombacher C., and Albrecht M. (2011). Anisotropy measurements of FePt thin films, *J. Appl. Phys.*, **109**, p. 07B713(3).

49. Siadou N., Androutsopoulos M., Panagiotopoulos I., Stoleriu L., Stancu A., Bakas T., and Alexandrakis V. (2011). Magnetization reversal in $[Ni/Pt]_6/Pt(x)/[Co/Pt]_6$ multilayers, *J. Magn. Magn. Mater.*, **323**, pp. 1671–1677.
50. Yang W., Lambeth D. N., and Laughlin D. E. (2000). Dependence of Co anisotropy constants on temperature, processing, and underlayer, *J. Appl. Phys.*, **87**, pp. 6884–6886.
51. Nemoto H., Araki R., and Hosoe Y. (2007). Magnetic anisotropy of perpendicular media: measurement and intermediate layer effect, *IEEE Trans. Magn.*, **43**, pp. 621–626.
52. Wu X. W., Asselin P. M., Veerdonk R. J., and Chantrell R. (2007). Evaluation of 45° method in characterizing the anisotropy of the perpendicular media, *J. Magn. Magn. Mater.*, **309**, pp. 139–143.
53. Zhou H. J., Chantrell R. W., and Weller D. (2003). Theory of rotational processes in perpendicular media and application to anisotropy measurement, *IEEE Trans. Magn.*, **39**, pp. 1902–1907.
54. Ju G., Wu X., Zhou H., Weller D., Chantrell R., Lu B., Kubota Y., and Yu J. (2003). Measurement of perpendicular media anisotropy field with accurate demagnetization correction, *J. Appl. Phys.*, **93**, p. 7846.
55. Hinata S., Saito S., Hasegawa D., and Takahashi M. (2011). Q-band ferromagnetic resonance for CoPt-based stacked perpendicular recording media with interlayer exchange coupling, *J. Appl. Phys.*, **109**, p. 083935(5).
56. Hinata S., Saito S., and Takahashi M. (2012). Ferromagnetic resonance analysis of internal effective field of classified grains by switching field for granular perpendicular recording media, *J. Appl. Phys.*, **111**, p. 07B722(3).
57. Harzer J. V., Hillebrands B., Stamps R. L., Güntherodt G., England C. D., and Falco C. M. (1991). Magnetic properties of Co/Pd multilayers determined by Brillouin light scattering and SQUID magnetometry, *J. Appl. Phys.*, **69**, pp. 2448–2454.
58. Bertram H. N., and Safonov V. L. (2001). The effect of uniaxial quartic anisotropy on thermal stability of magnetic nanograins, *Appl. Phys. Lett.*, **79**, p. 4402(3).
59. Shimizu Y., and Bertram H. N. (2003). Micromagnetic study of the transition parameter and position jitter in perpendicular recording, *IEEE Trans. Magn.*, **39**, pp. 1846–1850.
60. Berger A., Lengsfield B., and Ikeda Y. (2006). Determination of intrinsic switching field distributions in perpendicular recording media, *J. Appl. Phys.*, **99**, p. 08E705(6).

61. Voznyuk V. G., Misra A., Doyle W. D., Visscher P. B., and Chantrell R. W. (2005). Angular dispersion of the anisotropy in perpendicular media, *J. Appl. Phys.*, **97**, p. 10N508(3).
62. Churccmart J., Churccmart P., Evans R., Chantrcll R. W., and O'Grady K. (2011). Magnetic orientation in advanced recording media, *J. Phys. D. Appl. Phys.*, **44**, p. 455002(4).
63. Carter A., Fernandez-Outon L. E., Inaba Y., Greaves S. J., Muraoka H., and O'Grady K. (2008). Anisotropy dispersion in $(CoCrPt)_{1-x}(SiO_2)_x$ perpendicular recording media, *J. Magn. Magn. Mater.*, **320**, pp. 2269–2272.
64. Dong Y., and Victora R. H. (2012). Extracting intergranular exchange coupling in perpendicular magnetic recording media, *J. Magn. Magn. Mater.*, **324**, pp. 248–251.
65. Nemoto H., Takekuma I., Nakagawa H., Ichihara T., Araki R., and Hosoe Y. (2008). Designing magnetics of capped perpendicular media with minor-loop analysis, *J. Magn. Magn. Mater.*, **320**, pp. 3144–3150.
66. Berger A., Xu Y., Lengsfield B., Ikeda Y., and Fullerton E. E. (2005). $\Delta H(M,\Delta M)$ method of intrinsic switching field distributions in perpendicular media, *IEEE Trans. Magn.*, **41**, pp. 3178–3180.
67. Yuan E. (2004). Characterization of anisotropy field dispersion in perpendicular media, *IEEE Trans. Magn.*, **40**, pp. 2452–2454.
68. Liu Y., Hovorka O., Berger A., and Dahmen K. A. (2009). Effects of nonuniform exchange and magnetostatic interactions on the determination of intrinsic switching field distributions, *J. Appl. Phys.*, **105**, p. 123905(10).
69. Liu Y., Dahmen K. A., and Berger A. (2008). Determination of intrinsic switching field distributions in perpendicular recording media: numerical study of the $\Delta H(M,\Delta M)$ method, *Phys. Rev. B*, **77**, p. 054422(11).
70. Liu Y., Dahmen K. A., and Berger A. (2008). A comparative study of the $\Delta H(M,\Delta M)$ method reliability for square and triangular lattices, *J. Appl. Phys.*, **103**, p. 07F504(3).
71. Liu Y., Dahmen K. A., and Berger A. (2008). The role of dipolar interactions for the determination of intrinsic switching field distributions, *Appl. Phys. Lett.*, **92**, p. 222503(3).
72. Hovorka O., Pressesky J., Ju G., Berger A., and Chantrell R. W. (2012). Distribution of switching fields in magnetic granular materials, *Appl. Phys. Lett.*, **101**, p. 182405(5).

73. Hovorka O., Evans R. F. L., Chantrell R. W., and Berger A. (2010). Rate-dependence of the switching field distribution in nanoscale granular magnetic materials, *Appl. Phys. Lett.*, **97**, p. 062504(3).
74. Hovorka O., Evans R. F. L., Chantrell R. W., Liu Y., Dahmen K. A., and Berger A. (2010). Validation of ΔH(M,ΔM)-technique for identification of switching field distributions in the presence of thermal relaxation, *J. Appl. Phys.*, **108**, p. 123901(7).
75. Hovorka O., Liu Y., Dahmen K. A., and Berger A. (2010). On the ability to determine intrinsic switching field distributions from hysteresis loops in the partially correlated magnetization reversal regime, *J. Magn. Magn. Mater.*, **322**, pp. 459–468.
76. Hovorka O., Liu Y., Dahmen K. A., and Berger A. (2009). Simultaneous determination of intergranular interactions and intrinsic switching field distributions in magnetic materials, *Appl. Phys. Lett.*, **95**, p. 192504(3).
77. Tagawa I., and Nakamura Y. (1991). Relationship between high density recording performance and particle coercivity distribution, *IEEE Trans. Magn.*, **27**, pp. 4975–4977.
78. Hinata S., Saito S., Itagaki N., and Takahashi M. (2012). Effective switching field distribution evaluation using saturation field in minor loop*, *IEEE Trans. Magn.*, **48**, pp. 1248–1251.
79. Van de Veerdonk R. J. M., Wu X., and Weller D. (2003). Determination of switching field distributions for perpendicular recording media, *IEEE Trans. Magn.*, **39**, pp. 590–593.
80. Van de Veerdonk R. J. M., Wu X., and Weller D. (2002). Switching field distribution and ΔM measurements for perpendicular media, *IEEE Trans. Magn.*, **38**, pp. 2450–2452.
81. Dobrynin A. N., Gao T. R., Dempsey N. M., and Givord D. (2010). Experimental determination of the magnetization dependent part of the demagnetizing field in hard magnetic materials, *Appl. Phys. Lett.*, **97**, p. 192506(3).
82. Florez S. H., Boone C. T., Ikeda Y., Zhu F. Q., Takano K., and Terris B. D. (2012). Extraction of short time intrinsic switching field distributions in perpendicular media, *J. Appl. Phys.*, **111**, p. 07B703(3).
83. Winklhofer M., and Zimanyi G. T. (2006). Extracting the intrinsic switching field distribution in perpendicular media: a comparative analysis, *J. Appl. Phys.*, **99**, p. 08E710(3).
84. Wu X., and van de Veerdonk, R. J. M. (2008). Investigation on the mean field assumptions for switching field distribution measurements, *IEEE Trans. Magn.*, **44**, pp. 336–339.

85. Jung H. S., Ghaderi A., and Shi Z. (2011). Characterization of magnetic clustering and its effect on switching behavior and recording performance in perpendicular media, *J. Appl. Phys.*, **109**, p. 07B718(3).
86. Yin J., Zhang H., Hu F., Shen B., and Pan L. Q. (2009). First order reversal curve diagrams of perpendicular magnetic anisotropy films, *J. Appl. Phys.*, **106**, p. 103901(5).
87. Papusoi C., Srinivasan K., and Acharya R. (2011). Study of grain interactions in perpendicular magnetic recording media using first order reversal curve diagrams, *J. Appl. Phys.*, **110**, p. 083908(6).
88. Ju G., Zhou H., Chantrell R., Lu B., and Weller D. (2006). Direct probe of anisotropy field dispersion in perpendicular media, *J. Appl. Phys.*, **99**, p. 083902(4).
89. Srinivasan K., and Roddick E. (2012). Magnetic viscosity, hysteresis reptation, and their relationship with adjacent track interference in advanced perpendicular recording media, *J. Appl. Phys.*, **112**, p. 043911(13).
90. Pike C., Ross C., Scalettar R., and Zimanyi G. (2005). First-order reversal curve diagram analysis of a perpendicular nickel nanopillar array, *Phys. Rev. B*, **71**, p. 134407(12).
91. Pike C. R., Roberts A. P., and Verosub K. L. (1999). Characterizing interactions in fine magnetic particle systems using first order reversal curves, *J. Appl. Phys.*, **85**, pp. 6660–6667.
92. Harrison R. G. (2009). Physical theory of ferromagnetic first-order return curves, *IEEE Trans. Magn.*, **45**, pp. 1922–1939.
93. Harrison R. J., and Feinberg J. M. (2008). FORCinel: an improved algorithm for calculating first-order reversal curve distributions using locally weighted regression smoothing, *Geochem. Geophys.*, **9**, pp. 1–11.
94. Béron F., Carignan, Louis-Philippe Ménard D., and Yelon A. (2010). Extracting individual properties from global behaviour: first-order reversal curve method applied to magnetic nanowire arrays, in *Electrodeposited Nanowires and their Applications*, ed., Lupu N. (InTech).
95. Dobrota C.-I., and Stancu A. (2013). What does a first-order reversal curve diagram really mean? A study case: array of ferromagnetic nanowires, *J. Appl. Phys.*, **113**, p. 043928(11).
96. Valcu B. F., Gilbert D. A., and Liu K. (2011). Fingerprinting inhomogeneities in recording media using the first-order reversal curve method, *IEEE Trans. Magn.*, **47**, pp. 2988–2991.

97. Stancu A., Bissell P. R., and Chantrell R. W. (2000). Interparticle interactions in magnetic recording media as obtained from high-order measurements by a Preisach model, *J. Appl. Phys.*, **87**, pp. 8645–8652.
98. Gaur N., Piramanayagam S. N., Maurer S. L., Steen S. E., Yang H., and Bhatia C. S. (2012). First-order reversal curve investigations on the effects of ion implantation in magnetic media, *IEEE Trans. Magn.*, **48**, pp. 2753–2756.
99. Thomson T., Coffey K. R., and Thiele J.-U. (2003). Angle-dependent switching of granular and multilayer perpendicular media, *IEEE Trans. Magn.*, **39**, pp. 2314–2316.
100. Coffey K. R., Thomson T., and Thiele J.-U. (2002). Angular dependence of the switching field of thin-film longitudinal and perpendicular magnetic recording media, *J. Appl. Phys.*, **92**, p. 4553(7).
101. Tibus S., Strache T., Springer F., Makarov D., Rohrmann H., Schrefl T., Fassbender J., and Albrecht M. (2010). Magnetic properties of granular CoCrPt:SiO_2 films as tailored by Co^+ irradiation, *J. Appl. Phys.*, **107**, p. 093915(5).
102. Jung H. S., Kuo M., Malhotra S. S., and Bertero G. (2007). Origin of incoherent magnetic switching behavior in CoCrPt–SiO_2 perpendicular magnetic recording media, *Appl. Phys. Lett.*, **91**, p. 212502(3).
103. Morrison C., Saharan L., Ikeda Y., Takano K., Hrkac G., and Thomson T. (2013). Quantifying exchange coupling in segregated granular materials, *J. Phys. D. Appl. Phys.*, **46**, p. 475002(5).
104. Morrison C., Saharan L., Hrkac G., Schrefl T., Ikeda Y., Takano K., Miles J. J., and Thomson T. (2011). Inter/intra granular exchange and thermal activation in nanoscale granular magnetic materials, *Appl. Phys. Lett.*, **99**, p. 132507(3).
105. Saharan L., Morrison C., Ikeda Y., Takano K., Miles J. J., Thomson T., Schrefl T., and Hrkac G. (2013). Grain boundaries in granular materials: a fundamental limit for thermal stability, *Appl. Phys. Lett.*, **102**, p. 142402(4).
106. Saharan L., Morrison C., Miles J. J., Thomson T., Schrefl T., and Hrkac G. (2011). Angle dependence of the switching field of recording media at finite temperatures, *J. Appl. Phys.*, **110**, p. 103906(5).
107. Kondorsky (1940). Hysteresis in ferromagnetics, *J. Phys.*, **2**, pp. 161–181.
108. Schumacher F. (1991). On the modification of the Kondorsky function, *J. Appl. Phys.*, **70**, pp. 3184–3187.

109. Behler C., Neu V., Schultz L., and Fähler S. (2012). Magnetically and thermally induced switching processes in hard magnets, *J. Appl. Phys.*, **112**, p. 083919(6).

110. Kronmüller H., Durst K.-D., and Martinek G. (1987). Angular dependence of the coercive field in sitered $Fe_{77}Nd_{15}B_8$ magnets, *J. Magn. Magn. Mater.*, **69**, pp. 149–157.

111. Rahman M. T., Dumas R. K., Eibagi N., Shams N. N., Wu, Y.-C., Liu K., and Lai, C.-H. (2009). Controlling magnetization reversal in Co/Pt nanostructures with perpendicular anisotropy, *Appl. Phys. Lett.*, **94**, p. 042507(3).

112. Davies J., Hellwig O., Fullerton E., Denbeaux G., Kortright J., and Liu K. (2004). Magnetization reversal of Co/Pt multilayers: microscopic origin of high-field magnetic irreversibility, *Phys. Rev. B*, **70**, p. 224434(8).

113. Collocott S. J., and Neu V. (2012). Time-dependent magnetization in epitaxial hard magnetic thin films, *J. Phys. D. Appl. Phys.*, **45**, p. 035002(13).

114. Inoue D., Inaba Y., Komiyama K., and Shimatsu T. (2011). Characterization of coercivity at a recording speed of granular media for thermally assisted recording, *J. Appl. Phys.*, **109**, p. 07B727(3).

115. Shimatsu T., Uwazumi H., Muraoka H., and Nakamura Y. (2001). Thermal stability in perpendicular recording media, *J. Magn. Magn. Mater.*, **235**, pp. 273–280.

116. Piramanayagam S. N., and Ranjbar M. (2012). Influence of magnetic viscosity on the first order reversal curves of antiferromagnetically coupled perpendicular recording media, *J. Appl. Phys.*, **111**, p. 07B728(3).

117. Inaba Y., Shimatsu T., Muraoka H., Dutson J. D., and O'Grady K. (2005). Activation volumes in CoPtCr-SiO_2 perpendicular recording media, *IEEE Trans. Magn.*, **41**, pp. 3130–3132.

118. Misra A., Visscher P. B., and Apalkov D. M. (2003). Time-dependent remanent coercivity in nanosecond switching in perpendicular media, *J. Appl. Phys.*, **94**, pp. 6013–6016.

119. Plumer M. L., Leblanc M. D., Whitehead J. P., and van Ek J. (2012). Micromagnetic simulations of sweep-rate dependent coercivity in perpendicular recording media, *J. Appl. Phys.*, **111**, p. 123905(7).

120. Sharrock M. P. (1994). Time dependence of switching fields in magnetic recording media, *J. Appl. Phys.*, **76**, pp. 6413–6418.

121. Suess D., Eder S., Lee J., Dittrich R., Fidler J., Harrell J., Schrefl T., Hrkac G., Schabes M., Supper N., and Berger A. (2007). Reliability

of Sharrocks equation for exchange spring bilayers, *Phys. Rev. B*, **75**, p. 174430(11).

122. Harrell J. W. (2001). Orientation dependence of the dynamic coercivity of Stoner-Wohlfarth particles, *IEEE Trans. Magn.*, **37**, pp. 533–537.

123. Jung H. S., Kuo M., Malhotra S. S., and Bertero G. (2008). Effect of nonuniform microstructure on magnetic switching behavior in CoCrPt–SiO_2 perpendicular magnetic recording media, *J. Appl. Phys.*, **103**, p. 07F515(3).

124. Dutson J. D., O'Grady K., Lu B., Kubota Y., and Platt C. L. (2003). Determination of activation volumes of reversal in perpendicular media, *IEEE Trans. Magn.*, **39**, pp. 2344–2346.

125. Dutson J. D., Litvinov D., Gibbs M. R. J., Inaba Y., Muraoka H., and O'Grady K. (2006). Magnetisation reversal in media with perpendicular anisotropy, *J. Magn. Magn. Mater.*, **304**, pp. 51–54.

126. Dutson J. D., Hashimoto M., Inaba Y., Greaves S. J., Muraoka H., and O'Grady K. (2007). Magnetic and recording properties of $(CoCrPt)_{1-x}(SiO_2)_x$ media, *IEEE Trans. Magn.*, **43**, pp. 814–818.

127. El-Hilo M., O'Grady K., and Chantrell R. W. (2002). Fluctuation fields and reversal mechanisms in granular magnetic systems, *J. Magn. Magn. Mater.*, **248**, pp. 360–373.

128. El-Hilo M., and Bsoul I. (2007). Interaction effects on the coercivity and fluctuation field in granular powder magnetic systems, *Phys. B*, **389**, pp. 311–316.

129. Sbiaa R., Mochida M., Itoh Y., and Suzuki T. (2000). Relaxation time and thermal stability in magneto-optical recording media, *IEEE Trans. Magn.*, **36**, pp. 2279–2281.

130. Gaunt P. (1986). Magnetic viscosity and thermal activation energy, *J. Appl. Phys.*, **59**, pp. 4129–4132.

131. Igarashi M., Akagi F., and Sugita Y. (2008). Validity of thermal activation volume estimated by using Wohlfarth's equation in perpendicular recording hard disk media, *J. Appl. Phys.*, **103**, p. 07F529(3).

132. Igarashi M., Akagi F., and Sugita Y. (2006). Micromagnetic simulation of thermal activation volume in perpendicular recording media with dispersions of grain size, *IEEE Trans. Magn.*, **42**, pp. 2393–2395.

133. Kitano M., Miyashita E., Hayashi N., and Takenoiri S. (2007). Bit-percolation studies on CoPtCr-SiO_2 perpendicular magnetic-recording media, *IEEE Trans. Magn.*, **43**, pp. 2094–2096.

134. Wu X. W., Peng Y., and Wang B. (2008). Magnetic cluster size and cluster size distribution study on perpendicular media, *Appl. Phys. Lett.*, **93**, p. 152514(3).
135. Hashimoto M., Nakagawa H., Nakamoto K., Ide H., and Ichihara T. (2009). Magnetic cluster size distribution and its relation to read/write performance from spinstand testing, *IEEE Trans. Magn.*, **45**, pp. 3718–3721.
136. Suess D. (2007). Micromagnetics of exchange spring media: optimization and limits, *J. Magn. Magn. Mater.*, **308**, pp. 183–197.
137. Suess D. (2006). Multilayer exchange spring media for magnetic recording, *Appl. Phys. Lett.*, **89**, p. 113105(3).
138. Suess D., Schrefl T., Fähler S., Kirschner M., Hrkac G., Dorfbauer F., and Fidler J. (2005). Exchange spring media for perpendicular recording, *Appl. Phys. Lett.*, **87**, p. 012504(3).
139. Victora R. H. (2005). Composite media for perpendicular magnetic recording, *IEEE Trans. Magn.*, **41**, pp. 537–542.
140. Asti G., Ghidini M., Pellicelli R., Pernechele C., Solzi M., Albertini F., Casoli F., Fabbrici S., and Pareti L. (2006). Magnetic phase diagram and demagnetization processes in perpendicular exchange-spring multilayers, *Phys. Rev. B*, **73**, p. 094406(16).
141. Dobin A. Y., and Richter H. J. (2006). Domain wall assisted magnetic recording, *Appl. Phys. Lett.*, **89**, p. 062512(3).
142. Kronmüller H., and Goll D. (2011). Magnetic properties of exchange-spring composite films, *Phys. Status Solidi B*, **248**, pp. 2361–2367.
143. Skomski R., George T. A., and Sellmyer D. J. (2008). Nucleation and wall motion in graded media, *J. Appl. Phys.*, **103**, p. 07F531(3).
144. Zimanyi G. T. (2008). Graded media: optimization and energy barriers, *J. Appl. Phys.*, **103**, p. 07F543(4).
145. Casoli F., Albertini F., Nasi L., Fabbrici S., Cabassi R., Bolzoni F., and Bocchi C. (2008). Strong coercivity reduction in perpendicular FePt/Fe bilayers due to hard/soft coupling, *Appl. Phys. Lett.*, **92**, p. 142506(3).
146. Alexandrakis V., Speliotis T., Manios E., Niarchos D., Fidler J., Lee J., and Varvaro G. (2011). Hard/graded exchange spring composite media based on FePt, *J. Appl. Phys.*, **109**, p. 07B729(3).
147. Di Bona A., Luches P., Albertini F., Casoli F., Lupo P., Nasi L., D'Addato S., Gazzadi G. C., and Valeri S. (2013). Anisotropy-graded magnetic media obtained by ion irradiation of $L1_0$ FePt, *Acta Mater.*, **61**, pp. 4840–4847.

148. Thiele J.-U., Maat S., and Fullerton E. E. (2003). FeRh/FePt exchange spring films for thermally assisted magnetic recording media, *Appl. Phys. Lett.*, **82**, pp. 2859–2861.

149. Lubarda M. V., Li S., Livshitz B., Fullerton E. E., and Lomakin V. (2011). Reversal in bit patterned media with vertical and lateral exchange, *IEEE Trans. Magn.*, **47**, pp. 18–25.

150. Wang H., Zhao H., Rahman T., Isowaki Y., Kamata Y., Maeda T., Hieda H., Kikitsu A., and Wang, J.-P. (2013). Fabrication and characterization of FePt exchange coupled composite and graded bit patterned media, *IEEE Trans. Magn.*, **49**, pp. 707–712.

151. Wang H., Li W., Rahman M. T., Zhao H., Ding J., Chen Y., and Wang, J.-P. (2012). Characterization of $L1_0$-FePt/Fe based exchange coupled composite bit pattern media, *J. Appl. Phys.*, **111**, p. 07B914(3).

152. Bashir M. A., Schrefl T., Suess D., Dean J., Goncharov A., Hrkac G., Bance S., Allwood D. A., and Fidler J. (2009). Exchange coupled bit patterned media under the influence of RF-field pulses, *IEEE Trans. Magn.*, **45**, pp. 3851–3854.

153. Ikeda Y., Choe G., Florez S., Nemoto H., Lengsfield B., and Takano K. (2010). Incoherent switching amount analysis of exchange spring media by using initial minor loop slope measurement, *IEEE Trans. Magn.*, **46**, pp. 1852–1855.

154. Hauet T., Florez S., Margulies D., Ikeda Y., Lengsfield B., Supper N., Takano K., Hellwig O., and Terris B. D. (2009). Revealing the reversible rotation of magnetization in exchange-coupled composite media switching, *Appl. Phys. Lett.*, **95**, p. 222507(3).

155. Dumas R. K., Fang Y., Kirby B. J., Zha C., Bonanni V., Nogués J., and Åkerman J. (2011). Probing vertically graded anisotropy in FePtCu films, *Phys. Rev. B*, **84**, p. 054434(9).

156. Choi Y., Jiang J. S., Pearson J. E., Bader S. D., and Liu J. P. (2007). Origin of recoil hysteresis loops in Sm–Co/Fe exchange-spring magnets, *Appl. Phys. Lett.*, **91**, p. 022502(3).

157. Dumas R. K., Zha C., Fang Y., Bonanni V., Lau J. W., Nogués J., and Åkerman J. (2011). Graded anisotropy FePtCu films, *IEEE Trans. Magn.*, **47**, pp. 1580–1586.

158. Roy D., Sreenivasulu K. V., and Anil Kumar P. S. (2013). Investigation on non-exchange spring behaviour and exchange spring behaviour: a first order reversal curve analysis, *Appl. Phys. Lett.*, **103**, p. 222406(5).

159. Markou A., Panagiotopoulos I., Bakas T., Postolache P., Stoleriu L., and Stancu A. (2012). Magnetization reversal in graded anisotropy Co/Pt

multilayers: a first order reversal curve study, *J. Appl. Phys.*, **112**, p. 123914(7).

160. Suess D., Schrefl T., Dittrich R., Kirschner M., Dorfbauer F., Hrkac G., and Fidler J. (2005). Exchange spring recording media for areal densities up to 10Tbit/in^2, *J. Magn. Magn. Mater.*, **290-291**, pp. 551–554.
161. Jiang C. J., Chen J. S., Hu J. F., and Chow G. M. (2010). FePt–TiO_2 exchange coupled composite media with well-isolated columnar microstructure for high density magnetic recording, *J. Appl. Phys.*, **107**, p. 123915(7).
162. Feng C., Gong K., Xu C.-C., Yang M., Li X., Li B., and Yu G. (2014). Synthesis and property tunability of interparticle exchange-decoupled $L1_0$-FePt:Au/Fe perpendicular ECC films, *J. Alloys Compd.*, **590**, pp. 289–293.
163. Suess D., Fidler J., Zimanyi G., Schrefl T., and Visscher P. (2008). Thermal stability of graded exchange spring media under the influence of external fields, *Appl. Phys. Lett.*, **92**, p. 173111(3).
164. Griffiths R. A., Williams A., Oakland C., Roberts J., Vijayaraghavan A., and Thomson T. (2013). Directed self-assembly of block copolymers for use in bit patterned media fabrication, *J. Phys. D. Appl. Phys.*, **46**, p. 503001(29).
165. Hellwig O., Berger A., Thomson T., Dobisz E., Bandic Z. Z., Yang H., Kercher D. S., and Fullerton E. E. (2007). Separating dipolar broadening from the intrinsic switching field distribution in perpendicular patterned media, *Appl. Phys. Lett.*, **90**, p. 162516(3).
166. (a) Krone P., Makarov D., Cattoni A., Faini G., Haghiri-Gosnet, A.-M., KnittelI., Hartmann U., Schrefl T., and Albrecht M. (2011). Investigation of the magnetization reversal of a magnetic dot array of Co/Pt multilayers, *J. Nanopart. Res.*, **13**, pp. 5587–5593. (b) Thomson T., Hu G., and Terris B. (2006). Intrinsic distribution of magnetic anisotropy in thin films probed by patterned nanostructures, *Phys. Rev. Lett.*, **96**, p. 257204(4).
167. Ishio S., Takahashi S., Hasegawa T., Arakawa A., Sasaki H., Yan Z., Liu X., Kondo Y., Yamane H., Ariake J., Suzuki M., Kawamura N., and Mizumaki M. (2014). Switching field distribution and magnetization reversal process of FePt dot patterns, *J. Magn. Magn. Mater.*, **360**, pp. 205–210.
168. Bai J., Takahoshi H., Ito H., Saito H., and Ishio S. (2004). Dot-by-dot analysis of magnetization reversal in perpendicular patterned CoCrPt medium by using magnetic force microscopy, *J. Appl. Phys.*, **96**, pp. 1133–1137.

169. Li W. M., Chen Y. J., Huang T. L., Xue J. M., and Ding J. (2011). Calculation of individual bit island switching field distribution in perpendicular magnetic bit patterned media, *J. Appl. Phys.*, **109**, p. 07B758(3).

170. Springer F., Hellwig O., Dobisz E., Albrecht M., and Grobis M. (2011). Probing the time-dependent switching probability of individual patterned magnetic islands, *J. Appl. Phys.*, **109**, p. 07B905(3).

171. Chen Y., Ding J., Deng J., Huang T., Leong S. H., Shi J., Zong B., Ko H. Y. Y., Au C. K., Hu S., and Liu B. (2010). Switching probability distribution of bit islands in bit patterned media, *IEEE Trans. Magn.*, **46**, pp. 1990–1993.

172. Hu G., Thomson T., Rettner C. T., Raoux S., and Terris B. D. (2005). Magnetization reversal in Co/Pd nanostructures and films, *J. Appl. Phys.*, **97**, p. 10J702(3).

173. Weller D., Parker G., Mosendz O., Champion E., Stipe B., Wang X., Klemmer T., Ju G., and Ajan A. (2014). A HAMR media technology roadmap to an areal density of 4 Tb/in^2, *IEEE Trans. Magn.*, **50**, pp. 1–8.

174. Wang X., Gao K., Zhou H., Itagi A., Seigler M., and Gage E. (2013). HAMR recording limitations and extendibility, *IEEE Trans. Magn.*, **49**, pp. 686–692.

175. Shiimoto M., Igarashi M., Sugiyama M., Nishida Y., and Tagawa I. (2013). Effect of effective field distribution on recording performance in microwave assisted magnetic recording, *IEEE Trans. Magn.*, **49**, pp. 3636–3639.

176. Zhu J.-G., Zhu X., and Tang Y. (2008). Microwave assisted magnetic recording, *IEEE Trans. Magn.*, **44**, pp. 125–131.

177. Wang X., Gao K.-Z., Hohlfeld J., and Seigler M. (2010). Switching field distribution and transition width in energy assisted magnetic recording, *Appl. Phys. Lett.*, **97**, p. 102502(3).

178. Rong C. B., Li D., Nandwana V., Poudyal N., Ding Y., Wang Z. L., Zeng H., and Liu J. P. (2006). Size-dependent chemical and magnetic ordering in $L1_0$-FePt nanoparticles, *Adv. Mater.*, **18**, pp. 2984–2988.

179. Barber M. N. (1983). *Phase Transitions and Critical Phenomena,* eds., Domb C., and Lebowitz L. (Academic, New York).

180. Hovorka O., Devos S., Coopman Q., Fan W. J., Aas C. J., Evans R. F. L., Chen X., Ju G., and Chantrell R. W. (2012). The Curie temperature distribution of FePt granular magnetic recording media, *Appl. Phys. Lett.*, **101**, p. 052406(4).

181. Xu L., and Zhang S. (2012). Magnetization dynamics at elevated temperatures, *Phys. E*, **45**, pp. 72–76.

182. Pisana S., Jain S., Reiner J. W., Parker G. J., Poon C. C., Hellwig O., and Stipe B. C. (2014). Measurement of the Curie temperature distribution in FePt granular magnetic media, *Appl. Phys. Lett.*, **104**, p. 162407(5).

183. Thiele J.-U., Coffey K. R., Toney M. F., Hedstrom J. A., and Kellock A. J. (2002). Temperature dependent magnetic properties of highly chemically ordered $Fe_{55-x}Ni_xPt_{45}$ $L1_0$ films, *J. Appl. Phys.*, **91**, pp. 6595–6600.

184. Pisana S., Mosendz O., Parker G. J., Reiner J. W., Santos T. S., McCallum A. T., Richter H. J., and Weller D. (2013). Effects of grain microstructure on magnetic properties in FePtAg-C media for heat assisted magnetic recording, *J. Appl. Phys.*, **113**, p. 043910(6).

185. Kryder M. H., Gage E. C., Mcdaniel T. W., Challener W. A., Rottmayer R. E., Ju G., Hsia Y., and Erden M. F. (2008). Heat assisted magnetic recording, *Proc. IEEE*, **96**, pp. 1810–1835.

186. Farle M. (1998). Ferromagnetic resonance of ultrathin metallic layers, *Rep. Prog. Phys.*, **61**, pp. 755–826.

187. Heinrich B. (1994). *Ultrathin Magnetic Structures*, eds., Heinrich B., and Bland J. A. C. (Springer Verlag).

188. Bilzer C., Devolder T., Crozat P., Chappert C., Cardoso S., and Freitas P. P. (2007). Vector network analyzer ferromagnetic resonance of thin films on coplanar waveguides: comparison of different evaluation methods, *J. Appl. Phys.*, **101**, p. 074505(5).

189. Thirion C., Wernsdorfer W., and Mailly D. (2003). Switching of magnetization by nonlinear resonance studied in single nanoparticles, *Nat. Mater.*, **2**, pp. 524–527.

190. Woltersdorf G., and Back C. (2007). Microwave assisted switching of single domain $Ni_{80}Fe_{20}$ elements, *Phys. Rev. Lett.*, **99**, p. 227207(4).

191. Wannawong N., Tipcharoen W., and Kaewrawang A. (2014). Microwave assisted magnetization reversal on exchange coupled composite media, *Adv. Mater. Res.*, **931–932**, pp. 1265–1269.

192. Okamoto S., Furuta M., Kikuchi N., Kitakami O., and Shimatsu T. (2014). Theory and experiment of microwave-assisted magnetization switching in perpendicular magnetic nanodots, *IEEE Trans. Magn.*, **50**, pp. 83–88.

193. Boone C. T., Katine J. A., Marinero E. E., Pisana S., and Terris B. D. (2012). Demonstration of microwave assisted magnetic reversal in perpendicular media, *J. Appl. Phys.*, **111**, p. 07B907(3).

Chapter 9

New Trends in Magnetic Memories

Riccardo Bertacco and Matteo Cantoni
Dipartimento di Fisica, Politecnico di Milano, Via G. Colombo 81, 20133 Milano, Italy
matteo.cantoni@polimi.it

9.1 Introduction

Magnetoresistive random access memories (MRAMs) are a relatively new class of nonvolatile memories exploiting the hysteretic properties of ferromagnetic (FM) materials for data storage and magnetoresistive phenomena (giant magnetoresistance [GMR], tunneling magnetoresistance [TMR], anisotropic magnetoresistance [AMR]) for data reading. Briefly, an MRAM chip is a bidimensional array of magnetoresistive devices with stable remanent states (0 and 1), integrated on a silicon complementary metal–oxide semiconductor (CMOS) circuit allowing one to separately address each memory element.

Since the birth of the MRAM concept in the mid-1980s [1, 2] and the key improvement due to GMR discovery in 1988 [3, 4], leading to the Nobel Prize in 2007 for A. Fert and P. Grünberg, MRAMs have attracted great attention from many big memory producers. Just to name some of them, Motorola (later to become Freescale and

Ultrahigh-Density Magnetic Recording: Storage Materials and Media Designs
Edited by Gaspare Varvaro and Francesca Casoli

ISBN 978-981-4669-58-0 (Hardcover), 978-981-4669-59-7 (eBook)
www.panstanford.com

(DRAMs) and low writing speed (10 μs of flashes versus 10 ns of MRAMs).

Nowadays, new writing strategies beyond STT are under development, with the aim of further reducing power consumption and relatively high current density typical of STT. Furthermore, other architectures for increasing the memory density on chip are currently under evaluation. In 2008 IBM proposed a new concept, the so-called racetrack memory (RM), which is the evolution and on-chip implementation of the concept of magnetic tape recording. The motion of magnetic domain walls (DWs) in 3D conduits, which are then read by a unique magnetic sensor, is used to increase the number of bits stored per inch square of silicon chip. A second relevant and promising research line is that connected to the use of antiferromagnets as storing materials, the latter being much more stable against magnetic field perturbation and not producing stray fields which also limit the packing density in MRAMs. Despite antiferromagnet spintronics is definitely at its infancy, some companies, such as Hitachi, have recently turned their attention to this emerging field and are actively involved in some running research projects.

This chapter aims at providing an overview of new strategies currently under development to overcome major limitations of MRAMs that have impeded their spread in the consumer electronics market. The structure of the chapter is as follows. First, we present recent developments related to writing. After a brief presentation of the state of the art (the Stoner–Wolhfarth astroid (SWA) method, current lines, heat-assisted writing), new strategies still under development (STT, spin–orbit torque [SOT], magneto-electric coupling [MEC], optical writing) are discussed. Then, complementary strategies for increasing the density and stability of MRAMs are described, with reference to the RM and to the emerging field of antiferromagnet spintronics.

Finally, a brief overview of the new concept of neuromorphic computing and magnetic memristors is given. Even though this field is still in its infancy, bioinspired computing architectures beyond conventional digital von Neumann machines are expected to revolutionize the architecture of future computers, with a huge impact also on the field of magnetic memories.

9.2 State of the Art of Magnetic Random Access Memories

9.2.1 *Basics of MRAM Working Principles*

A conventional MRAM is based on a 2D array of memory cells (Fig. 9.1), which can be addressed individually because each cell is at the crosspoint of two lines (the bit line and the word line). Each cell is made of a magnetic heterostructure, composed of two FM layers, separated by a nonmagnetic layer: the latter can be conductive, as in GMR cells, or insulating, as in TMR cells. One of these layers (the reference, or pinned, layer) has its magnetization fixed in one direction, typically by exchange bias with an antiferromagnet. The magnetization of the second layer (the storage, or free, layer), instead, can be switched, and its relative direction with respect to the pinned layer (parallel or antiparallel) constitutes the bit of information.

The writing process of the bit in the cell consists of setting the magnetization of the free layer (FL) with respect to the pinned layer, parallel or antiparallel. A lot of effort has been put in the development of writing technologies allowing an increase in the

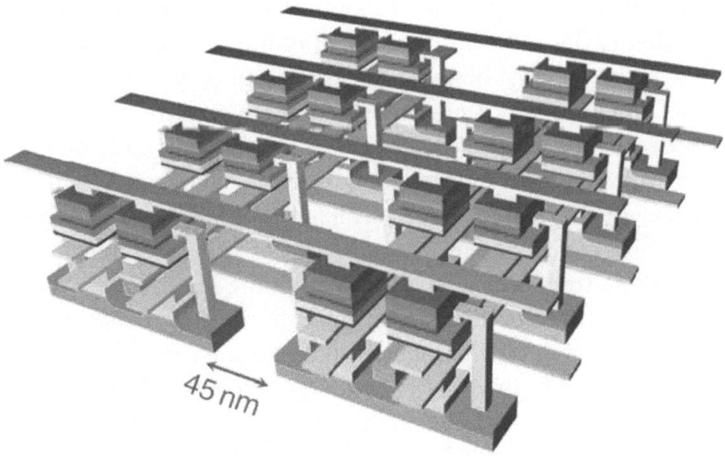

Figure 9.1 Array design of an MRAM device on a CMOS platform. Adapted by permission from Macmillan Publishers Ltd: *Nat. Commun.* (Ref. [18]), copyright (2011).

writing speed and a reduction in the power consumption. Some of them (current lines, heat-assisted writing, STT, SOT, MEC, optical writing) will be reviewed below due to the high relevance of this aspect for MRAMs.

The reading process of the bit consists in measuring the resistance of the heterostructure, whose value depends on the relative orientation of the pinned and free layers. Typically high (low) impedance states correspond to antiparallel (parallel) alignment. Since the discovery of GMR and TMR, the huge change in resistivity associated with the two magnetic alignments (up to 604% in the case of TMR [19] and 55% in the case of GMR [20], at room temperature) makes the reading process absolutely noncritical.

Because of its larger resistance variation, typically TMR-based cells are used. The TMR ratio is defined as

$$\mathrm{TMR} = \frac{R_{\mathrm{AP}} - R_{\mathrm{P}}}{R_{\mathrm{P}}} \tag{9.1}$$

where R_{P} and R_{AP} are the resistances in the parallel and antiparallel configurations, respectively, of the free and pinned layer magnetizations. The TMR effect for AlO_x tunneling barriers has been demonstrated in 1995 and typically does not exceed 70% [21], while with MgO a record value of 604% at room temperature has been achieved in 2008 [19]. This peculiar feature of MgO barriers is related to the different decay rates of Bloch states with different symmetries in crystalline MgO layers embedded in Fe/MgO/Fe tunnel junctions [22]. Only states with Δ_1 symmetry present a very slow decay rate and can then ensure high transmission. However, in Fe there exist only majority states with Δ_1 symmetry crossing the Fermi level, thus leading to very efficient spin filtering at low bias. In the parallel configuration, Δ_1 majority electrons are thus efficiently transmitted from one electrode through the MgO barrier and find available states with the same symmetry in the other electrode. On the contrary, in the antiparallel configuration Δ_1 majority spins will not find available final states with the same spin and symmetry, thus leading to a much lower transmission. As a consequence, the conductance in the parallel state is much higher than that in the antiparallel state, thus resulting in a very large value of TMR.

Different materials have been proposed for the role of FM electrodes [23]:

- 3*d* metals: Fe [18, 25], Co [26]
- Alloys of 3*d* metals: CoFe [26], NiFe [27], CoFeB [19, 28]
- Oxides: Fe_3O_4 [29], $La_{1-x}Sr_xMnO_3$ [30], $La_{1-x}Ca_xMnO_3$ [31], CrO_2 [32]
- Heusler alloys: NiMnSb [33], Co_2MnSi [34]
- Diluted magnetic semiconductors (DMSs): (GaMn)As [35]

Maximizing the TMR at room temperature is one of the key issues for realizing efficient (in terms of the signal-to-noise ratio) memory cells. It is widely recognized that the quality of the interfaces between the insulating layer and the FM electrodes plays a fundamental role. For example, single-crystalline high-quality MTJs based on the Fe/MgO/Fe template, grown by molecular beam epitaxy, lead to TMR values up to 200% at room temperature [24, 25], while a record value of 604% at room temperature can be achieved with the structure CoFeB/MgO/CoFeB, grown by magnetron sputtering [19]. Moreover, these structures are easily integrable on CMOS platforms by employing a buffer layer (Ta/Ru/Ta) grown directly on Si or SOI substrates. For these reasons, to date the CoFeB/MgO/CoFeB template is the leader technology for MRAM applications [6].

MTJs based on polycrystalline electrodes, such as the Co_2FeAl [36] and Co_2MnSi [37] Heusler alloys, and on FM oxide materials with complex stoichiometry, such as the half-metallic manganites [30, 31], have also been tested but with limited industrial interest up to now. While all these templates employ electrodes with in-plane (IP) magnetization, recently large interest has emerged toward magnetic layers with out-of-plane magnetization (see Section 9.3.1 for a more detailed discussion). The best candidates are actually ultrathin (~1 nm) CoFeB films, leading to TMR larger than 120% at room temperature and presenting a better stability and an easier fabrication process with respect to other candidates previously employed, such as multilayers based on Co/Pd and Fe/Pt bilayers [28].

9.2.2 *Current Writing*

In Fig. 9.2 an example of a cell pertaining to the first generation of MRAMs and exploiting current lines for writing is reported [38,

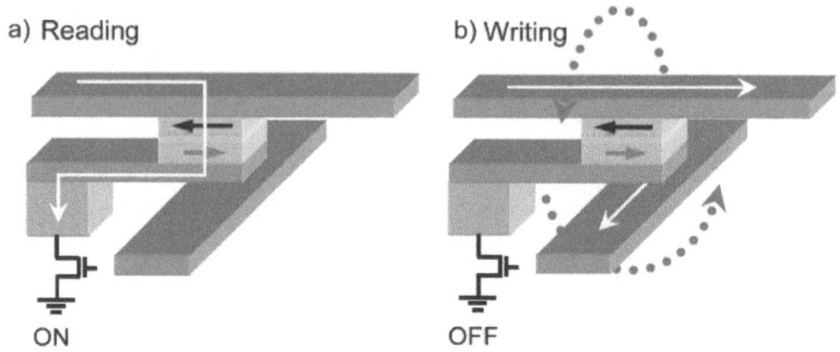

Figure 9.2 Basic processes of reading (a) and writing (b) in a CMOS-based MRAM cell. Adapted from Ref. [39].

39]. The memory cell combines a CMOS selection transistor with an MTJ. At reading (Fig. 9.2a), a low current pulse opens the transistor and addresses the cell. The cell resistance is measured by driving a current across the MTJ, from the top (bit line) to the bottom (word line) and measuring the corresponding voltage drop.

At writing (Fig. 9.2b), synchronized current pulses are sent to the word and bit lines, arranged in cross architecture, in order to generate magnetic fields able to switch the FL of the selected cell. This method is called field-induced magnetic switching (FIMS); alternative methods for writing will be presented in the next sections.

9.2.2.1 Stoner–Wolhfarth astroid

A problem with the FIMS method is that, because of the 2D geometry of the cell array, when a current pulse is injected along a word (or bit) line all the cells on the line will sense the magnetic field generated by the pulse.

To be able to address a single cell, that is, to change the magnetic state of the cell while leaving unchanged the remaining ones, the SWA property can be employed [40].

This method is based on the Stoner–Wolhfarth theory of coherent rotation in single-domain particles. In a single-domain magnetic structure with uniaxial magnetic anisotropy, the field parallel to the easy axis that is needed for reversing its magnetization ($\mathbf{H}_{\text{easy}}$) can be reduced by applying simultaneously a second field along the hard

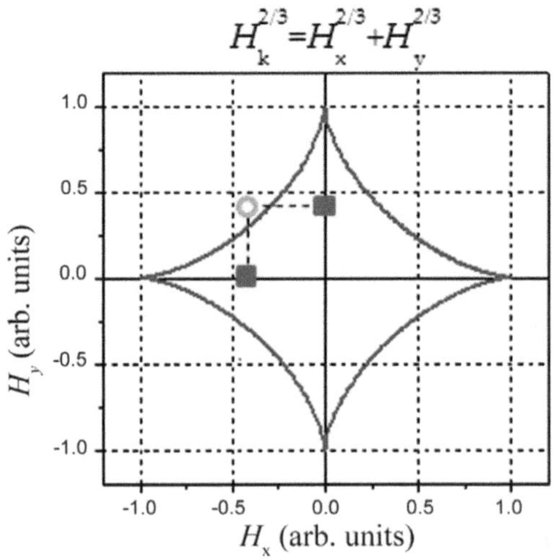

Figure 9.3 Stoner-Wolhfarth astroid: x and y are, respectively, the easy and the hard axis. A value of $H_K = 2K/M_S = 1$ has been used.

axis ($\mathbf{H}_{\text{hard}}$). The components of the switching fields are given by the astroid equation:

$$H_{\text{easy}}^{2/3} + H_{\text{hard}}^{2/3} = \left(\frac{2K}{M_S}\right)^{2/3} \quad (9.2)$$

where K is the effective anisotropy constant and M_S is the saturation magnetization [41]. This equation defines a curve in the ($\mathbf{H}_{\text{easy}}$, $\mathbf{H}_{\text{hard}}$) plane. Switching of the magnetic nanostructure occurs for any combination of $\mathbf{H}_{\text{easy}}$ and $\mathbf{H}_{\text{hard}}$ lying outside the astroid (dot in Fig. 9.3, e.g.) and does not occur for fields inside the astroid (squares in Fig. 9.3, e.g.).

Practically, two current pulses are sent along the bit and word lines, properly synchronized to arrive simultaneously at the cell. The two pulses generate two magnetic fields, $\mathbf{H}_{\text{easy}}$ and $\mathbf{H}_{\text{hard}}$ (as in Fig. 9.2b), in such a way that all the cells addressed by either $\mathbf{H}_{\text{easy}}$ or $\mathbf{H}_{\text{hard}}$ are not switched (squares in Fig. 9.3), because their switching fields are larger than both $\mathbf{H}_{\text{easy}}$ and $\mathbf{H}_{\text{hard}}$, while the cell simultaneously addressed by both $\mathbf{H}_{\text{easy}}$ and $\mathbf{H}_{\text{hard}}$ is switched (circle in Fig. 9.3) [42].

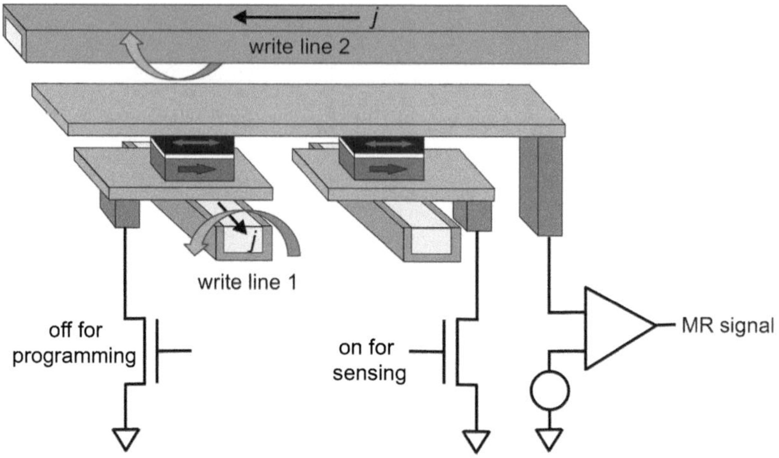

Figure 9.4 Cell writing by current lines.

The limits of this writing method appear when the cell dimension becomes smaller than 0.1 μm: larger switching fields are needed (their values are roughly inversely proportional to the cell size) so that the power dissipated by the current pulses increases, eventually introducing spurious thermal activation (see the following section) that reduces the long-term stability of the cell and introduces write errors.

9.2.2.2 Current lines

An alternative architecture is shown in Fig. 9.4 [43]. Here the current lines (write lines 1 and 2), generating the magnetic field pulses, are physically separated from the MTJ. This leads to the clear advantage that heating, and thus thermal activation, of the MTJ is avoided. Anyway, higher currents are needed, with respect to the architecture presented in the previous section, because the magnetic field sensed by the cell scales inversely to the distance from the current line generating it; this clearly results in larger power consumption.

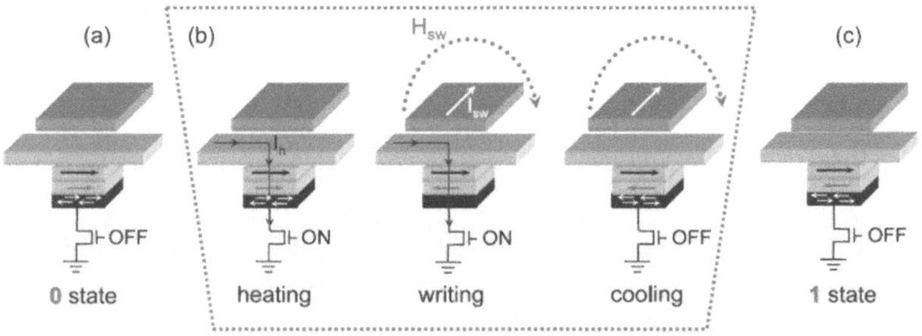

Figure 9.5 Thermally assisted cell writing. Adapted from Ref. [39]. © IOP Publishing. Reproduced by permission of IOP Publishing. All rights reserved.

9.2.2.3 Thermally assisted switching

To achieve writing of sub-0.1 μm cells with minimal power consumption, the thermally assisted switching (TAS) technique has been developed and successfully employed [39]. This approach relies on the temperature dependence of the magnetic switching field in magnetic nanostructures. As shown in Fig. 9.5, during the writing process the selection transistor is open and a current pulse is sent through the MTJ. The heat generated by the first pulse in the tunnel barrier (acting as a resistor) increases the temperature of the FL, thus reducing its switching field: as a consequence, current pulses with reduced intensities are needed for generating the external field required for achieving the magnetization switching.

The main advantage of this approach is that cell selection is made by two simultaneous pulses (or three if the SWA approach is used), of which the trigger one is the heating pulse: without it, the external field generated by the current pulse(s) is too small to influence the magnetic state of the MTJ, and no switching occurs. This results in fewer selection errors of the addressed cell, larger allowable writing speed, and larger packaging density (because the heating is strongly localized). Moreover, the power consumption of this method is lower than that of SWA and current lines. Despite the presence of the additional heating pulse, the overall power needed during the writing operation is smaller because the intensity of the pulse(s) generating the magnetic field is lower.

9.3 New Strategies for Writing Ferromagnetic Electrodes

9.3.1 *Spin Transfer Torque*

STT-MRAMs are considered one of the most promising candidates for the next generation of solid-state magnetic memories [19]. The STT effect has been predicted in 1996 [20, 46], while the first applications to MRAMs, with Al_2O_3 and MgO barriers respectively, date 2004 [47] and 2005 [48].

The typical structure of an STT-MRAM consists of a nonmagnetic layer sandwiched between two thin FM films (Fig. 9.6a) [49]. The magnetization of one of the FM layers (the pinned layer) is fixed ($\mathbf{M}_{\text{fixed}}$), while that of the second (FL) is free ($\mathbf{M}_{\text{free}}$). When a current is injected from the pinned layer to the FL, the carriers are spin-polarized with their spin aligned with $\mathbf{M}_{\text{fixed}}$. If $\mathbf{M}_{\text{fixed}}$ and $\mathbf{M}_{\text{free}}$ are not collinear, while passing through the FL the carriers' spins will align with $\mathbf{M}_{\text{free}}$ because of the exchange interaction, losing their transverse component. By conservation of angular momentum, this lost spin component is transferred to the FL in the form of torque (STT), with the effect of rotating $\mathbf{M}_{\text{free}}$ toward or away from $\mathbf{M}_{\text{fixed}}$, depending on the sign of the injected current. STT has two components, an out-of-plane component T_{OOP} and an IP

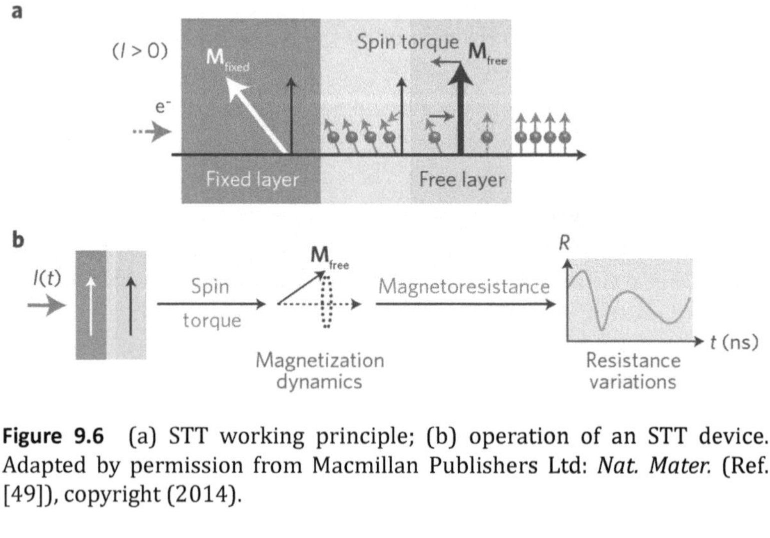

Figure 9.6 (a) STT working principle; (b) operation of an STT device. Adapted by permission from Macmillan Publishers Ltd: *Nat. Mater.* (Ref. [49]), copyright (2014).

component T_{IP}, where "out-of-plane" and "in-plane" refer to the plane defined by $\mathbf{M}_{fixed}$ and $\mathbf{M}_{free}$. T_{OOP} is generally much smaller than T_{IP} and most of STT devices use only T_{IP} as antidumping torque to destabilize the metastable states of the FL. Depending on the amplitude of the external field parallel to $\mathbf{M}_{fixed}$, the magnetization dynamics within the FL can be largely influenced, thus leading to precession, telegraphic switching or hystheretic switching, which can be read via magnetoresistance (see Fig. 9.6b). Relevant for MRAMs, at zero external field hysteretic reversible switching is achieved by reversing the sign of currents. The STT amplitude being proportional to the injected current density, current densities of about 10^7 A/cm^2 are typically needed to obtain full magnetization switching at zero field. However, the switching current scales with the device dimension. Recently, STT devices with 20 nm diameter, compatible with the 22 nm CMOS technology, showed full switching with very low current (<30 μA) [50].

Typical cell designs of STT-MRAMs are reported in Fig. 9.7 [44], with reference to the case of MTJs. In IP design, the magnetization of the magnetic layers lies in the plane of the device, while in perpendicular-to-plane (PP) design the magnetization is perpendicular. The FL is the storing element, whose magnetization can be parallel or antiparallel to the pinned layer. The reference layer (RL) is a synthetic antiferromagnet (SAF), made of two FM layers with opposite magnetizations, separated by a nonmagnetic layer

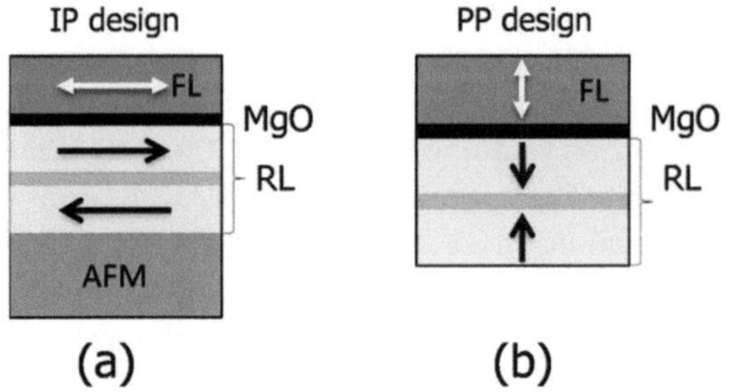

Figure 9.7 Basic STT-MRAMs cell designs: (a) in-plane configuration with SAF and antiferromagnet; (b) out-of-plane configuration with SAF. Adapted from Ref. [39]. © IOP Publishing. Reproduced by permission of IOP Publishing. All rights reserved.

that provides strong antiferromagnetic (AFM) interlayer coupling. This allows one to reduce the dipolar coupling between the RL and the FL. In the IP configuration, typically the RL magnetic robustness is enhanced via direct exchange bias coupling to an adjacent AFM layer.

Both IP and PP configurations offer peculiar advantages and there's not a clean winner in the competition between the two. This will be more clear after the discussion of the three figures of merit currently used for the evaluation of STT-MRAMs: (i) thermal stability, (ii) TMR and (iii) switching current.

Thermal stability is the main bottleneck in the route toward cell miniaturization and high storage density. By decreasing the cell volume (V), the energy barrier (E_b) for magnetization switching (that is KV, where K is the magnetic anisotropy energy density and V is the volume) proportionally decreases. If KV becomes comparable with $k_B T$, the thermal stability of the cell is mined as soon as the FL enters the superparamagnetic regime.

The thermal stability factor is usually defined as

$$\Delta = \frac{E_B}{k_B T} = \frac{KV}{k_B T} \tag{9.3}$$

In the case of a cell with a 40 nm diameter and 1.3 nm thick CoFeB film, [28], with a perpendicular anisotropy energy density $K = 2.1 \times 10^5$ J·m^{-3}, we obtain $\Delta = 83$ at room temperature. Because $\Delta > 25$, the conventional ratio assumed for defining the blocking temperature of a superparamagnetic system, this example shows how the high K of materials with perpendicular anisotropy can help in stabilizing MRAMs' nanosized cells at room temperature.

Δ depends on many parameters, such as the magnetization of the material, the crystalline anisotropy, the cell dimensions and aspect ratio (AR). The magnetization and AR determine the shape anisotropy contribution so that Δ scales with (AR-1) in the IP configuration and AR in the PP configuration, where AR is the cell-length-to-width ratio [15]. Crucial is the role of crystalline anisotropy, especially that of interfacial perpendicular anisotropy, which provides very high values of effective anisotropy constants for low thickness.

For the tunnel barrier a thin (few nanometers) layer of crystalline MgO is typically chosen, due to its high spin selectivity [22], which improves both STT (for writing) and TMR (for reading) effects. Values of TMR as high as 604% have been obtained in the CoFeB/MgO system [19] and TMR values higher than 100%–200% are currently used in applications. Moreover, the CoFeB/MgO template, with CoFeB thickness on the order of 1 nm, exhibits perpendicular magnetic anisotropy (PMA) [28] that is compatible with PP design.

The third key parameter of STT-RAMs is the critical current (I_C) for data writing. I_C represents the minimum current needed to reverse the magnetization of the FL due to the STT effect. When a current with density J enters the magnetic layer, the number of electrons per unit area and time is J/e, each one carrying an angular momentum $\eta\,(\hbar/2)$ on average. η is the STT efficiency parameter related to the polarization of the injected current. In general, η is a function of (i) the band structures of the electrodes and the barrier, (ii) the bias applied to the device, and (iii) the micromagnetic configuration. In the zero-order approximation, however, it can be considered a function of the electrode polarization (P) only. For inelastic tunneling in symmetric junctions with electrode polarization P, the STT efficiency is [44]

$$\eta = \frac{2P}{1+P^2} \tag{9.4}$$

In a very simple model, neglecting symmetry-based spin filtering, also the TMR can be related to the electrode polarization P via the Jullière formula:

$$\mathrm{TMR} = \frac{2P^2}{1-P^2} \tag{9.5}$$

The TMR efficiency and STT efficiency are then mutually related, as shown in Fig. 9.8 [44], with a high TMR implying a large STT. This is the reason for which structures with a high TMR, such as those based on MgO barriers [19, 22], are employed.

The net flow of magnetization per unit area is $\gamma\eta\,(\hbar/2)\,J/e$, where γ is the gyromagnetic ratio. The ratio of this quantity on the magnetic moment per unit area $M_S t$, where t is the FL thickness, gives the STT:

$$\mathbf{\Gamma}_{STT} = \gamma\eta\frac{\hbar}{2}\frac{J}{e}\frac{1}{M_S t}\mathbf{m}\times\mathbf{m}\times\mathbf{m}_{\mathrm{RL}} \tag{9.6}$$

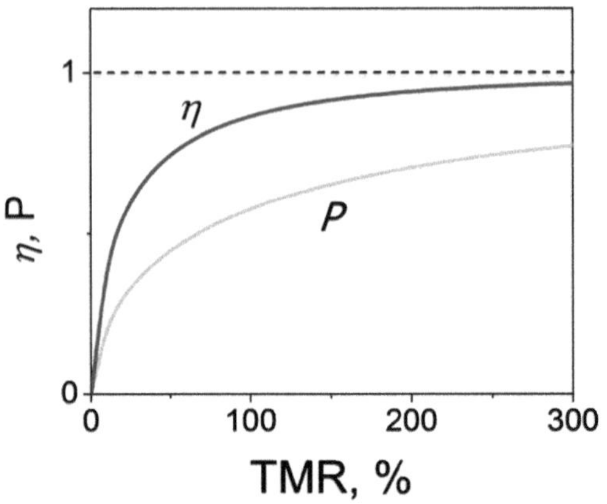

Figure 9.8 STT efficiency (η) and current polarization P versus TMR for a symmetrical junction. Adapted from Ref. [39]. © IOP Publishing. Reproduced by permission of IOP Publishing. All rights reserved.

where $\mathbf{m}$ and $\mathbf{m}_{\mathrm{RL}}$ are the magnetizations of the FL and RL, respectively. The angle-dependent term $\mathbf{m} \times \mathbf{m} \times \mathbf{m}_{\mathrm{RL}}$ accounts for the fact that only the transverse part of the incoming angular momentum can be absorbed by the FL. To produce a magnetization reversal of the FL, the STT must overcome the magnetization damping predicted by the Landau–Lifshitz–Gilbert equation: this happens over a critical current density J_{C0}. For PP cells this is given by

$$J_{\mathrm{C0}} = \frac{1}{\eta}\frac{2\alpha e}{\hbar} M_{\mathrm{S}} t H_{\mathrm{k}}^{*} \tag{9.7}$$

where α is the Gilbert damping constant and H_{k}^{*} is an effective anisotropy field. The corresponding critical current for PP cells can be written as

$$I_{\mathrm{C0}} = \left(\frac{4ek_{\mathrm{B}}T}{\hbar}\right)\frac{\alpha}{\eta}\Delta \tag{9.8}$$

For IP cells, I_{C0} must be multiplied by $(1 + 4\pi M_{\mathrm{eff}}/2H_{\mathrm{K}})$ (see Ref. [44] for further details). High values of the STT efficiency η (and hence of TMR) are then crucial for reducing the critical current and thus the power consumption of the cell. Note also that for the same η, α and Δ, the critical current is higher in the IP configuration due to the term $(1 + 4\pi M_{\mathrm{eff}}/2H_{\mathrm{K}})$, so PP should be more advantageous. Nevertheless, the IP configuration is still preferred in some applications because IP materials offer larger η, smaller α, and higher reproducibility of material parameters. Typical values of critical currents that flow across the device are quite high ($\sim 10^7$ A/cm^2 for full magnetization reversal). This is probably the main drawback of STT, reflecting in possible overheating that could deteriorate the device performance (e.g., the TMR of the junction).

Finally, note that STT-MRAMs are mature enough to compete with conventional MRAMs and other memories (SRAMs, DRAMs, flashes). The world's first STT-MRAM chip, with 64 Mb capacity, was announced in 2012 and realized in 2013 by Everspin, and the same company in 2014 announced that a new generation is forthcoming with 256 Mb capacity [51].

9.3.2 *Spin–Orbit Torque*

At variance with STT-based MRAMs, in SOT devices full magnetization reversal of a single magnetic layer can be obtained without the need of a separate RL. Breaking of inversion symmetry is required to induce net effects on the magnetization, which is usually realized by sandwiching an FM layer between two dissimilar layers (e.g., the Pt/Co/AlO_x heterostructure discussed later). When a current flows in the structure parallel to the interfaces, in a current-in-plane configuration, a nonequilibrium spin accumulation is generated, which eventually gives rise to a torque on the magnetization via the spin transfer between *s* and *d* electrons [52]. Usually SOT is ascribed to two different mechanisms, bulk spin Hall effect (SHE) and interface Rashba effect, both exploiting the coupling between electron spin and orbital motion. This is the reason for the SOT name to underline the link of these phenomena to spin–orbit interaction, which is the physical foundation of both the SHE and Rashba effect.

In Ref. [53], a current-driven SOT in a Pt/Co/AlO_x heterostructure is reported, and its origin is ascribed to the Rashba effect in the FM layer. Electrons moving in an asymmetric crystal field potential experience a net electric field. In the electron's rest frame, this field transforms into a magnetic field, which interacts with the electron spin. Electron motion is then strictly coupled to the electron spin, so it is possible to influence the latter by acting on the former: this is the principle on which SOT is based.

In a 2D system characterized by structural inversion asymmetry, the effective magnetic field (the Rashba field) produced by spin–orbit interaction on the conduction electrons is [54]

$$\mathbf{H}_{\mathrm{R}} = \alpha_{\mathrm{R}} \left(\mathbf{u}_{\mathrm{z}} \times \langle \mathbf{k} \rangle\right) \tag{9.9}$$

where α_R is a parameter that depends on the strength of the spin–orbit coupling (SOC) in the material, $\mathbf{u}_z$ is the unit vector parallel to the electric field, and $\langle \mathbf{k} \rangle$ is the average electron wavevector. For zero current, H_R cancels out because $\mathbf{k}$ and $-\mathbf{k}$ states are equally populated ($\langle \mathbf{k} \rangle = 0$). In the presence of a charge current, however, the electron distribution in the $\mathbf{k}$-space becomes asymmetric ($\langle \mathbf{k} \rangle \neq 0$), producing a net Rashba field and inducing a nonequilibrium spin density perpendicular to the current direction. This effect is common to nonmagnetic materials (such as semiconductors) and magnetic materials. In the latter case, as schematically shown in Fig. 9.9a, the action of $\mathbf{H}_R$ combines with the *s-d* exchange interaction that couples the conduction electron spin to the local magnetization. As a result of this interaction, a current passing through a uniformly magnetized FM layer exerts an exchange-mediated effective field $\mathbf{H}_{sd}$ on the magnetization $\mathbf{M}$, thereby producing a torque and eventually a reversal of $\mathbf{M}$.

In the device reported in Ref. [53], an ultrathin Co layer (0.6 nm thick) is sandwiched between AlO_x and Pt layers, as shown in Fig. 9.9b. Pt and AlO_x are present because spin–orbit effects at metal surfaces increase with the presence of heavy atoms and surface oxidation. The electron current (I) flows in the x direction and the electric field ($\mathbf{E}$) is applied in the z direction: the Rashba field ($\mathbf{H}_R$) and the effective field $\mathbf{H}_{sd}$, following Eq. 9.9, are along y.

The Co magnetization is out of plane (along z) because of the very small thickness and Pt proximity. The effective field ($\mathbf{H}_{sd}$) acts exactly as an external magnetic field, competing with the out-of-plane anisotropy field ($\mathbf{H}_K$) in determining the orientation of $\mathbf{M}$. As shown in Fig. 9.9c, the H_{sd}/H_K ratio determines the gradual rotation of the equilibrium magnetization (the minima in the energy curve) from PP ($\theta = 0$) to IP ($\theta = 90°$). Because H_{sd} scales linearly with the

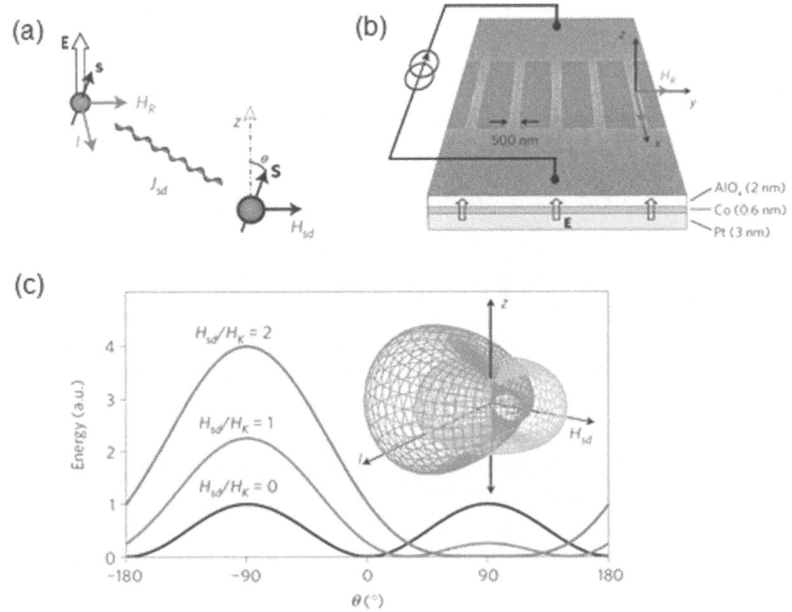

Figure 9.9 (a) Sketch of the s–d interaction: the Rashba field $\mathbf{H}_\mathrm{R}$ acting on free electrons with spin s leads to an effective field $\mathbf{H}_\mathrm{sd}$ on the localized moment **S**. (b) SOT device, with the current flowing in the x direction and the Co magnetization perpendicular to plane in the z direction. $\mathbf{H}_\mathrm{R}$ and $\mathbf{H}_\mathrm{sd}$ are parallel to y. (c) Magnetic energy as a function of the polar angle for different $H_\mathrm{sd}/H_\mathrm{K}$ ratios. Adapted by permission from Macmillan Publishers Ltd: *Nat. Mater.* (Ref. [53]), copyright (2010).

current density, it appears that a critical current exists on the order of 10^8 A/cm^2 (similar to STT currents), above which magnetization rotation is attained.

In a more recent paper [55] current-induced switching of the magnetic state of a Pt/CoFe/MgO heterostructure is explained by a combination of the SHE and the Slonczewski-like (SL) torque [45].

Whatever the physical interpretation of SOT effects, a crucial peculiarity is that the direction of the torque can be reverted by reversing the current, thereby enabling fully reversible control of the magnetization. An example of application to magnetic memories

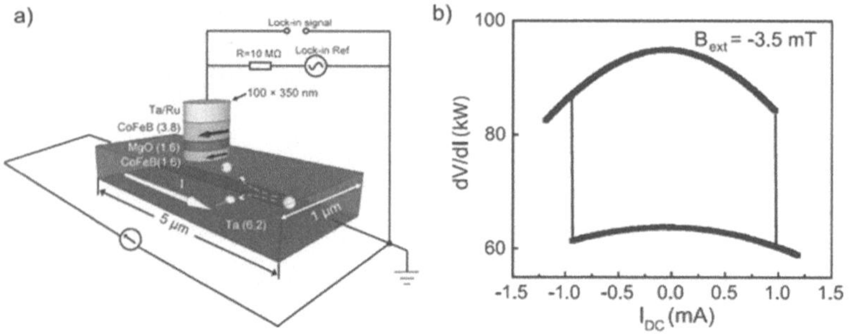

Figure 9.10 (a) Schematic of the three-terminal SOT-based device; (b) resistance of the device as a function of the current flowing in the bottom Ta layer. From Ref. [56]. Reprinted with permission from AAAS.

is reported in Ref. [56], where the switch of an MTJ by SOT is demonstrated, as shown in Fig. 9.10. The device stack is (from bottom to top) substrate/Ta(6.2)/CoFeB(1.6)/MgO(1.6)/CoFeB (3.8)/Ta(5)/Ru(5), where the thickness is in nanometers. For writing the MTJ state, a current is injected along the bottom Ta layer, shaped as a micrometric stripe, where the SHE separates spins and creates spin accumulation in the CoFeB adjacent film (Fig. 9.10b). Above a critical current of about 1 mA, corresponding to a current density of about 10^7 A/cm^2, the SHE reverses the magnetization of the bottom CoFeB layer, adjacent to Ta. Because the top CoFeB layer magnetization is fixed, this current-induced switch results in the abrupt change of the resistance, measured between the bottom Ta and the top CoFeB layer, as shown in Fig. 9.10b. While the employed current density is comparable with STT devices, we note that, here, this current is flowing not across the MTJ but in a separate layer (the Ta layer), thus avoiding thermal deterioration of the MTJ itself: this constitutes the main advantage of SOT- versus STT-based MRAMs.

9.3.3 *Electric Writing of Magnetic Information*

Another approach to reducing the power dissipation of MRAMs during writing and increasing the packing density is to control the magnetization orientation by an electric field.

In principle this can be achieved in two ways:

(i) Applying a volatile external field, for example using the pinned layer and the FL as electrodes of a capacitor where the barrier is the dielectric [57–59], and exploiting the phenomenon of the voltage-controlled magnetic anisotropy
(ii) Exploiting the coupling between dielectric polarization and magnetization in bulk multiferroics materials or in artificial multiferroics sytems [60]

Both methods are based on the accumulation of large charge surface densities at the interface between an insulating material and an FM one. The first method is highly appealing as it uses the very same structure of tunneling junctions without the introduction of new layers. On the other hand it usually requires some auxiliary external magnetic field to ensure the full reversibility of the switching or a precise timing of the writing pulses, especially in the case of coherent switching. The second method has the advantage that the effective electric field induced by the polarization change at the interface between the ferroelectric (FE) and the FM layers may be very high, resulting in a huge modification of the interfacial magnetic properties. Moreover, the remanent state of the FE layer allows one to achieve high stability of the written magnetic configuration, as well as the possibility of building up hybrid devices exploiting both the magnetic and the dielectric order as variables of state.

9.3.3.1 Voltage-controlled magnetic anisotropy

In Fig. 9.11 the operating principle of voltage-controlled magnetic anisotropy is sketched [57]. As any kind of electrically induced change of magnetic anisotropy is strongly confined within the Thomas–Fermi screening length of the metallic ferromagnet, ultrathin magnetic electrodes (about 1 nm thick) are typically used to maximize the impact on the magnetic behavior of the whole layer. For the typical FeCo and FeCoB materials used in conjunction with MgO barriers, this leads to a sizeable PMA. The FM FL thus presents remanent out-of-plane magnetization, while the coercive field is proportional to the interfacial anisotropy. As the strength of the anisotropy depends on the interface polarization induced by

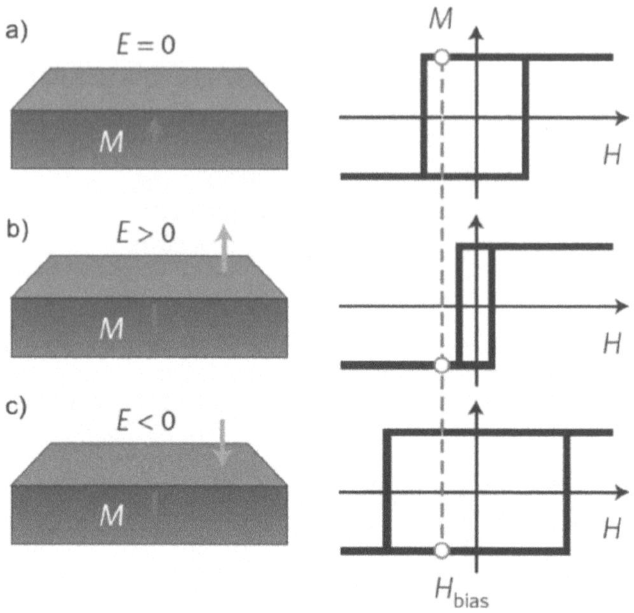

Figure 9.11 Operating principle of voltage-controlled MCA. Adapted by permission from Macmillan Publishers Ltd: *Nat. Mater.* (Ref. [57]), copyright (2012).

the electric field pulse, we then expect electric modulation of the magnetic hysteresis loop.

With reference to Fig. 9.11a, suppose that for a zero electric field ($E = 0$) the magnetization (M) is pointing up in the presence of a negative bias magnetic field (H_{bias}). If the application of $E>0$ (Fig. 9.11b) induces a decrease of interface magnetocrystalline anisotropy (MCA) the hysteresis loop becomes narrower. This leads to electrically induced magnetization switching from up do down. The opposite switching, however, cannot be achieved simply by reversing the electric field polarity. In fact, even though for $E<0$ (Fig. 9.11c) the interface MCA increases and the hysteresis loop becomes larger, M does not switch back unless the bias magnetic field is reversed, too.

In Fig. 9.12a,b the concept of the application of such an approach to an MTJ is sketched. Here the out-of-plane magnetization of the top layer is switched by application of an electric field across the barrier

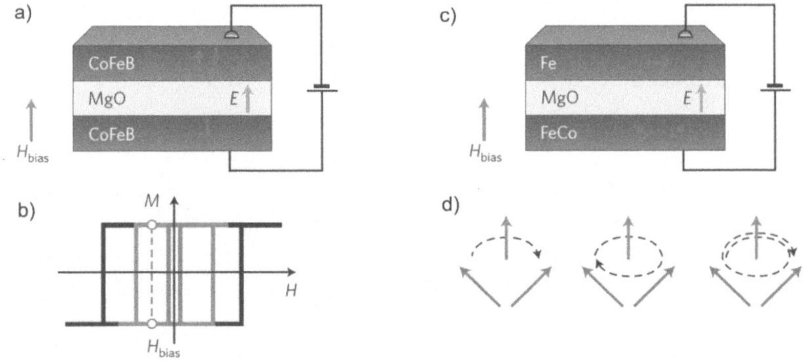

Figure 9.12 Two approaches for switching the magnetic state of an MTJ device: (a, b) voltage-controlled MCA; (c, d) voltage-controlled coherent precession of magnetization. Adapted by permission from Macmillan Publishers Ltd: *Nat. Mater.* (Ref. [57]), copyright (2012).

(writing) and then detected via the corresponding resistance change in the MTJ [58]. As pointed out earlier, this method does not allow the magnetization to be switched back without bias field reversal, as also shown in Fig. 9.11c. To overcome this problem, the spin torque effect can be used. By applying an electric field to reduce the interface MCA and the coercive field, the current needed for perform STT on the FL is 2 orders of magnitude smaller than in ordinary STT-MRAMs, thus reducing the power dissipation problem typical of STT devices.

A different approach to voltage-controlled magnetic anisotropy [59] is shown in Fig. 9.12c,d. Magnetization reversal of the FL in an MTJ is achieved via coherent rotation around the anisotropy field produced by electrical pulses of a specific duration. The IP film magnetization is tilted to about 45° by a bias field perpendicular to the plane. An electric field pulse of polarity that enhances the perpendicular anisotropy produces a coherent rotation of the magnetization around the bias field. Using pulses that produce half a revolution of the magnetization rotation, full switching is realized, which can be monitored by observing the MTJ resistance change. This method requires precise timing of the pulses in the nanosecond range, but there is no need of reversing the bias field in order to go back and forth between the antiparallel and the parallel magnetization alignments in the MTJ. Nevertheless a bias magnetic field is still needed in order to activate the electric toggle of the magnetization.

9.3.3.2 Magneto-electric coupling

MEC, that is, the coupling of magnetization (M) and polarization (P) in multiferroic materials, could permit one to build up a new generation of memory devices, as it would exploit the best aspects of ferroelectricity (low-cost electric writing) and magnetism (robustness and durability of the stored information). In spite of recent progress in spin torque [45], electric-field-assisted switching [58], and coherent magnetization switching [59], MEC represents an advantageous route toward low-power electrical control of magnetization.

Single-phase multiferroics [21–64] often present low Curie temperatures and weak MEC, thus hampering or limiting practical applications. Aiming at a large technological impact, multiferroic heterostructures, made of conventional ferromagnets and traditional ferroelectrics, are more promising [22, 66]. For the paradigmatic Fe/$BaTiO_3$ (BTO) system, sizable changes of the interfacial Fe magnetic moment upon reversal of the dielectric polarization of BTO have been predicted [23], and huge variations of magnetoresistance have been observed by reversing the BTO polarization in Fe/BTO/$La_{2/3}Sr_{1/3}MnO_3$ (LSMO) tunneling junctions [68].

Different mechanisms are currently explored in order to achieve electric field control of MEC: (i) strain-mediated coupling at FM/piezoelectric interfaces [24–71], (ii) exchange bias coupling at FM/FE interfaces [25–75], (iii) modulation of carrier density in FM layers [26–78], (iv) bond reconfiguration driven by ionic displacement at FM/FE interfaces [67], and (v) spin-dependent screening at FM/FE interfaces [79]. Recently the authors have demonstrated that a pure and reversible room-temperature electric switching of the interfacial magnetic order at the Fe/BTO interface can be achieved [80]. Upon reversal of the BTO dielectric polarization by ± 5 V pulses, the FM order within the interfacial Fe atomic layer can be switched on-off at room temperature as a result of the polarization-induced

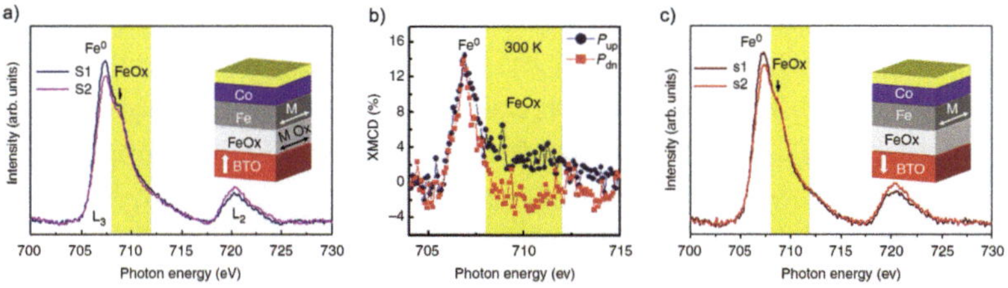

Figure 9.13 (a, c) XAS Fe–L2,3 spectra taken at room temperature on Au/Co/Fe/BTO/LSMO/STO capacitors at 300 K after polarization of BTO with $V+ = +5$ V ($E = +170$ kVcm^{-1}) (P_{up}) and $V- = -5$ V (P_{dn}), respectively. Insets in (a) and (c) are visual schemes of the Fe/BTO interface for P_{up} and P_{dn}: white vertical arrows correspond to the BTO polarization direction, while M and MOx indicate the net magnetization in metallic Fe and in the Fe oxide layer at the Fe/BTO interface. (b) XMCD signal in the Fe–L3 energy region for BTO polarization P_{up} (blue circles) or P_{dn} (red squares). The yellow shading indicates the region of the spectra corresponding to the oxidized Fe layer in contact with BTO. Adapted by permission from Macmillan Publishers Ltd: *Nat. Commun.* (Ref. [80]), copyright (2014).

bond reconfiguration. X-ray magnetic circular dichroism (XMCD) spectra show that the Fe oxide component of the peak (indicated with the yellow shading in Fig. 9.13 and ascribed to the interfacial Fe layer) displays a net magnetization when the BTO is in the up state (Fig. 9.13a), while no signal is present when the BTO is in the down state (i.e., with the opposite polarization state, Fig. 9.13c). These results suggest that the oxidized Fe layer undergoes a reversible transition (FM/nonmagnetic) driven by the switching of the electric state of the FE layer.

This new physical mechanism provides a record value of interfacial MEC, as the interfacial magnetization can be set or suppressed at will. Even though the effect is strongly confined within the Fe layer in contact with BTO, within a hybrid MTJ with a FE tunneling barrier this would not be an issue. In fact the TMR is mainly determined by the interfacial effective spin polarization, which can be very high when the FE polarization promotes the FM order and almost zero for the opposite polarity, which destroys magnetic order. In these hybrid devices the dielectric polarization then determines the overall impedance via two mechanisms, (i) the conventional tunneling electroresistance due to the barrier profile modification induced by polarization reversal and (ii) the on-off switching of the interfacial magnetization in one of the FM electrodes. The combination of these effects is highly appealing in view of the realization of novel nonconventional memory devices such as FE memristors [81].

To move MEC-based MRAMs toward industrial applications, integration of FE and magnetic oxides with CMOS technology should be accomplished. As a matter of fact, the high deposition temperatures (up to 1000 K for BTO and LSMO) needed for obtaining sharp interfaces with high crystal order [79] are not compatible with direct deposition on silicon. However, in recent years different buffer layers have been proposed in order to overcome this problem, such as $LaNiO_3/CeO_2/YSZ$ [82], MgO/TiN [83], $SrTiO_3/(Ba,Sr)O$ [84], $SrTiO_3/Sr$ [85], and $SrTiO_3/SiO_2$ [86]. Employing these templates, high-quality BTO and LSMO films can be grown by pulsed laser deposition; on top, epitaxial Fe layers can be obtained by molecular beam epitaxy [87], thus obtaining fully CMOS-integrated $Fe/BTO/La_{2/3}Sr_{1/3}MnO_3$ structures suitable for electric control of spin polarization [68].

9.3.4 *Optical Writing*

A nonelectrical approach for controlling the magnetization of the FL is optical writing, consisting of illuminating the cell with an ultrashort laser pulse that induces a reorientation of the magnetization without the need of any applied current or voltage. Noteworthy, this approach is characterized by a high writing speed due to the availability of pulsed laser sources in the femtosecond or subfemtosecond time scale. As a matter of fact, femtosecond laser pulses offer the possibility to probe and interact with a magnetic system on the time scale of the exchange interaction (10–100 fs), which is responsible for a magnetic order much shorter than the time scales of spin–orbit interactions (0.1–1 ps) and magnetic precession (1 ps–1 ns) [88].

Since the demonstration of subpicosecond demagnetization by a 60 fs laser pulse in 1996 [89], manipulating and controlling magnetization with ultrashort laser pulses have attracted increasing attention. Because the typical time scale for magnetization reversal is the magnetic precession time (magnetization is reversed in half a precessional period), the major issue of this method is understanding and controlling the angular momentum and energy transfer between the different thermodynamic reservoirs in a magnetic system, namely the magnetically ordered spins, the free electron system, and the lattice.

This is particularly relevant for magnetic recording and logic applications, where the high clock speed of the processor (up to terahertz) must be supported by storage elements with similar access speed.

When a laser pulse excites a magnetic medium, the latter is driven into a nonequilibrium state, where the conventional thermodynamic description fails. Pulse duration is comparable or shorter than the time scale of the spin–orbit interaction, so the magnetic anisotropy becomes a time-dependent parameter. Angular momentum is stored in different reservoirs (electrons, lattice, spins), as schematically shown in Fig. 9.14a, and their mutual exchange determines the response of the material to the laser pulse [88].

The possible effects of the pump laser on the magnetic system can be grouped into three classes:

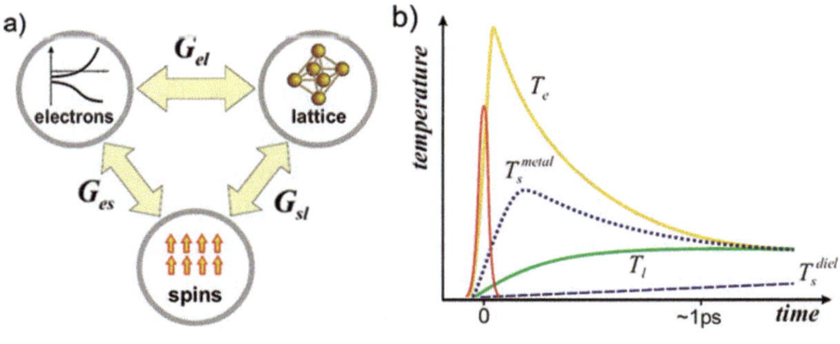

Figure 9.14 (a) Interacting reservoirs (carriers, spin, and lattice) in the three-temperature model; (b) temporal behavior of electron, lattice, and spin temperatures following excitation by a short laser pulse. Reprinted figure with permission from Ref. [88]. Copyright (2010) by the American Physical Society.

(a) Thermal effects. Because of the photon absorption, energy is directly pumped in the electron and phonon systems of the material and indirectly induces changes in the magnetization via internal equilibration processes such as electron–electron, electron–phonon, and electron–spin interactions, on a time scale that, in metals, can be very short ($\sim$50 fs). The lifetime of such thermal effects, instead, is determined by the device thermal conductivity and geometry.
(b) Nonthermal photomagnet effects. Photons are absorbed via electronic states that have a direct influence on magnetic parameters, such as the MCA. The rise time is very short and coincides with the rise time of the laser pulse. The lifetime of the excited states is that of the involved electronic states. However, as the parameter modulation causes a precessional motion of the magnetic moments, much longer than the electron states lifetimes, their dynamics is typically not visible.
(c) Nonthermal optomagnetic effects. Photons are not absorbed but produce optically coherent stimulated Raman scattering. The time scale is that of the SOC ($\sim$20 fs), which is the driving force behind the change of the magnetization, and the lifetime coincides with that of optical coherence (100–200 fs).

Depending on the cases, one or more of these effects are present. Thermal effects are always present to some extent and will be discussed in detail later.

In 1996, the transient reflectivity of a 22 nm thick Ni film, excited by 60 fs laser pulses, was measured by the linear magneto-optical Kerr effect (MOKE) (see Fig. 9.15) [89]. The electron thermalization time was about 260 fs and the electron temperature decay constant about 1 ps, both of them much faster than the spin–lattice relaxation time (about 30 ps). This seminal experiment showed that the excitation by the ultrashort laser pulse leads to ultrafast heating of the electronic system and to a subsequent decay of the magnetization following the electron temperature dependence.

To interpret the complex process of demagnetization, a phenomenological three-temperature model has been developed (see

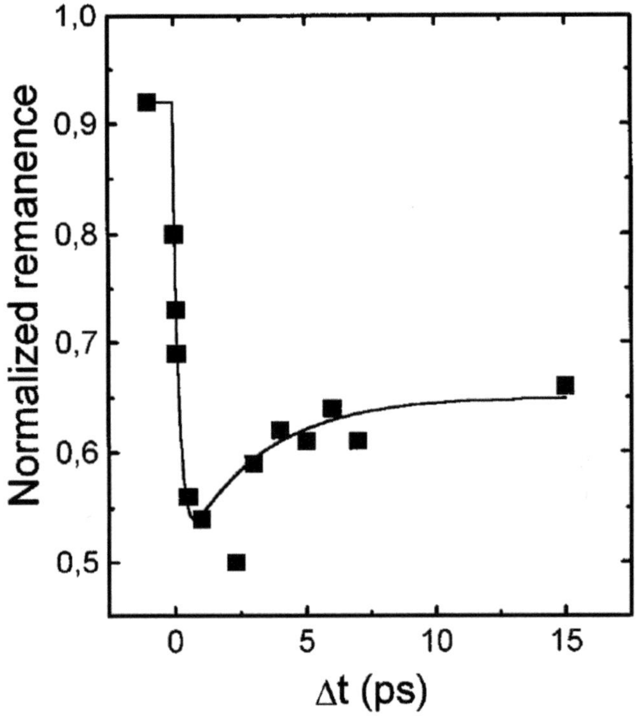

Figure 9.15 (a) Normalized remanence for a 22 nm Ni film excited by a 60 fs laser pulse. Reprinted figure with permission from Ref. [89]. Copyright (1996) by the American Physical Society.

Fig. 9.14a). This model contains three reservoirs (electron, lattice, and spin), separated but interacting via the coupling constants G_{ij}, each one defined by a proper temperature. The temporal evolution of the system can be described by three coupled differential equations [88, 90]:

$$
\begin{aligned}
C_e dT_e/dt &= -G_{el}(T_e - T_l) - G_{es}(T_e - T_s) + P(t) \\
C_s dT_s/dt &= -G_{es}(T_s - T_e) - G_{sl}(T_s - T_l) \\
C_l dT_l/dt &= -G_{el}(T_l - T_e) - G_{sl}(T_l - T_s) \qquad (9.10)
\end{aligned}
$$

C_i is the heat capacity and T_i the temperature of the i-th bath; G_{ij} represents the coupling between the i-th and j-th baths; $P(t)$ is the time-dependent optical input. In Fig. 9.14b the temporal behavior of electron (violet line), lattice (green line), and spin (yellow line) temperatures following excitation from a femtosecond laser pulse (red line) is reported. The spin temperature is shown for both metals and dielectrics.

The typical scenario after an ultrashort laser pulse is the following. At optical frequencies, only electrons are able to respond to electromagnetic excitations, absorbing or scattering the photons. Upon pulse absorption, the electronic system reaches a high equilibrium temperature within 50–500 fs by electron–electron interactions and then decays by electron–phonon interactions within 100 fs–1 ps, until the electron and lattice systems come in thermal equilibrium with each other, as shown in Fig. 9.14b. During this process, angular momentum must be conserved in the system, but part of the electron spin is transferred by spin flip events to the lattice via different mechanisms (Stoner excitations, inelastic electron–spin scattering, electron–phonon scattering), thus reducing the net sample magnetization, as shown in the experiment reported in Fig. 9.15.

Different strategies have been proposed to use short laser pulses for writing an MRAM cell. The first one corresponds to the so called heating-assisted magnetic writing technology, used also for hard disk writing. By heating the FL by a laser pulse, the magnetic anisotropy is strongly reduced and magnetic switching can be achieved by application of an external magnetic field smaller than the room-temperature coercive field of the layer. The femtosecond laser excitation first leads to the total demagnetization of the FL. During cooling a small external field opposite to the initial FM magnetization is enough to cause the recovery of the magnetic order in the opposite direction, that is, magnetic switching [91]. The time scale of this magnetization recovery is on the order of 700 fs, which represents the limiting factor for the writing speed of this technology.

A similar approach for optically induced magnetic switching is reported in Ref. [92]. A crystalline Fe film 8 nm thick, presenting a fourfold IP MCA with the easy axes oriented along the [100] and [010] directions, is initially magnetized along the [100] direction. An external field is then applied along the [110] hard axis, tilting the magnetization to an intermediate direction, with the main component along [100] and a small projection along [110]. When the sample is illuminated by a 50 fs laser pulse, it undergoes fast demagnetization within 100 fs and then recovers the saturation magnetization in ~100 ps. Because of the increase of the local temperature up to about 1000 K during this period, the magnetic anisotropy of the irradiated area strongly decreases, and consequently a gyroscopic precession of the magnetization vector around the external field is triggered [93]. For an external field of 130 Oe the authors demonstrate the magnetization switching from the [100] to the [010] direction, corresponding to half precession within the time in which the magnetic anisotropy is strongly depressed. Noteworthy, a second pulse can drive the magnetization back to [100] or, for an external field along [$\bar{1}$10], to the reversed state [$\bar{1}$00].

A different approach is that based on the use of circularly polarized light [94]. This has the advantage of not requiring any external magnetic field for magnetization reversal. The magnetic switching of a GdFeCo layer has been obtained by using a single 40 fs circularly polarized laser pulse, as a combined result of laser-induced heating of the magnetic system to just below the Curie temperature, and the magnetic circular dichroism (MCD) effect, that is, the differential absorption for left and right circularly polarized light by the magnetized material [95].

An application of optically induced ultrafast demagnetization to MgO-based MTJs, suitable for the implementation of MRAMs, is reported in Ref. [96]. The top layer of a CoFeB/MgO/CoFeB MTJ is illuminated by a 80 fs laser pulse, and the changes in magnetization are measured by recording the Kerr ellipticity as a function of time. As shown in Fig. 9.16, the laser-induced demagnetization decreases when a bias is applied to the junction (the maximum variation is about 40% in the investigated range), while the sample reflectivity is essentially unaffected. This effect is interpreted in terms of optically excited spin-polarized currents in the top CoFeB layer, which tunnel through the MgO barrier characterized by bias-dependent TMR. Because mainly majority carriers can flow across the barrier, this results in an increased degree of polarization of the photoinduced current moving from top to bottom CoFeB layers. Minority spins,

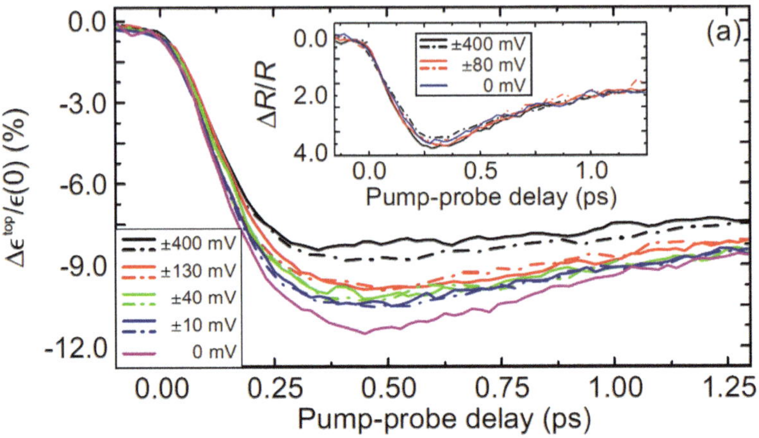

Figure 9.16 Ultrafast variation in the probe beam Kerr ellipticity and (inset) reflectivity as a function of the pump–probe delay and the bias voltage. Reprinted figure with permission from Ref. [96]. Copyright (2014) by the American Physical Society.

on the contrary, have a lower tunneling probability to the bottom layer due to the low density of states available, so they enhance the demagnetization effect in the top layer. When bias is applied to the MTJ, the TMR decreases. The difference of the tunneling probability for majority and minority electrons decreases, resulting in a lower unbalance between majority and minority electrons in the top layer, that is, smaller demagnetization.

This result may open new possibilities to control the magnetization state in MTJ-based devices. Extending this bias-modulated demagnetization effect to optically induced magnetization reversal, it is possible to envisage large MTJ arrays illuminated by a single laser pulse and separately addressed by applying the voltage to the cells one by one. In principle, this is an optical version of heat-assisted writing (see Section 9.2.2.3) but without the need of heating currents through the MTJs. Moreover, with this approach the addressing could be electrical and not optical, thus avoiding any limitation due to diffraction effects in case of rastering of a focused beam on the cell array.

9.4 Increasing Density and Stability

9.4.1 *Domain Wall–Based Memories*

Both conventional MRAMs and hard disk drives are essentially 2D arrays of magnetic cells, so that the only way to increase density relies on the reduction of the size of individual memory elements. An alternative approach to solve this problem is to develop 3D devices. In this context, a possible solution is given by racetrack memories (RMs), in which magnetic domains are used to store information in tall columns of magnetic material arranged perpendicularly to the surface of a silicon substrate [97].

Figure 9.17 shows a schematic view of a vertical RM, which is the solution offering, at present, the highest storage density. The racetrack is a U-shaped magnetic nanowire, normal to the plane of the substrate, with data encoded as a pattern of magnetic domains along a portion of the wire (different domains are indicated by different colors). Magnetic domain walls (DWs) are formed at the boundaries between magnetic domains magnetized in opposite directions (up or down) along the racetrack. The spacing between consecutive DWs, that is, the bit length, can be controlled by pinning

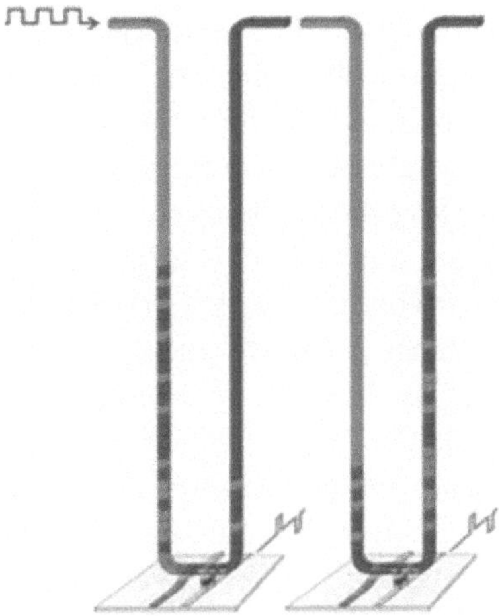

Figure 9.17 Vertical racetrack memory. From Ref. [97]. Reprinted with permission from AAAS.

sites fabricated along the racetrack. Different techniques have been explored: patterning of notches along the edges of the racetrack [98], modulation of the racetrack's size [99], and modulation of material properties or alternating materials along the racetrack [100]. In other implementations, the precise timing of transverse current pulses is used to write the DW structure of the nanowires, while longitudinal pulses are used to move the DW along the wire.

The motion of the DWs is induced by pulses of highly spin-polarized currents that shift the entire pattern of DWs coherently along the length of the wire. DW velocities up to 110 m/s with currents densities of 1.5×10^8 A/cm^2 can be currently achieved.

The write and read elements are placed at the bottom on the Si substrate. Reading is performed by measuring the TMR of MTJs in contact or very close to the racetrack. Writing can be accomplished in different ways: employing the fringing fields of a DW moved in a second FM nanowire oriented at right angles to the storage nanowire, using the self-field of a current passing along a neighboring metallic wire, or exploiting the STT effect produced by a current injected in the racetrack from external magnetic contacts.

With respect to conventional MRAMs, in addition to the higher storage density, RMs present the advantage of requiring a smaller number of reading sensors, thus simplifying the architecture. As a matter of fact, a single reading element in RM accesses 10–100 bits (essentially, all the bits stored along the U shape), while in MRAM each cell needs its proper MTJ. The average access time is 10–50 ns, comparable to MRAMs (about 10 ns).

Very recently, an alternative solution to FM RMs has been proposed by their inventor, S. S. P. Parkin [101]. To increase the racetrack storage density, closely packed DWs would be needed, but their intensity is limited by dipolar coupling from their fringing magnetic fields. These fields can be eliminated using an engineered synthetic antiferromagnet (SAF) structure composed of two magnetic sublayers, coupled by interlayer coupling via an ultrathin nonmagnetic spacer layer (e.g., Co/Ni/Ru/Ni/Co). Noteworthy, very recently it has been shown [101] that DWs in SAF racetracks can be moved by nanosecond-long current pulses with even higher speeds (up to 750 m/s) than in FM racetracks.

9.4.2 *Antiferromagnet Spintronics*

AFM materials represent the overwhelming majority of magnetically ordered materials. However, in current spin electronics they only play an ancillary role, namely that of pinning the magnetization of FM electrodes via the exchange bias effect. Anyway, very recently they have attracted increasing attention as active materials for the realization of memory devices where the bit of information is stored in the antiferromagnet's spin direction (e.g., parallel or perpendicular to the plane of the cell) [102]. The advantages with respect to conventional ferromagnet-based spintronic devices are twofold. In presence of an FM material, magnetic stray fields originate naturally and pose a physical limit to the packaging density: the minimum distance between cells should be enough to avoid any cross talk due to stray fields. If only antiferromagnets are employed, no stray fields exist and the packing density can be largely increased. Moreover, antiferromagnets are intrinsically more robust than ferromagnets to external perturbations from magnetic fields. Both in terms of density and in terms of stability, antiferromagnet-based RAMs are very promising for the development of a new generation of magnetic memories.

In antiferromagnet-based devices, information reading exploits relativistic SOC phenomena in AFM materials, in particular the tunneling anisotropic magnetoresistance (TAMR). In a tunneling junction with an AFM electrode and a nonmagnetic one, the conductance depends on the spin orientation of the two antiferromagnet sublattices with respect to the current direction. Indeed it turns out that, in many systems, the direction of the spin sublattices determines huge variations of the density of states (DOS) via the SOC, which in turns reflects in changes of the tunneling conductance, allowing one to electrically read the state of the antiferromagnet electrode. Also an antiferromagnet analogue of the AMR effect can be used, but the entity of the effect is much lower, below a few percent.

For writing the information, that is, for aligning the antiferromagnet spins in a given direction, the method typically employed is the exchange-spring effect. The antiferromagnet is put in contact with a ferromagnet, whose magnetization direction can be controlled by an external magnetic field. The FM magnetization and the antiferromagnet spin orientation are coupled by the interfacial exchange interaction (exchange bias), so the rotation of the FM magnetization drags the orientation of the antiferromagnet spins, in the so-called exchange-spring structure.

A typical structure used in such experiments is NiFe/IrMn/MgO/Pt, and a TAMR up to 160% has been observed in this system at low temperature [102]. With reference to the right panel of Fig. 9.18, initially the FM magnetization points to the right (red arrow) and the spins of the antiferromagnet are parallel to it. Then the FM magnetization is rotated by π (black arrow) in a continuous way by reversing the applied magnetic field. The exchange bias between the ferromagnet and the antiferromagnet acts as a spring, dragging the antiferromagnet orientation to be almost perpendicular to its original direction. This re-orientation reflects in the tunneling resistance, which increases by 160% from the parallel to the perpendicular configuration. The opposite process is described in the left panel of Fig. 9.18.

We note that, at variance with the classical approach of MTJs, in this case a single magnetic electrode is needed, making the device

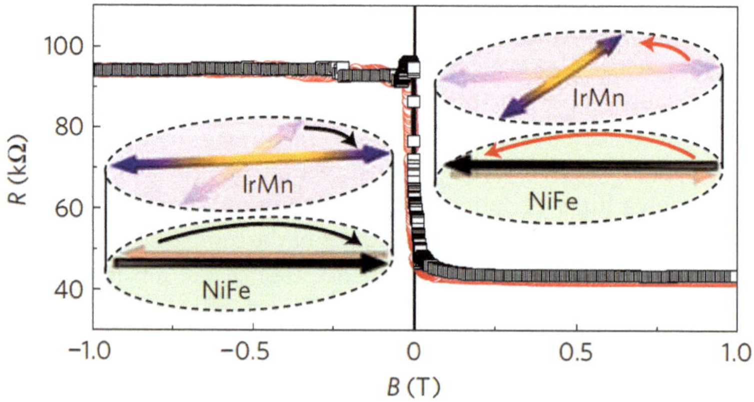

Figure 9.18 Exchange-spring method for storing information in the AFM configuration. The AFM state is read by measuring the voltage across the device. Adapted by permission from Macmillan Publishers Ltd: *Nat. Mater.* (Ref. [102]), copyright (2011).

fabrication easier. Furthermore, as no coherent transport is involved, the interface quality is less relevant.

Recently, an antiferromagnetic memory resistor [103], based on the antiferromagnetic material FeRh, has been proposed. This material presents a small but noticeable AMR value at room temperature (~0.3 %) and has the peculiarity of ordering ferromagnetically above 470 K and antiferromagnetically below it. The direction of Fe moments in the FM phase can be set via external magnetic fields. Upon cooling to room temperature, the direction of the antiferromagnet moments is determined by the magnetization direction in the FM phase, thus allowing one to write the information in the antiferromagnet.

A step forward in antiferromagneticic spintronics would be the elimination of any FM component from the device, leaving only the AFM and nonmagnetic elements (electrodes, barrier). This has been recently demonstrated in Ref. [104], employing an FM-free tunneling junction Ta/IrMn/MgO/Ta.

The reading of the device state was performed by TAMR, while writing was achieved by field cooling across the Néel temperature. By applying magnetic fields parallel ($\mathbf{H}_z$) and perpendicular ($\mathbf{H}_x$) to the current flow during the field cooling, distinct resistance states can be achieved, as reported in Fig. 9.19. The corresponding TAMR was about 10%, defined as the relative difference between the tunneling resistances obtained by application of $\mathbf{H}_z$ and $\mathbf{H}_x$. A key advantage of such device is the robustness of the stored information against external fields, as shown in the inset of Fig. 9.19. Upon field cooling in applied $\mathbf{H}_z$ down to 120 K (below the Néel temperature), the resistance of the metastable state was monitored while applying an external field up to 2 T along different directions with respect to the sample. The resistance state is unaffected by such relatively high fields, thus confirming the robustness of the information stored in antiferromagnet-based devices against external perturbations. Despite the low TAMR and operation temperature (well below room temperature), this fundamental result paves the way toward a "pure" antiferromagnet spintronics without any FM element.

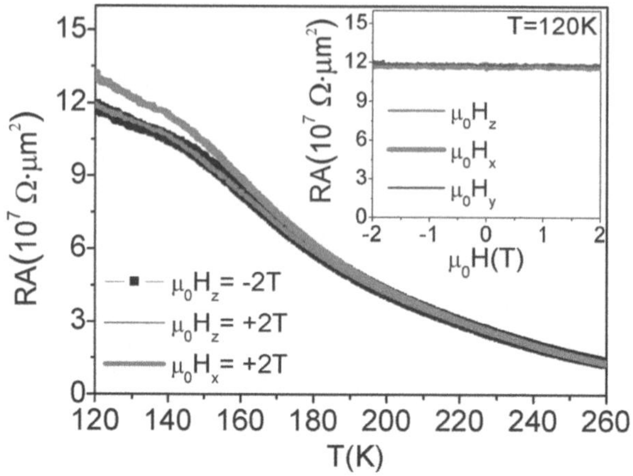

Figure 9.19 Resistance versus temperature during field cooling, with different external fields, of a Ta/IrMn/MgO/Ta multilayer. In the inset is reported the RA product, measured after field cooling with different external fields. Adapted with permission from Ref. [104]. Copyright 2013, AIP Publishing LLC.

9.5 Beyond Binary Static Memories

The classical von Neuman architecture of conventional digital computers has proved to be very efficient for solving heavy mathematical problems as well as for storing and processing large amounts of data. This architecture employs Boolean logic and is based on a sequential processing machine, with operations executed in sequence and synchronized by an internal clock. A control unit manages the process of moving data and programs in and out of the memory and executing programs following a one-at-a-time logic. Nevertheless the long-standing successes, since the first implementation in 1945 [105], the efficiency of such machines is getting worse and worse as their complexity is increasing. Even though digital computers can now emulate the brain functionalities of small animals like a spider, a mouse, and a cat [27, 107], the energy dissipation and physical space of computer networks developed for this scope exponentially increase with the intelligence of animals they intend to simulate. A Blue Gene/P super computer with 147456

CPUs and 144 TB of main memory is needed to simulate cortical processes in a cat. Despite all the efforts by many groups to emulate brain functionalities using von Neumann machines, the human brain still remains an unattainable, fantastic, compact apparatus able to perform very complex operations in a fast and energetically very efficient way. Indeed, the brain power consumption is only on the order of 10 W, the reason being that human brain uses a completely different architecture for computation and memory. Instead of Boolean logic, precise digital representations and clocked operations, the brain carries out robust and reliable computation using hybrid analog/digital unreliable components. Each neuron is connected to other neurons via 10^3–10^4 synapses in such a way as to implement a complex apparatus with about 10^{14} synapses capable of outperforming modern computers in tasks such as vision, audition, and remote control [108]. At variance with digital computers, the brain emphasizes distributed, event-driven, collective, chaotic, and parallel mechanisms and makes extensive use of adaptation, self-organization, and learning.

The basic unit of the biological neuronal network is made of a presynaptic neuron, a synapse, and a postsynaptic neuron. In response to external stimuli the firing activity of the two neurons influences the strength of the synaptic connection in a such a way that the network "learns" and the information is gradually stored in the synaptic connection via peculiar processes such as long-term potentiation (LTP), long-term depotentiation (LTD) [109], and spike-timing-dependent plasticity (STDP) [110]. A peculiar difference with respect to von Neumann machines is that there is no physical separation between logic and memory units, as the information is stored in the same network that process it. This is a very appealing feature that could revolutionize current computing architectures. Indeed, new bioinspired computing systems are definitely moving toward a new generation of neuronal networks where artificial neurons and synapses are used to implement both memory and computation. In this path toward new computing architectures, a crucial issue is to reach a high level of functionality integration within the same synaptic device, thus allowing one to achieve the high density level needed to compete with existing technologies. This requires the identification of novel and appropriate materials for nanometer-scale operation that have multifunctionality properties. Within this context a macroscopic order (ferromagnetism, ferroelectricity, crystallinity) is usually exploited for the memory operation, while its interplay with electrical conductivity provides the computational function. These peculiarities are found in a new class of devices, which have attracted renewed attention since 2008 [111]: the so-called *memristors*. Starting from the seminal paper by Leon Chua in 1971 [112], where the authors speculate about the existence of a fourth fundamental circuit element but resistor, capacitor and inductor, Strukov and coauthors [111] first proposed a model system displaying the peculiarities of a memristor, which, in the simplest form of a current-controlled memristor, is defined by the two following equations:

$$V = R(w)I \tag{9.11}$$

and

$$dw/dt = I \tag{9.12}$$

where V is the voltage, I the current, w the state variable of the device, and R a generalized resistance depending on the variable state. In short it behaves as a resistor with memory, that is, a nonlinear element with hysteresis, whose impedance depends on the value of the state variable w, which in turns depends on the current that flowed through it. It is clear why memristors have immediately attracted the attention of neuromorphic engineers: when placed between two artificial neurons, the memristor impedance changes in response to the peculiar sequence of firing pulses from neurons, thus implementing the learning functionality typical of a synapsis. In the search for novel computation and memory architectures beyond CMOS, memristors are rapidly assuming a leading role as suitable candidates for replacing conventional digital memories in neuromorphic systems. In the following section we will describe some relevant implementations of memristors that use a magnetic order parameter as state variable.

9.5.1 *Spin Transfer Torque Memristors*

As it has been explained before, the conductance of an MTJ depends on the relative alignment of the magnetization in the bottom and top electrodes. STT is nowadays widely used for storing the magnetic information in MTJs without the need for external magnetic fields, in a pure electrical way. Relevant for neuromorphic applications, the full switching of the magnetization can be achieved only for current pulses well above a characteristic threshold, corresponding to a current density on the order of 10^6–10^7Acm^{-2}. Below a characteristic current density threshold and for short enough current pulses, only incomplete magnetization reversal takes place, thus leading to intermediate impedance states that depend on the micromagnetic configuration assumed by the electrodes. The magnetization then assumes the role of a state variable, whose value depends on the sequence of current pulses sent though the MTJ and determines in turn the impedance value. This is the characteristic behaviour expected for a memristor.

In the literature there are some examples of implementation of this concept, essentially inspired by the seminal paper by Wang et al. [113], which first introduced theoretically the concept of spin torque memristors. A first proof-of-concept of the possibility of building up a spin torque memristor exploiting the spin torque–induced DW motion in the FL of an MTJ was reported in 2011 by A. Chanthbouala et al. [114]. The authors showed that in a c-shaped MTJ (Fig. 9.20a) a DW can be moved by perpendicular dc current injection in both directions, depending on the current sign. The magnetization of the RL is fixed and provides the spin polarization to the current pulses that enter the free electrode and exert a torque, causing the displacement of a magnetic DW.

Correspondingly, the device resistance changes according to the historic behavior of the current, thus implementing a memristive behavior (Fig. 9.20b). The DW position, or the average free electrode magnetization, can then be assumed as internal variables of state of the magnetic memristor. Even though this work deals with a strongly pinned DW so that the device impedance jumps between two stable states, the authors speculate about the possibility of achieving a precise control of the DW displacement by controlling the amplitude and pulse duration of the injected current in devices with weak DW pinning.

A more comprehensive analysis of an STT memristor can be found in the more recent paper by Krzysteczko et al. [115], where

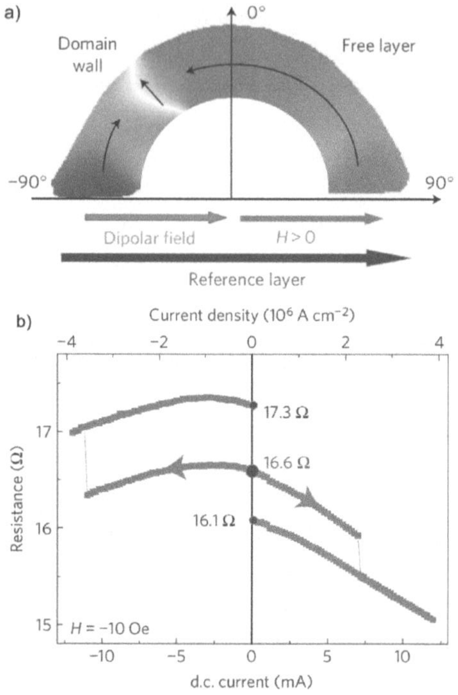

Figure 9.20 (a) Micromagnetic configuration of the c-shaped free layers showing a stable position of the DW. (b) Resistance vs. current curves obtained when sweeping the current starting from the pinned DW of panel (a). For high enough current in the device the STT induces the DW motion and this leads to a change of the MTJ impedance. Adapted by permission from Macmillan Publishers Ltd: *Nat. Phys.* (Ref. [114]), copyright (2011).

the authors present a detailed analysis of the memristive behaviour of nanoscopic MTJs. The synaptic behaviour has been investigated by monitoring the impedance variation of the MTJ upon stimulation with a train of voltage pulses that simulate the firing pulses in a neuronal network. The main result is reported in Fig. 9.21, where a clear change in the pre- and poststimulation impedance is seen. Noteworthy, the structure of the stimulus corresponds to N pulses enveloped by the positive part of a sinusoidal function $v(t)$ with amplitude v_{max}. The authors demonstrate that the impedance change is related only to the entity of the flux, as resulting from the integral over time of the voltage stimulation. The total flux then

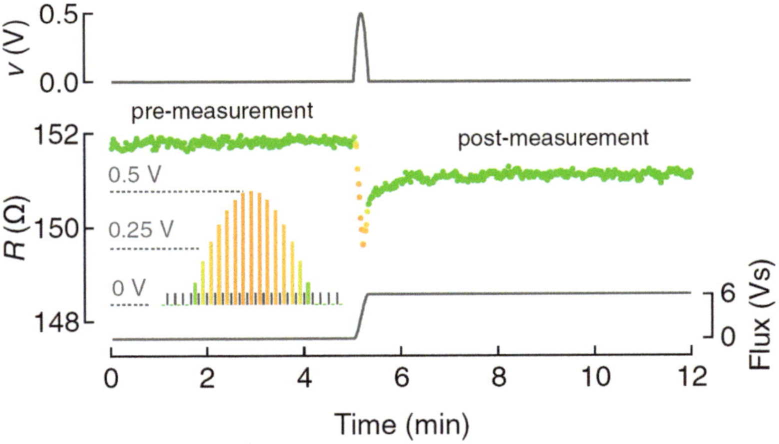

Figure 9.21 The resistance of the MTJ (green trace in the electronic version) decreases upon application of a sinus-shaped stimulus (top black signal) at about 5 min from the beginning of the experiment. The fine structure of the stimulus, which is made of 16 pulses, is shown in the inset. The bottom curve represents the total magnetic flux associated with the stimulus. From Ref. [115]. Copyright © 2012. Reproduced with permission of John Wiley & Sons, Inc.

appears to be the control variable of the system that sets the variable state represented by the micromagnetic configuration of the free electrode induced via STT.

Noteworthy, in the same paper it is also shown that the same nanoscopic MTJ can display a firing behavior similar to that displayed by biological neurons. This is shown in Fig. 9.22, where we see that at about 0.2 min the micromagnetic configuration of the free electrode was driven from a parallel state to an intermediate state by STT. In this case magnetic domains develop within the FL but the inherent instability of this configuration activates a back-hopping process with repeated switching between the intermediate and antiparallel state. This is analogous to the behavior of postsynaptic neurons, which generate spikes like those shown in the inset of Fig. 9.22 when properly stimulated.

Clearly these seminal papers are just proof-of-concept of magnetic memristors and neuron-like devices. Even though the investigation of neuromophic magnetic memories is still in its

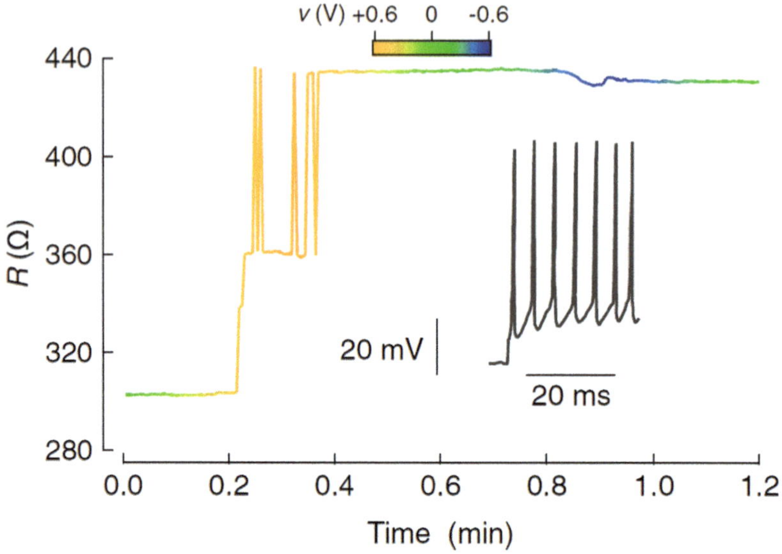

Figure 9.22 Resistance trace of the MTJ vs. time, colored according to the applied bias. Starting from a parallel alignment (low impedance state), for positive voltages (orange portion of the curve) the MTJ is driven in an intermediate metastable state. Then it switches form this intermediate state and the antiparallel state (back-hopping), displaying a behavior similar to the characteristic spiking of pyramidal neurons shown in the inset. From Ref. [115]. Copyright © 2012. Reproduced with permission of John Wiley & Sons, Inc.

infancy, the potential of this new research field is very high, especially if new materials with a high degree of multifunctionality will emerge in the near future, allowing for a more versatile implementation of neuromorphic behavior.

Acknowledgments

The authors acknowledge C. Rinaldi for valuable discussions and Fondazione Cariplo via the project "Magister" (Project No. 2013-0726).

References

1. Pohm A. V., Daughton J. M., Comstock C. S., Yoo H. Y., and Hur J. (1987). Threshold properties of 1, 2 and 4 μm multilayer magneto-resistive cells, *IEEE Trans. Magn.*, **23**, p. 2575.
2. Pohm A. V., Huang J. S. T., Daughton J. M., Krahn, D. R., and Mehra V. (1988). The design of a one megabit non-volatile M-R memory chip suing 1.5 × 5 μm cells, *IEEE Trans. Magn.*, **24**, p. 3117.
3. Baibich M. N., Broto J. M., Fert A., Nguyen Van Dau F., Petroff F., Etienne P., Creuzet G., Friederich A., and Chazelas J. (1988) Giant magnetoresistance of (001)Fe/(001)Cr magnetic superlattices, *Phys. Rev. Lett.*, **61**, p. 2472.
4. Binasch G., Grünberg P., Saurenbach F., and Zinn W. (1989). Enhanced magnetoresistance in layered magnetic structures with antiferromagnetic interlayer exchange, *Phys. Rev. B*, **39**, p. 4828(R).
5. Soutter W. (2013). Nanotechnology for universal memory. *AZoNano*, http://www.azonano.com/article.aspx?ArticleID=3064.
6. Stainer Q., Lombard L., Mackay K., Lee D., Bandiera S., Portemont C., Creuzet C., Sousa R. C., and Dieny B. (2014). Self-referenced multi-bit thermally assisted magnetic random access memories, *Appl. Phys. Lett.*, **105**, p. 032405.
7. Bedair S. M., Zavada J. M., and El-Masry N. (2010). Spintronic memories to revolutionize data storage, *IEEE Spectr.*, http://spectrum.ieee.org/semiconductors/memory/spintronic-memories-to-revolutionize-data-storage.
8. http://www.cnet.com/news/seagate-reaches-1tb-per-square-inch-hard-drive-to-reach-60tb-capacity/.
9. http://www.jcmit.com/disk2014.htm.
10. Seagate Cheetah 15K.5 datasheet.
11. Alowersson J., and Andersson, P. (1996). A 35 Gbit/s throughput 64 kbit CMOS buffer SRAM, in *Proceedings of the IEEE 1996 Conference: Custom Integrated Circuits Conference.*
12. Mosys 1T-SRAM-Q datasheet.
13. http://uw.physics.wisc.edu/~himpsel/memory.html.
14. http://www.mram-info.com/files/mraminfo/Yole-emerging-memories-potential-market.jpg.
15. Hruska J. (2014). *Everspin Teams up with GlobalFoundries to Build Commercial MRAMs*, www.extremetech.com/computing.

16. Jansen Ng. (2009). Micron announces world's first native 6Gbps SATA solid state drive, *DailyTech*. Retrieved 2009-12-02.
17. Hilson G. (2014). Magnetic RAM remains niche memory: for now, *EE Times.*
18. Hu J.-M., Li Z., Chen L.-Q., and Nan C.-W. (2011). High-density magnetoresistive random access memory operating at ultralow voltage at room temperature, *Nat. Commun.*, **2**, Article No. 553.
19. Ikeda S., Hayakawa J., Ashizawa Y., Lee Y. M., Miura K., Hasegawa H., Tsunoda M., Matsukura F., and Ohno H. (2008). Tunnel magnetoresistance of 604% at 300K by suppression of Ta diffusion in CoFeB/MgO/CoFeB pseudo-spin-valves annealed at high temperature, *Appl. Phys. Lett.*, **93**, p. 082508.
20. Bass J., and Prat W. P. (1999). Current-perpendicular (CPP) magnetoresistance in magnetic metallic multilayers, *J. Magn. Magn. Mater.*, **200**, p. 274.
21. Wang D. X., Nordman C., Daughton J. M., Qian Z. H., and Fink J. (2004). 70% TMR at room temperature for SDT sandwich junctions with CoFeB as free and reference layer, *IEEE Trans. Magn.*, **40**, p. 2269.
22. Butler W. H., Zhang, X.-G., Schulthess T. C., and MacLaren J. M. (2001). Spin-dependent tunneling conductance of Fe|MgO|Fe sandwiches, *Phys. Rev. B*, **63**, p. 054416.
23. Miyazaki T., and Jin H. (2012). *The Physics of Ferromagnetism* (Springer, London).
24. Yuasa S., Nagahama T., Fukushima A., Suzuki Y., and Ando K. (2004). Giant room-temperature magnetoresistance in single-crystal Fe/MgO/Fe magnetic tunnel junctions, *Nat. Mater.*, **3**, p. 868.
25. Parkin S. S., Kaiser C., Panchula A., Rice P. M., Hughes B., Samant M., and Yang S.-H. (2004). Giant tunneling magnetoresistance at room temperature with MgO(100) tunnel barriers, *Nat. Mater.*, **3**, p. 862.
26. Zhang X.-G., and Butler W. H. (2004). Large magnetoresistance in bcc Co/MgO/Co and FeCo/MgO/FeCo tunnel junctions, *Phys. Rev. B*, **70**, p. 172407.
27. Singh B. B., Chaudhary S., and Pandya D. K. (2012). Tunneling behavior in ion-assist ion-beam sputtered CoFe/MgO/NiFe magnetic tunnel junctions, *Mater. Res. Bull.*, **47**, p. 3786.
28. Ikeda S., Miura K., Yamamoto H., Mizunuma K., Gan H. D., Endo M., Kanai S., Hayakawa J., Matsukura F., and Ohno H. (2010). A perpendicular-anisotropy CoFeB-MgO magnetic tunnel junction, *Nat. Mater.*, **9**, p. 721.

29. Kado T. (2008). Large room-temperature inverse magnetoresistance in tunnel junctions with a Fe_3O_4 electrode, *Appl. Phys. Lett.*, **92**, p. 092502.
30. Bowen M., Bibes M., Barthelemy A., Contour J.-P., Anane A., Lemaitre Y., and Fert A. (2003). Nearly total spin polarization in $La_{2/3}Sr_{1/3}MnO_3$ from tunneling experiments, *Appl. Phys. Lett.*, **82**, p. 233.
31. Sefrioui A., Cros V., Barthelemy A., Pena V., Leon C., Santamaria J., Varela M., and Pennycook S. J. (2006). Tunnel magnetoresistance in $La_{0.7}Ca_{0.3}MnO_3/PrBa_2Cu_3O_7/La_{0.7}Ca_{0.3}MnO_3$, *Appl. Phys. Lett.*, **88**, p. 022512.
32. Barry A., Coey J. M. D., and Viret M. (2000). A CrO_2-based magnetic tunnel junction, *J. Phys.: Condens. Matter*, **12**, p. L173.
33. Liu J., Girgis E., Bach P., Ruster C., Gould C., Schmidt G., and Molenkamp L. W. (2006). Tunneling magnetoresistance in devices based on epitaxial NiMnSb with uniaxial anisotropy, *J. Appl. Phys.*, **99**, p. 036110.
34. Liu H.-X., Honda Y., Taira T., Matsuda K.-I., Arita M., Uemura T., and Yamamoto M. (2012). Giant tunneling magnetoresistance in epitaxial $Co_2MnSi/MgO/Co_2MnSi$ magnetic tunneling junctions by half-metallicity of Co_2MnSi and coherent tunneling, *Appl. Phys. Lett.*, **101**, p. 132418.
35. Ohya S., Muneta I., Hai P. N., and Tanaka M. (2009). GaMnAs-based magnetic tunnel junctions with an AlMnAs barrier, *Appl. Phys. Lett.*, **95**, p. 242503.
36. Wen Z., Sukegawa H., Kasai S., Inomata K., and Mitani S. (2014). Tunnel magnetoresistance and spin-transfer torque switching in polycrystalline Co_2FeAl full-Heusler-alloy magnetic tunnel junctions on amorphous Si/SiO_2 substrates, *Phys. Rev. Appl.*, **2**, p. 024009.
37. Schmalhorst J., Kammerer S., Reiss G., and Hutten A. (2005). Inelastic electron tunneling spectropscopy and bias voltage dependence of magnetic tunnel junctions with polycrystalline Co_2MnSi electrode, *Appl. Phys. Lett.*, **86**, p. 052501.
38. Nozières J. P., *Magnetic Random Access Memories (M-RAM): A Truly Universal Memory?* Internal report (Spintech, Grenoble).
39. Prejbeanu I. L., Kerekes M., Sousa R. C., Sibuet H., Redon O., Dieny B., and Nozières J. P. (2007). Thermally assisted MRAM, *J. Phys.: Condens. Matter*, **19**, p. 165218.
40. Stoner E. C., and Wohlfarth E. P. (1948). A mechanism of magnetic hysteresis in heterogeneous alloys, *Philos. Trans. R. Soc. London, A*, **240**, p. 599.

41. Sousa R. C., and Prejbeanu I. L. (2005). Non-volatile magnetic random access memories (MRAM), *C. R. Phys.*, **6**, p. 1013.

42. Zhu J.-G. (2008). Magnetoresistiverandom access memory: the path to competitiveness and scalability, *Proc. IEEE*, **96**, p. 1786.

43. Janesky J. (2007). *Impact of External Magnetic Fields on MRAM Products*. Application Note AN3525 (Freescale Semiconductor, Austin).

44. Khvalkovskiy A. V., Apalkov D., Watts S., Chepulskii R., Beach R. S., Ong A., Tang X., Driskill-Smith A., Butler W. H., Visscher P. B., Lottis D., Chen E., Nikitin V., and Krounbi M. (2013). Basic principles of STT-MRAM cell operation in memory arrays, *J. Phys. D: Appl. Phys.*, **46**, p. 074001.

45. Slonczewski J. C. (1996). Current-driven excitation of magnetic multilayers, *J. Magn. Magn. Mater.*, **159**, p. L1.

46. Berger L. (1996). Emission of spin waves by a magnetic multilayer traversed by a current, *Phys. Rev. B*, **54**, p. 9353.

47. Huai Y., Albert F., Nguyen P., Pakala M., and Valet T. (2004). Observation of spin transfer switching in deep submicron-sized and low resistance magnetic tunnel junctions, *Appl. Phys. Lett.*, **84**, p. 3118.

48. Diao Z., Apalkov D., Pakala M., Ding Y., Panchula A., and Huai Y. (2005). Spin transfer switching and spin polarization in magnetic tunnel junctions with MgO and AlO_x barriers, *Appl. Phys. Lett.*, **87**, p. 232502.

49. Locatelli N., Cros V., and Grollier J. (2014). Spin-torque building blocks, *Nat. Mater.*, **13**, p. 11.

50. Gajek M., Nowak J. J., Sun J. Z., Trouillod P. L., O'Sullivan E. J., Abraham D. W., Gaidis M. C., Hu G., Brown S., Zhu Y., Robertazzi R. P., Gallagher W. J., and Worledge D. C. (2012). Spin torque switching of 20nm magnetic tunnel junctions with perpendicular anisotropy, *Appl. Phys. Lett.*, **100**, p. 132408.

51. www.mram-info.com.

52. Garello K., Miron I. M., Avci C. O., Freimuth F., Mokrousov Y., Blügel S., Auffret S., Boulle O., Gaudin G., and Gambardella P. (2013). Symmetry and magnitude of spin-orbit torques in ferromagnetic heterostructures, *Nat. Nanotechnol.*, **8**, p. 587.

53. Miron I. M., Gaudin G., Auffret S., Rodmacq B., Schuhl A., Pizzini S., Vogel J., and Gambardella P. (2010). Current-driven spin torque induced by the Rashba effect in a ferromagnetic metal layer, *Nat. Mater.*, **9**, p. 230.

54. Bychkov Y. A., and Rashba E. I. (1984). Properties of a 2D electron gas with lifted spectral degeneracy, *JETP Lett.*, **39**, p. 78.

55. Emori S., Bauer U., Ahn S.-M., Martinez E., and Beach G. S. D. (2013). Current-driven dynamics of chiral ferromagnetic domain walls, *Nat. Mater.*, **12**, p. 611.

56. Liu L., Pai C.-F., Li Y., Tseng H. W., Ralph D. C., and Buhrman R. A. (2012). Spin-torque switching with the giant spin Hall effect of Tantalum, *Science*, **336**, p. 555.

57. Tsymbal E. Y. (2012). Electric toggling of magnets, *Nat. Mater.*, **11**, p. 12.

58. Wang W.-G., Li M., Hageman S., and Chien C. L. (2012). Electric-field-assisted switching in magnetic tunnel junctions, *Nat. Mater.*, **11**, p. 64.

59. Shiota Y., Nozaki T., Bonell F., Murakami S., Shinjo T., and Suzuki Y. (2012). Induction of coherent magnetization switching in a few atomic layers of FeCo using voltage pulses, *Nat. Mater.*, **11**, p.39.

60. Duan C.-G., Velev J. P., Sabirianov R. F., Mei W. N., Jaswal S. S., and Tsymbal E. Y. (2008). Tailoring magnetic anisotropy at the ferromagnetic/ferroelectric interface, *Appl. Phys. Lett.*, **92**, p. 122905.

61. Cheong S.-W., and Mostovoy M. (2007). Multiferroics: a magnetic twist for ferroelectricity, *Nat. Mater.*, **6**, p. 13.

62. Khomskii D. (2009). Classifying multiferroics: mechanisms and effects, *Physics*, **2**, p. 20.

63. Khomskii D. I. (2006). Multiferroics: different ways to combine magnetism and ferroelectricity, *J. Magn. Magn. Mater.*, **306**, p. 1.

64. Fiebig M. (2005). Revival of the magnetoelectric effect, *J. Phys. D: Appl. Phys.*, **38**, p. R123.

65. Zheng H., Wang J., Lofland S. E., Ma Z., Mohaddes-Ardabili L., Zhao T., Salamanca-Riba L., Shinde S. R., Ogale S. B., Bai F., Viehland D., Jia Y., Schlom D. G., Wuttig M., Roytburd A., and Ramesh R. (2004). Multiferroic $BaTiO_3$-$CoFe_2O_4$ nanostructures, *Science*, **303**, p. 661.

66. Zavaliche F., Zheng H., Mohaddes-Ardabili L., Yang S. Y., Zhan Q., Shafer P., Reilly E., Chopdekar R., Jia Y., Wright P., Schlom D. G., Suzuki Y., and Ramesh R. (2005). Electric field-induced magnetization switching in epitaxial columnar nanostructures, *Nano Lett.*, **5**, p. 1793.

67. Duan C.-G., Jaswal S. S., and Tsymbal E. Y. (2006). Predicted magnetoelectric effect in Fe/$BaTiO_3$ multilayers: ferroelectric control of magnetism, *Phys. Rev. Lett.*, **97**, p. 047201.

68. Garcia V., Bibes M., Bocher L., Valencia S., Kronast F., Crassous A., Moya X., Enouz-Vedrenne S., Gloter A., Imhoff D., Deranlot C., Mathur N. D., Fusil S., Bouzehouane K., and Barthélémy A. (2010). Ferroelectric control of spin polarization, *Science*, **327**, p. 1106.

69. Thiele C., Dörr K., Bilani O., Rödel J., and Schultz L. (2007). Influence of strain on the magnetization and magnetoelectric effect in $La_{0.7}A_{0.3}MnO_3$/PMN-PT(001) (A = Sr, Ca), *Phys. Rev. B*, **75**, p. 054408.
70. Eerenstein W., Wiora M., Prieto J. L., Scott J. L., and Mathur N. D. (2007). Giant sharp and persistent converse magnetoelectric effects in multiferroic epitaxial heterostructures, *Nat. Mater.*, **6**, p. 348.
71. Brivio S., Petti D., Bertacco R., Cezar J. C. (2011). Electric field control of magnetic anisotropies and magnetic coercivity in Fe/$BaTiO_3$ (001) heterostructures, *Appl. Phys. Lett.*, **98**, p. 092505.
72. Laukhin V., Skumryev V., Martí X., Hrabovsky D., Sánchez F., García-Cuenca M. V., Ferrater C., Varela M., Lüders U., Bobo J. F., and Fontcuberta J. (2006). Electric-field control of exchange bias in multiferroic epitaxial heterostructures, *Phys. Rev. Lett.*, **97**, p. 227201.
73. Skumryev V., Laukhin V., Fina I., Martí X., Sánchez F., Gospodinov M., and Fontcuberta J. (2011). Magnetization reversal by electric-field decoupling of magnetic and ferroelectric domain walls in multiferroic-based heterostructures, *Phys. Rev. Lett.*, **106**, p. 057206.
74. He X., Wang Y., Wu N., Caruso A. N., Vescovo E., Belashchenko K. D., Dowben P. A., and Binek C. (2010). Robust isothermal electric control of exchange bias at room temperature, *Nat. Mater.*, **9**, p. 579.
75. Wu S. M., Cybart S. A., Yu P., Rossell M. D., Zhang J. X., RameshR., and Dynes R. C. (2010). Reversible electric control of exchange bias in a multiferroic field-effect device, *Nat. Mater.*, **9**, p. 756.
76. Molegraaf H. J. A., Hoffman J., Vaz C. A. F., Gariglio S., van der Marel D., Ahn C. H., and Triscone J.-M. (2009). Magnetoelectriceffects in complex oxides with competing ground states, *Adv. Mater.*, **21**, p. 3470.
77. Maruyama T., Shiota Y., Nozaki T., Ohta K., Toda N., Mizuguchi M., Tulapurkar A. A., Shinjo T., Shiraishi M., Mizukami S., Ando Y., and Suzuki Y. (2009). Large voltage-induced magneticanisotropy change in a few atomic layers of iron, *Nat. Nanotechnol.*, **4**, p. 158.
78. Yin Y. W., Burton J. D., Kim Y.-M., Borisevich A. Y., Pennycook S. J., Yang S. M., Noh T. W., Gruverman A., Li X. G., Tsymbal E. Y., and Qi Li (2013). Enhanced tunneling electroresistance effect due to a ferroelectrically induced phase transition at a magnetic complex oxide interface, *Nat. Mater.*, **12**, p. 397.
79. Duan C.-G., Velev J. P., Sabirianov R. F., Zhu Z., Chu J., Jaswal S. S., and Tsymbal E. Y. (2008). Surface magnetoelectric effect in ferromagnetic metal films, *Phys. Rev. Lett.*, **101**, p. 137201.

80. Radaelli G., Petti D., Plekhanov E., Fina. I., Torelli P., Salles B. R., Cantoni M., Rinaldi C., Gutiérrez D., Panaccione G., Varela M., Picozzi S., Fontcuberta J., and Bertacco R. (2014). Electric control of magnetism at the Fe/$BaTiO_3$ interface, *Nat. Commun.*, **5**, p. 3404.

81. Chanthbouala A., Garcia V., Cherifi R. O., Bouzehouane K., Fusil S., Moya X., Xavier S., Yamada H., Deranlot C., Mathur N. D., Bibes M., Barthelemy A., and Grollier J. (2012). A ferroelectric memristor, *Nat. Mater.*, **1**, p. 860.

82. Scigaj M., Dix. N., fina I., Bachelet R., Warot-Fonrose B., Fontcuberta J., and Sanchez F. (2013). Ultra-flat $BaTiO_3$ films on si(001) with large out-of-plane polarization. *Appl. Phys. Lett.*, **102**, p. 112905.

83. Singamaneni S. R., Fan W., Prater J. T., and Narayan J. (2014). Magnetic properties of $BaTiO_3/La_{0.7}Sr_{0.3}MnO_3$ thin films integrated on Si(100), *J. Appl. Phys.*, **116**, p. 224104.

84. Marchiori C., Sousa M., Guiller A., Siegwart H., Locquet J.-P., Fompeyrine J., Norga G. J., and Seo J. W. (2006). Thermal stability of the $SrTiO_3$/(Ba, Sr)O stacks epitaxially grown on Si, *Appl. Phys. Lett.*, **88**, p. 072913.

85. Xiong C., Pernice W. H. P., Ngai J. H., Reiner J. W., Kumah D., Walker F. J., Ahn C. H., and Tang H. X. (2014). Active silicon integrated nanophotonics: ferroelectric $BaTiO_3$ devices, *Nano Lett.*, **14**, p. 1419.

86. Dubourdieu C., Bruley J., Arruda T. M., Posadas A., Jordan-Sweet J., Frank M. M., Cartier E., Franck D. J., Kalinin S. V., Demkov A. A., and Narayanan V. (2013). Switching of ferroelectric polarization in epitaxial $BaTiO_3$ films on silicon without a conducting bottom electrode, *Nat. Nanotechnol.*, **8**, p. 748.

87. Radaelli G., Cantoni M., Lijun L., Espahbodi M., and Bertacco R. (2014). Two dimensional growth of ultrathin Fe films on $BaTiO_3$ with sharp chemical interface, *J. Appl. Phys.*, **115**, p. 063501.

88. Kirilyuk A., Kimel A. V., and Rasing T. (2010). Ultrafast optical manipulation of magnetic order, *Rev. Mod. Phys.*, **82**, p. 2731.

89. Beaurepaire E., Merle J.-C., Daunois A., and Bigot J.-Y. (1996). Ultrafast spin dynamics in ferromagnetic nickel, *Phys. Rev. Lett.*, **76**, p. 4250.

90. Del Fatti N., Voisin C., Achermann M., Tzortzakis S., Christofilos D., and Vallèe F. (2000). Nonequlibrium electron dynamics in noble metals, *Phys. Rev. B*, **61**, p. 16956.

91. Hohlfeld J., Gerrits Th., Bilderbeek M., Rasing Th., Awano H., and Ohta N. (2001). Fast magnetization reversal of GdFeCo induced by femtosecond laser pulses, *Phys. Rev. B*, **65**, p. 012413.

92. Carpene E., Piovera C., Dallera C., Mancini E., and Puppin E. (2011). All-optical subnanosecond coherent spin switching in thin ferromagnetic layers, *Phys. Rev. B*, **84**, p. 134425.
93. Carpene E., Mancini E., Dallera C., Puppin E., and De Silvestri S. (2010). Three-dimensional magnetization evolution and the role of anisotropies in thin Fe/MgO films: static and dynamic measurements, *J. Appl. Phys.*, **108**, p. 063919.
94. Stanciu C. D., Hansteen F., Kimel A. V., Kirilyuk A., Tsukamoto A., Itoh A., and Rasing Th. (2007). All-optical magnetic recording with circularly polarized light, *Phys. Rev. Lett.*, **99**, p. 047601.
95. Khorsand A. R., Savoini M., Kirilyuk A., Kimel A. V., Tsukamoto A., Itoh A., and Rasing Th. (2012). Role of magnetic circular dichroism in all-optical magnetic recording, *Phys. Rev. Lett.*, **108**, p. 127205.
96. Savoini M., Piovera C., Rinaldi C., Albisetti E., Petti D., Khorsand A. R., Duò L., Dallera C., Cantoni M., Bertacco R., Finazzi M., Carpene E., Kimel A. V., Kirilyuk A., and Rasing Th. (2014). Bias-controlled ultrafast demagnetization in magnetic tunneling junctions, *Phys. Rev. B*, **89**, p. 140402(R).
97. Parkin S. S. P., Hayashi M., and Thomas L. (2008). Magnetic domain-wall racetrack memory, *Science*, **320**, p. 190.
98. Huang S.-H., and Lai C.-H. (2009). Domain-wall depinning by controlling its configuration at notch, *Appl. Phys. Lett.*, **95**, p. 032505.
99. Sekhar M. C., Goolaup S., Purnama I., and Lew W. S. (2014). Depinning assisted by domain wall deformation in cylindrical NiFe nanowires, *J. Appl. Phys.*, **115**, p. 083913.
100. Kim W. J., Lee, T. D., and Lee, K. J. (2007). Domain wall pinning by alternating materials in current-induced domain wall motion, *Phys. Status Solidi B*, **244**, p. 4439.
101. Parkin S. S. P. (2014). *EMRS 2014*, Fall meeting.
102. Park B. G., Wunderlich J., Martì X., Holy V., Kurosaki Y., Yamada M., Yamamoto H., Nishide A., Hayakawa J., Takahashi H., Shick A. B., and Jungwirth T. (2011). A spin-valve-like magneoresistance of an antiferromagnet-based tunnel junction, *Nat. Mater.*, **10**, p. 347.
103. Marti X., Fina I., Frontera C., Liu J., Wadley P., He Q., Paull R. J., Clarkson J. D., Kudrnovsky J., Turek I., Kunes J., Yi D., Chu J.-H., Nelson C. T., You L., Arenholz E., Salahuddin S., Fontcuberta J., Jungwirth T., and Ramesh R. (2014). Room temperature antiferromagnetic memory resistor, *Nat. Mater.*, **13**, p. 367.

104. Petti D., Albisetti E., Reichlova H., Gazquez J., Varela M., Molina-Ruiz M., Lopeandia A. F., Olejnik K., Novak V., Fina I., Dkhil B., Hayakawa J., Marti X., Wunderlich J., Jungwirth T., and Bertacco R. (2013). Storing magnetic information in IrMn/MgO/Ta tunnel junctions via field-cooling, *Appl. Phys. Lett.*, **102**, p. 192404.

105. Von Neumann J. (1945). *First Draft of a Report on the EDVAC*, http:// www.virtualtravelog.net/wp/wp-content/media/2003-08-TheFirstDraft.pdf.

106. Smith L. S. (2006). *Handbook of Nature Inspired and Innovative Computing: Integrating Classical Models with Emerging Technologies* (Springer, New York, USA).

107. Ananthanarayanan R., Esser S. K., Simon H. D., and Modha D. S. (2009). The cat is out of the bag: cortical simulations with 10^9 neurons, 10^{13} synapses, in *Proceedings of the 2009 IEEE/ACM Conference on High Performance Networking Computing*, Article No. 63.

108. Chicca E., Stefanini F., Bartolozzi C., and Indiveri G. (2014). Neuromorphic circuits for building autonomous cognitive systems, *Proc. IEEE*, **102**, p. 1367.

109. Lisman J., and Richard E. A. (2007). Long-term depression and depotentiation, *eLS*.

110. Feldman D. E. (2012). The spike-timing dependence of plasticity, *Neuron*, **75**, p. 556.

111. Strukov D. B., Snider G. S., Stewart D. R., and Williams R. S. (2008). The missing memristor found, *Nature*, **453**, p. 80.

112. Chua L. O. (1971). Memristor: the missing circuit element, *IEEE Trans. Circuit Theory*, **18**, p. 507.

113. Wang X., Chen Y., Xi H., Li H., and Dimitrov D. (2009). Spintronic memristor through spin-torque-induced magnetization motion, *IEEE Electron. Device Lett.*, **30**, p. 294.

114. Chanthbouala A., Matsumoto R., Grollier J., Cros V., Anane A., Fert A., Khvalkovskiy A. V., Zvezdin K. A., Nishimura K., Nagamine Y., Maehara H., Tsunekawa K., Fukushima A., and Yuasa S. (2011). Vertical-current-induced domain-wall motion in MgO-based magnetic tunnel junctions with low current densities, *Nat. Phys.*, **7**, p. 626.

115. Krzysteczko P., Münchenberger J., Schäfers M., Reiss G., and Thomas A. (2012). The memristive magnetic tunnel junction as a nanoscopic synapse-neuron system, *Adv. Mater.*, **24**, p. 762.

Index